STUDENT'S SOLUTIONS MANUAL
SINGLE VARIABLE

ROGER LIPSETT · MARK WOODARD

Brandeis University   *Furman University*

CALCULUS

William Briggs

University of Colorado, Denver

Lyle Cochran

Whitworth University

With the assistance of

Bernard Gillett

University of Colorado, Boulder

Addison-Wesley
is an imprint of

ISBN-13: 978-0-321-66521-8
ISBN-10: 0-321-66521-X

3 4 5 6 BRR 14 13 12 11

Addison-Wesley
is an imprint of

www.pearsonhighered.com

Contents

Chapter 1

Functions

1.1 Review of Functions

1.1.1 A function is a rule which assigns each domain element to a unique range element. The independent variable is associated with the domain, while the dependent variable is associated with the range.

1.1.3 The vertical line test is used to determine whether a given graph represents a function. (Specifically, it tests whether the variable associated with the vertical axis is a function of the variable associated with the horizontal axis.) If every vertical line which intersects the graph does so in exactly one point, then the given graph represents a function. If any vertical line $x = a$ intersects the curve in more than one point, then there is more than one range value for the domain value $x = a$, so the given curve does not represent a function.

1.1.5 Item (i) is true, since it is stipulated in the definition of function. However, item (ii) need not be true – for example, the function $f(x) = x^2$ has two different domain values associated with the one range value 4, since $f(2) = f(-2) = 4$.

1.1.7 $f(g(2)) = f(-2) = 2$ and $g(f(-2)) = g(2) = -2$.

1.1.9

The defining property for an even function is that $f(-x) = f(x)$, which ensures that the graph of the function is symmetric about the y-axis.

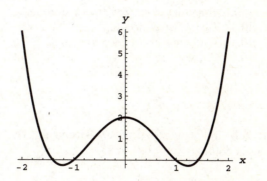

1.1.11 Graph A does not represent a function, while graph B does. Note that graph A fails the vertical line test, while graph B passes it.

1.1.13

The natural domain of this function is the set of a real numbers. The range is $[-10, \infty)$.

1.1.15

The natural domain of this function is $[-2, 2]$. The range is $[0, 2]$.

1.1.17

The natural domain and the range for this function are both the set of all real numbers.

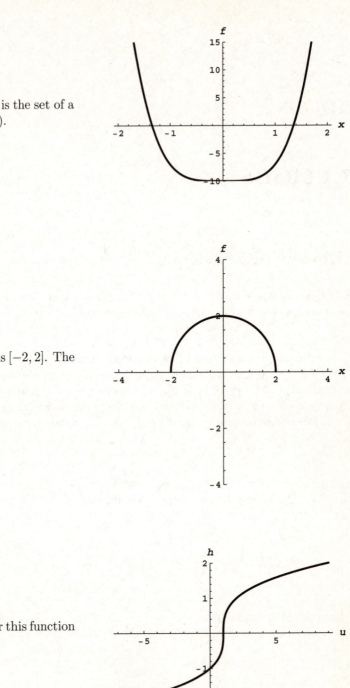

1.1.19 The independent variable t is time and the dependent variable d is distance above the ground. The domain in context is $[0, 8]$

1.1.21 $f(10) = 96$

1.1.23 $g(1/z) = (1/z)^3 = \frac{1}{z^3}$

1.1.25 $F(g(y)) = F(y^3) = \frac{1}{y^3 - 3}$

1.1.27 $g(f(u)) = g(u^2 - 4) = (u^2 - 4)^3$

1.1.29 $F(F(x)) = F\left(\dfrac{1}{x-3}\right) = \dfrac{1}{\frac{1}{x-3} - 3} = \dfrac{1}{\frac{1}{x-3} - \frac{3(x-3)}{x-3}} = \dfrac{1}{\frac{10-3x}{x-3}} = \dfrac{x-3}{10-3x}$

1.1.31 $g(x) = x^3 - 5$ and $f(x) = x^{10}$. The domain of h is the set of all real numbers.

1.1.33 $g(x) = x^4 + 2$ and $f(x) = \sqrt{x}$. The domain of h is the set of all real numbers.

1.1.35 $(f \circ g)(x) = f(g(x)) = f(x^2 - 4) = |x^2 - 4|$. The domain of this function is the set of all real numbers.

1.1.37 $(f \circ G)(x) = f(G(x)) = f\left(\dfrac{1}{x-2}\right) = \left|\dfrac{1}{x-2}\right|$. The domain of this function is the set of all real numbers except for the number 2.

1.1.39 $(G \circ g \circ f)(x) = G(g(f(x))) = G(g(|x|)) = G(x^2 - 4) = \dfrac{1}{x^2 - 4 - 2} = \dfrac{1}{x^2 - 6}$. The domain of this function is the set of all real numbers except for the numbers $\pm\sqrt{6}$.

1.1.41 Since $(x^2 + 3)^2 = x^4 + 6x^2 + 9$, it must be the case that $f(x) = x^2$.

1.1.43 Since $(x^2)^2 + 3 = x^4 + 3$, this expression results from squaring x^2 and adding 3 to it. Thus we must have $f(x) = x^2$.

1.1.45

    a. $f(g(2)) = f(2) = 4$ b. $g(f(2)) = g(4) = 1$ c. $f(g(4)) = f(1) = 3$

    d. $g(f(5)) = g(6) = 3$ e. $f(g(7)) = f(4) = 7$ f. $f(f(8)) = f(8) = 8$

1.1.47 This function is symmetric about the y-axis, since $f(-x) = (-x)^4 + 5(-x)^2 - 12 = x^4 + 5x^2 - 12 = f(x)$.

1.1.49 This function has none of the indicated symmetries. For example, note that $f(-2) = -26$, while $f(2) = 22$, so f is not symmetric about either the origin or about the y-axis, and is not symmetric about the x-axis because it is a function.

1.1.51 This curve (which is not a function) is symmetric about the x-axis, the y-axis, and the origin. Note that replacing either x by $-x$ or y by $-y$ (or both) yields the same equation. This is due to the fact that $(-x)^{2/3} = ((-x)^2)^{1/3} = (x^2)^{1/3} = x^{2/3}$, and a similar fact holds for the term involving y.

1.1.53 Function A is symmetric about the y-axis, so is even. Function B is symmetric about the origin, so is odd. Function C is also symmetric about the y-axis, so is even.

1.1.55

    a. True. A real number z corresponds to the domain element $z/2 + 19$, since $f(z/2 + 19) = 2(z/2 + 19) - 38 = z + 38 - 38 = z$.

    b. False. The definition of function does not require that each range element comes from a unique domain element, rather that each domain element is paired with a unique range element.

    c. True. $f(1/x) = \dfrac{1}{1/x} = x$, and $\dfrac{1}{f(x)} = \dfrac{1}{1/x} = x$.

    d. False. For example, suppose that f is the straight line through the origin with slope 1, so that $f(x) = x$. Then $f(f(x)) = f(x) = x$, while $(f(x))^2 = x^2$.

    e. False. For example, let $f(x) = x + 2$ and $g(x) = 2x - 1$. Then $f(g(x)) = f(2x - 1) = 2x - 1 + 2 = 2x + 1$, while $g(f(x)) = g(x + 2) = 2(x + 2) - 1 = 2x + 3$.

f. True. In fact, this is the definition of $f \circ g$.

g. True. If f is even, then $f(-z) = f(z)$ for all z, so this is true in particular for $z = ax$. So if $g(x) = cf(ax)$, then $g(-x) = cf(-ax) = cf(ax) = g(x)$, so g is even.

h. False. For example, $f(x) = x$ is an odd function, but $h(x) = x + 1$ isn't, since $h(2) = 3$, while $h(-2) = -1$ which isn't $-h(2)$.

i. True. If $f(-x) = -f(x) = f(x)$, then in particular $-f(x) = f(x)$, so $0 = 2f(x)$, so $f(x) = 0$ for all x.

1.1.57

We will make heavy use of the fact that $|x|$ is x if $x > 0$, and is $-x$ if $x < 0$. In the first quadrant where x and y are both positive, this equation becomes $x - y = 1$ which is a straight line with slope 1 and y-intercept -1. In the second quadrant where x is negative and y is positive, this equation becomes $-x - y = 1$, which is a straight line with slope -1 and y-intercept -1. In the third quadrant where both x and y are negative, we obtain the equation $-x - (-y) = 1$, or $y = x + 1$, and in the fourth quadrant, we obtain $x + y = 1$. Graphing these lines and restricting them to the appropriate quadrants yields the following curve:

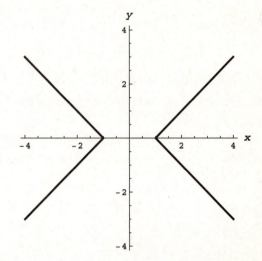

1.1.59 Since the composition of f with itself has first degree, we can assume that f has first degree as well, so let $f(x) = ax + b$. Then $(f \circ f)(x) = f(ax + b) = a(ax + b) + b = a^2x + (ab + b)$. Equating coefficients, we see that $a^2 = 9$ and $ab + b = -8$. If $a = 3$, we get that $b = -2$, while if $a = -3$ we have $b = 4$. So two possible answers are $f(x) = 3x - 2$ and $f(x) = -3x + 4$.

1.1.61 Let $f(x) = ax^2 + bx + c$. Then $(f \circ f)(x) = f(ax^2 + bx + c) = a(ax^2 + bx + c)^2 + b(ax^2 + bx + c) + c$. Expanding this expression yields $a^3x^4 + 2a^2bx^3 + 2a^2cx^2 + ab^2x^2 + 2abcx + ac^2 + abx^2 + b^2x + bc + c$, which simplifies to $a^3x^4 + 2a^2bx^3 + (2a^2c + ab^2 + ab)x^2 + (2abc + b^2)x + (ac^2 + bc + c)$. Equating coefficients yields $a^3 = 1$, so $a = 1$. Then $2a^2b = 0$, so $b = 0$. It then follows that $c = -6$, so the original function was $f(x) = x^2 - 6$.

1.1.63

b.

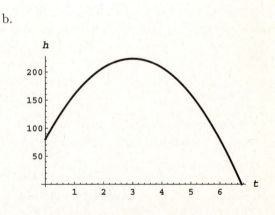

a. The formula for the height of the rocket is valid from $t = 0$ until the rocket hits the ground, which is the positive solution to $-16t^2 + 96t + 80 = 0$, which the quadratic formula reveals is $t = 3 + \sqrt{14}$. Thus, the domain is $[0, 3\sqrt{14}]$.

The maximum appears to occur at $t = 3$.
The height at that time would be 224.

1.1.65 This would not necessarily have either kind of symmetry. For example, $f(x) = x^2$ is an even function and $g(x) = x^3$ is odd, but the sum of these two is neither even nor odd.

1.1.67 This would be an odd function, so it would be symmetric about the origin. Suppose f is even and g is odd. Then $\frac{f}{g}(-x) = \frac{f(-x)}{g(-x)} = \frac{f(x)}{-g(x)} = -\frac{f}{g}(x)$.

1.1.69 This would be an even function, so it would be symmetric about the y-axis. Suppose f is even and g is even. Then $f(g(-x)) = f(g(x))$, since $g(-x) = g(x)$.

1.1.71 This would be an even function, so it would be symmetric about the y-axis. Suppose f is even and g is odd. Then $g(f(-x)) = g(f(x))$, since $f(-x) = f(x)$.

1.1.73

  a. $\dfrac{f(x) - f(a)}{x - a} = \dfrac{4x - 3 - (4a - 3)}{x - a} = \dfrac{4x - 4a}{x - a} = \dfrac{(4)(x - a)}{x - a} = 4.$

  b. $\dfrac{f(x + h) - f(x)}{h} = \dfrac{4(x + h) - 3 - (4x - 3)}{h} = \dfrac{4x + 4h - 3 - 4x + 3}{h} = \dfrac{4h}{h} = 4.$

1.1.75

  a. $\dfrac{f(x) - f(a)}{x - a} = \dfrac{\frac{1}{2x} - \frac{1}{2a}}{x - a} = \dfrac{\frac{a}{2ax} - \frac{x}{2ax}}{x - a} = \dfrac{a - x}{(2ax)(x - a)} = \dfrac{(-1)(x - a)}{(2ax)(x - a)} = \dfrac{-1}{2ax}.$

  b. $\dfrac{f(x + h) - f(x)}{h} = \dfrac{\frac{1}{(2)(x+h)} - \frac{1}{2x}}{h} = \dfrac{\frac{x}{(2)(x+h)(x)} - \frac{x+h}{(2)(x+h)(x)}}{h} = \dfrac{-h}{(2)(x + h)(x)(h)} = \dfrac{-1}{(2)(x + h)(x)}.$

1.2 Representing Functions

1.2.1 Functions can be defined and represented by a formula, through a graph, via a table, and by using words.

1.2.3 The domain of a rational function $\frac{p(x)}{q(x)}$ is the set of all real numbers for which $q(x) \neq 0$.

1.2.5

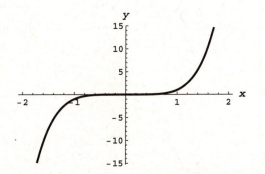

1.2.7 Compared to the graph of $f(x)$, the graph of $f(x + 2)$ will be shifted 2 units to the left.

1.2.9 Compared to the graph of $f(x)$, the graph of $f(3x)$ will be scaled horizontally by a factor of 3.

1.2.11 The slope of the line shown is $m = \frac{-3 - (-1)}{3 - 0} = -2/3$. The y-intercept is $b = -1$. Thus the function is given by $f(x) = (-2/3)x - 1$.

1.2.13 Using price as the independent variable p and the average number of units sold per day as the dependent variable d, we have the ordered pairs $(250, 12)$ and $(200, 15)$. The slope of the line determined by these points is $m = \frac{15-12}{200-250} = \frac{3}{-50}$. Thus the demand function has the form $d(p) = (-3/50)p + b$ for some constant b. Using the point $(200, 15)$, we find that $15 = (-3/50) \cdot 200 + b$, so $b = 27$. Thus the demand function is $d = (-3/50)p + 27$. While the natural domain of this linear function is the set of all real numbers, the formula is only likely to be valid for some subset of the interval $(0, 450)$, since outside of that interval either $p \leq 0$ or $d \leq 0$.

1.2.15 For $x < 0$, the graph is a line with slope 1 and y- intercept 3, while for $x > 0$, it is a line with slope $-1/2$ and y-intercept 3. Note that both of these lines contain the point $(0, 3)$. The function shown can thus be written

$$f(x) = \begin{cases} x + 3 & \text{if } x \leq 0; \\ (-1/2)x + 3 & \text{if } x > 0. \end{cases}$$

1.2.17

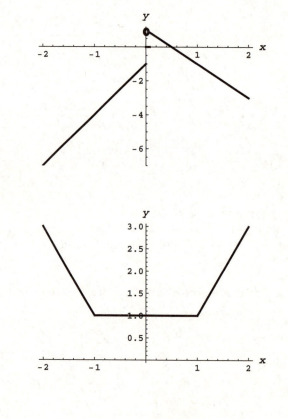

1.2.19

1.2.21

   a.

b. The function is a polynomial, so its domain is the set of all real numbers.

c. It has one peak near its y-intercept of $(0, 6)$ and one valley between $x = 1$ and $x = 2$. Its x-intercept is near $x = -4/3$.

1.2.23

a.

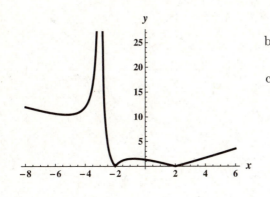

b. The domain of the function is the set of all real numbers except -3.

c. There is a valley near $x = -5.2$ and a peak near $x = -0.8$. The x-intercepts are at -2 and 2, where the curve does not appear to be smooth. There is a vertical asymptote at $x = -3$. The function is never below the x-axis. The y-intercept is $(0, 4/2)$.

1.2.25 The slope function is given by $s(x) = \begin{cases} 1 & \text{if } x < 0; \\ -1/2 & \text{if } x > 0. \end{cases}$

1.2.27

a. Since the area under consideration is that of a rectangle with base 2 and height 6, $A(2) = 12$.

b. Since the area under consideration is that of a rectangle with base 6 and height 6, $A(6) = 36$.

c. Since the area under consideration is that of a rectangle with base x and height 6, $A(x) = 6x$.

1.2.29 $f(x) = |x - 2| + 3$, since the graph of f is obtained from that of $|x|$ by shifting 2 units to the right and 3 units up.

$g(x) = -|x + 2| - 1$, since the graph of g is obtained from the graph of $|x|$ by shifting 2 units to the left, then reflecting about the x-axis, and then shifting 1 unit down.

1.2.31

a.

b.

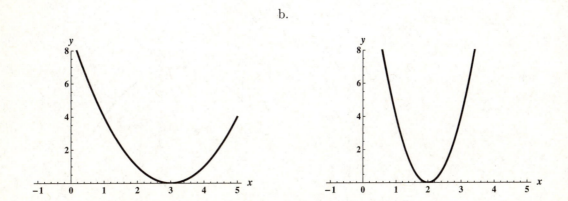

c.

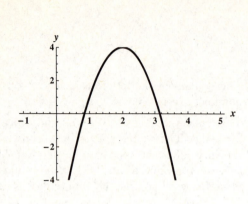

d.

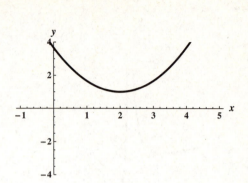

1.2.33

This function is $-3 \cdot f(x)$ where $f(x) = x^2$

1.2.35

This function is $2 \cdot f(x+3)$ where $f(x) = x^2$

1.2.37

By completing the square, we have that

$$h(x) = -4(x^2 + x - 3)$$
$$= -4\left(x^2 + x + \frac{1}{4} - \frac{1}{4} - 3\right)$$
$$= -4(x + (1/2))^2 + 13$$

So it is $-4f(x + (1/2)) + 13$ where $f(x) = x^2$.

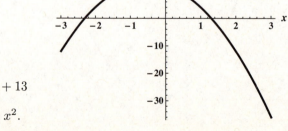

1.2.39

    a. True. A polynomial $p(x)$ can be written as the ratio of polynomials $\frac{p(x)}{1}$, so it is a rational function. However, a rational function like $\frac{1}{x}$ is not a polynomial.

    b. False. For example, if $f(x) = 2x$, then $(f \circ f)(x) = f(f(x)) = f(2x) = 4x$ is linear, not quadratic.

    c. True. In fact, if f is degree m and g is degree n, then the degree of the composition of f and g is $m \cdot n$, regardless of the order they are composed.

    d. False. The graph would be shifted two units to the left.

1.2.41 The points of intersection are found by solving $x^2 = -x^2 + 8x$. This yields the quadratic equation $2x^2 - 8x = 0$ or $(2x)(x - 4) = 0$. So the x-values of the points of intersection are 0 and 4. The actual points of intersection are $(0, 0)$ and $(4, 16)$.

1.2.43 $y = \sqrt{x} - 1$, since the y value is always 1 less than the square root of the x value.

1.2.45

$y = 5x$. The natural domain for the situation is $[0, h]$ where h represents the maximum number of hours that you can run at that pace before keeling over.

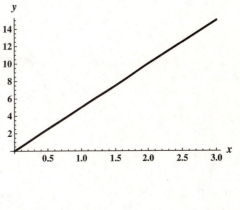

1.2.47

$y = \dfrac{3200}{x}$. Note that $\dfrac{x \text{ dollars per gallon}}{32 \text{ miles per gallon}} \cdot y \text{ miles}$ would represents the numbers of dollars, so this must be 100. So we have $\frac{xy}{32} = 100$, or $y = \frac{3200}{x}$. We certainly have $x > 0$, but unfortunately, there appears to be no no upper bound for x, so the domain is $(0, \infty)$.

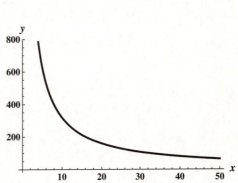

1.2.49

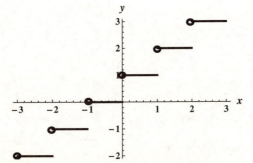

1.2.51

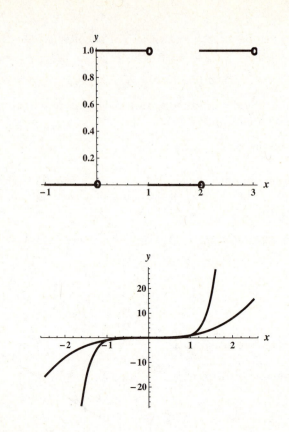

1.2.53

1.2.55

    a. By comparing various pairs of points, it appears that the slope of the line is about 328.3. At $t = 0$, the value of p is 1875. Therefore a line which reasonably approximates the data is $p(t) = 328.3t + 1875$.

    b. Using this line, we have that $p(9) = 4830$.

1.2.57

    a. Since you are paying \$350 per month, the amount paid after m months is $y = 350m + 1200$.

    b. After 4 years (48 months) you have paid $350 \cdot 48 + 1200 = 18000$ dollars. If you then buy the car for \$10,000, you will have paid a total of \$28,000 for the car instead of \$25,000. So you should buy the car instead of leasing it.

1.2.59

The function makes sense for $0 \le h \le 2$.

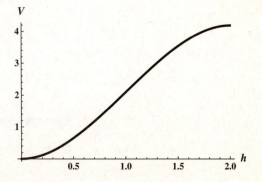

1.2.61

a. The volume of the box is x^2h, but since the box has volume 125 cubic feet, we have that $x^2h = 125$, so $h = \frac{125}{x^2}$. The surface area of the box is given by x^2 (the area of the base) plus $4 \cdot hx$, since each side has area hx. Thus $S = x^2 + 4hx = x^2 + \frac{4 \cdot 125 \cdot x}{x^2} = x^2 + \frac{500}{x}$.

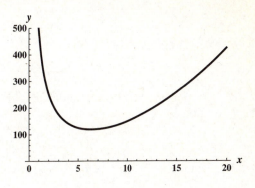

b. By inspection, it looks like the value of x which minimizes the surface area is about 6.3.

1.2.63 Suppose that the parabola f crosses the x-axis at a and b, with $a < b$. Then a and b are roots of the polynomial, so $(x - a)$ and $(x - b)$ are factors. Thus the polynomial must be $f(x) = c(x - a)(x - b)$ for some non-zero real number c. So $f(x) = cx^2 - c(a + b)x + abc$. Since the vertex always occurs at the x value which is $\frac{-\text{coefficient of } x}{2 \cdot \text{coefficient on } x^2}$ we have that the vertex occurs at $\frac{c(a + b)}{2c} = \frac{a + b}{2}$, which is halfway between a and b.

1.2.65

b.

a.

n	1	2	3	4	5
$n!$	1	2	6	24	120

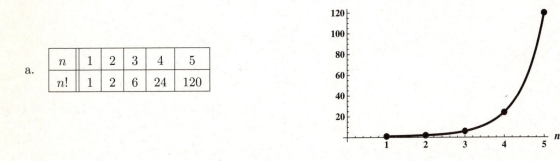

c. Using trial and error and a calculator yields that 10! is more than a million, but 9! isn't.

1.2.67

a.

n	1	2	3	4	5	6	7	8	9	10
$T(n)$	1	5	14	30	55	91	140	204	285	385

b. The domain of this function consists of the positive integers.

c. Using trial and error and a calculator yields that $T(n) > 1000$ for the first time for $n = 14$.

1.3 Trigonometric Functions and Their Inverses

1.3.1 Let O be the length of the side opposite the angle x, let A be length of the side adjacent to the angle x, and let H be the length of the hypotenuse. Then $\sin x = \frac{O}{H}$, $\cos x = \frac{A}{H}$, $\tan x = \frac{O}{A}$, $\csc x = \frac{H}{O}$, $\sec x = \frac{H}{A}$, and $\cot x = \frac{A}{O}$.

1.3.3 The radian measure of an angle θ is the length of the arc s on the unit circle associated with θ.

1.3.5 $\sin^2 x + \cos^2 x = 1$, $1 + \cot^2 x = \csc^2 x$, and $\tan^2 x + 1 = \sec^2 x$.

1.3.7 The tangent function is undefined where $\cos x = 0$, which is at all real numbers of the form $\frac{\pi}{2} + k\pi$, k an integer.

1.3.9 The point on the unit circle associated with $2\pi/3$ is $(-1/2, \sqrt{3}/2)$, so $\cos(2\pi/3) = -1/2$.

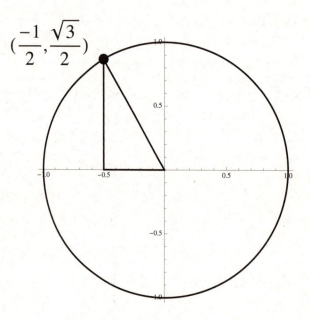

1.3.11

The point on the unit circle associated with $-3\pi/4$ is $(-\sqrt{2}/2, -\sqrt{2}/2)$, so

$$\tan(-3\pi/4) = 1.$$

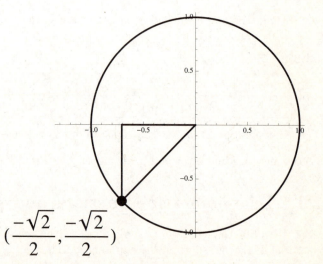

1.3.13

The point on the unit circle associated with $-13\pi/3$ is $(1/2, -\sqrt{3}/2)$, so

$$\cot(-13\pi/3) = -1/\sqrt{3} = -\sqrt{3}/3.$$

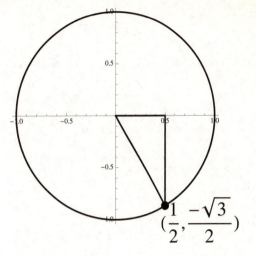

1.3.15

The point on the unit circle associated with $-17\pi/3$ is $(1/2, \sqrt{3}/2)$, so

$$\cot(-17\pi/3) = 1/\sqrt{3} = \sqrt{3}/3.$$

1.3.17 From our definitions of the trigonometric functions via a point $P(x,y)$ on a circle of radius $r = \sqrt{x^2 + y^2}$, we have

$$\tan^2\theta + 1 = \frac{y^2}{x^2} + 1 = \frac{y^2}{x^2} + \frac{x^2}{x^2} = \frac{r^2}{x^2} = \sec^2\theta.$$

1.3.19

Using the triangle pictured, we see that

$$\sec(\pi/2 - \theta) = \frac{c}{a} = \csc\theta.$$

This also follows from the sum identity

$$\cos(a + b) = \cos a \cos b - \sin a \sin b$$

as follows:

$$\sec(\pi/2 - \theta) = \frac{1}{\cos(\pi/2 + (-\theta))} = \frac{1}{\cos(\pi/2)\cos(-\theta) - \sin(\pi/2)\sin(-\theta)}$$

$$= \frac{1}{0 - (-\sin(\theta))} = \csc(\theta).$$

1.3.21 Using the fact that $\frac{\pi}{12} = \frac{\pi/6}{2}$ and the half-angle identity for cosine:

$$\cos^2(\pi/12) = \frac{1 + \cos(\pi/6)}{2} = \frac{1 + \sqrt{3}/2}{2} = \frac{2 + \sqrt{3}}{4}.$$

Thus, $\cos(\pi/12) = \sqrt{\frac{2 + \sqrt{3}}{4}}$.

1.3.23 First note that $\tan x = 1$ when $\sin x = \cos x$. Using our knowledge of the values of the standard angles between 0 and 2π, we recognize that the sine function and the cosine function are equal at $\pi/4$. Then, since we recall that the period of the tangent function is π, we know that $\tan(\pi/4 + k\pi) = \tan(\pi/4) = 1$ for every integer value of k. Thus the solution set is $\{\pi/4 + k\pi$, where k is an integer$\}$.

1.3.25 The equation $\sqrt{2}\sin(x) - 1 = 0$ can be written as $\sin x = \frac{1}{\sqrt{2}} = \frac{\sqrt{2}}{2}$. Standard solutions to this equation occur at $x = \pi/4$ and $x = 3\pi/4$. Since the sine function has period 2π the set of all solutions can be written as:

$$\{\pi/4 + 2k\pi, \text{where } k \text{ is an integer}\} \cup \{3\pi/4 + 2l\pi, \text{where } l \text{ is an integer}\}.$$

1.3.27 As in the previous problem, let $u = 3x$. Then we are interested in the solutions to $\cos u = \sin u$, for $0 \le u < 6\pi$.

This would occur for $u = 3x = \pi/4, 5\pi/4, 9\pi/4, 13\pi/4, 17\pi/4,$ and $21\pi/4$. Thus there are solutions for the original equation at

$$x = \pi/12, 5\pi/12, 9\pi/12, 13\pi/12, 17\pi/12, \text{and } 21\pi/12.$$

1.3.29

a. False. For example, $\sin(\pi/2 + \pi/2) = \sin(\pi) = 0 \neq \sin(\pi/2) + \sin(\pi/2) = 1 + 1 = 2$.

b. False. That equation has zero solutions, since the range of the cosine function is $[-1, 1]$.

c. False. It has infinitely many solutions of the form $\pi/6 + 2k\pi$, where k is an integer (among others.)

d. False. It has period $\frac{2\pi}{\pi/12} = 24$.

e. True. The others have a range of either $[-1, 1]$ or $(-\infty, -1] \cup [1, \infty)$.

1.3.31 If $\cos\theta = 5/13$, then the Pythagorean identity gives $|\sin\theta| = 12/13$. But if $0 < \theta < \pi/2$, then the sine of θ is positive, so $\sin\theta = 12/13$. Thus $\tan\theta = 12/5$, $\cot\theta = 5/12$, $\sec\theta = 13/5$, and $\csc\theta = 13/12$.

1.3.33 If $\csc\theta = 13/12$, then $\sin\theta = 12/13$, and the Pythagorean identity gives $|\cos\theta| = 5/13$. But if $0 < \theta < \pi/2$, then the cosine of θ is positive, so $\cos\theta = 5/13$. Thus $\tan\theta = 12/5$, $\cot\theta = 5/12$, and $\sec\theta = 13/5$.

1.3.35 The amplitude is 3, and the period is $\frac{2\pi}{1/3} = 6\pi$.

1.3.37 The amplitude is 3.6, and the period is $\frac{2\pi}{\pi/24} = 48$.

1.3.39

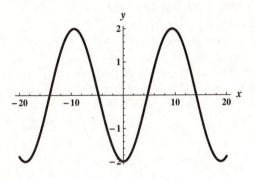

1.3.41

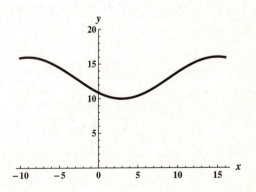

1.3.43

It is helpful to imagine first shifting the function horizontally so that the x intercept is where it should be, then stretching the function horizontally to obtain the correct period, and then stretching the function vertically to obtain the correct amplitude, and then shifting the whole graph up. Since the old x-intercept is at $x = 0$ and the new one should be at $x = 9$ (halfway between where the maximum and the minimum occur), we need to shift the function 9 units to the right. Then to get the right period, we need to multiply (before applying the sine function) by $\pi/12$ so that the new period is $\frac{2\pi}{\pi/12} = 24$. Finally, to get the right amplitude and to get the max and min at the right spots, we need to multiply on the outside by 3, and then shift the whole thing up 13 units. Thus, the desired function is:

$$f(x) = 3\sin((\pi/12)(x-9))+13 = 3\sin((\pi/12)x-3\pi/4)+13.$$

1.3.45 We are seeking a function with amplitude 10 and period 1.5, and value 10 at time 0, so it should have the form $10\cos(kt)$, where $\frac{2\pi}{k} = 1.5$. Solving for k yields $k = \frac{4\pi}{3}$, so the desired function is $d(t) = 10\cos(4\pi t/3)$.

1.3.47 There are various reasonable answers. One is $k + \sqrt{a^2 - h^2}$ (using the fact that the two legs of the 45–45–90 triangle are equal and using the Pythagorean theorem on the other triangle.)

1.3.49

To find $s(t)$ note that we are seeking a periodic function with period 365, and with amplitude 87.5 (which is half of the number of minutes between 7:25 and 4:30). We need to shift the function 4 days plus one fourth of 365, which is about 95 days so that the max and min occur at $t = 4$ days and at half a year later. Also, to get the right value for the maximum and minimum, we need to multiply by negative one and add 117.5 (which represents 30 minutes plus half the amplitude, since $s = 0$ corresponds to 4:00 AM.) Thus we have

$$s(t) = 117.5 - 87.5\sin\left(\frac{\pi}{182.5}(t - 95)\right).$$

A similar analysis leads to the formula

$$S(t) = 843.5 + 87.5\sin\left(\frac{\pi}{182.5}(t - 67)\right).$$

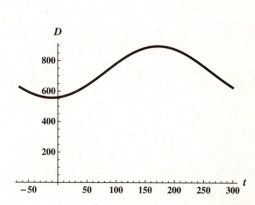

The graph pictured shows $D(t) = S(t) - s(t)$, the length of day function, which has its max at the summer solstice which is about the 172nd day of the year, and its min at the winter solstice.

1.3.51 Using the given diagram, drop a perpendicular from the point $(b\cos\theta, b\sin\theta)$ to the x axis, and consider the right triangle thus formed whose hypotenuse has length c. By the Pythagorean theorem, $(b\sin\theta)^2 + (a - b\cos\theta)^2 = c^2$. Expanding the binomial gives $b^2\sin^2\theta + a^2 - 2ab\cos\theta + b^2\cos^2\theta = c^2$. Now since $b^2\sin^2\theta + b^2\cos^2\theta = b^2$, this reduces to $a^2 + b^2 - 2ab\cos\theta = c^2$.

1.4   Chapter One Review

1.4.1

a. True. For example, $f(x) = x^2$ is such a function.

b. False. For example, $\cos(\pi/2 + \pi/2) = \cos(\pi) = -1 \neq \cos(\pi/2) + \cos(\pi/2) = 0 + 0 = 0$.

c. False. Consider $f(1 + 1) = f(2) = 2m + b \neq f(1) + f(1) = (m + b) + (m + b) = 2m + 2b$. (At least these aren't equal when $b \neq 0$.)

d. True. $f(f(x)) = f(1 - x) = 1 - (1 - x) = x$.

e. False. This set is the union of the disjoint intervals $(-\infty, -7)$ and $(1, \infty)$.

1.4.3

a.

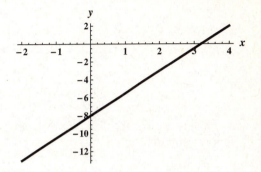

This line has slope $\frac{2-(-3)}{4-2} = 5/2$. Therefore the equation of the line is $y - 2 = \frac{5}{2}(x - 4)$, so $y = \frac{5}{2}x - 8$.

b.

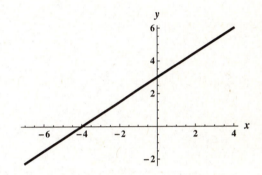

This line has the form $y = (3/4)x + b$, and since $(-4, 0)$ is on the line, $0 = (3/4)(-4) + b$, so $b = 3$. Thus the equation of the line is given by $y = (3/4)x + 3$.

c.

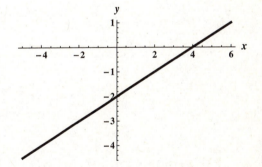

This line has slope $\frac{0-(-2)}{4-0} = \frac{1}{2}$, and the y-intercept is given to be -2, so the equation of this line is $y = (1/2)x - 2$.

1.4.5

Since $|x| = \begin{cases} -x & \text{if } x < 0; \\ x & \text{if } x \geq 0, \end{cases}$

we have

$2(x - |x|) = \begin{cases} 2(x - (-x)) = 4x & \text{if } x < 0; \\ 2(x - x) = 0 & \text{if } x \geq 0. \end{cases}$

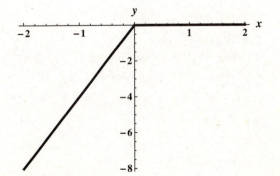

1.4.7

a.

This is a straight line with slope 2/3 and y-intercept 10/3.

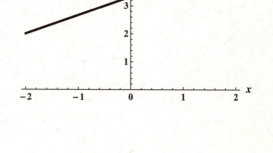

b.

Completing the square gives $y = (x^2 + 2x + 1) - 4$, or $y = (x+1)^2 - 4$, so this is the standard parabola shifted one unit to the left and down 4 units.

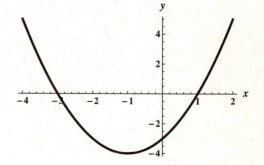

c.

Completing the square, we have $x^2 + 2x + 1 + y^2 + 4y + 4 = -1 + 1 + 4$, so we have $(x+1)^2 + (y+2)^2 = 4$, a circle of radius 2 centered at $(-1, -2)$.

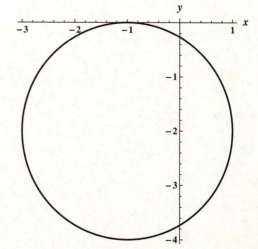

d.

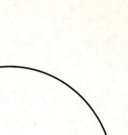

Completing the square, we have $x^2 - 2x + 1 + y^2 - 8y + 16 = -5 + 1 + 16$, or $(x-1)^2 + (y-4)^2 = 12$, which is a circle of radius $\sqrt{12}$ centered at $(1, 4)$.

1.4.9 The domain of $x^{1/7}$ is the set of all real numbers, as is its range. The domain of $x^{1/4}$ is the set of non-negative real numbers, as is its range.

1.4.11 We are looking for the line between the points $(0, 212)$ and $(6000, 180)$. The slope is $\frac{212-180}{0-6000} = -32/6000 = -2/375$. Since the intercept is given, we deduce that the line is $B = f(a) = (-2/375)(a) + 212$.

1.4.13

a.   b.

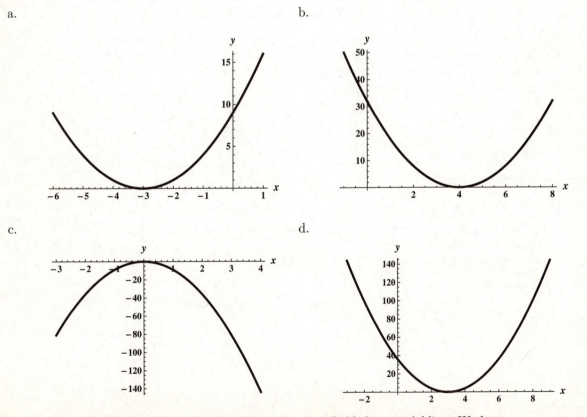

c.   d.

1.4.15

a. $h(g(\pi/2)) = h(1) = 1$

b. $h(f(x)) = h(x^3) = x^{3/2}$.

c. $f(g(h(x))) = f(g(\sqrt{x})) = f(\sin(\sqrt{x})) = (\sin(\sqrt{x}))^3$.

d. The domain of $g(f(x))$ is $(-\infty, \infty)$, since the domain of both functions is the set of all real numbers.

1.4.17

a. Since $f(-x) = \cos -3x = \cos 3x = f(x)$, this is an even function, and is symmetric about the y-axis.

b. Since $f(-x) = 3(-x)^4 - 3(-x)^2 + 1 = 3x^4 - 3x^2 + 1 = f(x)$, this is an even function, and is symmetric about the y-axis.

c. Since replacing x by $-x$ and/or replacing y by $-y$ gives the same equation, this represents a curve which is symmetric about the y-axis and about the origin and about the x-axis.

1.4.19

a. A $135°$ angle measures $135 \cdot (\pi/180)$ radians, which is about 2.356 radians.

b. A $4\pi/5$ radian angle measues $4\pi/5 \cdot (180/\pi)$ degrees, which is $144°$.

c. Since the length of the arc is the measure of the subtended angle (in radians) times the radius, this arc would be $4\pi/3 \cdot 10 = \frac{40\pi}{3}$ units long.

1.4.21

a. We need to scale the ordinary cosine function so that its period is 6, and then shift it 3 units to the right, and multiply it by 2. So the function we seek is $y = 2\cos((\pi/3)(t-3))$.

b. We need to scale the ordinary cosine function so that its period is 24, and then shift it to the right 6 units. We then need to change the amplitude to be half the difference between the maximum and minimum, which would be 5. Then finally we need to shift the whole thing up by 15 units. The function we seek is thus $y = 15 + 5\cos((\pi/12)(t-6))$.

1.4.23

a. $-\sin x$ is pictured in F.

b. $\cos 2x$ is pictured in E.

c. $\tan(x/2)$ is pictured in D.

d. $-\sec x$ is pictured in B.

e. $\cot 2x$ is pictured in C.

f. $\sin^2 x$ is pictured in A.

Chapter 2

Limits

2.1 The Idea of Limits

2.1.1 The average velocity of the object between time $t = a$ and $t = b$ is the change in position divided by the elapsed time: $v_{\mathrm{av}} = \frac{s(b)-s(a)}{b-a}$.

2.1.3 The slope of the secant line between points $(a, f(a))$ and $(b, f(b))$ is the ratio of the differences $f(b) - f(a)$ and $b - a$. Thus $m_{\mathrm{sec}} = \frac{f(b)-f(a)}{b-a}$.

2.1.5 Both problems involve the same mathematics, namely finding the limit as $t \to a$ of a quotient of differences of the form $\frac{g(t)-g(a)}{t-a}$ for some function g.

2.1.7

a. Over $[1, 4]$, we have $v_{\mathrm{av}} = \dfrac{s(4) - s(1)}{4 - 1} = \dfrac{256 - 112}{3} = 48.$

b. Over $[1, 3]$, we have $v_{\mathrm{av}} = \dfrac{s(3) - s(1)}{3 - 1} = \dfrac{240 - 112}{2} = 64.$

c. Over $[1, 2]$, we have $v_{\mathrm{av}} = \dfrac{s(2) - s(1)}{2 - 1} = \dfrac{192 - 112}{1} = 80.$

d. Over $[1, 1 + h]$, we have

$$v_{\mathrm{av}} = \frac{s(1 + h) - s(1)}{1 + h - 1} = \frac{-16(1 + h)^2 + 128(1 + h) - (112)}{h} = \frac{-16h^2 - 32h + 128h}{h}$$
$$= \frac{h(-16h + 96)}{h} = 96 - 16h.$$

2.1.9

Time Interval	$[1, 2]$	$[1, 1.5]$	$[1, 1.1]$	$[1, 1.01]$	$[1, 1.001]$
Average Velocity	80	88	94.4	95.84	95.984

The instantaneous velocity appears to be 96 ft/s.

2.1.11

Time Interval	$[2, 3]$	$[2.9, 3]$	$[2.99, 3]$	$[2.999, 3]$	$[2.9999, 3]$
Average Velocity	20	5.6	4.16	4.016	4.0016

The instantaneous velocity appears to be 4 ft/s.

2.1.13

Time Interval	$[3, 3.1]$	$[3, 3.01]$	$[3, 3.001]$	$[3, 3.0001]$
Average Velocity	-17.6	-16.16	-16.016	-16.002

The instantaneous velocity appears to be -16 ft/s.

2.1.15

Time Interval	[0, 0.1]	[0, 0.01]	[0, 0.001]	[0, 0.0001]
Average Velocity	79.4677	79.9947	79.9999	80.0000

The instantaneous velocity appears to be 80 ft/s.

2.1.17

x Interval	[2, 2.1]	[2, 2.01]	[2, 2.001]	[2, 2.0001]
Slope of Secant Line	8.2	8.02	8.002	8.0002

The slope of the tangent line appears to be 8.

2.1.19

x Interval	[1, 1.1]	[1, 1.01]	[1, 1.001]	[1, 1.0001]
Slope of the Secant Line	-2.1	-2.01	-2.001	-2.0001

The slope of the tangent line appears to be -2.

2.1.21

a. Note that the graph is a parabola with vertex $(2, -1)$.

b. At $(2, -1)$ the function has tangent line with slope 0.

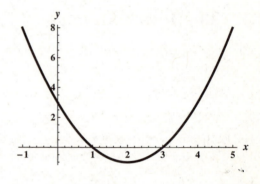

c.

x Interval	[2, 2.1]	[2, 2.01]	[2, 2.001]	[2, 2.0001]
Slope of the Secant Line	.1	.01	.001	.0001

The slope of the tangent line at $(2, -1)$ appears to be 0.

2.1.23

a. Note that the graph is a parabola with vertex $(4, 448)$.

b. At $(4, 448)$ the function has tangent line with slope 0, so $a = 4$.

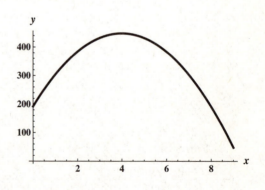

c.

x Interval	[4, 4.1]	[4, 4.01]	[4, 4.001]	[4, 4.0001]
Slope of the Secant Line	-1.6	$-.16$	$-.016$	$-.0016$

The slopes of the secant lines appear to be approaching zero.

d. On the interval $[0, 4)$ the instantaneous velocity of the projectile is positive.

e. On the interval $(4, 9]$ the instantaneous velocity of the projectile is negative.

2.1.25 For line AD, we have

$$m_{AD} = \frac{y_D - y_A}{x_D - x_A} = \frac{f(\pi) - f(\pi/2)}{\pi - (\pi/2)} = \frac{1}{\pi/2} \approx .63662.$$

For line AC, we have

$$m_{AC} = \frac{y_C - y_A}{x_C - x_A} = \frac{f(\pi/2 + .5) - f(\pi/2)}{(\pi/2 + .5) - (\pi/2)} = \frac{-\cos(\pi/2 + .5)}{.5} \approx .958851.$$

For line AB, we have

$$m_{AB} = \frac{y_B - y_A}{x_B - x_A} = \frac{f(\pi/2 + .05) - f(\pi/2)}{(\pi/2 + .05) - (\pi/2)} = \frac{-\cos(\pi/2 + .05)}{.05} \approx .999583.$$

Computing one more slope of a secant line:

$$m_{\text{sec}} = \frac{f(\pi/2 + .01) - f(\pi/2)}{(\pi/2 + .01) - (\pi/2)} = \frac{-\cos(\pi/2 + .01)}{.01} \approx .999983.$$

Conjecture: The slope of the tangent line to the graph of f at $x = \pi/2$ is 1.

2.2 Definitions of Limits

2.2.1 Suppose the function f is defined for all x near a except possibly at a. If $f(x)$ is arbitrarily close to a number L whenever x is sufficiently close to (but not equal to) a, then we write $\lim_{x \to a} f(x) = L$.

2.2.3 Suppose the function f is defined for all x near a but greater than a. If $f(x)$ is arbitrarily close to L for x sufficiently close to (but strictly greater than) a, then $\lim_{x \to a^+} f(x) = L$.

2.2.5 It must be true that $L = M$.

2.2.7

a. $h(2) = 5$

b. $\lim_{x \to 2} h(x) = 3$

c. $h(4)$ does not exist.

d. $\lim_{x \to 4} f(x) = 1$.

e. $\lim_{x \to 5} h(x) = 2$.

2.2.9

a. $f(1) = -1$

b. $\lim_{x \to 1} f(x) = 1$.

c. $f(0) = 2$.

d. $\lim_{x \to 0} f(x) = 2$.

2.2.11

a.

x	1.9	1.99	1.999	1.9999	2	2.0001	2.001	2.01	2.1
$f(x) = \frac{x^2-4}{x-2}$	3.9	3.99	3.999	3.9999	undefined	4.0001	4.001	4.01	4.1

b. $\lim\limits_{x\to 2} f(x) = 4$.

2.2.13

a.

t	8.9	8.99	8.999	9	9.001	9.01	9.1
$g(t) = \frac{t-9}{\sqrt{t}-3}$	5.98329	5.99833	5.99983	undefined	6.00017	6.00167	6.01662

b. $\lim\limits_{t\to 9} \dfrac{t-9}{\sqrt{t}-3} = 6$.

2.2.15

x	4.9	4.99	4.999	4.9999	5	5.0001	5.001	5.01	5.1
$f(x) = \frac{x^2-25}{x-5}$	9.9	9.99	9.999	9.9999	undefined	10.0001	10.001	10.01	10.1

$\lim\limits_{x\to 5^+} \dfrac{x^2-25}{x-5} = 10$, $\lim\limits_{x\to 5^-} \dfrac{x^2-25}{x-5} = 10$, and thus $\lim\limits_{x\to 5} \dfrac{x^2-25}{x-5} = 10$

2.2.17

a. $f(1) = 0$.

b. $\lim\limits_{x\to 1^-} f(x) = 1$.

c. $\lim\limits_{x\to 1^+} f(x) = 0$.

d. $\lim\limits_{x\to 1} f(x)$ does not exist, since the two one-sided limits aren't equal.

2.2.19

a. $f(1) = 3$

b. $\lim\limits_{x\to 1^-} f(x) = 2$.

c. $\lim\limits_{x\to 1^+} f(x) = 2$

d. $\lim\limits_{x\to 1} f(x) = 2$.

e. $f(3) = 2$.

f. $\lim\limits_{x\to 3^-} f(x) = 4$.

g. $\lim\limits_{x\to 3^+} f(x) = 1$.

h. $\lim\limits_{x\to 3} f(x)$ does not exist.

i. $f(2) = 3$.

j. $\lim\limits_{x\to 2^-} f(x) = 3$.

k. $\lim\limits_{x\to 2^+} f(x) = 3$.

l. $\lim\limits_{x\to 2} f(x) = 3$.

2.2.21

a.

x	$\frac{2}{\pi}$	$\frac{2}{3\pi}$	$\frac{2}{5\pi}$	$\frac{2}{7\pi}$	$\frac{2}{9\pi}$	$\frac{2}{11\pi}$
$f(x) = \sin(1/x)$	1	-1	1	-1	1	-1

If $x_n = \frac{2}{(2n+1)\pi}$, then $f(x_n) = (-1)^n$ where n is a positive integer.

b. As $x \to 0$, $1/x \to \infty$. So the values of $f(x)$ oscillate dramatically between -1 and 1.

c. $\lim\limits_{x\to 0} \sin(1/x)$ does not exist.

2.2.23

a. False. In fact $\lim\limits_{x\to 3} \dfrac{x^2-9}{x-3} = \lim\limits_{x\to 3} x + 3 = 6$.

b. False. For example, if $f(x) = \begin{cases} x^2 & \text{if } x \neq 0; \\ 5 & \text{if } x = 0 \end{cases}$ and if $a = 0$ then $f(a) = 5$ but $\lim\limits_{x\to a} f(x) = 0$.

c. False. For example, the limit in part a. of this problem existed, even though the corresponding function was undefined at $a = 3$.

2.2.25

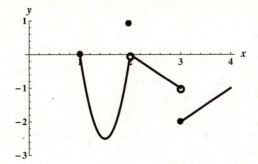

2.2.27

h	.01	.001	.0001	$-.01$	$-.001$	$-.0001$
$\tan(3h)/h$	3.0009	3.000009	3.00000009	3.0009	3.000009	3.00000009

$$\lim_{h \to 0} \frac{\tan(3h)}{h} = 3$$

2.2.29

h	.01	.001	.0001	$-.01$	$-.001$	$-.0001$
$(1 - \cos(h))/h$	.0049999	.0005	.00005	$-.0049999$	$-.0005$	$-.00005$

$$\lim_{h \to 0} \frac{1 - \cos(h)}{h} = 0$$

2.2.31

a. $\lim\limits_{x \to -1^-} \lfloor x \rfloor = -2$, $\lim\limits_{x \to -1^+} \lfloor x \rfloor = -1$, $\lim\limits_{x \to 2^-} \lfloor x \rfloor = 1$, $\lim\limits_{x \to 2^+} \lfloor x \rfloor = 2$.

b. $\lim\limits_{x \to 2.3^-} \lfloor x \rfloor = 2$, $\lim\limits_{x \to 2.3^+} \lfloor x \rfloor = 2$, $\lim\limits_{x \to 2.3} \lfloor x \rfloor = 2$.

c. In general, for an integer a, $\lim\limits_{x \to a^-} \lfloor x \rfloor = a - 1$ and $\lim\limits_{x \to a^+} \lfloor x \rfloor = a$.

d. In general, if a is not an integer, $\lim\limits_{x \to a^-} \lfloor x \rfloor = \lim\limits_{x \to a^+} \lfloor x \rfloor = \lfloor a \rfloor$.

e. $\lim\limits_{x \to a} \lfloor x \rfloor$ exists and is equal to $\lfloor a \rfloor$ for non-integers a.

2.2.33 $\lim\limits_{x \to 1} \dfrac{\sqrt{2x - x^4} - \sqrt[3]{x}}{1 - x^{3/4}} \approx 1.7777$.

2.2.35

a. Note that the function is piecewise constant.

b. $\lim\limits_{w \to 3.3} f(w) = .93$.

c. $\lim\limits_{w \to 1^+} f(w) = .59$ corresponds to the fact that for any piece of mail that weighs slightly over 1 ounce, the postage will cost 59 cents. $\lim\limits_{w \to 1^-} f(w) = .42$ corresponds to the fact that for any piece of mail that weighs slightly less than 1 ounce, the postage will cost 42 cents.

d. $\lim\limits_{w \to 4} f(w)$ does not exist because the two corresponding one-side limits don't exist. (The limit from the left is .93, while the limit from the right is 1.10.)

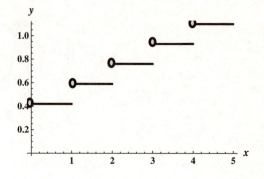

2.2.37

x	.1	.01	.001	.0001	.00001
$\frac{\sin(x^{20})}{x^{20}}$	1.	1.	1.	1.	1.

Yes, $\lim\limits_{x \to 0} \dfrac{\sin(x^{20})}{x^{20}} = 1$.

2.2.39

a. Because of the symmetry about the y axis, we must have $\lim\limits_{x \to -2^+} f(x) = 8$.

b. Because of the symmetry about the y axis, we must have $\lim\limits_{x \to -2^-} f(x) = 5$.

2.2.41

a.

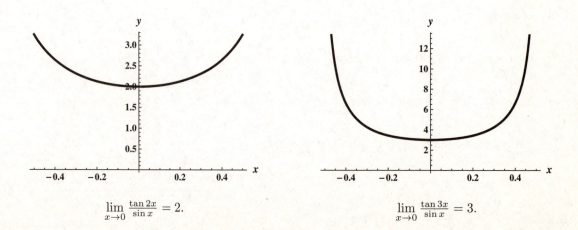

$$\lim\limits_{x \to 0} \frac{\tan 2x}{\sin x} = 2.$$ $$\lim\limits_{x \to 0} \frac{\tan 3x}{\sin x} = 3.$$

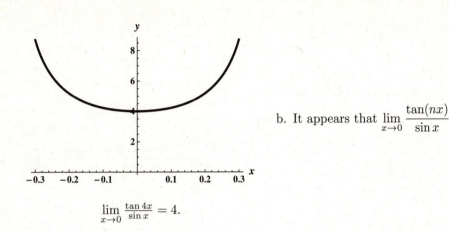

b. It appears that $\displaystyle\lim_{x\to 0}\frac{\tan(nx)}{\sin x} = n$.

$$\lim_{x\to 0}\frac{\tan 4x}{\sin x} = 4.$$

2.2.43

For $n = 8$ and $m = 2$, it appears that the limit is 4.

For $n = 12$ and $m = 3$, it appears that the limit is 4.

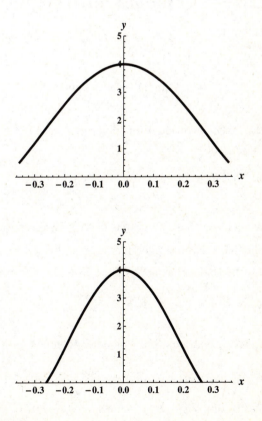

For $n = 4$ and $m = 16$, it appears that the limit is $1/4$.

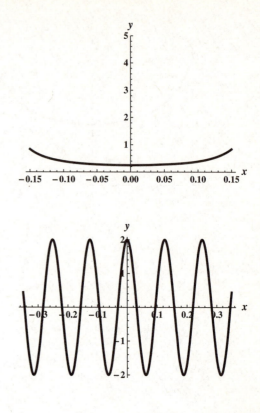

For $n = 100$ and $m = 50$, it appears that the limit is 2.

Conjecture: $\lim_{x \to 0} \dfrac{\sin nx}{\sin mx} = \dfrac{n}{m}$.

2.3   Techniques for Computing Limits

2.3.1 If $f(x) = a_n x^n + a_{n-1} x^{n-2} + \ldots + a_1 x + a_0$, then

$$\lim_{x \to a} f(x) = \lim_{x \to a} (a_n x^n + a_{n-1} x^{n-2} + \ldots + a_1 x + a_0)$$
$$= a_n (\lim_{x \to a} x)^n + a_{n-1} (\lim_{x \to a} x)^{n-1} + \ldots + a_1 \lim_{x \to a} x + \lim_{x \to a} a_0 = a_n a^n + a_{n-1} a^{na-1} + \ldots + a_1 a + a_0.$$

2.3.3 For a rational function $r(x)$, we have $\lim_{x \to a} r(x) = r(a)$ exactly for those numbers a which are in the domain of r.

2.3.5 Since

$$\frac{x^2 - 7x + 12}{x - 3} = \frac{(x - 3)(x - 4)}{x - 3} = x - 4 \text{ (for } x \neq 3),$$

we can see that the graphs of these two functions are the same except that one is undefined at $x = 3$ and the other is a straight line that is defined everywhere. Thus the function $\dfrac{x^2 - 7x + 12}{x - 3}$ is a straight line except that it has a "hole" at $(3, -1)$. The two functions therefore have the same limit as $x \to 3$, namely $\lim_{x \to 3} \dfrac{x^2 - 7x + 12}{x - 3} = \lim_{x \to 3} x - 4 = -1$.

2.3.7 If p and q are polynomials then $\lim_{x \to 0} \dfrac{p(x)}{q(x)} = \dfrac{\lim_{x \to 0} p(x)}{\lim_{x \to 0} q(x)} = \dfrac{p(0)}{q(0)}$. Since this quantity is given to be equal to 10, and $q(0) = 2$, we have $\frac{p(0)}{2} = 10$, so $p(0) = 20$.

2.3.9 $\lim_{x \to 5} \sqrt{x^2 - 9} = \sqrt{\lim_{x \to 5} x^2 - 9} = \sqrt{16} = 4$.

2.3.11 $\lim\limits_{x \to 4}(3x - 7) = 3\lim\limits_{x \to 4} x - 7 = 3 \cdot 4 - 7 = 5.$

2.3.13 $\lim\limits_{x \to -9}(5x) = 5\lim\limits_{x \to -9} x = 5 \cdot -9 = -45.$

2.3.15 $\lim\limits_{x \to 6} 4 = 4.$

2.3.17 $\lim\limits_{x \to 1} 4f(x) = 4\lim\limits_{x \to 1} f(x) = 4 \cdot 8 = 32.$ This follows from the Constant Multiple Law.

2.3.19 $\lim\limits_{x \to 1} \dfrac{f(x)g(x)}{h(x)} = \dfrac{\lim\limits_{x \to 1}(f(x)g(x))}{\lim\limits_{x \to 1} h(x)} = \dfrac{\lim\limits_{x \to 1} f(x) \cdot \lim\limits_{x \to 1} g(x)}{\lim\limits_{x \to 1} h(x)} = \dfrac{8 \cdot 3}{2} = 12.$ This follows from the Quotient and Product Laws.

2.3.21 $\lim\limits_{x \to 1}(h(x))^5 = \left(\lim\limits_{x \to 1} h(x)\right)^5 = (2)^5 = 32.$ The follows from the Power Law.

2.3.23

$$\lim_{x \to 1} 2x^3 - 3x^2 + 4x + 5 = \lim_{x \to 1} 2x^3 - \lim_{x \to 1} 3x^2 + \lim_{x \to 1} 4x + \lim_{x \to 1} 5$$
$$= 2(\lim_{x \to 1} x)^3 - 3(\lim_{x \to 1} x)^2 + 4(\lim_{x \to 1} x) + 5$$
$$= 2(1)^3 - 3(1)^2 + 4 \cdot 1 + 5 = 8.$$

2.3.25 $\lim\limits_{x \to 1} \dfrac{5x^2 + 6x + 1}{8x - 4} = \dfrac{\lim\limits_{x \to 1}(5x^2 + 6x + 1)}{\lim\limits_{x \to 1}(8x - 4)} = \dfrac{5(\lim\limits_{x \to 1} x)^2 + 6\lim\limits_{x \to 1} x + \lim\limits_{x \to 1} 1}{8\lim\limits_{x \to 1} x - \lim\limits_{x \to 1} 4} = \dfrac{5(1)^2 + 6 \cdot 1 + 1}{8 \cdot 1 - 4} = 3.$

2.3.27 $\lim\limits_{b \to 2} \dfrac{3b}{\sqrt{4b + 1} - 1} = \dfrac{\lim\limits_{b \to 2} 3b}{\lim\limits_{b \to 2}(\sqrt{4b + 1} - 1)} = \dfrac{3\lim\limits_{b \to 2} b}{\lim\limits_{b \to 2}\sqrt{4b + 1} - \lim\limits_{b \to 2} 1} = \dfrac{3 \cdot 2}{\sqrt{\lim\limits_{b \to 2} 4b + 1} - 1} = \dfrac{6}{3 - 1} = 3.$

2.3.29 $\lim\limits_{x \to 3} \dfrac{-5x}{\sqrt{4x - 3}} = \dfrac{\lim\limits_{x \to 3} -5x}{\lim\limits_{x \to 3}\sqrt{4x - 3}} = \dfrac{-5\lim\limits_{x \to 3} x}{\sqrt{\lim\limits_{x \to 3}(4x - 3)}} = \dfrac{-5 \cdot 3}{\sqrt{4\lim\limits_{x \to 3} x - \lim\limits_{x \to 3} 3}} = \dfrac{-15}{\sqrt{4 \cdot 3 - 3}} = -5.$

2.3.31

  a. $\lim\limits_{x \to -1^-} f(x) = \lim\limits_{x \to -1^-} x^2 + 1 = (-1)^2 + 1 = 2.$

  b. $\lim\limits_{x \to -1^+} f(x) = \lim\limits_{x \to -1^+} \sqrt{x + 1} = \sqrt{-1 + 1} = 0.$

  c. $\lim\limits_{x \to -1} f(x)$ does not exist.

2.3.33

  a. $\lim\limits_{x \to 2^+} \sqrt{x - 2} = \sqrt{2 - 2} = 0.$

  b. The domain of $f(x) = \sqrt{x - 2}$ is $[2, \infty)$. Thus, any question about this function that involves numbers less than 2 doesn't make any sense, since those numbers aren't in the domain of f.

2.3.35 Using the definition of $|x|$ given, we have $\lim\limits_{x \to 0^-} |x| = \lim\limits_{x \to 0^-} -x = -0 = 0.$ Also, $\lim\limits_{x \to 0^+} |x| = \lim\limits_{x \to 0^+} x = 0.$ Since the two one-sided limits are both 0, we also have $\lim\limits_{x \to 0} |x| = 0.$

2.3.37 $\lim\limits_{x \to 1} \dfrac{x^2 - 1}{x - 1} = \lim\limits_{x \to 1} \dfrac{(x+1)(x-1)}{x-1} = \lim\limits_{x \to 1} x + 1 = 2.$

2.3.39 $\lim\limits_{x \to 4} \dfrac{x^2 - 16}{4 - x} = \lim\limits_{x \to 4} \dfrac{(x+4)(x-4)}{-(x-4)} = \lim\limits_{x \to 4} -(x+4) = -8.$

2.3.41

$$\lim_{x \to b} \frac{(x-b)^{50} - x + b}{x - b} = \lim_{x \to b} \frac{(x-b)^{50} - (x-b)}{x - b} = \lim_{x \to b} \frac{(x-b)((x-b)^{49} - 1)}{x - b}$$
$$= \lim_{x \to b}(x-b)^{49} - 1 = -1.$$

2.3.43

$$\lim_{x \to -1} \frac{(2x-1)^2 - 9}{x + 1} = \lim_{x \to -1} \frac{(2x - 1 - 3)(2x - 1 + 3)}{x + 1} = \lim_{x \to -1} \frac{(2)(x-2)(2)(x+1)}{x + 1}$$
$$= \lim_{x \to -1} 4(x-2) = 4 \cdot (-3) = -12.$$

2.3.45 $\lim\limits_{x \to 9} \dfrac{\sqrt{x} - 3}{x - 9} = \lim\limits_{x \to 9} \dfrac{(\sqrt{x} - 3)(\sqrt{x} + 3)}{(x - 9)(\sqrt{x} + 3)} = \lim\limits_{x \to 9} \dfrac{x - 9}{(x - 9)(\sqrt{x} + 3)} = \lim\limits_{x \to 9} \dfrac{1}{\sqrt{x} + 3} = \dfrac{1}{6}.$

2.3.47

$$\lim_{h \to 0} \frac{\sqrt{16 + h} - 4}{h} = \lim_{h \to 0} \frac{(\sqrt{16+h} - 4)(\sqrt{16+h} + 4)}{h(\sqrt{16+h} + 4)} = \lim_{h \to 0} \frac{(16 + h) - 16}{h(\sqrt{16+h} + 4)}$$
$$= \lim_{h \to 0} \frac{h}{h(\sqrt{16+h} + 4)} = \lim_{h \to 0} \frac{1}{(\sqrt{16+h} + 4)} = \frac{1}{8}.$$

2.3.49

a. The statement we are trying to prove can be stated in cases as follows: For $x > 0$, $-x \le x \sin(1/x) \le x$, and for $x < 0$, $x \le x \sin(1/x) \le -x$.

Now for all $x \ne 0$, note that $-1 \le \sin(1/x) \le 1$ (since the range of the sine function is $[-1, 1]$). We will consider the two cases $x > 0$ and $x < 0$ separately, but in each case, we will multiply this inequality through by x, switching the inequalities for the $x < 0$ case.

For $x > 0$ we have $-x \le x \sin(1/x) \le x$, and for $x < 0$ we have $-x \ge x \sin(1/x) \ge x$, which are exactly the statements we are trying to prove.

b.

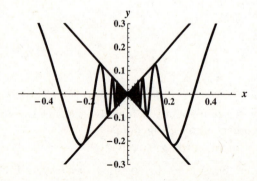

c. Since $\lim\limits_{x \to 0} -|x| = \lim\limits_{x \to 0} |x| = 0$, and since $-|x| \le x \sin(1/x) \le |x|$, the squeeze theorem assures us that $\lim\limits_{x \to 0} x \sin(1/x) = 0$ as well.

2.3.51

a.

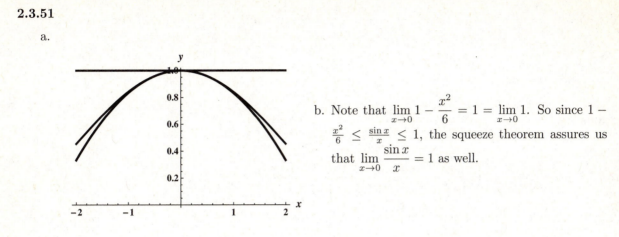

b. Note that $\lim\limits_{x\to0} 1 - \dfrac{x^2}{6} = 1 = \lim\limits_{x\to0} 1$. So since $1 - \dfrac{x^2}{6} \le \dfrac{\sin x}{x} \le 1$, the squeeze theorem assures us that $\lim\limits_{x\to0} \dfrac{\sin x}{x} = 1$ as well.

2.3.53

a. False. For example, if $f(x) = \begin{cases} x & \text{if } x \neq 1; \\ 4 & \text{if } x = 1, \end{cases}$ then $\lim\limits_{x\to1} f(x) = 1$ but $f(1) = 4$.

b. False. For example, if $f(x) = \begin{cases} x+1 & \text{if } x \le 1; \\ x-6 & \text{if } x > 1, \end{cases}$ then $\lim\limits_{x\to1^-} f(x) = 2$ but $\lim\limits_{x\to1^+} f(x) = -5$.

c. False. For example, if $f(x) = \begin{cases} x & \text{if } x \neq 1; \\ 4 & \text{if } x = 1, \end{cases}$ and $g(x) = 1$, then f and g both have limit 1 as $x \to 1$, but $f(1) = 4 \neq g(1)$.

d. False. For example $\lim\limits_{x\to2} \dfrac{x^2 - 4}{x - 2}$ exists and is equal to 4.

e. False. For example, it would be possible for the domain of f to be $[1, \infty)$, so that the one-sided limit exists but the two-sided limit doesn't even make sense. This would be true, for example, if $f(x) = x - 1$.

2.3.55 $\lim\limits_{x\to2}(5x - 6)^{3/2} = (5 \cdot 2 - 6)^{3/2} = 4^{3/2} = 2^3 = 8$.

2.3.57

$$\lim_{x\to1} \frac{\sqrt{10x - 9} - 1}{x - 1} = \lim_{x\to1} \frac{(\sqrt{10x - 9} - 1)(\sqrt{10x - 9} + 1)}{(x - 1)(\sqrt{10x - 9} + 1)} = \lim_{x\to1} \frac{(10x - 9) - 1}{(x - 1)(\sqrt{10x - 9} + 1)}$$

$$= \lim_{x\to1} \frac{10(x - 1)}{(x - 1)(\sqrt{10x - 9} + 1)} = \lim_{x\to1} \frac{10}{(\sqrt{10x - 9} + 1)} \overset{\cdot}{=} \frac{10}{2} = 5.$$

2.3.59 $\lim\limits_{h\to0} \dfrac{(5 + h)^2 - 25}{h} = \lim\limits_{h\to0} \dfrac{25 + 10h + h^2 - 25}{h} = \lim\limits_{h\to0} \dfrac{h(10 + h)}{h} = \lim\limits_{h\to0} 10 + h = 10.$

2.3.61 $\lim\limits_{w\to-k} \dfrac{w^2 + 5kw + 4k^2}{w^2 + kw} = \lim\limits_{w\to-k} \dfrac{(w + 4k)(w + k)}{(w)(w + k)} = \lim\limits_{w\to-k} \dfrac{w + 4k}{w} = \dfrac{-k + 4k}{-k} = -3.$

2.3.63 In order for $\lim\limits_{x\to-1} g(x)$ to exist, we need the two one-sided limits to exist and be equal. We have $\lim\limits_{x\to-1^-} g(x) = \lim\limits_{x\to-1^-} x^2 - 5x = 6$, and $\lim\limits_{x\to-1^+} g(x) = \lim\limits_{x\to-1^+} ax^3 - 7 = -a - 7$. So we need $-a - 7 = 6$, so we require that $a = -13$.

2.3.65 $\lim_{x \to 1} \dfrac{x^6 - 1}{x - 1} = \lim_{x \to 1} \dfrac{(x-1)(x^5 + x^4 + x^3 + x^2 + x + 1)}{x - 1} = \lim_{x \to 1} x^5 + x^4 + x^3 + x^2 + x + 1 = 6.$

2.3.67 $\lim_{x \to a} \dfrac{x^5 - a^5}{x - a} = \lim_{x \to a} \dfrac{(x-a)(x^4 + ax^3 + a^2 x^2 + a^3 x + a^4)}{x - a} = \lim_{x \to a} x^4 + ax^3 + a^2 x^2 + a^3 x + a^4 = 5a^4.$

2.3.69 $\lim_{x \to 1} \dfrac{\sqrt[3]{x} - 1}{x - 1} = \lim_{x \to 1} \dfrac{\sqrt[3]{x} - 1}{(\sqrt[3]{x} - 1)(\sqrt[3]{x^2} + \sqrt[3]{x} + 1)} = \lim_{x \to 1} \dfrac{1}{\sqrt[3]{x^2} + \sqrt[3]{x} + 1} = \dfrac{1}{3}.$

2.3.71 $\lim_{x \to 1} \dfrac{x - 1}{\sqrt{x} - 1} = \lim_{x \to 1} \dfrac{(x-1)(\sqrt{x} + 1)}{(\sqrt{x} - 1)(\sqrt{x} + 1)} = \dfrac{(x-1)(\sqrt{x} + 1)}{x - 1} = \lim_{x \to 1} \sqrt{x} + 1 = 2.$

2.3.73

$$\lim_{x \to 4} \frac{3(x-4)\sqrt{x+5}}{3 - \sqrt{x+5}} = \lim_{x \to 4} \frac{3(x-4)(\sqrt{x+5})(3 + \sqrt{x+5})}{(3 - \sqrt{x+5})(3 + \sqrt{x+5})} = \lim_{x \to 4} \frac{3(x-4)(\sqrt{x+5})(3 + \sqrt{x+5})}{9 - (x+5)}$$

$$= \lim_{x \to 4} \frac{3(x-4)(\sqrt{x+5})(3 + \sqrt{x+5})}{-(x-4)} = \lim_{x \to 4} -3(\sqrt{x+5})(3 + \sqrt{x+5}) = (-3)(3)(3+3) = -54.$$

2.3.75 Let $f(x) = x - 1$ and $g(x) = \frac{5}{x-1}$. Then $\lim_{x \to 1} f(x) = 0$, $\lim_{x \to 1} f(x)g(x) = \lim_{x \to 1} \dfrac{5(x-1)}{x-1} = \lim_{x \to 1} 5 = 5.$

2.3.77 Let $p(x) = x^2 + 2x - 8$. Then $\lim_{x \to 2} \dfrac{p(x)}{x-2} = \lim_{x \to 1} \dfrac{(x-2)(x+4)}{x-2} = \lim_{x \to 2} x + 4 = 6.$

The constants are unique. We know that 2 must be a root of p (otherwise the given limit couldn't exist), so it must have the form $p(x) = (x-2)q(x)$, and q must be a degree 1 polynomial with leading coefficient 1 (otherwise p wouldn't have leading coefficient 1.) So we have $p(x) = (x-2)(x+d)$, but since $\lim_{x \to 2} \dfrac{p(x)}{x-2} = \lim_{x \to 2} x + d = 2 + d = 6$, we are forced to realize that $d = 4$. Therefore, we have deduced that the only possibility for p is $p(x) = (x-2)(x+4) = x^2 + 2x - 8$.

2.3.79 $\lim_{S \to 0^+} r(S) = \lim_{S \to 0^+} (1/2)\left(\sqrt{100 + \dfrac{2S}{\pi}} - 10 \right) = 0.$ The radius of the circular cylinder approaches zero as the surface area approaches zero.

2.3.81 $\lim_{x \to 10} E(x) = \lim_{x \to 10} \dfrac{4.35}{x\sqrt{x^2 + 0.01}} = \dfrac{4.35}{10\sqrt{100.01}} \approx .0435.$

2.3.83 As $x \to 0^+$, $(1-x) \to 1^-$. So $\lim_{x \to 0^+} g(x) = \lim_{(1-x) \to 1^-} f(1-x) = \lim_{z \to 1^-} f(z) = 6.$ (Where $z = 1 - x$.)

As $x \to 0^-$, $(1-x) \to 1^+$. So $\lim_{x \to 0^-} g(x) = \lim_{(1-x) \to 1^+} f(1-x) = \lim_{z \to 1^+} f(z) = 4.$ (Where $z = 1 - x$.)

2.3.85

$$\lim_{x \to a} p(x) = \lim_{x \to a} (a_n x^n + a_{n-1} x^{n-1} + \ldots + a_1 x + a_0) = \lim_{x \to a} a_n x^n + \lim_{x \to a} a_{n-1} x^{n-1} + \ldots + \lim_{x \to a} a_1 x + \lim_{x \to a} a_0$$

$$= a_n \lim_{x \to a} x^n + a_{n-1} \lim_{x \to a} x^{n-1} + \ldots + a_1 \lim_{x \to a} x + a_0$$

$$= a_n (\lim_{x \to a} x)^n + a_{n-1} (\lim_{x \to a} x)^{n-1} + \ldots + a_1 (\lim_{x \to a} x) + a_0 = a_n a^n + a_{n-1} a^{n-1} + \ldots + a_1 a + a_0 = p(a).$$

2.4 Infinite Limits

2.4.1

$\lim\limits_{x \to a^+} f(x) = -\infty$ means that when x is very close to (but a little bigger than) a, the corresponding values for $f(x)$ are negative numbers whose absolute value is very large.

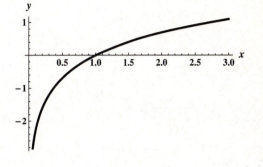

2.4.3 A vertical asymptote for a function f is a vertical line $x = a$ so that one or more of the following are true: $\lim\limits_{x \to a^-} f(x) = \pm\infty$, $\lim\limits_{x \to a^+} f(x) = \pm\infty$.

2.4.5 Since the numerator is approaching a non-zero constant while the denominator is approaching zero, the quotient of these numbers is getting big – at least the absolute value of the quotient is getting big. The quotient is actually always negative, since a number near 100 divided by a negative number is always negative. Thus $\lim\limits_{x \to 2} \dfrac{f(x)}{g(x)} = -\infty$.

2.4.7

x	$\frac{x+1}{(x-1)^2}$	x	$\frac{x+1}{(x-1)^2}$
1.1	210	.9	190
1.01	20,100	.99	19,900
1.001	2,001,000	.999	1,999,000
1.0001	200,010,000	.9999	199,990,000

From the data given, it appears that $\lim\limits_{x \to 1} f(x) = \infty$.

2.4.9

a. $\lim\limits_{x \to 1^-} f(x) = \infty$.

b. $\lim\limits_{x \to 1^+} f(x) = \infty$.

c. $\lim\limits_{x \to 1} f(x) = \infty$.

d. $\lim\limits_{x \to 2^-} f(x) = \infty$.

e. $\lim\limits_{x \to 2^+} f(x) = -\infty$.

f. $\lim\limits_{x \to 2} f(x)$ does not exist.

2.4.11

a. $\lim\limits_{x \to -2^-} h(x) = -\infty$.

b. $\lim\limits_{x \to -2^+} h(x) = -\infty$.

c. $\lim\limits_{x \to -2} h(x) = -\infty$.

d. $\lim\limits_{x \to 3^-} h(x) = \infty$.

e. $\lim\limits_{x \to 3^+} h(x) = -\infty$.

f. $\lim\limits_{x \to 3} h(x)$ does not exist.

2.4.13

a. $\displaystyle\lim_{x\to 0^-}\frac{1}{x^2-x}=\infty$.

b. $\displaystyle\lim_{x\to 0^+}\frac{1}{x^2-x}=-\infty$.

c. $\displaystyle\lim_{x\to 1^-}\frac{1}{x^2-x}=-\infty$.

d. $\displaystyle\lim_{x\to 1^+}\frac{1}{x^2-x}=\infty$.

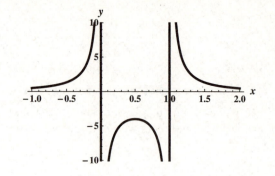

2.4.15

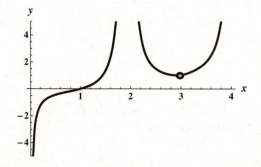

2.4.17

a. $\displaystyle\lim_{x\to 2^+}\frac{1}{x-2}=\infty$.   b. $\displaystyle\lim_{x\to 2^-}\frac{1}{x-2}=-\infty$.   c. $\displaystyle\lim_{x\to 2}\frac{1}{x-2}$ does not exist.

2.4.19 $\displaystyle\lim_{x\to 0}\frac{x^3-5x^2}{x^2}=\lim_{x\to 0}\frac{x^2(x-5)}{x^2}=\lim_{x\to 0}x-5=-5$.

2.4.21 $\displaystyle\lim_{x\to 1^+}\frac{x^2-5x+6}{x-1}=\lim_{x\to 1^+}\frac{(x-2)(x-3)}{x-1}=\infty$. (Note that as $x\to 1^+$, the numerator is near 2, while the denominator is near zero, but is positive. So the quotient is positive and large.)

2.4.23 $f(x)=\dfrac{x^2-9x+14}{x^2-5x+6}=\dfrac{(x-2)(x-7)}{(x-2)(x-3)}$. Note that $x=3$ is a vertical asymptote, while $x=2$ appears to be a candidate but isn't one. We have $\displaystyle\lim_{x\to 3^+}f(x)=\lim_{x\to 3^+}\frac{x-7}{x-3}=-\infty$ and $\displaystyle\lim_{x\to 3^-}f(x)=\lim_{x\to 3^-}\frac{x-7}{x-3}=\infty$, and thus $\displaystyle\lim_{x\to 3}f(x)$ doesn't exist. Note that $\displaystyle\lim_{x\to 2}f(x)=5$.

2.4.25 $f(x)=\dfrac{x+1}{x^3-4x^2+4x}=\dfrac{x+1}{x(x-2)^2}$. There are vertical asymptotes at $x=0$ and $x=2$. We have $\displaystyle\lim_{x\to 0^-}f(x)=\lim_{x\to 0^-}\frac{x+1}{x(x-2)^2}=-\infty$, while $\displaystyle\lim_{x\to 0^+}f(x)=\lim_{x\to 0^+}\frac{x+1}{x(x-2)^2}=\infty$, and thus $\displaystyle\lim_{x\to 0}f(x)$ doesn't exist.

Also we have $\displaystyle\lim_{x\to 2^-}f(x)=\lim_{x\to 2^-}\frac{x+1}{x(x-2)^2}=\infty$, while $\displaystyle\lim_{x\to 2^+}f(x)=\lim_{x\to 2^+}\frac{x+1}{x(x-2)^2}=\infty$, and thus $\displaystyle\lim_{x\to 2}f(x)=\infty$ as well.

2.4.27 $\lim\limits_{\theta \to 0^+} \csc \theta = \lim\limits_{\theta \to 0^+} \dfrac{1}{\sin \theta} = \infty.$

2.4.29 $\lim\limits_{x \to 0^+} -10 \cot x = \lim\limits_{x \to 0^+} \dfrac{-10 \cos x}{\sin x} = -\infty.$ (Note that as $x \to 0^+$, the numerator is near -10 and the denominator is near zero, but is positive. Thus the quotient is a negative number whose absolute value is large.)

2.4.31

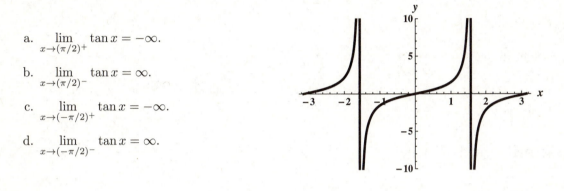

   a. $\lim\limits_{x \to (\pi/2)^+} \tan x = -\infty.$

   b. $\lim\limits_{x \to (\pi/2)^-} \tan x = \infty.$

   c. $\lim\limits_{x \to (-\pi/2)^+} \tan x = -\infty.$

   d. $\lim\limits_{x \to (-\pi/2)^-} \tan x = \infty.$

2.4.33

   a. False. $\lim\limits_{x \to 1^-} f(x) = \lim\limits_{x \to 1^+} f(x) = \lim\limits_{x \to 1} f(x) = \lim\limits_{x \to 1} \dfrac{(x-1)(x-6)}{(x-1)(x+1)} = \dfrac{-5}{2}.$

   b. True. For example, $\lim\limits_{x \to -1^+} f(x) = \lim\limits_{x \to -1^+} \dfrac{(x-1)(x-6)}{(x-1)(x+1)} = -\infty.$

   c. False. For example $g(x) = \frac{1}{x-1}$ has $\lim\limits_{x \to 1^+} g(x) = \infty$, but $\lim\limits_{x \to 1^-} g(x) = -\infty.$

2.4.35 One example is $f(x) = \frac{1}{x-6}.$

2.4.37 $f(x) = \dfrac{x^2 - 3x + 2}{x^{10} - x^9} = \dfrac{(x-2)(x-1)}{x^9(x-1)}.$ This has a vertical asymptote at $x = 0$, since $\lim\limits_{x \to 0^+} f(x) = -\infty$ (and $\lim\limits_{x \to 0^-} f(x) = \infty.$) Note that $\lim\limits_{x \to 1} f(x) = -1$, so there isn't a vertical asymptote at $x = 1$.

2.4.39 $h(x) = \dfrac{\cos x}{(x+1)^3}$ has a vertical asymptote at $x = -1$, since $\lim\limits_{x \to -1^+} \dfrac{\cos x}{(x+1)^3} = \infty$ (and $\lim\limits_{x \to -1^-} h(x) = -\infty.$)

2.4.41 $g(\theta) = \tan(\pi\theta/10) = \dfrac{\sin(\pi\theta/10)}{\cos(\pi\theta/10)}$ has a vertical asymptote at each $\theta = 10n + 5$ where n is an integer. This is due to the fact that $\cos(\pi\theta/10) = 0$ when $\pi\theta/10 = \pi/2 + n\pi$ where n is an integer, which is the same as $\{\theta : \theta = 10n + 5, n \text{ an integer}\}$. Note that at all of these numbers which make the denominator zero, the numerator isn't zero.

2.4.43 $f(x) = \dfrac{1}{\sqrt{x}\sec x} = \dfrac{\cos x}{\sqrt{x}}$ has a vertical asymptote at $x = 0$.

2.4.45

a. Note that the numerator of the given expression factors as $(x-3)(x-4)$. So if $a=3$ or if $a=4$ the limit would be a finite number. In fact, $\lim\limits_{x\to 3^+}\dfrac{(x-3)(x-4)}{x-3}=-1$ and $\lim\limits_{x\to 4}\dfrac{(x-3)(x-4)}{x-4}=1$.

b. For any number other than 3 or 4, the limit would be either $\pm\infty$. Since $x-a$ is always positive as $x\to a^+$, the limit would be $+\infty$ exactly when the numerator is positive, which is for a in the set $(-\infty,3)\cup(4,\infty)$.

c. The limit would be $-\infty$ for a in the set $(3,4)$.

2.4.47

a. The slope of the secant line is $\dfrac{f(h)-f(0)}{h}=\dfrac{h^{2/3}}{h}=h^{-1/3}$.

b. $\lim\limits_{h\to 0^+}\dfrac{1}{h^{1/3}}=\infty$, and $\lim\limits_{h\to 0^-}\dfrac{1}{h^{1/3}}=-\infty$. The tangent line is infinitely steep at the origin (i.e., it is a vertical line.)

2.5   Limits at Infinity

2.5.1

As $x<0$ becomes large in absolute value, the corresponding values of f level off near 10.

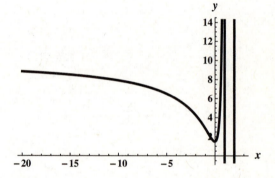

2.5.3 If $f(x)\to 100{,}000$ as $x\to\infty$ and $g(x)\to\infty$ as $x\to\infty$, then the ratio $\dfrac{f(x)}{g(x)}\to 0$ as $x\to\infty$. (Because *eventually* the values of f are small compared to the values of g.)

2.5.5 $\lim\limits_{x\to\infty}-2x^3=-\infty$, and $\lim\limits_{x\to-\infty}-2x^3=\infty$.

2.5.7
$$\lim_{x\to\infty}\frac{1-x}{2x}=-\frac{1}{2},\quad \lim_{x\to\infty}\frac{1-x}{x^2}=0,\quad \lim_{x\to\infty}\frac{1-x^2}{2x}=-\infty$$

2.5.9 $\lim\limits_{x\to\infty}(3+10/x^2)=3+\lim\limits_{x\to\infty}10/x^2=3+0=3$.

2.5.11 $\lim\limits_{\theta\to\infty}\dfrac{\cos\theta}{\theta^2}=0$. Note that $-1\le\cos\theta\le 1$, so $\dfrac{-1}{\theta^2}\le\dfrac{\cos\theta}{\theta^2}\le\dfrac{1}{\theta^2}$. The result now follows from the squeeze theorem.

2.5.13 $\lim\limits_{x\to-\infty}\dfrac{\cos x^5}{x}=0$. Note that $-1\le\cos x^5\le 1$, so $\dfrac{-1}{x}\le\dfrac{\cos x^5}{x}\le\dfrac{1}{x}$. The result now follows from the squeeze theorem.

2.5.15 $\lim\limits_{x\to\infty}3x^{12}-9x^7=\infty$.

2.5.17 $\lim\limits_{x\to-\infty} -3x^{16} + 2 = -\infty$.

2.5.19 $\lim\limits_{x\to\infty} -12x^{-5} = \lim\limits_{x\to\infty} \dfrac{-12}{x^5} = 0$.

2.5.21 $\lim\limits_{x\to\infty} \dfrac{(6x^2 - 9x + 8)}{(3x^2 + 2)} \cdot \dfrac{1/x^2}{1/x^2} = \lim\limits_{x\to\infty} \dfrac{6 - 9/x + 8/x^2}{3 + 2/x^2} = \dfrac{6 - 0 + 0}{3 + 0} = 2$. Similarly $\lim\limits_{x\to-\infty} f(x) = 2$. The line $y = 2$ is a horizontal asymptote.

2.5.23 $\lim\limits_{x\to\infty} \dfrac{(2x + 1)}{(3x^4 - 2)} \cdot \dfrac{1/x^4}{1/x^4} = \lim\limits_{x\to\infty} \dfrac{2/x^3 + 1/x^4}{3 - 2/x^4} = \dfrac{0 + 0}{3 - 0} = 0$. Similarly $\lim\limits_{x\to-\infty} f(x) = 0$. The line $y = 0$ is a horizontal asymptote.

2.5.25 $\lim\limits_{x\to\infty} \dfrac{(40x^5 + x^2)}{(16x^4 - 2x)} \cdot \dfrac{1/x^4}{1/x^4} = \lim\limits_{x\to\infty} \dfrac{40x + 1/x^2}{16 - 2/x^3} = \infty$. Similarly $\lim\limits_{x\to-\infty} f(x) = -\infty$. There are no horizontal asymptotes.

2.5.27 First note that $\sqrt{x^6} = x^3$ if $x > 0$, but $\sqrt{x^6} = -x^3$ if $x < 0$. We have

$$\lim_{x\to\infty} \frac{(4x^3 + 1)}{(2x^3 + \sqrt{16x^6 + 1})} \cdot \frac{1/x^3}{1/x^3} = \lim_{x\to\infty} \frac{4 + 1/x^3}{2 + \sqrt{16 + 1/x^6}} = \frac{4 + 0}{2 + \sqrt{16 + 0}} = \frac{2}{3}.$$

However,

$$\lim_{x\to-\infty} \frac{(4x^3 + 1)}{(2x^3 + \sqrt{16x^6 + 1})} \cdot \frac{1/x^3}{1/x^3} = \lim_{x\to-\infty} \frac{4 + 1/x^3}{2 - \sqrt{16 + 1/x^6}} = \frac{4 + 0}{2 - \sqrt{16 + 0}} = \frac{4}{-2} = -2.$$

So $y = \dfrac{2}{3}$ is a horizontal asymptote (as $x \to \infty$) and $y = -2$ is a horizontal asymptote (as $x \to -\infty$.)

2.5.29 First note that $\sqrt[3]{x^6} = x^2$ and $\sqrt{x^4} = x^2$ for all x (even when $x < 0$.) We have

$$\lim_{x\to\infty} \frac{\sqrt[3]{x^6 + 8}}{(4x^2 + \sqrt{3x^4 + 1})} \cdot \frac{1/x^2}{1/x^2} = \lim_{x\to\infty} \frac{\sqrt[3]{1 + 8/x^6}}{4 + \sqrt{3 + 1/x^4}} = \frac{1}{4 + \sqrt{3 + 0}} = \frac{4 - \sqrt{3}}{13}.$$

The calculation as $x \to -\infty$ is similar, so $y = \dfrac{4 - \sqrt{3}}{13}$ is a horizontal asymptote.

2.5.31

a.

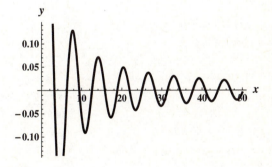

False. For example, the function $y = \frac{\sin x}{x}$ on the domain $[1, \infty)$ has a horizontal asymptote of $y = 0$, and it crosses the x axis infinitely many times.

b. False. If f is a rational function, and if $\lim\limits_{x\to\infty} f(x) = L$, then the degree of the polynomial in the numerator must equal the degree of the polynomial in the denominator. In this case, both $\lim\limits_{x\to\infty} f(x)$ and $\lim\limits_{x\to-\infty} f(x) = \dfrac{a_n}{b_n}$ where a_n is the leading coefficient of the polynomial in the numerator and b_n is the leading coefficient of the polynomial in the denominator.

c. True. There are only two directions which might lead to horizontal asymptotes: there could be one as $x \to \infty$ and there could be one as $x \to -\infty$, and those are the only possibilities.

2.5.33

a. $\lim\limits_{x\to\infty} \dfrac{2x^3 + 10x^2 + 12x}{x^3 + 2x^2} \cdot \dfrac{(1/x^3)}{(1/x^3)} = \lim\limits_{x\to\infty} \dfrac{2 + 10/x + 12/x^2}{1 + 2/x} = 2$. Similarly, $\lim\limits_{x\to-\infty} f(x) = 2$. Thus, $y = 2$ is a horizontal asymptote.

b. Note that $f(x) = \dfrac{2x(x+2)(x+3)}{x^2(x+2)}$. So we have $\lim\limits_{x\to 0^+} f(x) = \lim\limits_{x\to 0^+} \dfrac{2(x+3)}{x} = \infty$, and similarly, $\lim\limits_{x\to 0^-} f(x) = -\infty$. There is a vertical asymptote at $x = 0$. Note that there is no asymptote at $x = -2$ because $\lim\limits_{x\to -2} f(x) = -1$.

2.5.35

a. We have $\lim\limits_{x\to\infty} \dfrac{3x^4 + 3x^3 - 36x^2}{x^4 - 25x^2 + 144} \cdot \dfrac{(1/x^4)}{(1/x^4)} = \lim\limits_{x\to\infty} \dfrac{3 + 3/x - 36/x^2}{1 - 25/x^2 + 144/x^4} = 3$. Similarly, $\lim\limits_{x\to-\infty} f(x) = 3$. So $y = 3$ is a horizontal asymptote.

b. Note that $f(x) = \dfrac{3x^2(x+4)(x-3)}{(x+4)(x-4)(x+3)(x-3)}$. Thus, $\lim\limits_{x\to -3^+} f(x) = \infty$ and $\lim\limits_{x\to -3^-} f(x) = -\infty$. Also, $\lim\limits_{x\to 4^-} f(x) = -\infty$ and $\lim\limits_{x\to 4^+} f(x) = \infty$. Thus there are vertical asymptotes at $x = -3$ and $x = 4$.

2.5.37

a. $\lim\limits_{x\to\infty} \dfrac{x^2 - 9}{x^2 - 3x} \cdot \dfrac{(1/x^2)}{(1/x^2)} = \lim\limits_{x\to\infty} \dfrac{1 - 9/x^2}{1 - 3/x} = 1$. A similar result holds as $x \to -\infty$. So $y = 1$ is a horizontal asymptote.

b. Since $\lim\limits_{x\to 0^+} f(x) = \lim\limits_{x\to 0^+} \dfrac{x+3}{x} = \infty$ and $\lim\limits_{x\to 0^-} f(x) = -\infty$, there is a vertical asymptote at $x = 0$.

2.5.39

a. First note that

$$f(x) = \frac{\sqrt{x^2 + 2x + 6} - 3}{x - 1} \cdot \frac{\sqrt{x^2 + 2x + 6} + 3}{\sqrt{x^2 + 2x + 6} + 3} = \frac{x^2 + 2x + 6 - 9}{(x-1)(\sqrt{x^2 + 2x + 6} + 3)}$$

$$= \frac{(x-1)(x+3)}{(x-1)(\sqrt{x^2 + 2x + 6} + 3)}.$$

Thus

$$\lim\limits_{x\to\infty} f(x) = \lim\limits_{x\to\infty} \frac{x+3}{\sqrt{x^2 + 2x + 6} + 3} \cdot \frac{1/x}{1/x} = \lim\limits_{x\to\infty} \frac{1 + 3/x}{\sqrt{1 + 2/x + 6/x^2} + 3/x} = 1.$$

Using the fact that $\sqrt{x^2} = -x$ for $x < 0$, we have $\lim\limits_{x\to-\infty} f(x) = -1$. Thus the lines $y = 1$ and $y = -1$ are horizontal asymptotes.

b. f has no vertical asymptotes. Note that $\lim\limits_{x\to 1} f(x) = \dfrac{4}{6} = \dfrac{2}{3}$.

2.5.41

   a. Note that when $x > 1$, we have $|x| = x$ and $|x - 1| = x - 1$. Thus

$$f(x) = (\sqrt{x} - \sqrt{x-1}) \cdot \frac{\sqrt{x} + \sqrt{x-1}}{\sqrt{x} + \sqrt{x-1}} = \frac{1}{\sqrt{x} + \sqrt{x-1}}.$$

   Thus $\lim\limits_{x \to \infty} f(x) = 0$.

   When $x < 0$, we have $|x| = -x$ and $|x - 1| = 1 - x$. Thus

$$f(x) = (\sqrt{-x} - \sqrt{1-x}) \cdot \frac{\sqrt{-x} + \sqrt{1-x}}{\sqrt{-x} + \sqrt{1-x}} = \frac{-1}{\sqrt{-x} + \sqrt{1-x}}.$$

   Thus, $\lim\limits_{x \to -\infty} f(x) = 0$. There is a horizontal asymptote at $y = 0$.

   b. f has no vertical asymptotes.

2.5.43

One possible such graph is:

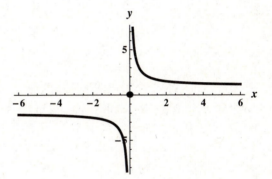

2.5.45 $\lim\limits_{x \to 0^+} \dfrac{\cos x + 2\sqrt{x}}{\sqrt{x}} = \infty$. $\lim\limits_{x \to \infty} \dfrac{\cos x + 2\sqrt{x}}{\sqrt{x}} = \lim\limits_{x \to \infty} 2 + \dfrac{\cos x}{\sqrt{x}} = 2$.
There is a vertical asymptote at $x = 0$ and a horizontal asymptote at $y = 2$.

2.5.47 $\lim\limits_{t \to \infty} p(t) = \lim\limits_{t \to \infty} \dfrac{3500t}{t + 1} = 3500$. The steady state exists. The steady state value is 3500.

2.5.49 $\lim\limits_{t \to \infty} a(t) = \lim\limits_{t \to \infty} 2\left(\dfrac{t + \sin t}{t}\right) = \lim\limits_{t \to \infty} 2\left(1 + \dfrac{\sin t}{t}\right) = 2$. The steady state exists. The steady state
value is 2.

2.5.51 $\lim\limits_{n \to \infty} f(n) = \lim\limits_{n \to \infty} \dfrac{n - 1}{n} = \lim\limits_{n \to \infty} 1 - (1/n) = 1$.

2.5.53 $\lim\limits_{n \to \infty} f(n) = \lim\limits_{n \to \infty} \dfrac{n + 1}{n^2} = \lim\limits_{n \to \infty} 1/n + 1/n^2 = 0$.

2.5.55

   a. $f(x) = \frac{x^2 - 3}{x + 6} = x - 6 + \frac{33}{x + 6}$. The oblique asymptote of f is $y = x - 6$.

b.

c.

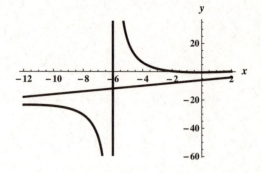

Since $\lim\limits_{x \to -6^+} f(x) = \infty$, there is a vertical asymptote at $x = -6$. Note also that $\lim\limits_{x \to -6^-} f(x) = -\infty$.

2.5.57

a. $f(x) = \frac{x^2 - 2x + 5}{3x - 2} = (1/3)x - 4/9 + \frac{37}{9(3x-2)}$. The oblique asymptote of f is $y = (1/3)x - 4/9$.

b.

c.

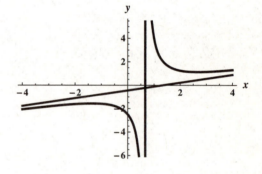

Since $\lim\limits_{x \to (2/3)^+} f(x) = \infty$, there is a vertical asymptote at $x = 2/3$. Note also that $\lim\limits_{x \to (2/3)^-} f(x) = -\infty$.

2.5.59

a. $f(x) = \frac{4x^3 + 4x^2 + 7x + 4}{1 + x^2} = 4x + 4 + \frac{3x}{1 + x^2}$. The oblique asymptote of f is $y = 4x + 4$.

b.

c.

There are no vertical asymptotes.

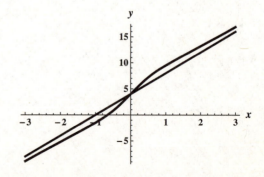

2.6 Continuity

2.6.1

   a. $a(t)$ is a continuous function during the time period from when she jumps from the plane and when she touches down on the ground, since her position is changing continuously with time.

   b. $n(t)$ is not a continuous function of time. The function "jumps" at the times when a quarter must be added.

   c. $T(t)$ is a continuous function, since temperature varies continuously with time.

   d. $p(t)$ is not continuous – it jumps by whole numbers when a player makes a shot.

2.6.3 A function f is continuous on an interval I if it is continuous at all points in the interior of I, and it must be continuous from the right at the left endpoint (if the left endpoint is included in I) and it must be left continuous at the right endpoint (if the right endpoint is included in I.)

2.6.5

   a. A function f is continuous from the left at $x = a$ if a is in the domain of f, and $\lim\limits_{x \to a^-} f(x) = f(a)$.

   b. A function f is continuous from the right at $x = a$ if a is in the domain of f, and $\lim\limits_{x \to a^+} f(x) = f(a)$.

2.6.7 The domain of $f(x) = \sqrt{1 - x^2}$ is those values of x where $1 - x^2 \geq 0$, i.e. $[-1, 1]$. The function is continuous on its domain.

2.6.9 f is discontinuous at $x = 1$, at $x = 2$, and at $x = 3$. At $x = 1$, f does not exist (so the first condition is violated.) At $x = 2$, $f(2)$ exists and $\lim\limits_{x \to 2} f(x)$ exists, but $\lim\limits_{x \to 2} f(x) \neq f(2)$ (so condition 3 is violated.). At $x = 3$, $\lim\limits_{x \to 3} f(x)$ does not exist (so condition 2 is violated.)

2.6.11 f is discontinuous at $x = 1$, at $x = 2$, and at $x = 3$. At $x = 1$, $\lim\limits_{x \to 1} f(x)$ does not exist, and f does not exist (so conditions 1 and 2 are violated.) At $x = 2$, $\lim\limits_{x \to 2} f(x)$ does not exist (so condition 2 is violated.) At $x = 3$, f does not exist (so condition 1 is violated.)

2.6.13 f is discontinuous at 1, since 1 is not in the domain of f.

2.6.15 f is discontinuous at 1, because $\lim\limits_{x \to 1} f(x) \neq f(1)$. In fact, $f(1) = 3$, but $\lim\limits_{x \to 1} f(x) = 2$.

2.6.17 f is discontinuous at 4, because 4 is not in the domain of f.

2.6.19 Since f is a polynomial, it is continuous on all of $\mathbb{R}$.

2.6.21 Since f is a rational function, it is continuous on its domain. Its domain is $(-\infty, -3) \cup (-3, 3) \cup (3, \infty)$.

2.6.23 Since f is a rational function, it is continuous on its domain. Its domain is $(-\infty, -2) \cup (-2, 2) \cup (2, \infty)$.

2.6.25 Since $f(x) = \left(x^8 - 3x^6 - 1\right)^{40}$ is a polynomial, it is continuous everywhere, including at 0. Thus $\lim\limits_{x \to 0} f(x) = f(0) = (-1)^{40} = 1$.

2.6.27 Since $f(x) = \left(\frac{x+5}{x+2}\right)^4$ is a rational function, it is continuous at all points in its domain, including at $x = 1$. Thus $\lim\limits_{x \to 1} f(x) = f(1) = 16$.

2.6.29 f is continuous on $[0, 1)$, on $(1, 2)$, on $(2, 3]$, and on $(3, 4]$.

2.6.31 f is continuous on $[0, 1)$, on $(1, 2)$, on $[2, 3)$, and on $(3, 5]$.

2.6.33

a. f is defined at 1. We have $f(1) = 1^2 + (3)(1) = 4$. To see whether or not $\lim\limits_{x \to 1} f(x)$ exists, we investigate the two one-sided limits. $\lim\limits_{x \to 1^-} f(x) = \lim\limits_{x \to 1^-} 2x = 2$, and $\lim\limits_{x \to 1^+} f(x) = \lim\limits_{x \to 1^+} x^2 + 3x = 4$, so $\lim\limits_{x \to 1} f(x)$ does not exist. Thus f is discontinuous at $x = 1$.

b. f is continuous from the right, since $\lim\limits_{x \to 1^+} f(x) = 4 = f(1)$.

c. f is continuous on $(-\infty, 1)$ and on $[1, \infty)$.

2.6.35 f is continuous on $(-\infty, -\sqrt{8}]$ and on $[\sqrt{8}, \infty)$.

2.6.37 Since f is the composition of two functions which are continuous everywhere, it is continuous everywhere.

2.6.39 Since f is the composition of two functions which are continuous everywhere, it is continuous everywhere.

2.6.41 $\lim\limits_{x \to 2} \sqrt{\dfrac{4x + 10}{2x - 2}} = \sqrt{\dfrac{18}{2}} = 3$.

2.6.43 $\lim\limits_{x \to 3} \sqrt{x^2 + 7} = \sqrt{9 + 7} = 4$.

2.6.45 $f(x) = \csc x$ isn't defined at $x = k\pi$ where k is an integer, so it isn't continuous at those points. So it is continuous on intervals of the form $(k\pi, (k+1)\pi)$ where k is an integer. $\lim\limits_{x \to \pi/4} \csc x = \sqrt{2}$. $\lim\limits_{x \to 2\pi^-} \csc x = -\infty$.

2.6.47 f isn't defined for any number of the form $\pi/2 + k\pi$ where k is an integer, so it isn't continuous there. It is continuous on intervals of the form $(\pi/2 + k\pi, \pi/2 + (k+1)\pi)$, where k is an integer.

$\lim\limits_{x \to \pi/2^-} f(x) = \infty$. $\lim\limits_{x \to 4\pi/3} f(x) = \dfrac{1 - \sqrt{3}/2}{-1/2} = \sqrt{3} - 2$.

2.6.49

a. Since A is a continuous function of r on $[0, .08]$, and since $A(0) = 5000$ and $A(.08) \approx 11098.2$, (and 7000 is an intermediate value between these two numbers) the Intermediate Value Theorem guarantees a value of r between 0 and .08 where $A(r) = 7000$.

b.

Solving $5000(1 + (r/12))^{120} = 7000$ for r, we see that $(1 + (r/12))^{120} = 7/5$, so $1 + r/12 = \sqrt[120]{7/5}$, so $r = 12(\sqrt[120]{7/5} - 1) \approx .03369$.

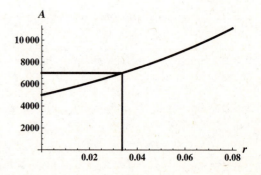

2.6.51

a. Note that $f(x) = 2x^3 + x - 2$ is continuous everywhere, so in particular it is continuous on $[-1, 1]$. Note that $f(-1) = -5 < 0$ and $f(1) = 1 > 0$. Since 0 is an intermediate value between $f(-1)$ and $f(1)$, the Intermediate Value Theorem guarantees a number c between -1 and 1 where $f(c) = 0$.

b.

c.

Using a graphing calculator and a computer algebra system, we see that the root of f is about $.835$.

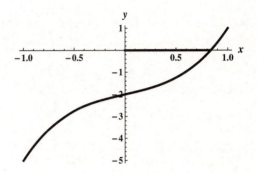

2.6.53

a. Note that $f(x) = x^3 - 5x^2 + 2x$ is continuous everywhere, so in particular it is continuous on $[-1, 5]$. Note that $f(-1) = -8 < -1$ and $f(5) = 10 > -1$. Since -1 is an intermediate value between $f(-1)$ and $f(5)$, the Intermediate Value Theorem guarantees a number c between -1 and 5 where $f(c) = -1$.

b.

c.

Using a graphing calculator and a computer algebra system, we see that there are actually three different values of c between -1 and 5 for which $f(c) = -1$. They are $c \approx -.285$, $c \approx .778$, and $c \approx 4.50$.

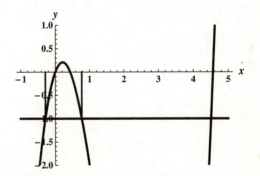

2.6.55

a. True. If f is right continuous at a, then $f(a)$ exists and the limit from the right at a exists and is equal to $f(a)$. Since it is left continuous, the limit from the left exists — so we now know that the limit as $x \to a$ of $f(x)$ exists, since the two one-sided limits are both equal to $f(a)$.

b. True. If $\lim\limits_{x \to a} f(x) = f(a)$, then $\lim\limits_{x \to a^+} f(x) = f(a)$ and $\lim\limits_{x \to a^-} f(x) = f(a)$.

c. False. The statement would be true if f were continuous. However, if f isn't continuous, then the statement doesn't hold. For example, suppose that $f(x) = \begin{cases} 0 & \text{if } 0 \leq x < 1; \\ 1 & \text{if } 1 \leq x \leq 2, \end{cases}$ Note that $f(0) = 0$ and $f(2) = 1$, but there is no number c between 0 and 2 where $f(c) = 1/2$.

d. True. The number $\frac{f(a)+f(b)}{2}$ is between $f(a)$ and $f(b)$ (since it is the average of these two numbers.) So the Intermediate Value Theorem guarantees a c between a and b with $f(c) = \frac{f(a)+f(b)}{2}$.

2.6.57 Since $f(x) = x^3 + 3x - 18$ is a polynomial, it is continuous on $(-\infty, \infty)$, and since the absolute value function is continuous everywhere, $|f(x)|$ is continuous everywhere.

2.6.59 Let $f(x) = \frac{1}{\sqrt{x}-4}$. Then f is continuous on $[0, 16) \cup (16, \infty)$. So $h(x) = |f(x)|$ is continuous on this set as well.

2.6.61 $\lim\limits_{x \to \pi} \dfrac{\cos^2 x + 3\cos x + 2}{\cos x + 1} = \lim\limits_{x \to \pi} \dfrac{(\cos x + 1)(\cos x + 2)}{\cos x + 1} = \lim\limits_{x \to \pi} \cos x + 2 = 1.$

2.6.63 $\lim\limits_{x \to \pi/2} \dfrac{\sin x - 1}{\sqrt{\sin x} - 1} = \lim\limits_{x \to \pi/2} \sqrt{\sin x} + 1 = 2.$

2.6.65 $\lim\limits_{x \to 0} \dfrac{\cos x - 1}{\sin^2 x} = \lim\limits_{x \to 0} \dfrac{\cos x - 1}{1 - \cos^2 x} = \lim\limits_{x \to 0} \dfrac{\cos x - 1}{(1 - \cos x)(1 + \cos x)} = \lim\limits_{x \to 0} \dfrac{-1}{1 + \cos x} = \dfrac{-1}{2}.$

2.6.67

The graph shown isn't drawn correctly at the integers. At an integer a, the value of the function is 0, whereas the graph shown appears to take on all the values from 0 to 1.

Note that in the correct graph, $\lim\limits_{x \to a^-} f(x) = 1$ and $\lim\limits_{x \to a^+} f(x) = 0$ for every integer a.

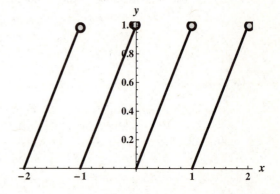

2.6.69 With slight modifications, we can use the examples from the previous two problems.

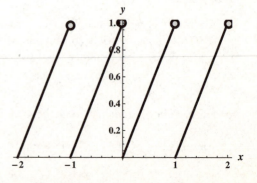

a. The function $y = x - \lfloor x \rfloor$ is defined at $x = 1$ but isn't continuous there.

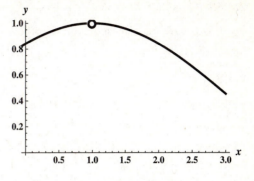

b. The function $y = \frac{\sin(x-1)}{x-1}$ has a limit at $x = 1$, but isn't defined there, so isn't continuous there.

2.6.71

a. In order for g to be continuous from the left at $x = 1$, we must have $\lim_{x \to 1^-} g(x) = g(1) = a$. We have $\lim_{x \to 1^-} g(x) = \lim_{x \to 1^-} x^2 + x = 2$. So we must have $a = 2$.

b. In order for g to be continuous from the right at $x = 1$, we must have $\lim_{x \to 1^+} g(x) = g(1) = a$. We have $\lim_{x \to 1^+} g(x) = \lim_{x \to 1^+} 3x + 5 = 8$. So we must have $a = 8$.

c. Since the limit from the left and the limit from the right at $x = 1$ don't agree, there is no value of a which will make the function continuous at $x = 1$.

2.6.73 Let $f(x) = 70x^3 - 87x^2 + 32x - 3$. Note that $f(0) < 0$, $f(.2) > 0$, $f(.55) < 0$, and $f(1) > 0$. Since the given polynomial is continuous everywhere, the Intermediate Value Theorem guarantees us a root on $(0, .2)$, at least one on $(.2, .55)$, and at least one on $(.55, 1)$. Since there can be at most 3 roots and there are at least 3 roots, there must be exactly 3 roots. The roots are $x_1 = 1/7$, $x_2 = 1/2$ and $x_3 = 3/5$.

2.6.75

a. Note that $A(.01) \approx 2615.55$ and $A(.1) \approx 3984.35$. By the Intermediate Value Theorem, there must be a number r_0 between .01 and .1 so that $A(r_0) = 3500$.

b. The desired value is $r_0 \approx 0.073$.

2.6.77 We can argue essentially like the previous problem, or we can imagine an identical twin to the original monk, who takes an identical version of the original monk's journey up the winding path while the monk is taking the return journey down. Because they must pass somewhere on the path, that point is the one we are looking for.

2.6.79 The discontinuity is not removable, since $\lim_{x \to a} f(x)$ does not exist. The discontinuity pictured is a jump discontinuity.

2.6.81 Note that $\lim_{x \to 2} \frac{x^2 - 7x + 10}{x - 2} = \lim_{x \to 2} \frac{(x - 2)(x - 5)}{x - 2} = \lim_{x \to 2} x - 5 = -3$. Since this limit exists, the discontinuity is removable.

2.6.83

a. Note that $-1 \le \sin(1/x) \le 1$ for all $x \ne 0$, so $-x \le x\sin(1/x) \le x$ (for $x > 0$. For $x < 0$ we would have $x \le x\sin(1/x) \le -x$.) Since both $x \to 0$ and $-x \to 0$ as $x \to 0$, the squeeze theorem tells us that $\lim_{x \to 0} x\sin(1/x) = 0$ as well. Since this limit exists, the discontinuity is removable.

b. Note that as $x \to 0^+$, $1/x \to \infty$, and thus $\lim_{x \to 0^+} \sin(1/x)$ does not exist.

2.6.85 Note that $h(x) = \dfrac{x^3 - 4x^2 + 4x}{x(x-1)} = \dfrac{(x)(x-2)^2}{(x)(x-1)}$. Thus $\lim\limits_{x \to 0} h(x) = -4$, and the discontinuity at $x = 0$ is removable. However, $\lim\limits_{x \to 1} h(x)$ does not exist, and the discontinuity at $x = 1$ is not removable (it is infinite.)

2.6.87

a. Consider $g(x) = x + 1$ and $f(x) = \frac{|x-1|}{x-1}$. Note that both g and f are continuous at $x = 0$. However $f(g(x)) = f(x+1) = \frac{|x|}{x}$ is not continuous at 0.

b. The previous theorem says that the composition of f and g is continuous at a if g is continuous at a and f is continuous at $g(a)$. It does not say that if g and f are both continuous at a that the composition is continuous at a.

2.6.89

a. Using the hint, we have

$$\sin x = \sin(a + (x - a)) = \sin a \cos(x - a) + \sin(x - a)\cos(a).$$

Note that as $x \to a$, we have that $\cos(x - a) \to 1$ and $\sin(x - a) \to 0$.

So, $\lim\limits_{x \to a} \sin x = \lim\limits_{x \to a} \sin(a + (x - a)) = \lim\limits_{x \to a} \sin a \cos(x - a) + \sin(x - a)\cos(a) = \sin a \cdot 1 + 0 \cdot \cos a = \sin a$.

b. Using the hint, we have

$$\cos x = \cos(a + (x - a)) = \cos a \cos(x - a) - \sin a \sin(x - a).$$

So, $\lim\limits_{x \to a} \cos x = \lim\limits_{x \to a} \cos(a + (x - a)) = \lim\limits_{x \to a} \cos a \cos(x - a) - \sin a \sin(x - a) = \cos a \cdot 1 - \sin a \cdot 0 = \cos a$.

2.7 Precise Definitions of Limits

2.7.1 Note that all the numbers in the interval $(1, 3)$ are within 1 unit of the number 2. So $|x - 2| < 1$ is true for all numbers in that interval, so that $\delta = 1$ is the smallest possible value for δ. In fact, $\{x : 0 < |x - 2| < 1\}$ is exactly the set $(1, 3)$ with $x \neq 2$.

2.7.3

$(3, 8)$ has center 5.5, so it is not symmetric about the number 5.

$(1, 9)$ and $(4, 6)$ and $(4.5, 5.5)$ are symmetric about the number 5.

2.7.5 $\lim\limits_{x \to a} f(x) = L$ if for any arbitrarily small positive number ϵ, there exists a number δ, so that $f(x)$ is within ϵ units of L for any number x within δ units of a (but not including a itself.)

2.7.7 We are given that $|f(x) - 5| < .1$ for values of x in the interval $(0, 5)$, so we need to ensure that the set of x values we are allowing fall in this interval.

Note that the number 0 is 2 units away from the number 2 and the number 5 is three units away from the number 2. In order to be sure that we are talking about numbers in the interval $(0, 5)$ when we write $|x - 2| < \delta$, we would need to have $\delta = 2$ (or a number less than 2.) In fact, the set of numbers for which $|x - 2| < 2$ is the interval $(0, 4)$ which is a subset of $(0, 5)$.

If we were to allow δ to be any number greater than 2, then the set of all x so that $|x - 2| < \delta$ would include numbers less than 0, and those numbers aren't on the interval $(0, 5)$.

2.7.9

a. In order for f to be within 2 units of 5, it appears that we need x to be within 1 unit of 2. So $\delta = 1$.

b. In order for f to be within 1 unit of 5, it appears that we would need x to be within $1/2$ unit of 2. So $\delta = .5$.

2.7.11

a. In order for f to be within 3 units of 6, it appears that we would need x to be within 2 units of 3. So $\delta = 2$.

b. In order for f to be within 1 unit of 6, it appears that we would need x to be within 1/2 unit of 3. So $\delta = 1/2$.

2.7.13

a.

If $\epsilon = 1$, we need $|x^3 + 3 - 3| < 1$. So we need $|x| < \sqrt[3]{1} = 1$ in order for this to happen. Thus $\delta = 1$ will suffice.

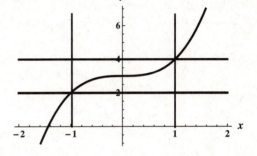

b.

If $\epsilon = .5$, we need $|x^3 + 3 - 3| < .5$. So we need $|x| < \sqrt[3]{.5}$ in order for this to happen. Thus $\delta = \sqrt[3]{.5} \approx .79$ will suffice.

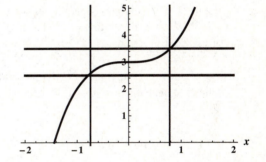

2.7.15

a. For $\epsilon = 1$, the required value of δ would also be 1. A larger value of δ would work to the right of 2, but this is the largest one that would work to the left of 2.

b. For $\epsilon = 1/2$, the required value of δ would also be 1/2.

c. It appears that for a given value of ϵ, it would be wise to take $\delta = \epsilon$. This assures that the desired inequality is met on both sides of 2.

2.7.17

a. For $\epsilon = 1$, it appears that a value of $\delta = .2$ would work.

b. For $\epsilon = .5$, it appears that a value of $\delta = .1$ would work.

c. For an arbitrary ϵ, a value of $\delta = \epsilon/5$ or smaller appears to suffice. At least this appears to work for small values of ϵ, so we could let $\delta = \min(1, \epsilon/5)$.

2.7.19 For any $\epsilon > 0$, let $\delta = \epsilon/8$. Then if $0 < |x - 1| < \delta$, we would have $|x - 1| < \epsilon/8$. Then $|8x - 8| < \epsilon$, so $|(8x + 5) - 13| < \epsilon$. This last inequality has the form $|f(x) - L| < \epsilon$, which is what we were attempting to show. Thus, $\lim_{x \to 1} 8x + 5 = 13$.

2.7.21 First note that if $x \neq 4$, $f(x) = \dfrac{x^2 - 16}{x - 4} = x + 4$.

Now if $\epsilon > 0$ is given, let $\delta = \epsilon$. Now suppose $0 < |x - 4| < \delta$. Then $x \neq 4$, so the function $f(x)$ can be described by $x + 4$. Also, since $|x - 4| < \delta$, we have $|x - 4| < \epsilon$. Thus $|(x + 4) - 8| < \epsilon$. This last inequality has the form $|f(x) - L| < \epsilon$, which is what we were attempting to show. Thus, $\lim_{x \to 4} \dfrac{x^2 - 16}{x - 4} = 8$.

2.7.23 Let $\epsilon > 0$ be given. Let $\delta = \sqrt{\epsilon}$. Then if $0 < |x - 0| < \delta$, we would have $|x| < \sqrt{\epsilon}$. But then $|x^2| < \epsilon$, which has the form $|f(x) - L| < \epsilon$. Thus, $\lim_{x \to 0} f(x) = 0$.

2.7.25 Let $\epsilon > 0$ be given.

Since $\lim_{x \to a} f(x) = L$, we know that there exists a $\delta_1 > 0$ so that $|f(x) - L| < \epsilon/2$ when $0 < |x - a| < \delta_1$. Also, since $\lim_{x \to a} g(x) = M$, there exists a $\delta_2 > 0$ so that $|g(x) - M| < \epsilon/2$ when $0 < |x - a| < \delta_2$.

Now let $\delta = \min(\delta_1, \delta_2)$.

Then if $0 < |x - a| < \delta$, we would have $|f(x) - g(x) - (L - M)| = |(f(x) - L) + (M - g(x))| \leq |f(x) - L| + |M - g(x)| = |f(x) - L| + |g(x) - M| \leq \epsilon/2 + \epsilon/2 = \epsilon$. Note that the key inequality in this sentence follows from the triangle inequality.

2.7.27

    a. Let $\epsilon > 0$ be given. It won't end up mattering what δ is, so let $\delta = 1$. Note that the statement $|f(x) - L| < \epsilon$ amounts to $|c - c| < \epsilon$, which is true for any positive number ϵ, without any restrictions on x. So $\lim_{x \to a} c = c$.

    b. Let $\epsilon > 0$ be given. Let $\delta = \epsilon$. Note that the statement $|f(x) - L| < \epsilon$ has the form $|x - a| < \epsilon$, which follows whenever $0 < |x - a| < \delta$ (since $\delta = \epsilon$.) Thus $\lim_{x \to a} x = a$.

2.7.29 Let $N > 0$ be given. Let $\delta = 1/\sqrt{N}$. Then if $0 < |x - 4| < \delta$, we have $|x - 4| < 1/\sqrt{N}$. Taking the reciprocal of both sides, we have $\frac{1}{|x-4|} > \sqrt{N}$, and squaring both sides of this inequality yields $\left|\frac{1}{(x-4)^2}\right| > N$. Thus $\lim_{x \to 4} f(x) = \infty$.

2.7.31 Let $N > 1$ be given. Let $\delta = 1/\sqrt{N - 1}$. Suppose that $0 < |x - 0| < \delta$. Then $|x| < 1/\sqrt{N - 1}$, and taking the reciprocal of both sides, we see that $1/|x| > \sqrt{N - 1}$. Then squaring both sides yields $\frac{1}{x^2} > N - 1$, so $\frac{1}{x^2} + 1 > N$. Thus $\lim_{x \to 0} f(x) = \infty$.

2.7.33

    a. False. In fact, if the statement is true for a specific value of δ_1, then it would be true for any value of $\delta < \delta_1$. This is because if $0 < |x - a| < \delta$, it would automatically follow that $0 < |x - a| < \delta_1$.

    b. False. This statement is not equivalent to the definition – note that it says "for an arbitrary δ there exists an ϵ" rather than "for an arbitrary ϵ there exists a δ."

    c. True. This is the definition of $\lim_{x \to a} f(x) = L$.

    d. True. Both inequalities describe the set of x's which are within δ units of a.

2.7.35 Assume $|x - 3| < 1$, as indicated in the hint. Then $2 < x < 4$, so $\frac{1}{4} < \frac{1}{x} < \frac{1}{2}$, and thus $\left|\frac{1}{x}\right| < \frac{1}{2}$.

Also note that the expression $\left|\frac{1}{x} - \frac{1}{3}\right|$ can be written as $\left|\frac{x-3}{3x}\right|$.

The Proof:

Let $\epsilon > 0$ be given. Let $\delta = \min(6\epsilon, 1)$. Now assume that $0 < |x - 3| < \delta$. Then

$$|f(x) - L| = \left|\frac{x - 3}{3x}\right| < \left|\frac{x - 3}{6}\right|$$

$$\leq \frac{6\epsilon}{6} = \epsilon.$$

Thus we have established that $\left|\frac{1}{x} - \frac{1}{3}\right| < \epsilon$ whenever $0 < |x - 3| < \delta$.

2.7.37 Assume $|x - (1/10)| < (1/20)$, as indicated in the hint. Then $1/20 < x < 3/20$, so $\frac{20}{3} < \frac{1}{x} < \frac{20}{1}$, and thus $\left|\frac{1}{x}\right| < 20$.

Also note that the expression $\left|\frac{1}{x} - 10\right|$ can be written as $\left|\frac{10x-1}{x}\right|$.

The Proof:

Let $\epsilon > 0$ be given. Let $\delta = \min(\epsilon/200, 1/20)$. Now assume that $0 < |x - (1/10)| < \delta$. Then

$$|f(x) - L| = \left|\frac{10x - 1}{x}\right| < |(10x - 1)(20)|$$

$$\leq |x - (1/10)|(200) < \frac{\epsilon}{200} \cdot 200 = \epsilon.$$

Thus we have established that $\left|\frac{1}{x} - 10\right| < \epsilon$ whenever $0 < |x - (1/10)| < \delta$.

2.7.39 Since we are approaching a from the right, we are only considering values of x which are close to, but a little larger than a. The numbers x to the right of a which are within δ units of a satisfy $0 < x - a < \delta$.

2.7.41

a. Let $\epsilon > 0$ be given. let $\delta = \epsilon/2$. Suppose that $0 < x < \delta$. Then $0 < x < \epsilon/2$ and

$$|f(x) - L| = |2x - 4 - (-4)| = |2x| = 2|x|$$

$$= 2x < \epsilon.$$

b. Let $\epsilon > 0$ be given. let $\delta = \epsilon/3$. Suppose that $0 < 0 - x < \delta$. Then $-\delta < x < 0$ and $-\epsilon/3 < x < 0$, so $\epsilon > -3x$. We have

$$|f(x) - L| = |3x - 4 - (-4)| = |3x| = 3|x|$$

$$= -3x < \epsilon.$$

c. Let $\epsilon > 0$ be given. Let $\delta = \epsilon/3$. Since $\epsilon/3 < \epsilon/2$, we can argue that $|f(x) - L| < \epsilon$ whenever $0 < |x| < \delta$ exactly as in the previous two parts of this problem.

2.7.43 Let $\epsilon > 0$ be given, and let $\delta = \epsilon^2$. Suppose that $0 < x < \delta$, which means that $x < \epsilon^2$, so that $\sqrt{x} < \epsilon$. Then we have

$$|f(x) - L| = |\sqrt{x} - 0| = \sqrt{x} < \epsilon.$$

as desired.

2.7.45

a. We say that $\lim\limits_{x \to a^+} f(x) = \infty$ if for each positive number N, there exists $\delta > 0$ such that

$$f(x) > N \quad \text{whenever} \quad a < x < a + \delta.$$

b. We say that $\lim\limits_{x \to a^-} f(x) = -\infty$ if for each negative number N, there exists $\delta > 0$ such that

$$f(x) < N \quad \text{whenever} \quad a - \delta < x < a.$$

c. We say that $\lim\limits_{x \to a^-} f(x) = \infty$ if for each positive number N, there exists $\delta > 0$ such that

$$f(x) > N \quad \text{whenever} \quad a - \delta < x < a.$$

2.7.47 Let $N > 0$ be given. Let $\delta = 1/N$, and suppose that $1 - \delta < x < 1$. Then $\frac{N-1}{N} < x < 1$, so $\frac{1-N}{N} > -x > -1$, and therefore $1 + \frac{1-N}{N} > 1 - x > 0$, which can be written as $\frac{1}{N} > 1 - x > 0$. Taking reciprocals yields the inequality $N < \frac{1}{1-x}$, as desired.

2.7.49 Let $M < 0$ be given. Let $\delta = \sqrt[4]{-10/M}$. Suppose that $0 < |x + 2| < \delta$. Then $(x+2)^4 < -10/M$, so $\frac{1}{(x+2)^4} > \frac{M}{-10}$, and $\frac{-10}{(x+2)^4} < M$, as desired.

2.7.51 Let $\epsilon > 0$ be given. Let $N = 1/\epsilon$. Suppose that $x > N$. Then $\frac{1}{x} < \epsilon$, and so $|f(x) - L| = |2 + \frac{1}{x} - 2| < \epsilon$.

2.7.53 Let $M > 0$ be given. Let $N = M - 1$. Suppose that $x > N$. Then $x > M - 1$, so $x + 1 > M$, and thus $\frac{x^2 + x}{x} > M$, as desired.

2.7.55 Let $\epsilon > 0$ be given. Let $N = \lfloor (1/\epsilon) \rfloor + 1$. By assumption, there exists an integer $M > 0$ so that $|f(x) - L| < 1/N$ whenever $|x - a| < 1/M$. Let $\delta = 1/M$.

Now assume $0 < |x - a| < \delta$. Then $|x - a| < 1/M$, and thus $|f(x) - L| < 1/N$. But then

$$|f(x) - L| < \frac{1}{\lfloor (1/\epsilon) \rfloor + 1} < \epsilon,$$

as desired.

2.7.57 Note that $f(x) = |x|/x = \begin{cases} 1 & \text{if } x > 0; \\ -1 & \text{if } x < 0. \end{cases}$

Thus $\lim_{x \to 0^+} f(x) = 1$, and $\lim_{x \to 0^-} f(x) = -1$, and therefore $\lim_{x \to 0} f(x)$ does not exist.

2.7.59 Since f is continuous at a, we know that $\lim_{x \to a} f(x)$ exists and is equal to $f(a) > 0$. Let $\epsilon = f(a)/3$. Then there is a number $\delta > 0$ so that $|f(x) - f(a)| < f(a)/3$ whenever $|x - a| < \delta$. Then whenever x lies in the interval $(a - \delta, a + \delta)$ we have $-f(a)/3 \le f(x) - f(a) \le f(a)/3$, so $2f(a)/3 \le f(x) \le 4f(a)/3$, so f is positive in this interval.

2.8 Chapter Two Review

2.8.1

a. False. Since $\lim_{x \to 1} \frac{x - 1}{x^2 - 1} = \lim_{x \to 1} \frac{1}{x + 1} = \frac{1}{2}$, f doesn't have a vertical asymptote at $x = 1$.

b. False. In general, these methods are too imprecise to produce accurate results.

c. False. For example, the function $f(x) = \begin{cases} 2x & \text{if } x < 0; \\ 1 & \text{if } x = 0; \\ 4x & \text{if } x > 0 \end{cases}$ has a limit of 0 as $x \to 0$, but $f(0) = 1$.

d. True. When we say that a limit exists, we are saying that there is a real number L that the function is approaching. If the limit of the function is ∞, it is still the case that there is no real number that the function is approaching. (There is no real number called "infinity.")

e. False. It could be the case that $\lim_{x \to a^-} f(x) = 1$ and $\lim_{x \to a^+} f(x) = 2$.

f. False. For example, the function $f(x) = \begin{cases} 2 & \text{if } 0 < x < 1; \\ 3 & \text{if } 1 \le x < 2, \end{cases}$ is continuous on $(0, 1)$, and on $[1, 2)$, but isn't continuous on $(0, 2)$.

g. True. $\lim_{x \to a} f(x) = f(a)$ if and only if f is continuous at a.

2.8.3 This function is discontinuous at $x = -1$, at $x = 1$, and at $x = 3$. At $x = -1$ it is discontinuous because $\lim_{x \to -1} f(x)$ does not exist. At $x = 1$, it is discontinuous because $\lim_{x \to 1} f(x) \neq f(1)$. At $x = 3$, it is discontinuous because $f(3)$ does not exist, and because $\lim_{x \to 3} f(x)$ does not exist.

2.8.5

a.

x	$.9\pi/4$	$.99\pi/4$	$.999\pi/4$	$.9999\pi/4$
$f(x)$	1.4098	1.4142	1.4142	1.4142

x	$1.1\pi/4$	$1.01\pi/4$	$1.001\pi/4$	$1.0001\pi/4$
$f(x)$	1.4098	1.4142	1.4142	1.4142

b. $\displaystyle \lim_{x \to \pi/4} \frac{\cos 2x}{\cos x - \sin x} = \lim_{x \to \pi/4} \frac{\cos^2 x - \sin^2 x}{\cos x - \sin x} = \lim_{x \to \pi/4} \cos x + \sin x = \sqrt{2}.$

2.8.7

There are infinitely many different correct functions which you could draw. One of them is:

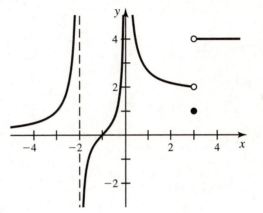

2.8.9 $\displaystyle \lim_{x \to 1} \sqrt{5x + 6} = \sqrt{11}.$

2.8.11 $\displaystyle \lim_{x \to 1} \frac{x^3 - 7x^2 + 12x}{4 - x} = \lim_{x \to 1} \frac{1 - 7 + 12}{4 - 1} = \frac{6}{3} = 2.$

2.8.13 $\displaystyle \lim_{x \to 1} \frac{1 - x^2}{x^2 - 8x + 7} = \lim_{x \to 1} \frac{(1 - x)(1 + x)}{(x - 7)(x - 1)} = \lim_{x \to 1} \frac{-(x + 1)}{x - 7} = \frac{1}{3}.$

2.8.15

$$\lim_{x \to 3} \frac{1}{x - 3} \left(\frac{1}{\sqrt{x + 1}} - \frac{1}{2} \right) = \lim_{x \to 3} \frac{2 - \sqrt{x + 1}}{2(x - 3)\sqrt{x + 1}} \cdot \frac{(2 + \sqrt{x + 1})}{(2 + \sqrt{x + 1})}$$

$$= \lim_{x \to 3} \frac{4 - (x + 1)}{2(x - 3)(\sqrt{x + 1})(2 + \sqrt{x + 1})}$$

$$= \lim_{x \to 3} \frac{-(x - 3)}{2(x - 3)(\sqrt{x + 1})(2 + \sqrt{x + 1})}$$

$$= \lim_{x \to 3} \frac{-1}{2\sqrt{x + 1}(2 + \sqrt{x + 1})} = \frac{-1}{16}.$$

2.8.17 $\displaystyle \lim_{x \to 3} \frac{x^4 - 81}{x - 3} = \lim_{x \to 3} \frac{(x - 3)(x + 3)(x^2 + 9)}{x - 3} = \lim_{x \to 3} (x + 3)(x^2 + 9) = 108.$

2.8.19 $\lim\limits_{x\to 81} \dfrac{\sqrt[4]{x}-3}{x-81} = \lim\limits_{x\to 81} \dfrac{\sqrt[4]{x}-3}{(\sqrt{x}+9)(\sqrt[4]{x}+3)(\sqrt[4]{x}-3)} = \lim\limits_{x\to 81} \dfrac{1}{(\sqrt{x}+9)(\sqrt[4]{x}+3)} = \dfrac{1}{108}.$

2.8.21 $\lim\limits_{x\to \pi/2} \dfrac{\frac{1}{\sqrt{\sin x}}-1}{x+\pi/2} = \dfrac{0}{\pi} = 0.$

2.8.23

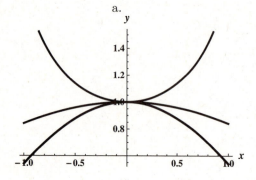

a.

b. Since $\lim\limits_{x\to 0}\cos x = \lim\limits_{x\to 0}\dfrac{1}{\cos x} = 1$, the squeeze theorem assures us that $\lim\limits_{x\to 0}\dfrac{\sin x}{x} = 1$ as well.

2.8.25 $\lim\limits_{x\to 5}\dfrac{x-7}{x(x-5)^2} = -\infty.$

2.8.27 $\lim\limits_{x\to 3^-}\dfrac{x-4}{x^2-3x} = \lim\limits_{x\to 3^-}\dfrac{x-4}{x(x-3)} = \infty.$

2.8.29 $\lim\limits_{x\to 0^-}\dfrac{2}{\tan x} = -\infty.$

2.8.31 $\lim\limits_{x\to\infty}\dfrac{2x-3}{4x+10} = \lim\limits_{x\to\infty}\dfrac{2-(3/x)}{4+(10/x)} = \dfrac{2}{4} = \dfrac{1}{2}.$

2.8.33 $\lim\limits_{x\to-\infty} -3x^3+5 = \infty.$

2.8.35 $\lim\limits_{x\to\infty}\dfrac{\sqrt{25x^2+8}}{x+2} = \lim\limits_{x\to\infty}\dfrac{\sqrt{25+8/x^2}}{1+2/x} = 5$

2.8.37 $\lim\limits_{x\to\infty}\dfrac{4x^3+1}{1-x^3} = \lim\limits_{x\to\infty}\dfrac{4+(1/x^3)}{(1/x^3)-1} = \dfrac{4+0}{0-1} = -4.$ A similar result holds as $x\to-\infty$. Thus, $y=-4$ is a horizontal asymptote as $x\to\infty$ and as $x\to-\infty$.

2.8.39 $\lim\limits_{x\to\pm\infty}\dfrac{12x^2}{\sqrt{16x^4+7}} = \lim\limits_{x\to\pm\infty}\dfrac{12}{\sqrt{16+7/x^4}} = 3$

2.8.41 $\lim\limits_{x\to\pm\infty}\dfrac{x^2-x}{x^2-1} = \lim\limits_{x\to\pm\infty}\dfrac{x}{x+1} = \lim\limits_{x\to\pm\infty}\dfrac{1}{1+1/x} = 1$, so that $f(x)$ has a horizontal asymptote at $y=1$. There is a vertical asymptote when $x+1=0$, i.e. for $x=-1$.

2.8.43 f is discontinuous at 5, since $f(5)$ does not exist, and also because $\lim\limits_{x\to 5} f(x)$ does not exist

2.8.45 h is continuous from the right at 3 since $\lim\limits_{x\to 3^+} h(x) = 0 = h(3)$.

2.8.47 The domain of f is $(-\infty, -\sqrt{5}] \cup [\sqrt{5}, \infty)$, and f is continuous on that domain.

2.8.49 The domain of h is $(-\infty, -5)\cup(-5,0)\cup(0,5)\cup(5,\infty)$, and like all rational functions, it is continuous on its domain.

2.8.51 In order for g to be left continuous at 1, it is necessary that $\lim_{x\to 1^-} g(x) = g(1)$, which means that $a = 3$. In order for g to be right continuous at 1, it is necessary that $\lim_{x\to 1^+} g(x) = g(1)$, which means that $a + b = 3 + b = 3$, so $b = 0$.

2.8.53

One such possible graph is pictured to the right.

2.8.55

a. Any such rectangle with length x has width $y = 100/x$, so its perimeter is

$$P(x) = 2x + 2y = 2x + \frac{200}{x}.$$

b. $P(x)$ is continuous for $x > 0$. $P(10) = 40$ while $P(30) = 60 + \frac{200}{3} > 60$. Thus by the intermediate value theorem, there must be some x with $10 \le x \le 30$ and $P(x) = 50$.

c. The rectangle with perimeter 50 can be determined by solving $P(x) = 50$.

$$2x + \frac{200}{x} = 50$$
$$2x^2 - 50x + 200 = 0$$
$$x = 5, 20$$

So a perimeter of 50 occurs for a 5×20 or a 20×5 rectangle.

d. If there were a rectangle with perimeter 30, it would be a solution to $P(x) = 30$, so

$$2x + \frac{200}{x} = 30, \qquad 2x^2 - 30x + 200 = 0$$

The discriminant of this quadratic is $900 - 1600 < 0$, so the quadratic has no (real) solution.

e. Plotting $P(x)$ for various viewing windows seems to show that the smallest possible perimeter is 40, for a 10×10 rectangle.

2.8.57 Let $\epsilon > 0$ be given. Let $\delta = \epsilon$. Now suppose that $0 < |x - 5| < \delta$.
Then

$$|f(x) - L| = \left| \frac{x^2 - 25}{x - 5} - 10 \right| = \left| \frac{(x-5)(x+5)}{x-5} - 10 \right| = |x + 5 - 10|$$
$$= |x - 5| < \epsilon.$$

2.8.59 Let $N > 0$ be given. Let $\delta = 1/\sqrt[4]{N}$. Suppose that $0 < |x - 2| < \delta$.
Then $|x - 2| < \frac{1}{\sqrt[4]{N}}$, so $\frac{1}{|x-2|} > \sqrt[4]{N}$, and $\left| \frac{1}{(x-2)^4} \right| > N$, as desired.

Chapter 3

Derivatives

3.1 Introducing the Derivative

3.1.1 The secant line through the points $(a, f(a))$ and $(x, f(x))$ for x near a, of the graph of f, is given by $m_{\text{sec}} = \dfrac{f(x) - f(a)}{x - a}$. As x approaches a, we obtain the limit $m_{\text{tan}} = \lim\limits_{x \to a} \dfrac{f(x) - f(a)}{x - a} = \lim\limits_{x \to a} m_{\text{sec}}$.

3.1.3 The average rate of change of f over $[a, x]$ is the slope of the secant line $m_{\text{sec}} = \frac{f(x) - f(a)}{x - a}$. As x approaches a, the length of the interval $x - a$ goes to zero, and in the limit we obtain the instantaneous rate of change of f at a given by $m_{\text{tan}} = \lim\limits_{x \to a} \dfrac{f(x) - f(a)}{x - a}$.

3.1.5 $f'(a)$ is the value of the derivative of f at a. Also, $f'(a)$ is the slope of the tangent line to the graph of f at $(a, f(a))$. Furthermore, $f'(a)$ is the instantaneous rate of change of f at a.

3.1.7 The letter Δ is the Greek letter for d and stands for difference or change. dy and dx are used to represent small changes in y and x, respectively. Since the derivative is the limit of the ratio of small changes in y and x, it makes sense to write it as $\frac{dy}{dx}$.

3.1.9 No, there are continuous functions which are not differentiable. For example $f(x) = |x|$ is continuous everywhere but the graph of f has a corner at $a = 0$, and thus f is not differentiable at $a = 0$.

3.1.11

c.

a.

$$m_{\text{tan}} = \lim_{x \to 3} \frac{x^2 - 5 - 4}{x - 3} = \lim_{x \to 3} \frac{x^2 - 9}{x - 3}$$
$$= \lim_{x \to 3} \frac{(x - 3)(x + 3)}{x - 3} = \lim_{x \to 3} (x + 3) = 6.$$

b. Using the point-slope form of the equation of a line, we obtain $y - 4 = 6(x - 3)$.

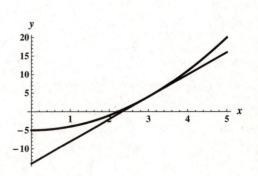

3.1.13

c.

a.

$$m_{\tan} = \lim_{x \to 1} \frac{-5x + 1 + 4}{x - 1} = \lim_{x \to 1} \frac{-5x + 5}{x - 1}$$
$$= \lim_{x \to 1} -5 \frac{x - 1}{x - 1} = -5.$$

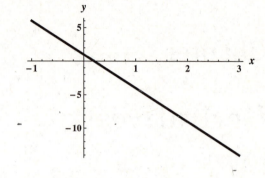

b. Using the point-slope form of the equation of a line, we get $y + 4 = -5(x - 1)$ which equals $y = -5x + 1$, the function itself.

3.1.15

c.

a.

$$m_{\tan} = \lim_{x \to -1} \frac{\frac{1}{x} + 1}{x + 1} = \lim_{x \to -1} \frac{\frac{1+x}{x}}{x + 1}$$
$$= \lim_{x \to -1} \frac{1}{x} = -1.$$

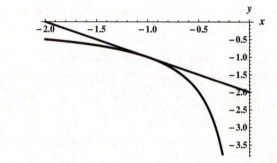

b. $y - (-1) = -1(x + 1)$, or $y = -x - 2$.

3.1.17

a. $m_{\tan} = \lim_{h \to 0} \frac{2(0 + h) + 1 - 1}{h} = \lim_{h \to 0} \frac{2h}{h} = 2.$

b. $y - 1 = 2x.$

3.1.19

a. $m_{\tan} = \lim_{h \to 0} \frac{(-1 + h)^4 - 1}{h} = \lim_{h \to 0} \frac{1 - 4h + 6h^2 - 4h^3 + h^4 - 1}{h} = \lim_{h \to 0} (-4 + 6h - 4h^2 + h^3) = -4.$

b. $y - 1 = -4(x - (-1)).$

3.1.21

a. $m_{\tan} = \lim_{h \to 0} \frac{\frac{1}{3 - 2(h-1)} - \frac{1}{5}}{h} = \lim_{h \to 0} \frac{\frac{5 - (3 - 2h + 2)}{15 - 10(h-1)}}{h} = \lim_{h \to 0} \frac{2}{15 - 10(h - 1)} = \frac{2}{25}.$

b. $y - \frac{1}{5} = \frac{2}{25}(x + 1).$

3.1.23

a. $f'(-3) = \lim_{h \to 0} \frac{8(-3 + h) + 24}{h} = \lim_{h \to 0} \frac{8h}{h} = 8.$

b. $y - (-24) = 8(x + 3).$

3.1.25

a.

$$f'(-2) = \lim_{h \to 0} \frac{4(-2+h)^2 + 2(-2+h) - 12}{h} = \lim_{h \to 0} \frac{16 - 16h + 4h^2 - 4 + 2h - 12}{h}$$

$$= \lim_{h \to 0} \frac{-14h + 4h^2}{h} = -14$$

b. $y - 12 = -14(x+2)$.

3.1.27

a.

$$f'\left(\frac{1}{4}\right) = \lim_{h \to 0} \frac{\frac{1}{\sqrt{\frac{1}{4}+h}} - 2}{h} = \lim_{h \to 0} \frac{1 - 2\sqrt{\frac{1}{4}+h}}{h\sqrt{\frac{1}{4}+h}} = \lim_{h \to 0} \frac{\left(1 - 2\sqrt{\frac{1}{4}+h}\right)\left(1 + 2\sqrt{\frac{1}{4}+h}\right)}{h\sqrt{\frac{1}{4}+h}\left(1 + 2\sqrt{\frac{1}{4}+h}\right)}$$

$$= \lim_{h \to 0} \frac{1 - 4\left(\frac{1}{4}+h\right)}{h\sqrt{\frac{1}{4}+h}\left(1 + 2\sqrt{\frac{1}{4}+h}\right)} = \lim_{h \to 0} \frac{-4}{\sqrt{\frac{1}{4}+h}\left(1 + 2\sqrt{\frac{1}{4}+h}\right)} = -4.$$

b. $y - 2 = -4\left(x - \frac{1}{4}\right)$.

3.1.29

a.

$$f'(x) = \lim_{h \to 0} \frac{3(x+h)^2 + 2(x+h) - 10 - (3x^2 + 2x - 10)}{h}$$

$$= \lim_{h \to 0} \frac{3x^2 + 6xh + 3h^2 + 2x + 2h - 10 - 3x^2 - 2x + 10}{h}$$

$$= \lim_{h \to 0} \frac{6xh + 2h + 3h^2}{h} = \lim_{h \to 0} (6x + 2 + 3h) = 6x + 2.$$

c.

b. We have $f'(1) = 8$, and the tangent line is given by $y + 5 = 8(x - 1)$.

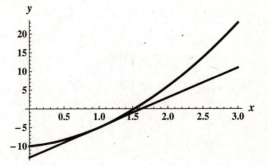

3.1.31

a.

$$f'(x) = \lim_{h \to 0} \frac{5(x+h)^2 - 6(x+h) + 1 - (5x^2 - 6x + 1)}{h} = \lim_{h \to 0} \frac{5x^2 + 10xh + 5h^2 - 6x - 6h - 5x^2 + 6x}{h}$$

$$= \lim_{h \to 0} \frac{10xh + 5h^2 - 6h}{h} = \lim_{h \to 0} (10x + 5h - 6) = 10x - 6.$$

c.

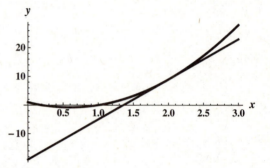

b. We have $f'(2) = 14$, so the tangent line is given by $y - 9 = 14(x - 2)$.

3.1.33

a.

$$\frac{d}{dx}\left(ax^2 + bx + c\right) = \lim_{h \to 0} \frac{a(x+h)^2 + b(x+h) + c - \left(ax^2 + bx + c\right)}{h}$$

$$= \lim_{h \to 0} \frac{ax^2 + 2axh + ah^2 + bx + bh + c - ax^2 - bx - c}{h}$$

$$= \lim_{h \to 0} \frac{2axh + ah^2 + bh}{h} = \lim_{h \to 0} (2ax + ah + b) = 2ax + b.$$

b. With $a = 4, b = -3, c = 10$ we have $\frac{d}{dx}(4x^2 - 3x + 10) = 2 \cdot 4 \cdot x + (-3) = 8x - 3.$

3.1.35 $m_{\tan} = \lim\limits_{h \to 0} \dfrac{\frac{1}{1+h+1} - \frac{1}{2}}{h} = \lim\limits_{h \to 0} \dfrac{2 - (2+h)}{h(2+h)2} = \lim\limits_{h \to 0} \dfrac{-1}{(2+h)2} = -\dfrac{1}{4}.$

3.1.37 $m_{\tan} = \lim\limits_{h \to 0} \dfrac{2\sqrt{25+h} - 1 - (2\sqrt{25} - 1)}{h} = \lim\limits_{h \to 0} \dfrac{2(\sqrt{25+h} - \sqrt{25})}{h} =$

$\lim\limits_{h \to 0} \dfrac{2(\sqrt{25+h} - \sqrt{25})(\sqrt{25+h} + \sqrt{25})}{h(\sqrt{25+h} + \sqrt{25})} = \lim\limits_{h \to 0} \dfrac{2(25+h-25)}{h(\sqrt{25+h} + \sqrt{25})} = \lim\limits_{h \to 0} \dfrac{2}{(\sqrt{25+h} + \sqrt{25})} = \dfrac{1}{5}.$

3.1.39

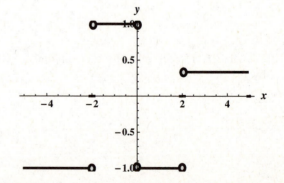

The function f is not differentiable at $x = -2, 0, 2$, so f' is not defined at those points. Elsewhere, the slope is constant.

3.1.41

a. The function has non-negative slope everywhere, and as there is a horizontal tangent at $x = 0$, so the derivative has to be zero at zero. The graph of the derivative has to be above the x-axis and touching it at $x = 0$, so (D) is the graph of the derivative.

b. The graph of this function has three horizontal tangent lines, at $x = -1, 0, 1$, and the matching graph of the derivative with three zeros is (C).

c. The function has negative slope on $(-1, 0)$, and positive slope on $(0, 1)$ and has a horizontal tangent at $x = 0$, so the derivative has to be negative on $(-1, 0)$, positive on $(0, 1)$ and zero at $x = 0$; the graph is (B).

d. The function has negative slope everywhere so the graph of the derivative has to be negative everywhere, which is graph (A).

3.1.43

The function always has non-negative slope, so the derivative is never below the x axis. However, it does have slope zero at about $x = 2$.

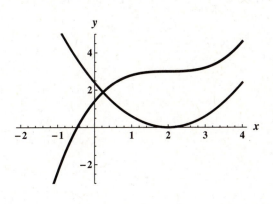

3.1.45

a. The function f is not continuous at $x = 1$, since the graph has a jump there.

b. The function f is not differentiable at $x = 1$ since it is not continuous at that point (Theorem 3.1 Alternate Version), and it is also not differentiable at $x = 2$ since the graph has a corner there.

c.

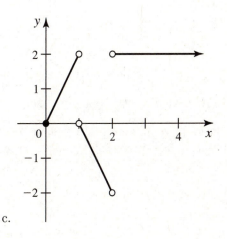

3.1.47

a. True. Since the graph is a line, any secant line has the same graph as the function and thus the same slope.

b. False. For example, take $f(x) = x^2$, $P = (0,0)$ and $Q = (1,1)$. Then the secant line has slope $m_{\text{sec}} = \frac{1-0}{1-0} = 1$, but the the graph has a horizontal tangent at P so $m_{\text{tan}} = 0$ and $m_{\text{sec}} > m_{\text{tan}}$.

c. True. Since $m_{\text{sec}} = \frac{(x+h)^2 - x^2}{h} = \frac{2xh + h^2}{h} = 2x + h$ while $m_{\text{tan}} = \lim_{h \to 0} (2x + h) = 2x$. Since we assume that $h > 0$, we have $m_{\text{sec}} = 2x + h > 2x = m_{\text{tan}}$.

d. True. If f is differentiable at a number a, then it is continuous at a.

3.1.49

a.

$$f'(x) = \lim_{h \to 0} \frac{\sqrt{3(x+h)+1} - \sqrt{3x+1}}{h}$$

$$= \lim_{h \to 0} \frac{\sqrt{3(x+h)+1} - \sqrt{3x+1}}{h} \cdot \frac{\sqrt{3x+3h+1} + \sqrt{3x+1}}{\sqrt{3x+3h+1} + \sqrt{3x+1}}$$

$$= \lim_{h \to 0} \frac{3x+3h+1 - 3x - 1}{h(\sqrt{3x+3h+1} + \sqrt{3x+1})} = \lim_{h \to 0} \frac{3}{(\sqrt{3x+3h+1} + \sqrt{3x+1})} = \frac{3}{2\sqrt{3x+1}}.$$

b. We have $f'(8) = \frac{3}{10}$. Using the point-slope form, we get that the tangent line has equation $y - 5 = \frac{3}{10}(x - 8)$.

3.1.51

a.

$$f'(x) = \lim_{h \to 0} \frac{\frac{2}{3(x+h)+1} - \frac{2}{3x+1}}{h} = \lim_{h \to 0} \frac{6x + 2 - (6x + 6h + 2)}{h(3x+1)(3x+3h+1)}$$

$$= \lim_{h \to 0} \frac{-6h}{h(3x+1)(3x+3h+1)} = \frac{-6}{(3x+1)^2}.$$

b. We have $f'(-1) = \frac{-3}{2}$. Using the point-slope form, we get that the tangent line has equation $y + 1 = -\frac{3}{2}(x + 1)$.

3.1.53

a. At C and D, the slope of the tangent line (and thus of the curve) is negative.

b. At A, B, and E, the slope of the curve is positive.

c. The graph is in its steepest ascent at A followed by B. At E it barely increases, at D it slightly decreases and at C it is decreasing the most, so the points in decreasing order of slope are A, B, E, D, C.

3.1.55

Since $f'(x) = x$ is negative for $x < 0$ and positive for $x > 0$, we have that the graph of f has to have negative slope on $(-\infty, 0)$ and positive slope on $(0, \infty)$ and has to have a horizontal tangent at $x = 0$. Since f' only gives us the slope of the tangent line and not the actual value of f, there are infinitely many graphs possible, they all have the same shape, but are shifted along the y-axis.

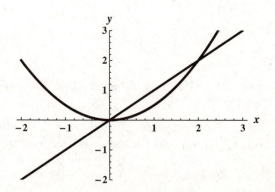

3.1.57

a. From the graph we approximate the derivative by the slope of a secant line: For example we see that $E(6) = 250\,\text{kWh}$ and $E(18) = 350$ kWh, so the power after 10 hours is approximately the slope of the secant line through these points, so $P(10) \approx m_{sec} = \frac{E(18)-E(6)}{18-6} = \frac{350\,\text{kWh}-250\,\text{kWh}}{12\text{h}} \approx 8.3\,\text{kW}$. Similarly, after 20 hours, using 18 hours and 25 hours, that $P(20) \approx m_{\text{sec}} = \frac{E(22)-E(18)}{22-18} = \frac{325\,\text{kWh}-350\,\text{kWh}}{4\text{h}} \approx -6.25\text{kW}$.

b. The power is zero where the graph of $E(t)$ has a horizontal tangent line, which happens approximately at $t = 6$ hours and $t = 18$ hours.

c. The power has a maximum where the graph of $E(t)$ has the steepest increase, which is approximately at $t = 12$ hours.

3.1.59

a.

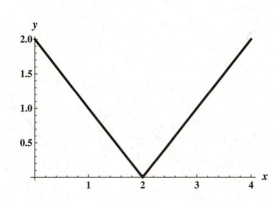

b.

$$f'_+(2) = \lim_{h \to 0^+} \frac{|2+h-2|-0}{h} = \lim_{h \to 0^+} \frac{h}{h} = 1,$$

since for $h > 0$, we have $|h| = h$. Similarly,

$$f'_-(2) = \lim_{h \to 0^-} \frac{|2+h-2|-0}{h} = \lim_{h \to 0^-} \frac{-h}{h} = -1,$$

since for $h < 0$, we have $|h| = -h$.

c. Since f is defined at $a = 2$ and the graph of f does not jump, f is continuous at $a = 2$. Since the left-hand and right-hand derivatives are not equal, f is not differentiable at $a = 2$.

3.1.61

a. The graph has a vertical tangent at $x = 2$.

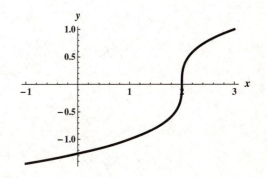

b.   The graph has a vertical tangent at $x = -1$.

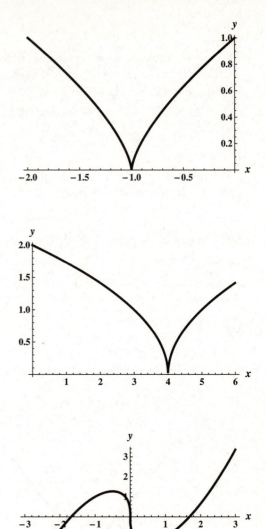

c.   The graph has a vertical tangent at $x = 4$.

d.   The graph has a vertical tangent at $x = 0$.

3.1.63 $f'(x) = \lim\limits_{h \to 0} \dfrac{h^{1/3}}{h} = \lim\limits_{h \to 0} \dfrac{1}{h^{2/3}} = +\infty$ regardless of $h > 0$ or $h < 0$, since h only appears with a power of $\frac{2}{3}$. Thus the graph of f has a vertical tangent at $x = 0$.

3.1.65 Consider $a = 2$ and $f(x) = \frac{1}{x+1}$. Then $f'(2) = \lim\limits_{x \to 2} \dfrac{\frac{1}{x+1} - \frac{1}{3}}{x - 2}$ as desired. We have $f'(2) =$

$\lim\limits_{x \to 2} \dfrac{\frac{1}{x+1} - \frac{1}{3}}{x - 2} = \lim\limits_{x \to 2} \dfrac{3 - (x+1)}{(x-2)3(x+1)} = \lim\limits_{x \to 2} \dfrac{-(x-2)}{(x-2)3(x+1)} = \lim\limits_{x \to 2} \dfrac{-1}{3(x+1)} = -\dfrac{1}{9}.$

3.1.67 Consider $a = 2$ and $f(x) = x^4$. Then $f'(2) = \lim\limits_{h \to 0} \dfrac{(2+h)^4 - 16}{h}$ as desired. We have

$$f'(2) = \lim_{h \to 0} \frac{(2+h)^4 - 16}{h} = \lim_{h \to 0} \frac{16 + 32h + 24h^2 + 8h^3 + h^4 - 16}{h}$$

$$= \lim_{h \to 0} \frac{h\left(32 + 24h + 8h^2 + h^3\right)}{h} = \lim_{h \to 0} \left(32 + 24h + 8h^2 + h^3\right) = 32.$$

3.1.69 It is not differentiable at $x = 2$. The denominator of f is zero when $x = 2$, so f is not defined and is not continuous at $x = 2$, and by Theorem 3.1 (alternate) it cannot be differentiable there.

3.1.71 In order for f to be differentiable at $x = 1$, it would need to be continuous there. Thus, $\lim\limits_{x \to 1^-} f(x) = \lim\limits_{x \to 1^-} 2x^2 = 2 = \lim\limits_{x \to 1^+} f(x) = \lim\limits_{x \to 1^+} ax - 2 = a - 2$, so the only possible value for a is 4. Now checking the differentiability at 1, we have

$$f'_-(1) = \lim_{x \to 1^-} \frac{f(x) - f(1)}{x - 1} = \lim_{x \to 1^-} \frac{2x^2 - 2}{x - 1} = \lim_{x \to 1^-} 2(x + 1) = 4.$$

Also

$$f'_+(1) = \lim_{x \to 1^+} \frac{f(x) - f(1)}{x - 1} = \lim_{x \to 1^+} \frac{4x - 2 - 2}{x - 1} = \lim_{x \to 1^-} 4 = 4,$$

so f is differentiable at 1 for $a = 4$.

3.2 Rules of Differentiation

3.2.1 Often the limit definition of f' is difficult to compute, especially for functions which are reasonably complicated. The rules for differentiation allow us to easily compute the derivatives of complex functions.

3.2.3 By the constant multiple rule, the derivative of the function cf where c is a constant and f is a function is cf'. That is, the derivative of a constant times a function is that same constant times the derivative of the function.

3.2.5 $\dfrac{d}{dx}\left(\dfrac{1}{2}x^6 - 3x^4 + 101x + 7\right) = 3x^5 - 12x^3 + 101$ by applying the Sum Rule and then applying the Power Rule to each term.

3.2.7 By the power rule, $y' = 5x^{5-1} = 5x^4$.

3.2.9 By the constant rule, $f'(x) = 0$.

3.2.11 By the power rule $h'(t) = 1t^{1-1} = t^0 = 1$.

3.2.13 By the constant multiple rule and the power rule, $f'(x) = 5 \cdot \frac{d}{dx}x^3 = 5 \cdot 3x^2 = 15x^2$.

3.2.15 By the constant multiple and power rules, $p'(x) = 8 \cdot \frac{d}{dx}x = 8 \cdot 1 = 8$.

3.2.17 By the constant multiple and power rules, $g'(t) = 100\frac{d}{dt}t^2 = 100 \cdot 2t = 200t$.

3.2.19

$$f'(x) = \frac{d}{dx}(3x^4 + 7x) = \frac{d}{dx}(3x^4) + \frac{d}{dx}(7x) = 12x^3 + 7.$$

3.2.21 $f'(x) = \dfrac{d}{dx}\left(10x^4 - 32x + \dfrac{1}{2}\right) = \dfrac{d}{dx}(10x^4) - \dfrac{d}{dx}(32x) + \dfrac{d}{dx}\left(\dfrac{1}{2}\right) = 40x^3 - 32 + 0 = 40x^3 - 32.$

3.2.23 $g'(w) = \frac{d}{dw}(2w^3 + 3w) = \frac{d}{dw}(2w^3) + \frac{d}{dw}(3w) = 6w^2 + 3.$

3.2.25 Expanding the product yields $f(x) = 6x^3 + 3x^2 + 4x + 2$. So

$$f'(x) = \frac{d}{dx}(6x^3 + 3x^2 + 4x + 2) = \frac{d}{dx}(6x^3) + \frac{d}{dx}(3x^2) + \frac{d}{dx}(4x) + \frac{d}{dx}(2)$$
$$= 18x^2 + 6x + 4.$$

3.2.27 Expanding the product yields $h(x) = x^4 + 2x^2 + 1$. So

$$h'(x) = \frac{d}{dx}(x^4 + 2x^2 + 1) = \frac{d}{dx}(x^4) + \frac{d}{dx}(2x^2) + \frac{d}{dx}(1)$$
$$= 4x^3 + 4x.$$

3.2.29 f simplifies as $f(w) = w^2 - 1$, so $f'(w) = 2w$ for $w \neq 0$.

3.2.31 g simplifies as $g(x) = \dfrac{(x-1)(x+1)}{x-1} = x + 1$. Thus $g'(x) = 1$ for $x \neq 1$.

3.2.33 y simplifies as $y = \dfrac{(\sqrt{x} - \sqrt{a})(\sqrt{x} + \sqrt{a})}{\sqrt{x} - \sqrt{a}} = \sqrt{x} + \sqrt{a}$. Thus $\dfrac{dy}{dx} = \dfrac{1}{2\sqrt{x}}$ for $x \neq a$.

3.2.35

b.

a. $y' = -6x$, so the slope of the tangent line at $a = 1$ is -6. Thus, the tangent line at the point $(1, -1)$ is $y + 1 = -6(x - 1)$, or $y = -6x + 5$.

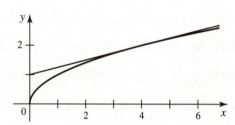

3.2.37

b.

a. $y' = 1/(2\sqrt{x})$, so the slope of the tangent line at $x = 4$ is $y'(4) = \frac{1}{4}$. At $x = 4$, $y = \sqrt{4} = 2$, so the equation of the tangent line is

$$y - 2 = \frac{1}{4}(x - 4), \qquad y = \frac{1}{4}x + 1$$

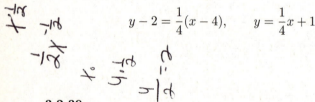

3.2.39

a. $f'(x) = 2x - 6$, so the slope is zero when $2x - 6 = 0$, which is at $x = 3$.

b. The slope is 2 when $2x - 6 = 2$ which is at $x = 4$.

3.2.41

a. The slope of the tangent line is given by $f'(x) = 6x^2 - 6x - 12$, and this quantity is zero when $x^2 - x - 2 = 0$, or $(x - 2)(x + 1) = 0$. The two solutions are thus $x = -1$ and $x = 2$, so the points on the graph are $(-1, 11)$ and $(2, -16)$.

b. The slope of the tangent line is 60 when $6x^2 - 6x - 12 = 60$, which is when $6x^2 - 6x - 72 = 0$. Simplifying this quadratic expression yields the equation $x^2 - x - 12 = 0$, which has solutions $x = -3$ and $x = 4$, so the points on the graph are $(-3, -41)$, and $(4, 36)$.

3.2.43 $f'(x) = 20x^3 + 30x^2 + 3$, $f''(x) = 60x^2 + 60x$, and $f^{(3)}(x) = 120x + 60$.

3.2.45 f simplifies as $f(x) = \frac{(x-8)(x+1)}{x+1} = x - 8$. So for $x \neq -1$, $f'(x) = 1$, $f''(x) = 0$, and $f^{(3)}(x) = 0$.

3.2.47

a. False. 10^5 is a constant, so the constant rule assures us that $\frac{d}{dx}10^5 = 0$.

b. True. This follows because the slope is given by $4 \neq 0$ for all x.

c. False. We have $\frac{d}{dx}(5x^3 + 2x + 5) = 15x^2 + 2$. Thus we have $\frac{d^2}{dx^2}(5x^3 + 2x + 5) = 30x$, and $\frac{d^3}{dx^3}(5x^3 + 2x + 5) = 30$. It is true that $\frac{d^n}{dx^n}(5x^3 + 2x + 5) = 0$ for $n \geq 4$.

3.2.49 First note that since the slope of $4x + 1$ is 4, it must be the case that $f'(2) = 4$. Also, at $x = 2$, we have $y = 4 \cdot 2 + 1 = 9$, so $f(2) = 9$. Similarly, the slope of $y = 3x - 2$ is 3, so $g'(2) = 3$. Also, at $x = 2$, we hve $y = 3 \cdot 2 - 2 = 4$, so $g(2) = 4$.

a. $y'(2) = f'(2) + g'(2) = 4 + 3 = 7$. The line contains the point $(2, f(2) + g(2)) = (2, 13)$. Thus, the equation of the tangent line is $y - 13 = 7(x - 2)$, or $y = 7x - 1$.

b. $y'(2) = f'(2) - 2g'(2) = 4 - 2 \cdot 3 = -2$. The line contains the point $(2, f(2) - 2g(2)) = (2, 1)$. Thus, the equation of the tangent line is $y - 1 = -2(x - 2)$, or $y = -2x + 5$.

c. $y'(2) = 4f'(2) = 4 \cdot 4 = 16$. The line contains the point $(2, 4f(2)) = (2, 36)$. Thus, the equation of the tangent line is $y - 36 = 16(x - 2)$, or $y = 16x + 4$.

3.2.51 $G'(2) = 3f'(2) - g'(2) = 3(-3) - 1 = -10$.

3.2.53 $G'(5) = 3f'(5) - g'(5) = 3 \cdot 1 - (-1) = 4$.

3.2.55 $\frac{d}{dx}[1.5f(x)]_{x=2} = 1.5f'(2) = 1.5 \cdot 5 = 7.5$.

3.2.57

a. Let $f(x) = \sqrt{x}$ and $a = 9$. Then $\lim_{h \to 0} \frac{f(a + h) - f(a)}{h} = \lim_{h \to 0} \frac{\sqrt{9 + h} - \sqrt{9}}{h} = f'(9)$.

b. Since $f'(x) = \frac{1}{2\sqrt{x}}$, we have $f'(9) = \frac{1}{6}$, so this is the value of the original limit.

3.2.59

a. Let $f(x) = x^{100}$ and $a = 1$. Then $\lim_{x \to 1} \frac{f(x) - f(1)}{x - 1} = f'(1)$.

b. Since $f'(x) = 100x^{99}$, we have $f'(1) = 100$, so this is the value of the original limit.

3.2.61

a. $d'(t) = 32t$ is the velocity of the stone after t seconds, measured in feet per second.

b. The stone travels $d(6) = 16 \cdot 6^2 = 576$ feet and strikes the ground with a velocity of $32 \cdot 6 = 192$ feet per second. Converting to miles per hour, we have $192 \cdot \frac{3600}{5280} \approx 130.9$ miles per hour.

3.2.63

a. $\frac{dD}{dg} = .1g + 35$ miles per gallon. It measures the rate of change of the car's distance travelled with respect to the amount of gas consumed. We usually call this "gas mileage."

b. $\frac{dD}{dg}(0) = 35$ miles per gallon. $\frac{dD}{dg}(5) = 35.5$ miles per gallon. $\frac{dD}{dg}(10) = 36$ miles per gallon. The gas mileage gets better as the tank gets emptier.

c. $D(12) = 0.05 \cdot 12^2 + 35 \cdot 12 \approx 427.2$ miles.

3.2.65

$$\frac{d}{dx}x^n = \lim_{h \to 0} \frac{(x+h)^n - x^n}{h} = \lim_{h \to 0} \left(\frac{x^n + nx^{n-1}h + \frac{n(n-1)}{2}x^{n-2}h^2 + \cdots + nxh^{n-1} + h^n - x^n}{h} \right)$$

$$= \lim_{h \to 0} \left(nx^{n-1} + \frac{n(n-1)}{2}x^{n-2}h + \cdots + nxh^{n-2} + h^{n-1} \right) = nx^{n-1} + 0 + 0 + \cdots + 0 = nx^{n-1}$$

3.2.67

a. $\frac{d}{dx}\left(\sqrt{x}\right) = \frac{d}{dx}x^{1/2} = \frac{1}{2} \cdot x^{-1/2} = \frac{1}{2\sqrt{x}}$.

b.

$$\frac{d}{dx}x^{3/2} = \lim_{h \to 0} \frac{(x+h)^{3/2} - x^{3/2}}{h} = \lim_{h \to 0} \frac{((x+h)^{3/2} - x^{3/2})((x+h)^{3/2} + x^{3/2})}{(h)((x+h)^{3/2} + x^{3/2})}$$

$$= \lim_{h \to 0} \frac{(x+h)^3 - x^3}{(h)((x+h)^{3/2} + x^{3/2})} = \lim_{h \to 0} \frac{x^3 + 3x^2h + 3xh^2 + h^3 - x^3}{(h)((x+h)^{3/2} + x^{3/2})}$$

$$= \lim_{h \to 0} \frac{3x^2 + 3xh + h^2}{((x+h)^{3/2} + x^{3/2})} = \frac{3x^2 + 0 + 0}{x^{3/2} + x^{3/2}} = \frac{3x^2}{2x^{3/2}}$$

$$= \frac{3}{2}x^{1/2}$$

c.

$$\frac{d}{dx}x^{5/2} = \lim_{h \to 0} \frac{(x+h)^{5/2} - x^{5/2}}{h} = \lim_{h \to 0} \frac{((x+h)^{5/2} - x^{5/2})((x+h)^{5/2} + x^{5/2})}{(h)((x+h)^{5/2} + x^{5/2})}$$

$$= \lim_{h \to 0} \frac{(x+h)^5 - x^5}{(h)((x+h)^{5/2} + x^{5/2})} = \lim_{h \to 0} \frac{x^5 + 5x^4h + 10x^3h^2 + 10x^2h^3 + 5xh^4 + h^5 - x^5}{(h)((x+h)^{5/2} + x^{5/2})}$$

$$= \lim_{h \to 0} \frac{5x^4 + 10x^3h + 10x^2h^2 + 5xh^3 + h^4}{((x+h)^{5/2} + x^{5/2})} = \frac{5x^4 + 0 + 0 + 0 + 0}{x^{5/2} + x^{5/2}} = \frac{5x^4}{2x^{5/2}} = \frac{5}{2}x^{3/2}$$

d. It appears that $\frac{d}{dx}x^{n/2} = \frac{n}{2} \cdot x^{(n/2)-1}$.

3.3 The Product and Quotient Rules

3.3.1 The derivative of the product fg with respect to x is given by $f'(x)g(x) + f(x)g'(x)$.

3.3.3 $\frac{d}{dx}(x^n) = nx^{n-1}$ for all integers n.

3.3.5

$$\frac{d}{dx}(x^n \cdot x^{-n}) = x^{-n}\frac{d}{dx}x^n + x^n\frac{d}{dx}x^{-n} = nx^{-n}x^{n-1} - nx^nx^{-n-1} = nx^{-1} - nx^{-1} = 0$$

3.3.7 $f'(x) = 12x^3(2x^2 - 1) + 3x^4 \cdot 4x = 24x^5 - 12x^3 + 12x^5 = 36x^5 - 12x^3$.

3.3.9 $h'(x) = (35x^6 + 5)(6x^3 + 3x^2 + 3) + (5x^7 + 5x)(18x^2 + 6x) = 300x^9 + 135x^8 + 105x^6 + 120x^3 + 45x^2 + 15$.

3.3.11 $\frac{d}{dw}(w^3 + 4)(w^3 - 1) = 3w^2(w^3 - 1) + 3w^2(w^3 + 4) = 6w^5 + 9w^2 = 3w^2(2w^3 + 3)$

3.3.13

  a. $f'(x) = 1(3x + 4) + (x - 1) \cdot 3 = 6x + 1.$

  b. $f'(x) = \frac{d}{dx}(3x^2 + x - 4) = 6x + 1.$

3.3.15

  a. $g'(y) = (12y^3 - 2y)(y^2 - 4) + (3y^4 - y^2) \cdot 2y = 18y^5 - 52y^3 + 8y.$

  b. $g'(y) = \frac{d}{dy}(3y^6 - 13y^4 + 4y^2) = 18y^5 - 52y^3 + 8y.$

3.3.17 $f'(x) = \dfrac{(x+1) \cdot 1 - x \cdot 1}{(x+1)^2} = \dfrac{1}{(x+1)^2}.$

3.3.19 $y' = \dfrac{d}{dt}\left(\dfrac{3t-1}{2t-2}\right) = \dfrac{(2t-2) \cdot 3 - (3t-1) \cdot 2}{(2t-2)^2} = \dfrac{-4}{(2t-2)^2} = \dfrac{-1}{(t-1)^2}.$

3.3.21

$$\frac{d}{dx}\left(\frac{x^4+1}{x^2-1}\right) = \frac{(x^2-1)4x^3 - (x^4+1)(2x)}{(x^2-1)^2} = \frac{4x^5 - 4x^3 - 2x^5 - 2x}{(x^2-1)^2}$$

$$= \frac{2x^5 - 4x^3 - 2x}{(x^2-1)^2} = \frac{2x(x^4 - 2x^2 - 1)}{(x^2-1)^2}$$

3.3.23

  a. $f'(w) = \dfrac{w(3w^2 - 1) - (w^3 - w) \cdot 1}{w^2} = \dfrac{2w^3}{w^2} = 2w.$

  b. $f'(w) = \dfrac{d}{dw}(w^2 - 1) = 2w.$

3.3.25

  a.

$$y' = \frac{(\sqrt{x} - \sqrt{a}) \cdot 1 - (x - a)\frac{1}{2\sqrt{x}}}{(\sqrt{x} - \sqrt{a})^2} = \left(\frac{(\sqrt{x} - \sqrt{a}) \cdot 1 - (x - a)\frac{1}{2\sqrt{x}}}{(\sqrt{x} - \sqrt{a})^2}\right) \cdot \frac{2\sqrt{x}}{2\sqrt{x}}$$

$$= \frac{2x - 2\sqrt{ax} - x + a}{2\sqrt{x}(\sqrt{x} - \sqrt{a})^2} = \frac{x - 2\sqrt{ax} + a}{2\sqrt{x}(\sqrt{x} - \sqrt{a})^2} = \frac{(\sqrt{x} - \sqrt{a})^2}{2\sqrt{x}(\sqrt{x} - \sqrt{a})^2} = \frac{1}{2\sqrt{x}}.$$

  b. $y' = \dfrac{d}{dx}\left(\dfrac{(\sqrt{x} - \sqrt{a})(\sqrt{x} + \sqrt{a})}{\sqrt{x} - \sqrt{a}}\right) = \dfrac{d}{dx}(\sqrt{x} - \sqrt{a}) = \dfrac{1}{2\sqrt{x}}.$

3.3.27

  b.

  a. $y' = \frac{(x-1)-(x+5)}{(x-1)^2} = \frac{-6}{(x-1)^2}.$ At $a = 3$ we have $y' = \frac{-6}{4} = \frac{-3}{2}$ and $y = 4$, so the equation of the tangent line is $y - 4 = \frac{-3}{2} \cdot (x - 3).$

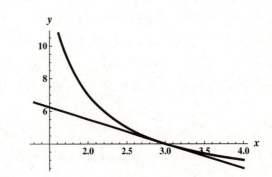

3.3.29

b.

a.

$$y' = (2x^{-2}+1)+x(-4x^{-3}) = -2x^{-2}+1 = 1-\frac{2}{x^2}$$

so that for $x = -1$, $y'(-1) = 1 - 2 = -1$. At $x = -1$, $y(-1) = (-1)(2+1) = -3$, so the tangent line is $y + 3 = -(x + 1)$, or $y = -x - 4$.

3.3.31 $f'(x) = (-9) \cdot 3 \cdot x^{-9-1} = -27x^{-10}$.

3.3.33 $g'(t) = \frac{d}{dt}(3t^2 + 6t^{-7}) = 6t - 42t^{-8}$.

3.3.35 $g'(t) = \frac{d}{dt}(1 + 3t^{-1} + t^{-2}) = -3t^{-2} - 2t^{-3}$.

3.3.37

e.

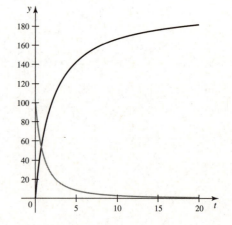

a. $p'(t) = \dfrac{(t + 2)200 - 200t}{(t + 2)^2} = \dfrac{400}{(t + 2)^2}$.

b. $p'(5) = \dfrac{400}{49} \approx 8.16$.

c. The value of p' is as large as possible when its denominator is as small as possible, which is when $t = 0$. The value of $p'(0)$ is 100.

d. $\displaystyle\lim_{t \to \infty} p'(t) = \lim_{t \to \infty} \frac{400}{(t + 2)^2} = 0$. This means that the population growth rate approaches 0 over time, which means that the population tends to stabilize at a long-term value.

3.3.39

a.

$$f'(x) = \frac{(2x^2 + 1)(1 - 2x) - (x - x^2)(4x)}{(2x^2 + 1)^2} = \frac{-2x^2 - 2x + 1}{(2x^2 + 1)^2}$$

$f'(x)$ is zero when the numerator is zero; the roots of $2x^2 + 2x - 1$ are $(-1 \pm \sqrt{3})/2$.

b. This means that the slope of the tangent to $f(x)$ at $x = (-1 \pm \sqrt{3})/2$ is zero.

3.3.41 $g'(x) = \dfrac{[(3-x)+x(-1)]2x^2 - 4x\,[x(3-x)]}{(2x^2)^2} = \dfrac{6x^2 - 4x^3 - 12x^2 + 4x^3}{4x^4} = \dfrac{-6x^2}{4x^4} = \dfrac{-3}{2x^2}.$

3.3.43

$$g'(x) = \frac{(x^2+x)(1-x)(4) - 4x((x^2+x)(-1)+(2x+1)(1-x))}{(x^2+x)^2(1-x)^2}$$

$$= \frac{8x^3}{(x^2+x)^2(1-x)^2} = \frac{8x}{(1-x^2)^2}$$

3.3.45

a. True. Note that $f'(x) = -nx^{-(n+1)}$, $f''(x) = (-n)(-(n+1))x^{-(n+2)} = n(n+1)x^{-(n+2)}$, and that in general, after taking an even number of derivatives, the sign of the coefficient of x will be positive, so that evaluating $f^{(8)}(1)$ gives a positive answer.

b. False. It is certainly a reasonable way to proceed, but one could also write the given quantity as $x + 3 + 2x^{-1}$, and then proceed using the sum rule and the power rule and the extended power rule.

c. False. $\frac{d}{dx}\frac{1}{5}\cdot x^{-4} = \frac{-4}{5}\cdot x^{-5} = \frac{-4}{5x^5}.$

3.3.47

$$f'(x) = 2x(2+x^{-3}) + x^2(-3x^{-4}) = 4x - x^{-2}$$
$$f''(x) = 4 + 2x^{-3}$$
$$f'''(x) = -6x^{-4}$$

3.3.49

$$f'(x) = \frac{d}{dx}\left(\frac{x^2-7x}{x+1}\right) = \frac{(x+1)(2x-7)-(x^2-7x)\cdot 1}{(x+1)^2} = \frac{x^2+2x-7}{x^2+2x+1}$$

$$f''(x) = \frac{d}{dx}\left(\frac{x^2+2x-7}{x^2+2x+1}\right) = \frac{(x^2+2x+1)(2x+2)-(x^2+2x-7)(2x+2)}{(x+1)^4}$$
$$= \frac{(2x^2+4x+2)-(2x^2+4x-14)}{(x+1)^3} = \frac{16}{(x+1)^3}$$

$$f'''(x) = \frac{d}{dx}\left(\frac{16}{x^3+3x^2+3x+1}\right) = \frac{(x^3+3x^2+3x+1)\cdot 0 - 16(3x^2+6x+3)}{(x+1)^6}$$
$$= \frac{(-48)(x^2+2x+1)}{(x+1)^6} = \frac{-48}{(x+1)^4}$$

3.3.51 $f'(x) = \dfrac{d}{dx}\left(4x^2 - \dfrac{2x}{5x+1}\right) = 8x - \dfrac{(5x+1)2 - (2x)(5)}{(5x+1)^2} = 8x - \dfrac{2}{(5x+1)^2}.$

3.3.53

$$h'(r) = \frac{(r+1)(-1-\frac{1}{2\sqrt{r}}) - (2-r-\sqrt{r})\cdot 1}{(r+1)^2} = \frac{-r-\frac{\sqrt{r}}{2}-1-\frac{1}{2\sqrt{r}}-2+r+\sqrt{r}}{(r+1)^2}$$
$$= \frac{\frac{\sqrt{r}}{2}-\frac{1}{2\sqrt{r}}-3}{(r+1)^2}\cdot\frac{2\sqrt{r}}{2\sqrt{r}} = \frac{r-1-6\sqrt{r}}{2\sqrt{r}(r+1)^2}.$$

3.3.55

a.

b.

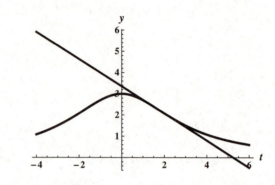

$y' = \dfrac{-54x}{(x^2+9)^2}$. At $x=2$, $y' = \dfrac{-108}{169}$ and $y = \dfrac{27}{13}$.

Thus the tangent line is given by

$$y - \frac{27}{13} = \frac{-108}{169}(x-2).$$

3.3.57 $\dfrac{d}{dx}\left[\dfrac{f(x)}{g(x)}\right]\Bigg|_{x=2} = \dfrac{g(2)f'(2) - f(2)g'(2)}{(g(2))^2} = \dfrac{2\cdot 5 - 4\cdot 4}{4} = \dfrac{-3}{2}.$

3.3.59 $\dfrac{d}{dx}\left[\dfrac{f(x)}{x+2}\right]\Bigg|_{x=4} = \dfrac{(4+2)f'(4) - f(4)}{36} = \dfrac{6-2}{36} = \dfrac{1}{9}.$

3.3.61 $\dfrac{d}{dx}\left[\dfrac{f(x)g(x)}{x}\right]\Bigg|_{x=4} = \dfrac{4(f'(4)g(4) + f(4)g'(4)) - f(4)g(4)}{16} = \dfrac{4(1\cdot 3 + 2\cdot 1) - (2\cdot 3)}{16} = \dfrac{14}{16} = \dfrac{7}{8}.$

3.3.63

a. The instantaneous rate of change is $\frac{d}{dx}F(x) = \frac{-2kQq}{x^3}$ Newtons per meter.

b. $\left[\dfrac{d}{dx}F(x)\right]\Bigg|_{x=0.001} = \dfrac{-2(9\times 10^9)}{(0.001)^3} = \dfrac{-18\times 10^9}{10^{-9}} = -18\times 10^{18} = -1.8\times 10^{19}$ Newtons per meter.

c. Since the distance x appears in the denominator of $F'(x)$, the absolute value of the instantaneous rate of change decreases with the separation.

3.3.65

a. The tangent line at $x=a$ is $y - a^2 = 2a(x-a)$ and at $x=b$ is $y - b^2 = 2b(x-b)$. These intersect when $a^2 + 2ax - 2a^2 = b^2 + 2bx - 2b^2$, or $(2a-2b)x = a^2 - b^2$, which is met when $x = \frac{a+b}{2}$. So $c = \frac{a+b}{2}$.

b. The tangent line at $x=a$ is $y - \sqrt{a} = \frac{1}{2\sqrt{a}}(x-a)$ and at $x=b$ is $y - \sqrt{b} = \frac{1}{2\sqrt{b}}(x-b)$. These intersect when $\sqrt{a} + \frac{1}{2\sqrt{a}}(x-a) = \sqrt{b} + \frac{1}{2\sqrt{b}}(x-b)$, or $\left(\frac{1}{2\sqrt{a}} - \frac{1}{2\sqrt{b}}\right)x = \frac{\sqrt{b}-\sqrt{a}}{2}$, which is met when $x = \sqrt{ab}$. So $c = \sqrt{ab}$.

c. The tangent line at $x=a$ is $y - \frac{1}{a} = \frac{-1}{a^2}(x-a)$ and at $x=b$ is $y - \frac{1}{b} = \frac{-1}{b^2}(x-b)$. These intersect when $\frac{1}{a} + \frac{-1}{a^2}(x-a) = \frac{1}{b} + \frac{-1}{b^2}(x-b)$, or $\left(\frac{2}{a} - \frac{x}{a^2}\right) = \left(\frac{2}{b} - \frac{x}{b^2}\right)$, which is met when $x\left(\frac{1}{b^2} - \frac{1}{a^2}\right) = \frac{2}{b} - \frac{2}{a}$, or $x\cdot\left(\frac{a^2-b^2}{a^2b^2}\right) = \frac{2(a-b)}{ab}$. Thus we arrive at $x = \frac{2ab}{a+b}$. So $c = \frac{2ab}{a+b}$.

d. The tangent line at $x=a$ is $y - f(a) = f'(a)(x-a)$ and at $x=b$ is $y - f(b) = f'(b)(x-b)$. These intersect when $f(a) + f'(a)(x-a) = f(b) + f'(b)(x-b)$, or $(f'(a) - f'(b))x = f(b) - f(a) - f'(b)b + f'(a)a$. Solving for x yields $x = \dfrac{f(b) - f(a) - f'(b)b + f'(a)a}{f'(a) - f'(b)}$ provided $f'(a) \neq f'(b)$ (which occurs when the tangent lines are parallel and don't intersect.)

3.3.67

$$\frac{d^2}{dx^2}(f(x)g(x)) = \frac{d}{dx}(f'(x)g(x) + f(x)g'(x)) = f''(x)g(x) + f'(x)g'(x) + f'(x)g'(x) + f(x)g''(x)$$
$$= f''(x)g(x) + 2f'(x)g'(x) + f(x)g''(x).$$

3.3.69

a.

$$\frac{d}{dx}[(f(x)g(x))h(x)] = \frac{d}{dx}[f(x)g(x)] \cdot h(x) + f(x)g(x) \cdot \frac{d}{dx}h(x)$$
$$= [f'(x)g(x) + f(x)g'(x)]h(x) + f(x)g(x)h'(x)$$
$$= f'(x)g(x)h(x) + f(x)g'(x)h(x) + f(x)g(x)h'(x).$$

b. $\frac{d}{dx}[x(x-1)(x+3)] = (x-1)(x+3) + x(x+3) + x(x-1).$

3.4 Derivatives of Trigonometric Functions

3.4.1 A direct substitution would yield the quotient of zero with itself, which isn't defined

3.4.3 Since $\tan x = \frac{\sin x}{\cos x}$, and $\cot x = \frac{\cos x}{\sin x}$, we can use the quotient rule to compute these derivatives, since we know the derivatives of $\sin x$ and of $\cos x$.

3.4.5 $f'(x) = \cos x$ and $f'(\pi) = \cos \pi = -1.$

3.4.7 $\lim\limits_{x \to 0} \frac{\sin 3x}{x} = \lim\limits_{x \to 0} \frac{3 \sin 3x}{3x} = 3 \lim\limits_{x \to 0} \frac{\sin 3x}{x} = 3 \cdot 1 = 3.$

3.4.9 $\lim\limits_{x \to 0} \frac{\tan 5x}{x} = \lim\limits_{x \to 0} \frac{5 \sin 5x}{5x \cos 5x} = 5 \lim\limits_{x \to 0} \frac{\sin 5x}{5x} \cdot \lim\limits_{x \to 0} \frac{1}{\cos 5x} = 5 \cdot 1 \cdot 1 = 5.$

3.4.11 $\lim\limits_{x \to 0} \frac{\tan 7x}{\sin x} = \lim\limits_{x \to 0} \frac{\sin 7x}{\cos 7x \cdot \sin x} = \lim\limits_{x \to 0} \left(\frac{1}{\cos 7x} \cdot \frac{x}{\sin x} \cdot \frac{7 \sin 7x}{7x} \right) =$
$7 \cdot \lim\limits_{x \to 0} \frac{1}{\cos 7x} \cdot \lim\limits_{x \to 0} \frac{x}{\sin x} \cdot \lim\limits_{x \to 0} \frac{\sin 7x}{7x} = 7 \cdot 1 \cdot 1 \cdot 1 = 7.$

3.4.13 $\lim\limits_{x \to 2} \frac{\sin(x-2)}{x^2 - 4} = \lim\limits_{x \to 2} \left(\frac{1}{x+2} \right) \cdot \left(\frac{\sin(x-2)}{x-2} \right) = \lim\limits_{x \to 2} \frac{1}{x+2} \cdot \lim\limits_{x \to 0} \frac{\sin(x-2)}{x-2} = \frac{1}{4} \cdot 1 = \frac{1}{4}.$

3.4.15 $y' = \cos x - \sin x.$

3.4.17 $y' = 12x^3 \sin x + 3x^4 \cos x.$

3.4.19 $y' = \cos x \cos x + \sin x \cdot (-\sin x) = \cos^2 x - \sin^2 x = \cos(2x).$

3.4.21 $y' = -\sin x \cos x + \cos x(-\sin x) = -2 \sin x \cos x = -\sin(2x).$

3.4.23 $\frac{d}{dx} \cot x = \frac{d}{dx} \frac{\cos x}{\sin x} = \frac{\sin x(-\sin x) - \cos x(\cos x)}{\sin^2 x} = \frac{-(\sin^2 x + \cos^2 x)}{\sin^2 x} = \frac{-1}{\sin^2 x} = -\csc^2 x.$

3.4.25 $\frac{d}{dx} \csc x = \frac{d}{dx} \frac{1}{\sin x} = \frac{0 - \cos x}{\sin^2 x} = \frac{-1}{\sin x} \cdot \frac{\cos x}{\sin x} = -\csc x \cot x.$

3.4.27 $y' = \sec x \tan x - \csc x \cot x.$

3.4.29

$$y' = \frac{(1 + \csc x)(-\csc^2 x) - \cot x(-\csc x \cot x)}{(1 + \csc x)^2} = \frac{-\csc^2 x - \csc^3 x + \csc x(\csc^2 x - 1)}{(1 + \csc x)^2}$$

$$= \frac{-\csc x(1 + \csc x)}{(1 + \csc x)^2} = \frac{-\csc x}{1 + \csc x}$$

3.4.31

$$y' = \frac{0 - (\sec z \tan z \csc z - \sec z \csc z \cot z)}{\sec^2 z \csc^2 z} = \frac{\sec z \csc z(\cot z - \tan z)}{\sec^2 z \csc^2 z}$$

$$= \frac{\cot z - \tan z}{\sec z \csc z} = \cos^2 z - \sin^2 z = \cos(2z).$$

3.4.33 $y' = -\csc^2 x$ and $y'' = -((-\csc x \cot x)\csc x + \csc x(-\csc x \cot x)) = 2 \cot x \csc^2 x.$

3.4.35

$$y' = \sec x \tan x \csc x - \sec x \csc x \cot x = \sec x \csc x(\tan x - \cot x) = \sec^2 x - \csc^2 x$$

$$y'' = \sec x(\sec x \tan x) + (\sec x \tan x)\sec x - ((-\csc x \cot x)\csc x + \csc x(-\csc x \cot x))$$

$$= 2 \sec^2 x \tan x + 2 \csc^2 x \cot x$$

3.4.37

a. False. $\frac{d}{dx} \sin^2 x = \sin x \cos x + \cos x \sin x = 2 \sin x \cos x \neq \cos^2 x.$

b. False. $\frac{d^2}{dx^2} \sin x = \frac{d}{dx} \cos x = -\sin x \neq \sin x.$

c. True. $\frac{d^4}{dx^4} \cos x = \frac{d^3}{dx^2}(-\sin x) = \frac{d^2}{dx^2}(-\cos x) = \frac{d}{dx} \sin x = \cos x.$

d. True. In fact, $\pi/2$ isn't even in the domain of $\sec x$.

3.4.39 $\lim\limits_{x \to 0} \dfrac{\sin ax}{\sin bx} = \lim\limits_{x \to 0} \dfrac{a \sin ax}{ax} \cdot \dfrac{bx}{b \sin bx} = \dfrac{a}{b} \lim\limits_{x \to 0} \dfrac{\sin ax}{ax} \cdot \lim\limits_{x \to 0} \dfrac{bx}{\sin bx} = \dfrac{a}{b} \cdot 1 \cdot 1 = \dfrac{a}{b}.$

3.4.41 $\lim\limits_{x \to 0} \dfrac{3 \sec^5 x}{x^2 + 4} = \dfrac{3 \sec^5(0)}{4} = \dfrac{3}{4}.$

3.4.43 $\lim\limits_{x \to \pi/4} 3 \csc(2x)\cot(2x) = \lim\limits_{x \to \pi/4} 3\dfrac{1}{\sin 2x}\dfrac{\cos 2x}{\sin 2x} = 3\dfrac{\cos \pi/2}{(\sin \pi/2)^2} = 3 \cdot \dfrac{0}{1} = 0.$

3.4.45 $\frac{dy}{dx} = \cos x \sin x + x(-\sin x)\sin x + x \cos x \cos x = \sin x \cos x - x \sin^2 x + x \cos^2 x.$

3.4.47 $\dfrac{dy}{dx} = \dfrac{(1 + \sin x)(-2 \sin x) - 2 \cos x \cdot \cos x}{(1 + \sin x)^2} = -2 \cdot \dfrac{\sin x + 1}{(1 + \sin x)^2} = \dfrac{-2}{1 + \sin x}.$

3.4.49 $\dfrac{dy}{dx} = \dfrac{(1 + \cos x)\sin x - (1 - \cos x)(-\sin x)}{(1 + \cos x)^2} = \dfrac{2 \sin x}{(1 + \cos x)^2}.$

3.4.51

b.

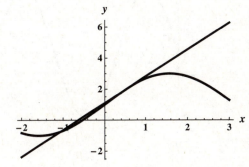

a. $y' = 2\cos x$, so $y'(\pi/6) = \sqrt{3}$. $y(\pi/6) = 2$. The tangent line is thus given by $y - 2 = \sqrt{3}(x - \pi/6)$.

3.4.53

b.

a.

$$y' = \frac{(1 - \cos x)(-\sin x) - \cos x \sin x}{(1 - \cos x)^2}$$

$$= -\frac{\sin x}{(1 - \cos x)^2},$$

so $y'(\pi/3) = -2\sqrt{3}$. $y(\pi/3) = 1$. The tangent line is thus given by $y - 1 = -2\sqrt{3}(x - \pi/3)$.

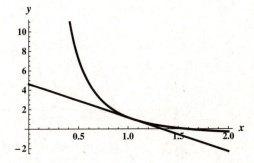

3.4.55 For a horizontal tangent line we need $f'(x) = 1 + 2\sin x = 0$, or $\sin x = \frac{-1}{2}$. This occurs for $x = \frac{12n+7}{6} \cdot \pi$ where n is any integer, or for $x = \frac{12n+11}{6} \cdot \pi$ where n is any integer.

3.4.57

a.

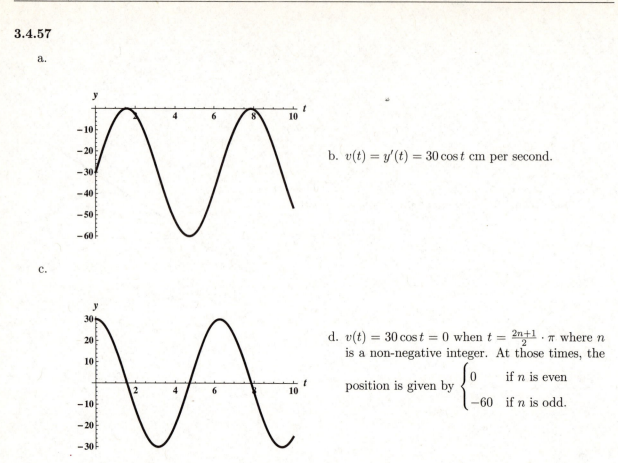

b. $v(t) = y'(t) = 30\cos t$ cm per second.

c.

d. $v(t) = 30\cos t = 0$ when $t = \frac{2n+1}{2} \cdot \pi$ where n is a non-negative integer. At those times, the position is given by $\begin{cases} 0 & \text{if } n \text{ is even} \\ -60 & \text{if } n \text{ is odd.} \end{cases}$

e. The maximum velocity is 30 cm per second since $|\cos t| \le 1$ for all t. We have $\cos t = 1$ for $t = 2n\pi$ for a positive integer n. At those times, $y(2n\pi) = -30$.

f.

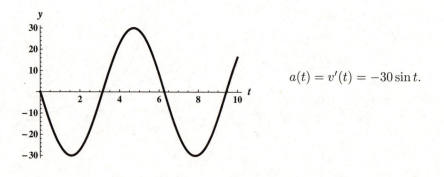

$a(t) = v'(t) = -30\sin t.$

3.4.59

a. $y'(t) = A\cos t$, $y''(t) = -A\sin t$, so $y''(t) + y(t) = -A\sin t + A\sin t = 0$ for all A and all t.

b. $y'(t) = -B\sin t$, $y''(t) = -B\cos t$, so $y''(t) + y(t) = -B\cos t + B\cos t = 0$ for all B and all t.

c. $y' = A\cos t - B\sin t$, $y'' = -A\sin t - B\cos t$, so $y''(t) + y(t) = -A\sin t - B\cos t + A\sin t + B\cos t = 0$ for all A, B, t.

3.4.61

$$\lim_{x \to 0} \frac{\cos x - 1}{x} = \lim_{x \to 0} \frac{(\cos x - 1)(\cos x + 1)}{x(\cos x + 1)} = \lim_{x \to 0} \frac{\cos^2 x - 1}{x(\cos x + 1)} = \lim_{x \to 0} \frac{-\sin^2 x}{x(\cos x + 1)}$$

$$= \lim_{x \to 0} \frac{\sin x}{x} \cdot \lim_{x \to 0} \frac{-\sin x}{(\cos x + 1)} = 1 \cdot \frac{0}{2} = 0.$$

3.4.63

$$\frac{d}{dx} \cos x = \lim_{h \to 0} \frac{\cos(x + h) - \cos x}{h} = \lim_{h \to 0} \frac{\cos x \cos h - \sin x \sin h - \cos x}{h}$$

$$= \cos x \left(\lim_{h \to 0} \frac{\cos h - 1}{h} \right) - \sin x \left(\lim_{h \to 0} \frac{\sin h}{h} \right) = \cos x \cdot 0 - \sin x \cdot 1 = -\sin x.$$

3.4.65 g is continuous at 0 if and only if $\lim_{x \to 0} g(x) = g(0)$. Since $\lim_{x \to 0} g(x) = \lim_{x \to 0} \frac{1 - \cos x}{2x} = \frac{1}{2} \cdot 0 = 0$, we require $a = 0$ in order for g to be continuous.

3.4.67

a. $\dfrac{d}{dx} \sin^2 x = \sin x \cos x + \cos x \sin x = 2 \sin x \cos x.$

b. $\dfrac{d}{dx} \sin^3 x = \dfrac{d}{dx} (\sin^2 x)(\sin x) = (2 \sin x \cos x) \sin x + \sin^2 x \cdot \cos x = 3 \sin^2 x \cos x.$

c. $\dfrac{d}{dx} \sin^4 x = \dfrac{d}{dx} (\sin^3 x)(\sin x) = (3 \sin^2 x \cos x)(\sin x) + (\sin^3 x)(\cos x) = 4 \sin^3 x \cos x.$

d. We guess that $\dfrac{d}{dx} \sin^n x = n \sin^{n-1} x \cos x$. We have already seen that the claim is valid for $n = 2$. Suppose our guess is valid for a given positive integer n. Then

$$\frac{d}{dx} \sin^{n+1}(x) = \frac{d}{dx} (\sin^n(x))(\sin x) = (n \sin^{n-1}(x) \cos(x))(\sin x) + \sin^n(x) \cos x = (n+1) \sin^n(x) \cos(x).$$

Thus by induction, the result holds for all n.

3.4.69

a. $f(x) = \sin x$, $a = \pi/6$.

b. $\lim_{h \to 0} \dfrac{\sin(\pi/6 + h) - (1/2)}{h} = f'(\pi/6) = \cos(\pi/6) = \sqrt{3}/2.$

3.4.71

a. $f(x) = \cot x$, $a = \pi/4$.

b. $\lim_{x \to \pi/4} \dfrac{\cot(x) - 1}{x - \pi/4} = f'(\pi/4) = -\csc^2(\pi/4) = -2.$

3.5   Derivatives as Rates of Change

3.5.1

The average rate of change over the interval $[a, a + \Delta x]$ is the slope of the line through $(a, f(a))$ and $(a + \Delta x, f(a + \Delta x))$, given by

$$m_{\text{avg}} = \frac{f(a + \Delta x) - f(a)}{\Delta x}.$$

The instantaneous rate of change is the limit of this expression as $\Delta x \to 0$, which is the slope of the tangent line at $(a, f(a))$.

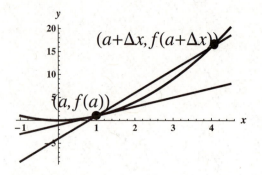

3.5.3 If $\frac{dy}{dx}$ is small, then small changes in x will result in relatively small changes in the value of y.

3.5.5 Acceleration is the instantaneous rate of change of the velocity, that is, if $s(t)$ is the position of an object at time t, then $s''(t) = \frac{d}{dt}v(t) = a(t)$ is the acceleration of the object at time t.

3.5.7 Each of the first 200 stoves cost on average \$70 to produce, while the 201st stove costs \$65 to produce.

3.5.9

a. $v_{\text{avg}} = \dfrac{f(0.75) - f(0)}{0.75} = \dfrac{30 - 0}{0.75} = 40$ mph.

b. $v_{\text{avg}} = \dfrac{f(0.75) - f(0.25)}{0.75 - 0.25} = \dfrac{30 - 10}{0.5} = 40$ mph. This is a pretty good estimate, since the graph is nearly linear over that time interval.

c. $v_{\text{avg}} = \dfrac{f(2.25) - f(1.75)}{2.25 - 1.75} = \dfrac{-14 - 16}{0.5} = -60$ mph. At 11 a.m. the velocity is $v(2) \approx -60$ mph. The car is moving south with a speed of approximately 60 mph.

d. From 9 a.m. until about 10:08 a.m., the car moves north, away from the station. then it moves south, passing the station at approximately 11:02 a.m., and continues south until about 11:40 a.m. Then the car drives north until 12:00 noon stopping south of the station.

3.5.11

a.

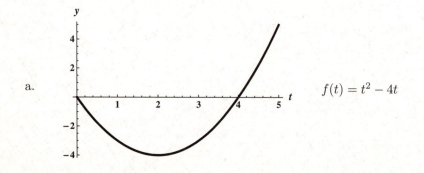

$f(t) = t^2 - 4t$

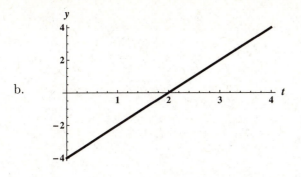

b. $f'(t) = 2t - 4$. $f'(t) = 0$ when $t = 2$ – that is when the object is stationary. For $0 \le t < 2$ we have $f'(t) < 0$ so the object is moving to the left. For $2 < t \le 5$ we have $f'(t) > 0$ so the object is moving to the right.

c. $f'(1) = -2$ ft/sec and $f''(t) = 2$ ft/sec^2, so in particular, $f''(1) = 2$ ft/sec^2.

d. $f'(t) = 0$ when $t = 2$ and $f''(2) = 2$ ft/sec^2.

3.5.13

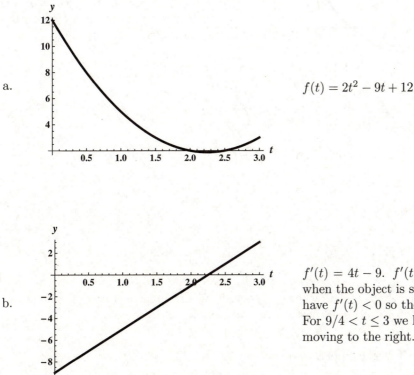

a.

$f(t) = 2t^2 - 9t + 12$

b. $f'(t) = 4t - 9$. $f'(t) = 0$ when $t = 9/4$ – that is when the object is stationary. For $0 \le t < 9/4$ we have $f'(t) < 0$ so the object is moving to the left. For $9/4 < t \le 3$ we have $f'(t) > 0$ so the object is moving to the right.

c. $f'(1) = -5$ ft/sec and $f''(t) = 4$ ft/sec^2, so in particular, $f''(1) = 4$ ft/sec^2.

d. $f'(t) = 0$ when $t = 9/4$ and $f''(9/4) = 4$ ft/sec^2.

3.5.15

a.

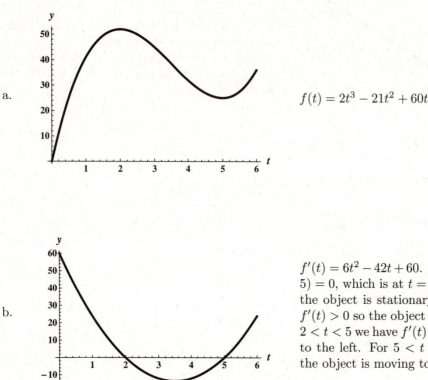

$f(t) = 2t^3 - 21t^2 + 60t$

b.

$f'(t) = 6t^2 - 42t + 60$. $f'(t) = 0$ when $6(t - 2)(t - 5) = 0$, which is at $t = 2$ and $t = 5$ – that is when the object is stationary. For $0 \leq t < 2$ we have $f'(t) > 0$ so the object is moving to the right. For $2 < t < 5$ we have $f'(t) < 0$ so the object is moving to the left. For $5 < t \leq 8$ we have $f'(t) > 0$, so the object is moving to the right again.

c. $f'(1) = 24$ ft/sec and $f''(t) = 12t - 42$, so $f''(1) = -30$ ft/sec^2.

d. $f'(t) = 0$ when $t = 2$ and $t = 5$. We have $f''(2) = -18$ ft/sec^2 and $f''(5) = 18$ ft/sec^2.

3.5.17

a. $v(t) = s'(t) = -32t + 64$ ft/sec.

b. The highest point is reached at the instant when the stone changes from moving upward (where $v > 0$) to moving downward (where $v < 0$), so it must occur when $v = 0$, which is at $t = 2$.

c. The height of the stone at its highest point is $s(2) = -16 \cdot 4 + 64 \cdot 2 + 32 = 96$ feet.

d. The stone strikes the ground when $s(t) = 0$ for $t > 0$. Using the quadratic formula we see that this occurs when $t = 2 + \sqrt{6} \approx 4.45$ seconds.

e. The velocity when the stone hits the ground is $v(2 + \sqrt{6}) = -32(2 + \sqrt{6}) + 64 = -32\sqrt{6} \approx -78.38$ feet per second.

3.5.19

c.

a. The average growth rate from 1995 to 2005 is
$\frac{p(10) - p(0)}{10 - 0} = \frac{8038 - 7055}{10} = 98.3$ thousand
people/year.

b. The instantaneous growth rate is $p'(t) = -0.54t + 101$. In 1997 we have $p'(2) = 99.92$ thousand people per year and in 2005 we have $p'(10) = 95.6$ thousand people per year.

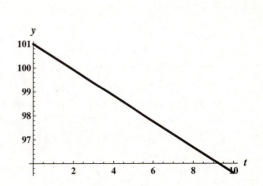

The population was growing but the rate was slowing over this time interval.

3.5.21

a. The average cost function is given by $\overline{C}(x) = \frac{C(x)}{x} = \frac{1000}{x} + .1$. The marginal cost function is given by $M(x) = C'(x) = .1$.

b. At $a = 2000$ we have $\overline{C}(2000) = \frac{1000}{2000} + .1 = .6$, and $M(2000) = .1$.

c. The average cost per item when producing 2000 items is \$0.60. The cost of producing the next item is \$0.10.

3.5.23

a. The average cost function is given by $\overline{C}(x) = \frac{C(x)}{x} = \frac{100}{x} + 40 - 0.01x$. The marginal cost function is given by $M(x) = C'(x) = 40 - 0.02x$.

b. At $a = 1000$ we have $\overline{C}(1000) = \frac{100}{1000} + 40 - (.01)(1000) = 30.1$, and $M(1000) = 20$.

c. The average cost per item when producing 1000 items is \$30.10. The cost of producing the next item is \$20.00.

3.5.25

a. False. For example, when a ball is thrown up in the air near the surface of the earth, its acceleration is constant (due to gravity) but its velocity changes during its trip.

b. True. If the rate of change of velocity is zero, then velocity must be constant.

c. False. If the velocity is constant over an interval, then the average velocity is equal to the instantaneous velocity over the interval.

d. True. For example, a ball dropped from a tower has negative acceleration and increasing speed as it falls toward the earth.

3.5.27 In each case, the bullet reaches its maximum height when its velocity is zero. On Mars, this occurs when $v(t) = s'(t) = 1200 - 12t = 0$, or when $t = 100$ seconds. So the maximum height on Mars is $s(100) = 60,000$ feet. On Earth, this occurs when $v(t) = s'(t) = 1200 - 32t = 0$, or when $t = 37.5$ seconds. So the maximum height on Earth is $s(37.5) = 22,500$ feet. The bullet will travel $60,000 - 22,500 = 37,500$ feet higher on Mars.

3.5.29

c.

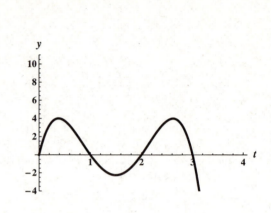

a. The velocity is zero at $t = 1, 2,$ and 3.

b. The object is moving in the positive direction when the slope of s is positive, so from $t = 0$ to $t = 1$, and from $t = 2$ to $t = 3$. It is moving in the negative direction from $t = 1$ to $t = 2$.

3.5.31

a. $P(x) = xp(x) - C(x) = 100x + 0.02x^2 - 50x - 100 = 0.02x^2 + 50x - 100$.

b. The average profit is $\overline{P}(x) = \frac{P(x)}{x} = 0.02x + 50 - \frac{100}{x}$. The marginal profit is $P'(x) = .04x + 50$.

c. $\overline{P}(500) = 59.8$. $P'(500) = 70$.

d. The average profit for the first 500 items sold is \$59.80, while the profit on the 501st item is \$70.00.

3.5.33

a. $P(x) = xp(x) - C(x) = 100x + 0.04x^2 - 800$.

b. The average profit is $\overline{P}(x) = \frac{P(x)}{x} = .04x + 100 - \frac{800}{x}$. The marginal profit is $P'(x) = .08x + 100$.

c. $\overline{P}(1000) = 139.2$. $P'(1000) = 180$.

d. The average profit for the first 1000 items sold is \$139.20, while the profit on the 1001st item is \$180.00.

3.5.35

a. Since the graph represents the growth rate, the slowest rate corresponds to the minimum, which is at about $t = 30$, which corresponds to the year 1930.

b. The largest growth rate corresponds to the maximum at $t = 60$, so the year 1960 at the largest growth rate of about 2.8 million per year.

c. Since $p'(t) > 0$ for all t shown on the graph, $p(t)$ is never decreasing.

d. The population growth rate $p'(t)$ is increasing from about 1905 to 1915, from 1930 to 1960, and from 1980 to 1990.

3.5.37

a.

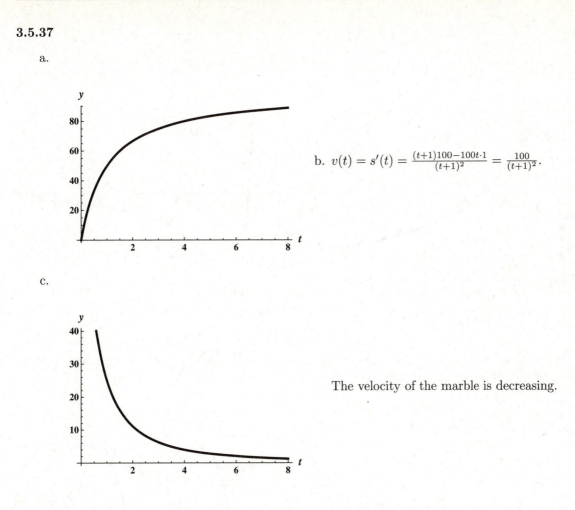

b. $v(t) = s'(t) = \frac{(t+1)100 - 100t \cdot 1}{(t+1)^2} = \frac{100}{(t+1)^2}$.

c.

The velocity of the marble is decreasing.

d. $s(t) = 80$ when $\frac{100t}{t+1} = 80$, or $100t = 80t + 80$, which occurs when $t = 4$ seconds.

e. $v(t) = 50$ when $\frac{100}{(t+1)^2} = 50$, or $(t+1)^2 = 2$. This occurs for $t = \sqrt{2} - 1$ seconds.

3.5.39

a. The average cost function is
$\overline{C}(x) = \frac{C(x)}{25000} = 50 + \frac{5000}{x} + 0.00006x$.
The marginal cost function is
$C'(x) = \frac{-125000000}{x^2} + 1.5$.
The average cost decreases to about 50 per unit as
the batch size increases, while the marginal cost is
negative but increases.

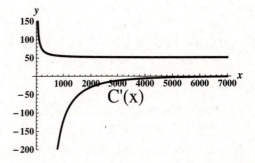

b. $\overline{C}(5000) = 51.3$, $C'(5000) = -3.5$.

c. If the batch size is 5000, then the average cost of producing 25000 items is \$51.30. If the batch size is
increased from 5000 to 5001, then the cost of producing 25000 items would decrease by about \$3.50.

3.5.41

a.

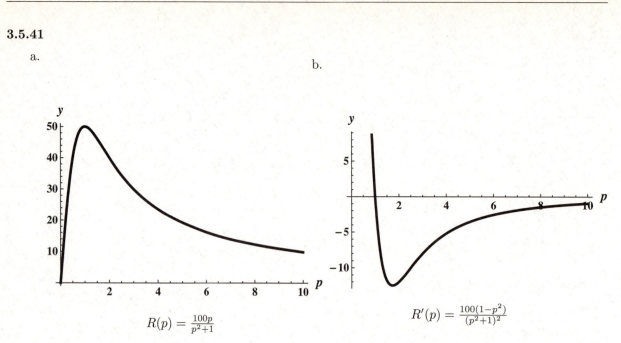

$$R(p) = \frac{100p}{p^2+1}$$

b.

$$R'(p) = \frac{100(1-p^2)}{(p^2+1)^2}$$

c. $R'(p)$ is zero at $p = 1$, and the maximum of $R(p)$ occurs at this same value of p, so that is the price to charge in order to maximize revenue. The revenue at this price is $50.00.

3.5.43

a.   The mass oscillates about the equilibrium point.

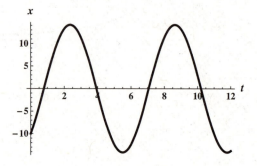

b. $\frac{dx}{dt} = 10\cos t + 10\sin t$ is the velocity of the mass at time t.

c. $\frac{dx}{dt} = 0$ when $\sin t = -\cos t$, which occurs when $t = \frac{4n+3}{4}\cdot\pi$ where n is any positive integer.

d. The model is unrealistic as it ignores the effects of friction and gravity. In reality, the amplitude would decrease as the mass oscillates.

3.5.45

a. Juan starts out faster, but slows toward the end, while Jean starts slower but increases her speed toward the end.

b. Since both start and finish at the same time, they finish with the same average angular velocity.

c. It is a tie.

d. Jean's angular velocity is given by $\theta'(t) = \frac{\pi t}{4}$. At $t = 2$, $\theta'(2) = \frac{\pi}{2}$ radians per minute. Her angular velocity is greatest at $t = 4$.

e. Juan's angular velocity is given by $\phi'(t) = \pi - \frac{\pi t}{4}$. At $t = 2$, $\phi'(2) = \frac{\pi}{2}$ radians per minute as well. His angular velocity is greatest at $t = 0$.

3.5.47

a.

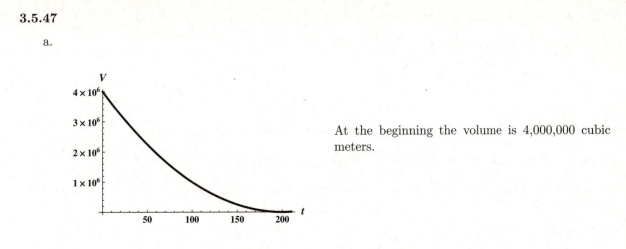

At the beginning the volume is 4,000,000 cubic meters.

b. The tank is empty when $V(t) = 100(200 - t)^2 = 0$, which occurs when $t = 200$.

c.

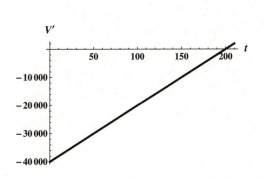

Since $V(t)$ can be written as $V(t) = 4,000,000 - 40,000t + 100t^2$, the flow rate is $V'(t) = -40,000 + 200t$ cubic meters per minute.

d. The flow rate is largest (in absolute value) when $t = 0$ and smallest when $t = 200$.

3.5.49

a. $T'(t) = 160 - 80x$, so $T'(1) = 80$, so the heat flux at 1 is -80. At $x = 3$ we have $T'(3) = -80$, so the heat flux at 3 is 80.

b. The heat flux $-T'(x)$ is negative for $0 \leq x < 2$ and positive for $2 < x \leq 4$.

c. At any point other than the midpoint of the rod, heat flows toward the closest end of the rod, and "out the end."

3.6   The Chain Rule

3.6.1 If $y = f(x)$ and $u = g(x)$ then $\dfrac{dy}{dx} = \dfrac{dy}{du} \cdot \dfrac{du}{dx}$. Alternatively, we have $\dfrac{d}{dx}f(g(x)) = f'(g(x))g'(x)$.

3.6.3 The derivative of $f(g(x))$ equals f' evaluated at $g(x)$ multiplied by g' evaluated at x.

3.6.5 The inner function is $x^2 + 10$ and the outer function is u^{-5}, so with $y = f(u)$ and $u = g(x)$, we have $f(u) = u^{-5}$ and $g(x) = x^2 + 10$. Then $y = (x^2 + 10)^{-5}$.

3.6.7 With $u = 3x + 7$ and $y = u^{10}$ we have $\dfrac{dy}{dx} = \dfrac{dy}{du} \cdot \dfrac{du}{dx} = 10u^9 \cdot 3 = 30(3x + 7)^9$.

3.6.9 With $u = x^2 + 1$ and $y = \sqrt{u}$ we have $\dfrac{dy}{dx} = \dfrac{dy}{du} \cdot \dfrac{du}{dx} = \dfrac{1}{2\sqrt{u}} \cdot (2x) = \dfrac{x}{\sqrt{x^2+1}}$.

3.6.11 With $u = 5x^2$ and $y = \tan u$ we have $\frac{dy}{dx} = \frac{dy}{du} \cdot \frac{du}{dx} = \sec^2 u \cdot (10x) = 10x \sec^2(5x^2)$.

3.6.13 With $u = \cos x$ and $y = \sqrt{u}$ we have

$$\frac{dy}{dx} = \frac{dy}{du} \cdot \frac{du}{dx} = \frac{1}{2\sqrt{u}} \cdot (-\sin x) = -\frac{\sin x}{2\sqrt{\cos x}}.$$

3.6.15 With $u = x^4$ and $y = \tan u$ we have $\frac{dy}{dx} = \frac{dy}{du} \cdot \frac{du}{dx} = (\sec^2 u)(4x^3) = 4x^3 \sec^2(x^4)$.

3.6.17 With $g(x) = 3x^2 + 7x$ and $f(u) = u^{10}$ we have $\frac{d}{dx}[f(g(x))] = f'(g(x))g'(x) = 10(3x^2 + 7x)^9(6x + 7)$.

3.6.19 With $g(x) = 7x^3 + 1$ and $f(u) = 5u^{-3}$ we have $\frac{d}{dx}[f(g(x))] = f'(g(x))g'(x) = -15(7x^3 + 1)^{-4}(21x^2) = -315(7x^3 + 1)^{-4} \cdot x^2$.

3.6.21 With $f(u) = \tan u$ and $g(x) = 3x + 1$,

$$\frac{d}{dx}[f(g(x))] = f'(g(x))g'(x) = 3\sec^2(3x + 1)$$

3.6.23 With $g(x) = 4x^3 + 3x + 1$ and $f(u) = \sin u$ we have $\frac{d}{dx}[f(g(x))] = f'(g(x))g'(x) = \cos(u) \cdot (12x^2 + 3) = (12x^2 + 3) \cdot \cos(4x^3 + 3x + 1)$.

3.6.25 With $g(\theta) = 5\theta$ and $f(u) = \sec u$ we have $\frac{d}{d\theta} \sec(5\theta) = f'(g(\theta)) \cdot g'(\theta) = \sec(5\theta)\tan(5\theta) \cdot 5$. Now using the product rule, we have $\frac{dy}{d\theta} = (2\theta)\sec(5\theta) + \theta^2(5\sec(5\theta)\tan(5\theta)) = \theta\sec(5\theta)(2 + 5\theta\tan(5\theta))$.

3.6.27 With $g(x) = \sec x + \tan x$ and $f(u) = u^5$ we have $\frac{d}{dx}[f(g(x))] = f'(g(x))g'(x) = 5u^4 \cdot (\sec x \tan x + \sec^2 x) = 5(\sec x + \tan x)^4(\sec x \tan x + \sec^2 x) = 5\sec x(\sec x + \tan x)^5$.

3.6.29

a. $u = g(x) = \cos x$, $y = f(u) = u^3$. So $\frac{dy}{dx} = \frac{dy}{du} \cdot \frac{du}{dx} = 3\cos^2 x \cdot (-\sin x) = -3\cos^2 x \sin x$.

b. $u = g(x) = x^3$, $y = f(u) = \cos u$. So $\frac{dy}{dx} = \frac{dy}{du} \cdot \frac{du}{dx} = -\sin(x^3) \cdot 3x^2 = -3x^2 \sin(x^3)$.

3.6.31

a. $h'(3) = f'(g(3))g'(3) = f'(1) \cdot 20 = 5 \cdot 20 = 100$.

b. $h'(2) = f'(g(2))g'(2) = f'(5) \cdot 10 = (-10)(10) = -100$.

c. $p'(4) = g'(f(4))f'(4) = g'(1) \cdot (-8) = 2 \cdot (-8) = -16$.

d. $p'(2) = g'(f(2))f'(2) = g'(3) \cdot (2) = (20)(2) = 40$.

e. $h'(5) = f'(g(5))g'(5) = f'(2) \cdot 20 = 2 \cdot 20 = 40$.

3.6.33 Take $g(x) = 2x^6 - 3x^3 + 3$, and $n = 25$. Then $y' = n(g(x))^{n-1}g'(x) = 25(2x^6 - 3x^3 + 3)^{24}(12x^5 - 9x^2)$.

3.6.35 Take $g(x) = 1 + 2\tan x$, and $n = 15$. Then $y' = n(g(x))^{n-1}g'(x) = 15(1 + 2\tan x)^{14}(2\sec^2 x) = 30(1 + 2\tan x)^{14} \sec^2 x$.

3.6.37

$$\frac{d}{dx}\sqrt{1 + \cot^2 x} = \frac{1}{2\sqrt{1 + \cot^2 x}} \cdot \frac{d}{dx}(1 + \cot^2 x) = \frac{1}{2\sqrt{1 + \cot^2 x}} \cdot 2\cot x \cdot \frac{d}{dx}\cot x$$

$$= \frac{1}{2\sqrt{1 + \cot^2 x}} \cdot 2\cot x \cdot (-\csc^2 x) = -\frac{\cot x \csc^2 x}{\sqrt{1 + \cot^2 x}}.$$

3.6.39

$$\frac{d}{dx}\sin^5(\cos(3x)) = 5\sin^4(\cos(3x)) \cdot \frac{d}{dx}(\sin(\cos(3x)) = 5\sin^4(\cos(3x)) \cdot \cos(\cos(3x)) \cdot \frac{d}{dx}\cos(3x)$$
$$= 5\sin^4(\cos(3x)) \cdot \cos(\cos(3x)) \cdot (-\sin(3x)) \cdot 3 = -15\sin^4(\cos(3x))\cos(\cos(3x))\sin(3x).$$

3.6.41 Let $k(x) = \tan x$, $g(x) = \sqrt{x}$, $h(x) = \sec x$. Then $f(x) = k(g(h(x)))$, so

$$\frac{d}{dx}f(x) = k'(g(h(x)))(g(h(x)))' = k'(g(h(x)))g'(h(x))h'(x)$$

$$= \sec^2(\sqrt{\sec x})\left(\frac{1}{2\sqrt{\sec x}}\right)\sec x \tan x = \frac{\sin x\sqrt{\sec x}}{2\cos^2(\sqrt{\sec x})\cos x}$$

3.6.43 $\frac{d}{dx}\sqrt{x+\sqrt{x}} = \frac{1}{2\sqrt{x+\sqrt{x}}} \cdot \frac{d}{dx}(x+\sqrt{x}) = \frac{1}{2\sqrt{x+\sqrt{x}}} \cdot \left(1 + \frac{1}{2\sqrt{x}}\right).$

3.6.45 $\frac{d}{dx}f(g(x^2)) = f'(g(x^2)) \cdot \frac{d}{dx}(g(x^2)) = f'(g(x^2)) \cdot g'(x^2) \cdot 2x.$

3.6.47

a. True. The product rule alone will suffice.

b. False. You could write the expression as $\dfrac{1}{(x^2+10)^2}$ and use the quotient rule.

c. True. The derivative of the composition $f(g(x))$ is the product of $f'(g(x))$ with $g'(x)$, so it is the product of two derivatives.

d. False. In fact, $\frac{d}{dx}P(Q(x)) = P'(Q(x))Q'(x)$.

3.6.49 $\frac{d^2}{dx^2}\sin(x^2) = \frac{d}{dx}(2x\cos(x^2)) = 2(\cos(x^2) - 2x^2\sin(x^2)) = 2\cos(x^2) - 4x^2\sin(x^2)$. Note that in the middle of this calculation we used a result from the middle of the previous problem – namely the derivative of $x\cos(x^2)$.

3.6.51

$$\frac{d^2y}{dx^2}(x^2+1)^{-2} = \frac{dy}{dx}(-2(x^2+1)^{-3}(2x)) = -4\frac{dy}{dx}(x(x^2+1)^{-3})$$

$$= -4((x^2+1)^{-3} - 6x^2(x^2+1)^{-4}) = 4(x^2+1)^{-4}(6x^2-x^2-1) = \frac{4(5x^2-1)}{(x^2+1)^4}$$

3.6.53 $\frac{d}{dx}\sqrt{f(x)} = \frac{1}{2\sqrt{f(x)}} \cdot f'(x).$

3.6.55

$$y' = \frac{(x^3-6x-1)(2)(x^2-1)(2x) - (x^2-1)^2(3x^2-6)}{(x^3-6x-1)^2},$$

so

$$y'(3) = \frac{(27-18-1)(2)(8)(6) - (64)(21)}{64}$$

$$= \frac{768-1344}{64} = \frac{-576}{64} = -9.$$

The equation of the tangent line is thus $y - 8 = -9(x-3)$.

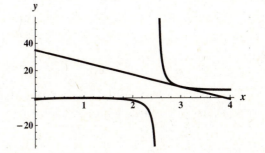

3.6.57

a. $g'(4) = 3$, $g(4) = 3 \cdot 4 - 5 = 7$. $f'(7) = -2$, $f(7) = -2 \cdot 7 + 23 = 9$. Thus, $h(4) = f(g(4)) = f(7) = 9$, and $h'(4) = f'(g(4))g'(4) = f'(7) \cdot 3 = -2 \cdot 3 = -6$.

b. The tangent line to h at $(4, 9)$ is given by $y - 9 = -6(x - 4)$.

3.6.59

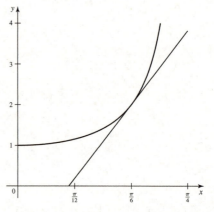

$y' = 2\sec(2x)\tan(2x)$; $y(\pi/6) = \sec \pi/3 = 2$ and $y'(\pi/6) = 2\sec(\pi/3)\tan(\pi/3) = 4\sqrt{3}$, so the tangent line at $\pi/6$ is

$$y - 2 = 4\sqrt{3}\left(x - \frac{\pi}{6}\right), \qquad y = 4\sqrt{3}x + \frac{6 - 2\pi\sqrt{3}}{3}$$

3.6.61 First, note that $g'(x) = \cos(\pi f(x)) \cdot \pi f'(x)$.

a. $g'(0) = \cos(\pi \cdot f(0)) \cdot \pi f'(0) = \cos(-3\pi) \cdot 3\pi = -3\pi$.

b. $g'(1) = \cos(\pi \cdot f(1)) \cdot \pi f'(1) = \cos(3\pi) \cdot 5\pi = -5\pi$.

3.6.63

a. $\frac{d^2 y}{dt^2} = \frac{d}{dt}\left(-y_0 \sqrt{\frac{k}{m}} \sin\left(t\sqrt{\frac{k}{m}}\right)\right) = -y_0 \cdot \frac{k}{m} \cdot \cos\left(t\sqrt{\frac{k}{m}}\right)$.

b. $-\frac{k}{m}y = -\frac{k}{m}\left(y_0 \cos\left(t\sqrt{\frac{k}{m}}\right)\right) = \frac{d^2 y}{dt^2}$.

3.6.65

a. Assuming the year is not a leap year, March 1st corresponds to $t = 59$. We have $D(59) = 12 - 3\cos\left(\frac{2\pi(69)}{365}\right) \approx 10.88$ hours.

b. $\frac{d}{dt}D(t) = 3 \cdot \frac{2\pi}{365} \sin\left(\frac{2\pi(t+10)}{365}\right)$ hours per day.

c. $D'(59) \approx 0.048$ hours per day ≈ 2 minutes and 52 seconds per day. This means that on March 1st, the days are getting longer by just shy of 3 minutes per day.

d.

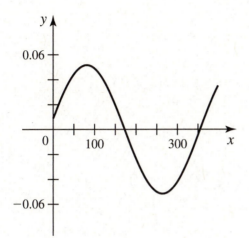

e. The largest increase in the length of the days appears to be at about $t = 81$, and the largest decrease at about $t = 265$. These correspond to March 22nd and to September 22nd. The least rapid changes occur at about $t = 172$ and $t = 355$. These correspond to June 21st and December 21st.

3.6.67

a. $E'(t) = 400 + 200\cos\left(\frac{\pi t}{12}\right)$ MW.

b. Since the maximum value of $\cos\theta$ is 1, the maximum value of $E'(t)$ will be 600 MW, where $\cos\left(\frac{\pi t}{12}\right) = 1$, which is where $t = 0$, which corresponds to noon.

c. Since the minimum value of $\cos\theta$ is -1, the minimum value of $E'(t)$ will be 200 MW, where $\cos\left(\frac{\pi t}{12}\right) = -1$, which is where $\frac{\pi t}{12} = \pi$, or $t = 12$, which corresponds to midnight.

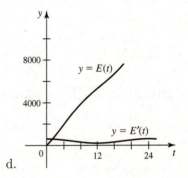

d.

3.6.69

a. $f'(x) = \frac{d}{dx}(\cos^2 x + \sin^2 x) = 2(\cos x)(-\sin x) + 2\sin x \cos x = 0.$

b. If $f(x)$ is a constant, then the output value must be the same at any input value, so we choose to evaluate f at a nice value like $x = 0$. We see that $f(0) = \cos^2(0) + \sin^2(0) = 1^2 + 0^2 = 1$, so we must have $\cos^2 x + \sin^2 x = 1$ for all x.

3.6.71

$$\frac{d}{dx}\left(f(x)\right) \cdot (g(x))^{-1}) = f'(x) \cdot (g(x))^{-1} + f(x) \cdot (-(g(x))^{-2} \cdot g'(x)) = \frac{f'(x)}{g(x)} - \frac{f(x)g'(x)}{(g(x))^2}$$
$$= \frac{g(x)f'(x) - f(x)g'(x)}{(g(x))^2}.$$

3.6.73

a. $h(x) = (x^2 - 3)^5$, $a = 2$.

b. $h'(x) = 5(x^2 - 3)^4(2x) = 10x(x^2 - 3)^4$, so the value of this limit is $h'(2) = 20$.

3.6.75

a. $h(x) = \sin x^2$, $a = \frac{\pi}{2}$.

b. $h'(x) = (\cos x^2)(2x)$, so the value of this limit is $h'\left(\frac{\pi}{2}\right) = \pi \cdot \cos\left(\frac{\pi^2}{4}\right) \approx -2.45$.

3.6.77 $\lim\limits_{x \to 5} \dfrac{f(x^2) - f(25)}{x - 5} = \dfrac{d}{dx}\left[f(x^2)\right]_{x=5} = 2 \cdot 5 \cdot f'(25) = 10f'(25)$.

3.6.79

a. $\lim\limits_{v \to u} H(v) = \lim\limits_{v \to u}\left(\dfrac{f(v) - f(u)}{v - u} - f'(u)\right) = \lim\limits_{v \to u}\left(\dfrac{f(v) - f(u)}{v - u}\right) - f'(u) = f'(u) - f'(u) = 0$.

b. Suppose $u = v$. Then clearly both sides of the given expression are 0, so they are equal. Suppose $u \neq v$. Then $H(v) = \frac{f(v)-f(u)}{v-u} - f'(u)$, so $H(v) + f'(u) = \frac{f(v)-f(u)}{v-u}$, so the result holds by multiplying both sides of this equation by $v - u$.

c.
$$h'(a) = \lim\limits_{x \to a} \frac{f(g(x)) - f(g(a))}{x - a} = \lim\limits_{x \to a} \frac{H(g(x)) + f'(g(a))}{x - a} \cdot (g(x) - g(a))$$
$$= \lim\limits_{x \to a} \left[(H(g(x)) + f'(g(a))) \cdot \frac{g(x) - g(a)}{x - a}\right]$$

d. $h'(a) = \lim\limits_{x \to a} \left[(H(g(x)) + f'(g(a))) \cdot \dfrac{g(x) - g(a)}{x - a}\right] = (0 + f'(g(a))) \cdot g'(a) = f'(g(a))g'(a)$.

3.7 Implicit Differentiation

3.7.1 Solving for y involves computing $\sqrt{y^2} = |y|$ which then typically requires an analysis by cases, which can be messy.

3.7.3 Often, an implicit function doesn't represent y as a function of x in the sense of a unique y value for each input x value. For example, the "implicit function" $x^2 + y^2 = 1$ represents the unit circle which fails the vertical line test. Implicit Differentiation can tell us the slope of the tangent line at both of the points $(1/2, \sqrt{3}/2)$ and $(1/2, -\sqrt{3}/2)$, but if one were to just specify $x = 1/2$, we wouldn't know which point was under consideration.

3.7.5

a. $2yy' = 4$, so $y' = \frac{2}{y}$.

b. $y'(1, 2) = \frac{2}{2} = 1$.

3.7.7

a. $y' \cos y = 20x^3$, so $y' = \frac{20x^3}{\cos y}$.

b. $y'(1, \pi) = \frac{20}{\cos \pi} = -20$.

3.7.9

a. $-y' \sin y = 1$, so $y' = -\frac{1}{\sin y} = -\csc y$.

b. $y'(0, \pi/2) = -\csc(\pi/2) = -1$.

3.7.11 $(y + xy')\cos(xy) = 1 + y'$, so $y\cos(xy) + xy'\cos(xy) = 1 + y'$. If we rearrange terms in order to have the terms with a factor of y' all on the same side, we obtain $y\cos(xy) - 1 = y' - xy'\cos(xy)$. Factoring out the y' factor gives $y\cos(xy) - 1 = y'(1 - x\cos(xy))$, so $y' = \dfrac{y\cos(xy) - 1}{1 - x\cos(xy)}$.

3.7.13 $-\sin(y^2)(2yy') + 1 = 2yy'$, so $2yy'(1 + \sin(y^2)) = 1$ and

$$y' = \frac{1}{2y(1 + \sin(y^2))}.$$

3.7.15

$$3x^2 = \frac{(x-y)(1+y') - (x+y)(1-y')}{(x-y)^2}$$
$$3x^2(x-y)^2 = x + xy' - y - yy' - x + xy' - y + yy'$$
$$3x^2(x-y)^2 + 2y = 2xy'$$
$$\frac{3x^2(x-y)^2 + 2y}{2x} = y'$$

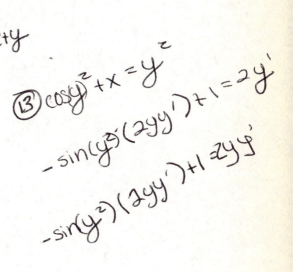

3.7.17

$$18x^2 + 21y'y^2 = 13(y + xy')$$
$$21y'y^2 - 13xy' = 13y - 18x^2$$
$$y' = \frac{13y - 18x^2}{21y^2 - 13x}.$$

3.7.19

$$\frac{4x^3 + 2yy'}{2\sqrt{x^4 + y^2}} = 5 + 6y'y^2$$
$$yy' - 6y'y^2\sqrt{x^4 + y^2} = 5\sqrt{x^4 + y^2} - 2x^3$$
$$y' = \frac{5\sqrt{x^4 + y^2} - 2x^3}{y - 6y^2\sqrt{x^4 + y^2}}.$$

3.7.21

a. $2^2 + 2 \cdot 1 + 1^2 = 7$, so the point $(2, 1)$ does lie on the curve.

b. $2x + y + xy' + 2yy' = 0$, so at the given point we have $4 + 1 + 2y' + 2y' = 0$, so $y' = \frac{-5}{4}$. The equation of the tangent line is therefore $y - 1 = \frac{-5}{4}(x - 2)$.

3.7.23

a. $\sin\pi + 5\left(\frac{\pi^2}{5}\right) = \pi^2$, so the point $(\pi^2/5, \pi)$ does lie on the curve.

b. $y'\cos y + 5 = 2yy'$, so at the given point we have $-y' + 5 = 2\pi y'$, so $y' = \frac{5}{1+2\pi}$. The equation of the tangent line is therefore $y - \pi = \frac{5}{1+2\pi}\left(x - \pi^2/5\right)$.

3.7.25

a. $\cos(\pi/2 - \pi/4) + \sin(\pi/4) = (\sqrt{2}/2) + (\sqrt{2}/2) = \sqrt{2}$, so the point $(\pi/2, \pi/4)$ does lie on the curve.

b. $(1 - y')(-\sin(x - y)) + y'\cos y = 0$, so at the given point we have $(1 - y')(-\sqrt{2}/2) + y'(\sqrt{2}/2) = 0$, so $\sqrt{2}y' = \sqrt{2}/2$, so $y' = 1/2$. The equation of the tangent line is therefore $y - (\pi/4) = (1/2)(x - \pi/2)$.

3.7.27 $1 + 2yy' = 0$, so $y' = \frac{-1}{2y}$. Differentiating again, we obtain

$$y'' = \frac{-1}{2} \cdot \frac{-y'}{y^2} = \frac{y'}{2y^2} = \left(\frac{-1}{2y}\right) \cdot \frac{1}{2y^2} = \frac{-1}{4y^3}.$$

3.7.29 $\frac{y'}{2\sqrt{y}} + y + xy' = 0$, so $y' + 2x\sqrt{y}y' = -2y\sqrt{y}$, so $y' = \frac{-2y\sqrt{y}}{2x\sqrt{y} + 1} = \frac{-2y^{(3/2)}}{2x\sqrt{y} + 1}$. Differentiating again we obtain

$$y'' = -\frac{(2\sqrt{y}x + 1)(3\sqrt{y}y') - 2y^{3/2}\left(2\sqrt{y} + \frac{xy'}{\sqrt{y}}\right)}{(2x\sqrt{y} + 1)^2} = \frac{-(2\sqrt{y}x + 1)(3\sqrt{y}y') + 4y^2 + 2xyy'}{(2x\sqrt{y} + 1)^2}$$

$$= \left(\frac{-(2\sqrt{y}x + 1)(3\sqrt{y})\left(\frac{-2y\sqrt{y}}{2x\sqrt{y}+1}\right) + 4y^2 + 2xy\left(\frac{-2y\sqrt{y}}{1+2x\sqrt{y}}\right)}{(2x\sqrt{y} + 1)^2}\right) \cdot \frac{2x\sqrt{y} + 1}{2x\sqrt{y} + 1}$$

$$= \frac{(2x\sqrt{y} + 1)(6y^2) + 4y^2(1 + 2x\sqrt{y}) - 4xy^{5/2}}{(2x\sqrt{y} + 1)^3} = \frac{10y^2 + 16xy^2\sqrt{y}}{(2x\sqrt{y} + 1)^3}.$$

3.7.31 $y'\cos y + 1 = y'$, so that $y' = \frac{1}{1-\cos y}$. Differentiating again, we get

$$y'' = -(1 - \cos y)^{-2}(y'\sin y) = -\frac{\sin y}{(1 - \cos y)^3}$$

3.7.33 $\frac{dy}{dx} = \frac{5}{4}x^{\frac{5}{4}-1} = \frac{5}{4}x^{\frac{1}{4}}$.

3.7.35 $\frac{dy}{dx} = 5 \cdot \frac{2}{3}(5x + 1)^{\frac{-1}{3}} = \frac{10}{3(5x + 1)^{\frac{1}{3}}}$.

3.7.37 $\frac{dy}{dx} = \frac{1}{4}\left(\frac{2x}{4x - 3}\right)^{\frac{-3}{4}} \cdot \frac{2(4x - 3) - 2x \cdot 4}{(4x - 3)^2} = \frac{-3}{2}\left(\frac{4x - 3}{2x}\right)^{\frac{3}{4}} \cdot \frac{1}{(4x - 3)^2}$.

3.7.39

$$\frac{dy}{dx} = (x^2 + 5x + 1)^{\frac{1}{3}} + \frac{1}{3}x(x^2 + 5x + 1)^{\frac{-2}{3}} \cdot (2x + 5)$$

$$= \left((x^2 + 5x + 1)^{\frac{1}{3}} + \frac{1}{3}x(x^2 + 5x + 1)^{\frac{-2}{3}} \cdot (2x + 5)\right) \cdot \frac{3(x^2 + 5x + 1)^{\frac{2}{3}}}{3(x^2 + 5x + 1)^{\frac{2}{3}}}$$

$$= \frac{3x^2 + 15x + 3 + 2x^2 + 5x}{3(x^2 + 5x + 1)^{\frac{2}{3}}} = \frac{5x^2 + 20x + 3}{3(x^2 + 5x + 1)^{\frac{2}{3}}}.$$

3.7.41 $\frac{1}{3}x^{\frac{-2}{3}} + \frac{4}{3}y^{\frac{1}{3}}y' = 0$, so at the given point we have $\frac{1}{3} + \frac{4}{3}y' = 0$, so $y' = \frac{-1}{4}$.

3.7.43 $y^{\frac{1}{3}} + \frac{1}{3}xy^{\frac{-2}{3}}y' + y' = 0$, so at the given point we have $2 + \frac{1}{3} \cdot 1 \cdot \frac{1}{4}y' + y' = 0$, so $\frac{13}{12}y' = -2$, so $y' = \frac{-24}{13}$.

3.7.45 $y + xy' + \frac{3}{2}x^{\frac{1}{2}}y^{\frac{-1}{2}} - \frac{1}{2}x^{\frac{3}{2}}y^{\frac{-3}{2}}y' = 0$, so at the given point we have $1 + y' + \frac{3}{2} - \frac{1}{2}y' = 0$, so $\frac{1}{2}y' = \frac{-5}{2}$, so $y' = -5$.

3.7.47

a. False. For example, the equation $y\cos(xy) = x$, cannot be solved explicitly for y in terms of x.

b. True. We have $2x + 2yy' = 0$, and the result follows by solving for y'.

c. False. The equation $x = 1$ doesn't represent any sort of function – it is either just a number, or perhaps a vertical line, but it doesn't represent a differentiable function.

d. False. $y + xy' = 0$, so $y' = \frac{-y}{x}$, $x \neq 0$.

3.7.49

b.

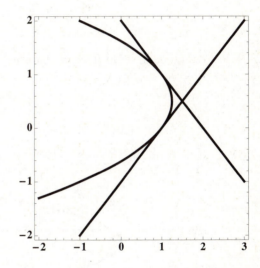

a. There are two points on the curve associated with $x = 1$. When $x = 1$, we have $1 + y^2 - y = 1$, so $y(y - 1) = 0$. The two points are thus $(1, 0)$ and $(1, 1)$. Differentiating yields $1 + 2yy' - y' = 0$, so $y' = \frac{1}{1-2y}$. At $(1, 0)$, we have $y' = 1$, so the tangent line is given by $y = x - 1$. At $(1, 1)$, we have $y' = -1$, so the tangent line is given by $y - 1 = -1(x - 1)$.

3.7.51

a. $y(2x) + (x^2 + 4)y' = 0$, so $y' = \frac{-2xy}{x^2+4}$.

b. At $y = 1$ we have $x^2 + 4 = 8$, so $x = \pm 2$. At the point $(2, 1)$ we have $y' = \frac{-4}{8} = \frac{-1}{2}$. At the point $(-2, 1)$ we have $y' = \frac{4}{8} = \frac{1}{2}$. Thus, the equations of the tangent lines are given by $y - 1 = \frac{-1}{2}(x - 2)$ and $y - 1 = \frac{1}{2}(x + 2)$.

c. $y = \frac{8}{x^2+4}$, so $y' = \frac{0-8\cdot 2x}{(x^2+4)^2} = \frac{-16x}{(x^2+4)^2}$.

d. $y' = \frac{-16x}{(x^2+4)^2} = \frac{-2x}{x^2+4} \cdot \frac{8}{x^2+4} = \frac{-2x}{x^2+4} \cdot y = \frac{-2xy}{x^2+4}$.

3.7.53

a. From Exercise 49, we have that $y' = \frac{1}{1-2y}$. A vertical tangent would occur at a point whose y value would make $1 - 2y$ equal to zero. So we are looking for where $2y = 1$ or $y = \frac{1}{2}$. If $y = \frac{1}{2}$, then $x + \frac{1}{4} - \frac{1}{2} = 1$, so $x = \frac{5}{4}$, and there is a vertical tangent at $\left(\frac{5}{4}, \frac{1}{2}\right)$.

b. Since y' is never zero, there are no horizontal tangent lines.

3.7.55

a. $1 + 3y^2y' - y - xy' = 0$, so $y' = \frac{y-1}{3y^2-x}$.

b.

$$y^3 - 1 = x(y-1)$$
$$(y-1)(y^2+y+1) = x(y-1)$$
$$y^2 + y + 1 = x$$
$$y^2 + y + (1-x) = 0,$$

so by the quadratic formula we have $y = \frac{-1 \pm \sqrt{4x-3}}{2}$. Note that this means that $\pm\sqrt{4x-3} = 2y+1$ (we will use this in part d. below.)

c.

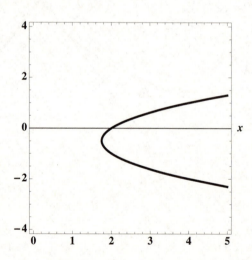

d. From part b.: $y' = \pm\frac{1}{\sqrt{4x-3}}$. Suppose $y \neq 1$. Then (from part (b)) $x = y^2 + y + 1$. So $y' = \frac{y-1}{3y^2-x} = \frac{y-1}{3y^2-y^2-y-1} = \frac{y-1}{2y^2-y-1} = \frac{1}{2y+1} = \pm\frac{1}{\sqrt{4x-3}}$.

3.7.57

a. $4x^3 = 4x - 4yy'$, so $y' = \frac{x-x^3}{y}$.

b. $y = \pm\sqrt{x^2 - \dfrac{x^4}{2}}$.

c.

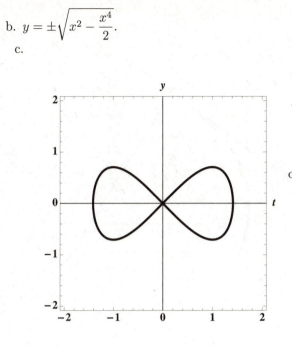

d. From part a. we have that $y' = \frac{x-x^3}{y} = \frac{2x-2x^3}{2y}$. Differentiating our result in part b. we have $y' = \pm\frac{1}{2}\left(x^2 - \frac{x^4}{2}\right)^{-1/2} \cdot (2x - 2x^3) = \frac{2x-2x^3}{2y}$.

3.7.59

The slope of the normal line is the negative reciprocal of the slope of the tangent line. From 21: $y' = \frac{-5}{4}$, so the slope of the normal line is $\frac{4}{5}$. At the point $(2, 1)$ we have the line $y - 1 = \frac{4}{5}(x - 2)$.

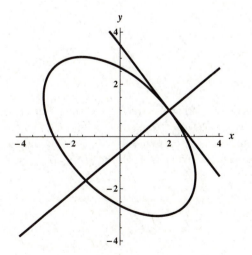

3.7.61

The slope of the normal line is the negative reciprocal of the slope of the tangent line. From 23: $y' = \frac{5}{2\pi+1}$, so the slope of the normal line is $\frac{-2\pi-1}{5}$. At the point $\left(\frac{\pi^2}{5}, \pi\right)$ we have the line $y - \pi = \frac{-2\pi-1}{5}\left(x - \frac{\pi^2}{5}\right)$.

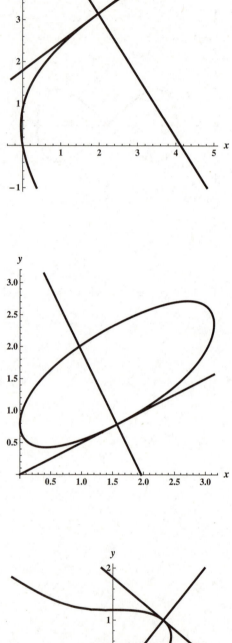

3.7.63

The slope of the normal line is the negative reciprocal of the slope of the tangent line. From 25: $y' = \frac{1}{2}$, so the slope of the normal line is -2. At the point $\left(\frac{\pi}{2}, \frac{\pi}{4}\right)$ we have the line $y - \frac{\pi}{4} = -2\left(x - \frac{\pi}{2}\right)$.

3.7.65

We have $9x^2 + 21y^2 y' = 10y'$, so at the point $(1,1)$ we have $9 + 21y' = 10y'$, so $y' = \frac{-9}{11}$. Thus, the tangent line is given by $y - 1 = \frac{-9}{11}(x - 1)$, and the normal line is given by $y - 1 = \frac{11}{9}(x - 1)$.

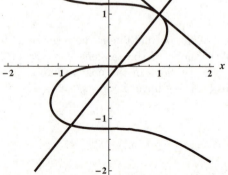

3.7.67

We have $2(x^2+y^2-2x)(2x+2yy'-2)=4x+4yy'$, so at the point $(2,2)$ we have $2(4+4-4)(4+4y'-2)=8+8y'$, so $16+32y'=8+8y'$, so $y'=\frac{-1}{3}$. Thus, the tangent line is given by $y-2=\frac{-1}{3}(x-2)$, and the normal line is given by $y-2=3(x-2)$.

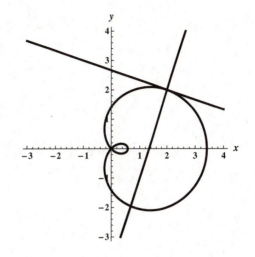

3.7.69

a. $1280=40L^{1/3}K^{2/3}$, so $0=\frac{40}{3}L^{-2/3}K^{2/3}+\frac{80}{3}L^{1/3}K^{-1/3}\cdot\frac{dK}{dL}$. Multiplying both sides by $\frac{3}{40}L^{2/3}K^{1/3}$ yields $0=K+2L\frac{dK}{dL}$, so $\frac{dK}{dL}=\frac{-K}{2L}$.

b. With $L=8$ and $K=64$, $\frac{dK}{dL}=\frac{-64}{16}=-4$.

3.7.71

a. $V=\frac{\pi h^2(3r-h)}{3}=\frac{5\pi}{3}$. So
$$\frac{1}{3}[2\pi h(3r-h)+\pi h^2(3r'-1)]=0,$$
$$6rh-2h^2+3h^2r'-h^2=0,$$
so $r'=1-\frac{2r}{h}$.

b. At $r=2$ and $h=1$, we have $r'=1-(4)=-3$.

3.7.73 For $y=mx$, $y'=m$, and for $x^2+y^2=a^2$, $y'=\frac{-x}{y}$. Let (a,b) be a point on both curves. Then $b=ma$, so the point has the form (a,ma). A normal line to the circle at the point (a,ma) would have slope $\frac{y}{x}=\frac{ma}{a}=m$, which is the slope of the tangent line to the curve $y=mx$. Thus the two curves are orthogonal at any point of intersection.

3.7.75 For $xy=a$ we have $xy'+y=0$, so $y'=\frac{-y}{x}$. For $x^2-y^2=b$, we have $2x-2yy'=0$, so $y'=\frac{x}{y}$. Let (c,d) be a point on both curves. Then the slope of the normal line to the first curve is $\frac{c}{d}$, but that is the slope of the tangent line to the second curve. Thus the two curves are orthogonal at any points of intersection.

3.8 Related Rates

3.8.1 The area of a circle of radius r is $A(r)=\pi r^2$. If the radius $r=r(t)$ changes with time, then the area of the circle is a function of r and r is a function of t, so ultimately A is a function of t. If the radius changes at rate $\frac{dr}{dt}$, then the area changes at rate $2\pi r\frac{dr}{dt}$.

3.8.3 Since area is width times height, if one increases, the other must decrease at the same rate in order for the area to remain constant.

3.8.5 $A(x) = x^2$, $\frac{dx}{dt} = 2$ meters per second.

c.

a. $\frac{dA}{dt} = 2x\frac{dx}{dt}$, so at $x = 10$ meters we have $\frac{dA}{dt} = 2 \cdot 10\text{m} \cdot 2\text{m/s} = 40\text{m}^2/\text{s}$.

b. At $x = 20$m we have $\frac{dA}{dt} = 2 \cdot 20\text{m} \cdot 2\text{m/s} = 80\text{m}^2/\text{s}$.

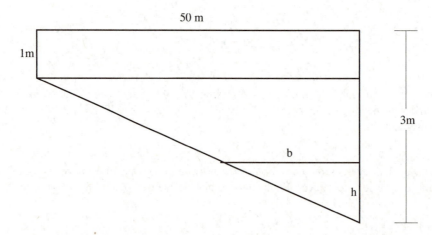

3.8.7 $A(x) = \pi x^2$, so $\frac{dA}{dt} = 2\pi x \frac{dx}{dt}$. At $x = 10\,\text{ft}$ and $\frac{dx}{dt} = -2\,\text{ft/min}$ we have $\frac{dA}{dt} = 2\pi \cdot 10 \cdot (-2) = -40\pi\,\text{ft}^2/\text{min}$.

3.8.9 $V(r) = \frac{4}{3}\pi r^3$, so $\frac{dV}{dt} = 4\pi r^2 \frac{dr}{dt} = 15\,\text{in}^3/\text{min}$. At $r = 10$ inches we have $4\pi(10\,\text{in})^2 \frac{dr}{dt} = 15\,\text{in}^3/\text{min}$. Thus, $\frac{dr}{dt} = \frac{3}{80\pi}\,\text{in/min} \approx 0.012\,\text{in/min}$.

3.8.11 $V(r) = \frac{4}{3}\pi r^3$, and $S(r) = 4\pi r^2$. $\frac{dV}{dt} = 4\pi r^2 \frac{dr}{dt} = k \cdot 4\pi r^2$, so $\frac{dr}{dt} = k$, the constant of proportionality.

3.8.13   By similar triangles, $\frac{2}{50} = \frac{h}{b}$, so $b = 25h$. Also, $A = \frac{1}{2}bh = 12.5h^2$, so the volume for $0 \le h \le 2$ is $V(h) = 12.5 \cdot h^2 \cdot 20 = 250h^2$. For $2 < h \le 3$, $V(h) = 250 \cdot 2^2 + 50 \cdot 20 \cdot (h - 2) = 1000h - 1000$. When $t = 250$ minutes, then $V = 250\,\text{min} \cdot 1\,\text{m}^3/\text{min} = 250\,\text{m}^3$. So $V(h) = 250h^2 = 250$, so $h = 1\,\text{m}$. At that time $\frac{dV}{dt} = 500h\frac{dh}{dt} = 500 \cdot 1 \cdot \frac{dh}{dt} = 1\,\text{m}^3/\text{min}$. So $\frac{dh}{dt} = \frac{1}{500}\,\text{m/min} = 0.002\,\text{m/min} = 2\,\text{mm/min}$.

Fill time: The volume of the entire swimming pool is 2000 cubic meters, so at 1 cubic meter per minute, it will take 2000 minutes.

3.8.15   Let x be the distance the surface ship has traveled and D the depth of the submarine. We have $\frac{dx}{dt} = 10\,\text{km/hr}$. Note that $\frac{D}{x} = \tan 20°$, so $D = x \cdot \tan 20° \approx 0.364x$. We have $\frac{dD}{dt} = 0.364\frac{dx}{dt} = 3.64\,\text{km/hr}$. The depth of the submarine is increasing at a rate of 3.64 km/hr.

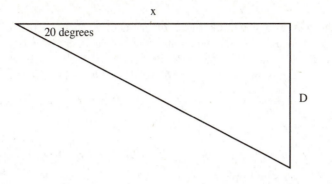

3.8.17 By the Pythagorean Theorem, we have that $x^2 + h^2 = 169$. Thus, $2x\frac{dx}{dt} + 2h\frac{dh}{dt} = 0$, so $\frac{dh}{dt} = \frac{-x}{h}\frac{dx}{dt}$, and we are given that $\frac{dx}{dt} = 0.5$ feet per second. At $x = 5$ we have $h = \sqrt{169 - 25} = 12$ feet. Thus, $\frac{dh}{dt} = \frac{-5}{12} \cdot \frac{1}{2} = \frac{-5}{24}$ feet per second. So the top of the ladder slides down the wall at $\frac{5}{24}$ feet per second.

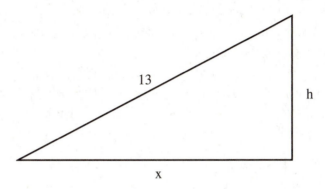

3.8.19 By similar triangles, $\frac{x+y}{20} = \frac{y}{5}$, so $x + y = 4y$, so $x = 3y$, and $\frac{dx}{dt} = 3\frac{dy}{dt}$. Since we are given that $\frac{dx}{dt} = -8$, we have $\frac{dy}{dt} = \frac{-8}{3}$ feet per second, which is the rate of change of the length of her shadow. The tip of her shadow is therefore moving at $-8 - \frac{8}{3} = \frac{-32}{3}$ feet per second (or $\frac{32}{3}$ feet per second towards the light).

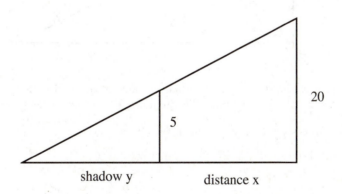

3.8.21 $V = \frac{1}{3}\pi r^2 h$ where $r = 3h$, so $V = 3\pi h^3$. We have that $\frac{dV}{dt} = 9\pi h^2 \frac{dh}{dt}$, and we given that $\frac{dh}{dt} = 2$ at the moment when $h = 12$, so at that time, $\frac{dV}{dt} = 9\pi \cdot 144\,\text{cm}^2 \cdot 2\,\text{cm/sec} = 2592\pi\,\text{cm}^3/\text{s}$. This is the rate at which the volume of the sandpile is increasing, so it must also be the rate at which the sand is leaving the bin, since there is no other sand involved.

3.8.23 Let h be the depth of the water in the tank at time t, and let r be the radius of the cone-shaped water at time t. By similar triangles, we have that $\frac{h}{r} = \frac{12}{6}$, so $h = 2r$. The volume of the water in the tank is given by $V = \frac{1}{3}\pi r^2 h = \frac{1}{3}\pi \frac{h^2}{4} \cdot h = \frac{\pi h^3}{12}$. Thus, $\frac{dV}{dt} = \frac{\pi h^2}{4}\frac{dh}{dt}$, and so when $h = 3$ we have $-2\,\text{ft}^3/\text{s} = \frac{9\pi\,\text{ft}^2}{4}\frac{dh}{dt}$, so $\frac{dh}{dt} = \frac{-8}{9\pi}$ ft/s. So the depth of the water is decreasing at a rate of $8/(9\pi)$ feet per second.

3.8.25

a. The volume of the water in the tank (as a function of h – the depth of the water in the tank) is given by 5 times the area of the segment of water in a cross-sectional circle. For a tank of radius 1, the formula for such a segment is $\cos^{-1}(1 - h) - (1 - h)\sqrt{2h - h^2}$. Thus the volume of the water in the tank is given by $V = 5(\cos^{-1}(1 - h) - (1 - h)\sqrt{2h - h^2})$. We have $\frac{dV}{dt} = 5 \cdot \left(\frac{-1}{\sqrt{1-(1-h)^2}} \cdot \left(-\frac{dh}{dt}\right) + \frac{dh}{dt}\sqrt{2h - h^2} - \frac{(1-h)^2}{\sqrt{2h-h^2}}\frac{dh}{dt} \right)$. When $h = .5$, we have $-\frac{3}{2} = \frac{dV}{dt} = 5\sqrt{3}\frac{dh}{dt}$, so $\frac{dh}{dt} = \frac{-\sqrt{3}}{10}$ meters per hr.

b. The surface area of the water is given by $S = 5 \cdot 2\sqrt{2h - h^2}$. So $\frac{dS}{dt} = 10 \cdot \frac{2-2h}{2\sqrt{2h-h^2}} \cdot \frac{dh}{dt}$, so at $h = .5$, we have $\frac{5}{\sqrt{3/4}} \cdot \frac{-\sqrt{3}}{10} = -1$ square meter per hr.

3.8.27 Let x be the distance the motorcycle has traveled since the instant it went under the balloon, and let y be the height of the balloon above the ground t seconds after the motorcycle went under it. We have $x^2 + y^2 = D^2$ where D is the distance between the motorcycle and the balloon. Thus, $2x\frac{dx}{dt} + 2y\frac{dy}{dt} = 2D\frac{dD}{dt}$, and we are given that $\frac{dy}{dt} = 10$ feet per second, and $\frac{dx}{dt} = 40\,\text{mph} = \frac{176}{3}$ ft/s. After 10 seconds have passed, we have that $y = 150 + 100 = 250\,\text{ft}$, $x = \frac{1760}{3}$ ft and $D = \sqrt{250^2 + \left(\frac{1760}{3}\right)^2} \approx 638\,\text{ft}$. Thus, $\frac{dD}{dt} \approx \frac{1}{638}\left(\frac{1760}{3} \cdot \frac{176}{3} + 2500\right) \approx 57.86$ feet per second.

3.8.29 Let x be the distance between the fish and the feet of the fisherman, and let D be the distance between the fish and the tip of the pole. Then $D^2 = x^2 + 144$, so $2D\frac{dD}{dt} = 2x\frac{dx}{dt}$. Note that $\frac{dD}{dt} = -\frac{1}{3}$ ft/s, so when $x = 20$, we have $\frac{dx}{dt} = \frac{\sqrt{400+144}}{20} \cdot \frac{-1}{3} \approx -0.39$ feet per second. The fish is moving toward the fisherman at about $0.39 \cdot 12 = 4.66$ inches per second.

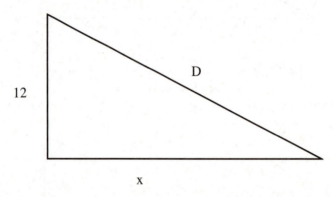

3.8.31 Let D be the length of the rope from the boat to the capstan, and let x be the horizontal distance from the boat to the dock. By the Pythagorean Theorem, $x^2 + 25 = D^2$, so $2x\frac{dx}{dt} = 2D\frac{dD}{dt}$, so $\frac{dx}{dt} = \frac{D}{x}\frac{dD}{dt}$. We are given that $\frac{dD}{dt} = -3$ feet per second, so when $x = 10$, we have $\frac{dx}{dt} = \frac{\sqrt{125}}{10} \cdot (-3) = \frac{-3\sqrt{5}}{2}$ feet per second. The boat is approaching the dock at $\frac{3\sqrt{5}}{2}$ feet per second.

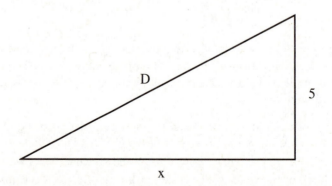

3.8.33 Let x be the distance the westbound airliner has traveled between noon and t hours after 1:00, and let y be the distance the northbound airliner has traveled t hours after 1:00, and let D be the distance between the planes. We have $D^2 = x^2 + y^2$, so $2D\frac{dD}{dt} = 2x\frac{dx}{dt} + 2y\frac{dy}{dt}$. We are given that $\frac{dx}{dt} = 500$ mph and $\frac{dy}{dt} = 550$ mph. At 2:30, we have that $x = 500 + 500 \cdot 1.5 = 1250$, and $y = 550 \cdot 1.5 = 825$ miles. $D = \sqrt{2243125} \approx 1497.7$ miles. Thus $\frac{dD}{dt} \approx \frac{1250 \cdot 500 + 825 \cdot 550}{1497.7} \approx 720.27$ miles per hour.

3.8.35 Let θ be the angle between the hands of the clock, and D the distance between the tips of the hands. By the law of cosines, $D^2 = 2.5^2 + 3^2 - 15\cos\theta$. So $2D\frac{dD}{dt} = 15\sin\theta\frac{d\theta}{dt}$. At 9:00 AM, we have $D^2 = 6.25 + 9$, so $D = \sqrt{15.25}$. Also, $\theta = \pi/2$ so $\sin\theta = 1$. Thus, $\frac{dD}{dt} = \frac{15}{2\sqrt{15.25}}\frac{d\theta}{dt}$. Now $\frac{d\theta}{dt} = \frac{d\theta_1}{dt} - \frac{d\theta_2}{dt}$ where $\frac{d\theta_1}{dt}$ is the angular change of the minute hand and $\frac{d\theta_2}{dt}$ is the angular change of the hour hand. We have $\frac{d\theta_1}{dt} = \frac{\pi}{30}$ radians per minute and $\frac{d\theta_2}{dt} = \frac{\pi}{360}$ radians per minute, so $\frac{d\theta}{dt} = \frac{11\pi}{360}$ radians per minute. Thus $\frac{dD}{dt} = \frac{15}{2\sqrt{15.25}} \cdot \frac{11\pi}{360} \approx .18436$ meters per minute, or about 11.06 meters per hour.

3.8.37

a. Let A be the point where the dragster started, let B be the point where camera 1 is located and let $C = y(t)$ be the position of the car at time t. Let θ be angle ABC. Note that $\tan\theta = \frac{y}{50}$, so $\sec^2\theta \cdot \frac{d\theta}{dt} = \frac{1}{50}\frac{dy}{dt}$. At time $t = 2$, we have that $\tan^2\theta = 4$, so $\sec^2\theta = \tan^2\theta + 1 = 5$. So $\frac{dy}{dt} = 5 \cdot 50 \cdot .75 = 187.5$ feet per second.

b. Let D be the point where camera 2 is located, and let ϕ be angle ADC. The $\phi = \tan^{-1}\left(\frac{y}{100}\right)$, so $\frac{d\phi}{dt} = \frac{1}{100\left(1 + \left(\frac{y}{100}\right)^2\right)} \cdot \frac{dy}{dt}$. After 2 seconds, we know that $y = 100$ and $\frac{dy}{dt} = 187.5$. Thus $\frac{d\phi}{dt} = \frac{100}{20,000} \cdot 187.5 = .9375$ radians per second.

3.8.39 By the Law of Sines, $\frac{\sin\theta}{s} = \frac{\sin\left(\frac{3\pi}{4} - \theta\right)}{2}$, so $2\sin\theta = s \cdot \sin\left(\frac{3\pi}{4} - \theta\right) = s\left(\sin\left(\frac{3\pi}{4}\right)\cos\theta - \cos\left(\frac{3\pi}{4}\right)\sin\theta\right)$. We have

$$2\sin\theta = \frac{\sqrt{2}}{2}s(\sin\theta + \cos\theta)$$

$$2\tan\theta = \frac{\sqrt{2}}{2}s(\tan\theta + 1)$$

$$\tan\theta = \frac{(\sqrt{2}/2) \cdot s}{2 - (\sqrt{2}/2)s} = \frac{\sqrt{2}s}{4 - \sqrt{2}s}$$

$$\theta = \tan^{-1}\left(\frac{\sqrt{2}s}{4 - \sqrt{2}s}\right).$$

Thus, $\frac{d\theta}{dt} = \frac{\sqrt{2} \cdot \frac{ds}{dt}}{4 - 2\sqrt{2}s + s^2}$. When $\frac{ds}{dt} = 15$ and $s = 7.5$ we arrive at $\frac{d\theta}{dt} = 0.54$ radians per hour.

3.8.41 We have $\frac{h}{20} = \tan\theta$, so $\frac{1}{20}\frac{dh}{dt} = \sec^2\theta\frac{d\theta}{dt}$. We are given that $\frac{dh}{dt} = 5\,\text{m/s}$. At $h = -10$, we have $\tan\theta = -.5$, so $\sec^2\theta = 1 + \tan^2\theta = 1 + (.5)^2 = 1.25$. So $\frac{d\theta}{dt} = \frac{1}{20 \cdot 1.25} \cdot 5 = \frac{1}{5}$ radian per second. When $h = 20$, we have that $\tan\theta = 1$, so $\sec^2\theta = 1 + 1^2 = 2$, and thus $\frac{d\theta}{dt} = \frac{1}{20 \cdot 2} \cdot 5 = \frac{1}{8}$ radian per hour.

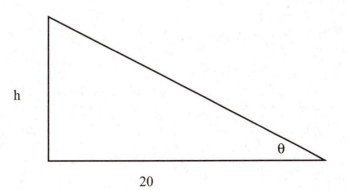

3.8.43 Let x be the distance the eastbound boat has traveled at time t and let s be the distance the northeastbound boat has traveled. Note the diagram shown. By the Law of Sines, $\dfrac{\sin\left(\frac{\pi}{2}-\theta\right)}{s}=\dfrac{\sin\left(\frac{\pi}{4}+\theta\right)}{x}$. Thus,

$$x\left(\sin\left(\frac{\pi}{2}\right)\cos\theta-\cos\left(\frac{\pi}{2}\right)\sin\theta\right)=s\left(\sin\left(\frac{\pi}{4}\right)\cos\theta+\cos\left(\frac{\pi}{4}\right)\sin\theta\right)$$

So

$$x\cos\theta=\frac{\sqrt{2}}{2}\cdot s\cdot\cos\theta+\frac{\sqrt{2}}{2}\cdot s\cdot\sin\theta,\qquad x=\frac{\sqrt{2}}{2}\cdot s+\frac{\sqrt{2}}{2}\cdot s\cdot\tan\theta,$$

and thus $\tan\theta=\dfrac{x-\frac{\sqrt{2}}{2}s}{\frac{\sqrt{2}}{2}s}=\dfrac{\sqrt{2}x-s}{s}$, and therefore $\theta=\tan^{-1}\left(\dfrac{\sqrt{2}x-s}{s}\right)$.

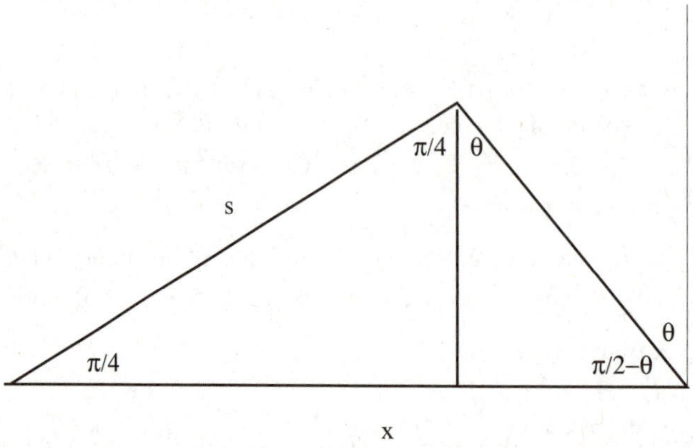

We have

$$\frac{d\theta}{dt}=\frac{1}{1+\left(\frac{\sqrt{2}x-s}{s}\right)^2}\cdot\frac{\left(\sqrt{2}\left(\frac{dx}{dt}\right)-\left(\frac{ds}{dt}\right)\right)\cdot s-\left(\sqrt{2}x-s\right)\cdot\frac{ds}{dt}}{s^2}.$$

After 30 minutes we have $\frac{dx}{dt}=12$ and $\frac{ds}{dt}=15$, and $x=6$ and $s=7.5$. Evaluating this expression at these values gives $\frac{d\theta}{dt}=0$. After 2 hours, we have $x=24$ and $s=30$. Again, evaluating $\frac{d\theta}{dt}$ at these values yields a rate of 0.

3.8.45 Let α be the angle between the line of sight to the bottom of the screen and the line of sight to the point 3 feet below where the floor and the wall meet. Note that $\cot\alpha=\frac{x}{3}$ and $\cot(\alpha+\theta)=\frac{x}{10}$, so $\alpha=\cot^{-1}\left(\frac{x}{3}\right)$ and $\alpha+\theta=\cot^{-1}\left(\frac{x}{10}\right)$. Thus, $\theta=\cot^{-1}\left(\frac{x}{10}\right)-\cot^{-1}\left(\frac{x}{3}\right)$. So

$$\frac{d\theta}{dt}=\frac{-10x'}{100+x^2}+\frac{3x'}{9+x^2},$$

and at $x=30$ feet, and with $\frac{dx}{dt}=3$ feet per second, we have $\dfrac{d\theta}{dt}=\dfrac{-30}{1000}+\dfrac{9}{909}\approx-0.0201$ radians per second.

3.9   Chapter Three Review

3.9.1

a. False. This function is not differentiable at $x=\frac{-1}{2}$. It is possible for a function to be continuous at a point and not differentiable at that point.

b. False. For example, $f(x)=x^2+3$ and $g(x)=x^2+100$ have the same derivative, but aren't the same function.

c. False. For example, $\frac{d}{dx}|e^{-x}|=\frac{d}{dx}e^{-x}=-e^{-x}\neq|-e^{-x}|$.

d. False. For example, the function $f(x) = |x|$ has no derivative at 0, but there is no vertical tangent there.

e. True. For example, a ball dropping from a high tower has acceleration due to gravity which is negative, but it is speeding up as it falls because the velocity (which is negative also) is in the same direction as the acceleration.

3.9.3

a.

$$f'(1) = \lim_{h \to 0} \frac{5(1+h)^3 + (1+h) - 6}{h} = \lim_{h \to 0} \frac{5(1 + 3h^2 + 3h + h^3) + 1 + h - 6}{h}$$

$$= \lim_{h \to 0} \frac{5 - 15h^2 + 15h + 5h^3 + h - 5}{h} = \lim_{h \to 0} \frac{5h^3 - 15h^2 + 16h}{h} = \lim_{h \to 0} (5h^2 - 15h + 16) = 16.$$

b. The tangent line at $(1,6)$ is given by $y - 6 = 16(x - 1)$.

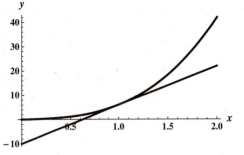

3.9.5

a.

$$f'(0) = \lim_{h \to 0} \frac{\frac{1}{2\sqrt{3h+1}} - \frac{1}{2}}{h} = \lim_{h \to 0} \frac{1 - \sqrt{3h+1}}{2\sqrt{3h+1} \cdot h} = \lim_{h \to 0} \frac{(1 - \sqrt{3h+1})(1 + \sqrt{3h+1})}{2\sqrt{3h+1} \cdot h(1 + \sqrt{3h+1})}$$

$$= \lim_{h \to 0} \frac{1 - (3h+1)}{2\sqrt{3h+1} \cdot h(1 + \sqrt{3h+1})} = \lim_{h \to 0} \frac{-3}{2\sqrt{3h+1}(1 + \sqrt{3h+1})} = \frac{-3}{4}.$$

b. The tangent line at $\left(0, \frac{1}{2}\right)$ is given by $y - \frac{1}{2} = \frac{-3}{4}(x)$.

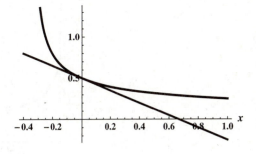

3.9.7

a. Average growth is $\frac{p(60) - p(50)}{10} = 2.7$ million people per year.

b. The curve is pretty straight between $t = 50$ and $t = 60$, so the secant line between these two points is approximately as steep as the tangent line at a point in between.

c. A reasonable estimate to the instantaneous grow rate at 1985 would be the slope of the secant line between $t = 80$ and $t = 90$. This is $\frac{p(90) - p(80)}{10} = 2.217$ million people per year.

3.9.9

a. $v(15) \approx \frac{400 - 200}{5} = 40$ meters per second.

b. Since the graph is a straight line for $t \geq 30$, $v(70) = \frac{D(90) - D(60)}{30} = \frac{1600 - 1400}{30} = \frac{20}{3}$ meters per second. The points at 60 and 90 were chosen because it is easier to detect the function values at those points using the given grid.

c. The average velocity is $\frac{D(90) - D(20)}{70} \approx \frac{1600 - 550}{70} = 15$ meters per second.

d.

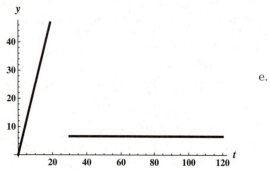

e. The parachute was deployed.

3.9.11

$$g'(x) = \lim_{h \to 0} \frac{\sqrt{2(x+h) - 3} - \sqrt{2x - 3}}{h} = \lim_{h \to 0} \frac{\sqrt{2(x+h) - 3} - \sqrt{2x - 3}}{h} \cdot \frac{\sqrt{2(x+h) - 3} + \sqrt{2x - 3}}{\sqrt{2(x+h) - 3} + \sqrt{2x - 3}}$$

$$= \lim_{h \to 0} \frac{2(x+h) - 3 - (2x - 3)}{h(\sqrt{2(x+h) - 3} + \sqrt{2x - 3})} = \lim_{h \to 0} \frac{2}{\sqrt{2(x+h) - 3} + \sqrt{2x - 3}} = \frac{2}{2\sqrt{2x - 3}} = \frac{1}{\sqrt{2x - 3}}.$$

3.9.13

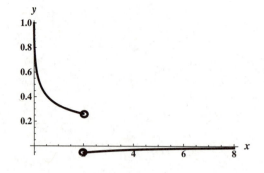

3.9.15 $f'(x) = 2x^2 + 2\pi x + 7.$

3.9.17 $f'(t) = 10t \sin t + 5t^2 \cos t.$

3.9.19 $f'(\theta) = (4 \sec^2(\theta^2 + 3\theta + 2)) \cdot (2\theta + 3).$

3.9.21 $f'(u) = \dfrac{(8u + 1)(8u + 1) - (4u^2 + u)(8)}{(8u + 1)^2} = \dfrac{64u^2 + 16u + 1 - 32u^2 - 8u}{(8u + 1)^2} = \dfrac{32u^2 + 8u + 1}{(8u + 1)^2}.$

3.9.23 $f'(\theta) = \sec^2(\sin\theta) \cdot \cos\theta$.

3.9.25 $f'(x) = 2(\sin x)\sqrt{3x-1} + 2x(\cos x)\sqrt{3x-1} + \dfrac{3x\sin x}{\sqrt{3x-1}}$.

3.9.27 $y' = \dfrac{(1+\sin x)(-y'\sin y) - (\cos y)(\cos x)}{(1+\sin x)^2}$ so that $y' = -\dfrac{\cos x\cos y}{(1+\sin x)^2 + \sin y(1+\sin x)}$.

3.9.29 $y'\sqrt{x^2+y^2} + y\cdot\dfrac{x+yy'}{\sqrt{x^2+y^2}} = 0$, and thus $y'\left(\sqrt{x^2+y^2} + \dfrac{y^2}{\sqrt{x^2+y^2}}\right) = \dfrac{-xy}{\sqrt{x^2+y^2}}$. This can be

written as $y'\left(\dfrac{x^2+2y^2}{\sqrt{x^2+y^2}}\right) = \dfrac{-xy}{\sqrt{x^2+y^2}}$, so $y' = \dfrac{-xy}{x^2+2y^2}$.

3.9.31 $y' = 9x^2 + \cos x$. At $x = 0$, $y' = 1$. So the tangent line is given by $y - 0 = 1(x-0)$, or $y = x$.

3.9.33 $y' + \dfrac{y+xy'}{2\sqrt{xy}} = 0$. At the point $(1,4)$, we have $y' + \dfrac{4+y'}{4} = 0$, so $y' = \frac{-4}{5}$. The tangent line is given by $y - 4 = \frac{-4}{5}(x-1)$.

3.9.35 We are looking for values of x so that $y'(x) = 0$. We have $y' = \sqrt{6-x} - \frac{x}{2\sqrt{6-x}}$, and this quantity is zero when $2(6-x) - x = 0$, or $12 - 3x = 0$, so when $x = 4$. So at the point $(4, 4\sqrt{2})$ there is a horizontal tangent line.

3.9.37

$$y' = \frac{1}{2}x^{-1/2}\cos\sqrt{x}$$

$$y'' = \frac{-1}{4}x^{-3/2}\cos\sqrt{x} + \frac{-1}{4}x^{-1}\sin\sqrt{x}$$

$$y''' = \frac{3}{8}x^{-5/2}\cos\sqrt{x} + \frac{1}{8}x^{-2}\sin\sqrt{x} + \frac{1}{4}x^{-2}\sin\sqrt{x} - \frac{1}{8}x^{-3/2}\cos\sqrt{x}$$

$$= \frac{3}{8}x^{-5/2}\cos\sqrt{x} + \frac{3}{8}x^{-2}\sin\sqrt{x} - \frac{1}{8}x^{-3/2}\cos\sqrt{x}.$$

3.9.39 $\dfrac{d}{dx}[x^2 f(x)] = 2xf(x) + x^2 f'(x)$.

3.9.41 $\dfrac{d}{dx}\left(\dfrac{xf(x)}{g(x)}\right) = \dfrac{(f(x) + xf'(x))g(x) - xf(x)g'(x)}{(g(x))^2}$.

3.9.43

a. $\frac{d}{dx}[f(x) + 2g(x)]_{x=3} = f'(3) + 2g'(3) = 9 + 2\cdot 9 = 27$.

b. $\dfrac{d}{dx}\left[\dfrac{xf(x)}{g(x)}\right]_{x=1} = \dfrac{g(1)(1\cdot f'(1) + f(1)) - 1\cdot f(1)\cdot g'(1)}{(g(1))^2} = \dfrac{9\cdot[7+3] - 15}{81} = \dfrac{25}{27}$.

c. $\frac{d}{dx}f(g(x^2))\big|_{x=3} = f'(g(9))\cdot g'(9)\cdot 2\cdot 3 = f'(1)\cdot 7\cdot 6 = 7\cdot 42 = 294$.

3.9.45 Let $a = 5$ and $f(x) = \tan(\pi\sqrt{3x-11})$. Note that $f'(5) = \dfrac{3\pi\sec^2(2\pi)}{2} = \dfrac{3\pi}{4}$. So $\lim\limits_{x\to 5}\dfrac{f(x) - f(5)}{x-5} =$

$\lim\limits_{x\to 5}\dfrac{\tan(\pi\sqrt{3x-11}) - 0}{x-5} = f'(5) = \dfrac{3\pi}{4}$.

3.9.47

a. The average cost is $\frac{C(3000)}{3000} = \frac{1025000}{3000} \approx \$341.67$. The marginal cost is $C'(3000) = -0.04(3000) + 400 = \$280$.

b. The average cost of producing 3000 lawnmowers is \$341.67 per mower. The cost of producing the 3001st lawnmower is approximately \$280.

3.9.49

a. The average growth rate is $\frac{p(50) - p(0)}{50} = \frac{407500 - 80000}{50} = 6550$ people per year.

b. The growth rate in 1990 is $p'(40) = -5.1(40^2) + 144 \cdot 40 + 7200 = 4800$ people per year.

3.9.51 Let x be the distance the eastbound boat has traveled, and y the distance the southbound boat has traveled. By the Pythagorean Theorem, $D^2 = x^2 + y^2$, so $2D\frac{dD}{dt} = 2x\frac{dx}{dt} + 2y\frac{dy}{dt}$, so $\frac{dD}{dt} = \frac{x \cdot x' + y \cdot y'}{D}$. We are given that $x' = 40$, $y' = 30$, and at $t = .5$ hours, we have $x = 20$, $y = 15$, and $D = 25$. Thus, $\frac{dD}{dt} = \frac{20 \cdot 40 + 30 \cdot 15}{25} = 50$ mph.

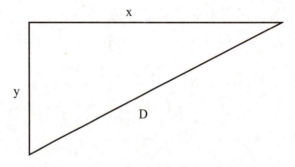

3.9.53 Let h be the elevation of the balloon, and s the length of the rope. We have $h = s\sin(65°)$, so $h' = s'\sin(65°) = -5 \cdot \sin(65°) \approx -4.53$ feet per second.

3.9.55 Let x be the distance the jet has flown since it went over the spectator. Let θ be the angle of elevation between the ground and the line from the spectator to the jet. Note that θ is also the angle pictured, and that $\cot\theta = \frac{x}{500}$. Thus, $\theta = \cot^{-1}\left(\frac{x}{500}\right)$. We are given that $x' = 450\,\text{mph} = 660\,\text{ft/sec}$. $\theta' = \frac{-x'}{500 \cdot \left(1 + \left(\frac{x}{500}\right)^2\right)} = \frac{-500x'}{250,000 + x^2}$. After 2 seconds, $x = 1320$ feet, so at this time $\theta' = \frac{-500 \cdot 660}{250,000 + (1320)^2} \approx -0.166$ radians per second.

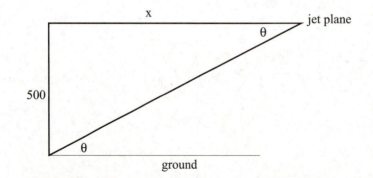

Chapter 4

Applications of the Derivative

4.1   Maxima and Minima

4.1.1 A number $M = f(c)$ where $c \in [a,b]$ with the property that $f(x) \leq M$ for all $x \in [a,b]$ is an absolute maximum for f on $[a,b]$, and a number $m = f(d)$ where $d \in [a,b]$ with the property that $f(x) \geq m$ for all $x \in [a,b]$ is an absolute minimum for f on $[a,b]$.

4.1.3 The function must be a continuous function defined on a closed interval.

4.1.5

The function shown has no absolute minimum on $[0,3]$ because $\lim\limits_{x \to 0^-} f(x) = -\infty$. It has an absolute maximum near $x = 1$ and a local minimum near $x = 2.5$.

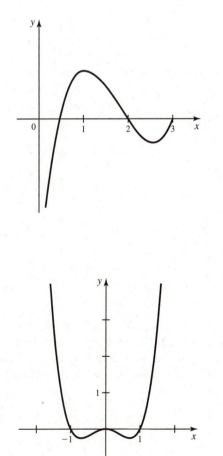

4.1.7

Note the existence of a horizontal tangent line at $x = 0$ where the maximum occurs.

107

4.1.9 First find all the critical points by seeking all points x in the domain of f so that $f'(x) = 0$ or $f'(x)$ doesn't exist. Now compare the y-values of all of these points, together with the y-values of the endpoints. The largest y-value from among these is the maximum, and the smallest is the minimum.

4.1.11 $y = h(x)$ has an absolute maximum at $x = b$ and an absolute minimum at $x = c_2$.

4.1.13 $y = g(x)$ has no absolute maximum, but has an absolute minimum at $x = a$.

4.1.15 $y = f(x)$ has an absolute maximum at $x = b$ and an absolute minimum at $x = a$. It has local maxima at $x = p$ and $x = r$, and local minima at $x = q$ and $x = s$.

4.1.17 $y = g(x)$ has an absolute minimum at $x = b$ and an absolute maximum at $x = p$. It has local maxima at $x = p$ and $x = r$. It has a local minimum at $x = q$.

4.1.19

Note the horizontal tangent lines at 1 and 2, and the minimum at 0 and the maximum at 4.

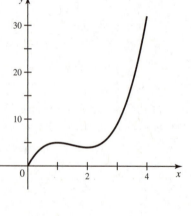

4.1.21

Note the horizontal tangent line at $x = 2$, and the "corners" at $x = 1$ and $x = 3$. Also note the absolute maximum at $x = 3$ and the absolute minimum at $x = 4$.

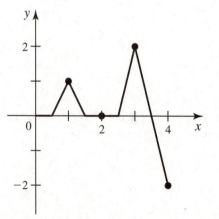

4.1.23

a. $f'(x) = 6x - 4$, which is zero when $x = 2/3$.

b. At $x = 2/3$ there is a local minimum.

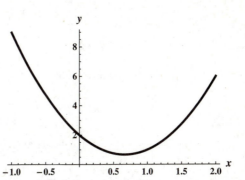

4.1.25

a.

$$f'(x) = \frac{4(x^2 + 1) - (4x - 3)(2x)}{(x^2 + 1)^2}$$

$$= \frac{-2(2x^2 - 3x - 2)}{(x^2 + 1)^2}$$

Thus $f'(x) = 0$ exactly when

$$0 = 2x^2 - 3x - 2 = (2x + 1)(x - 2),$$

so when $x = -1/2$ or $x = 2$.

b. From the graph, there is a minimum at $x = -1/2$ and a maximum at $x = 2$.

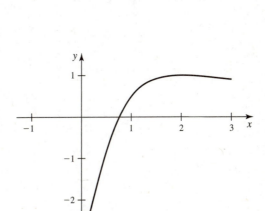

4.1.27

a. $f'(x) = -2\sin 2x + 2\sqrt{3}\cos 2x$; thus $f'(x) = 0$ when $\tan 2x = \sqrt{3}$, thus $2x = \pi/3, 4\pi/3$; thus on $[0, \pi]$, for $x = \pi/6, 2\pi/3$.

b. From the graph, there is a maximum at $x = \pi/6$ and a minimum at $x = 2\pi/3$.

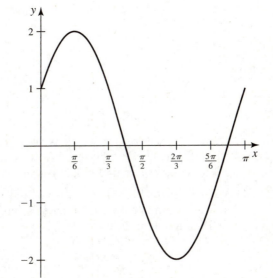

4.1.29

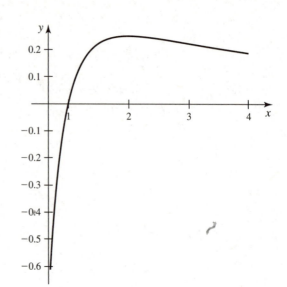

a. $f'(x) = -\frac{1}{x^2} + \frac{2}{x^3} = \frac{2-x}{x^3}$ so that $f'(x) = 0$ when $x = 2$.

b. From the graph, there is a maximum at $x = 2$.

4.1.31

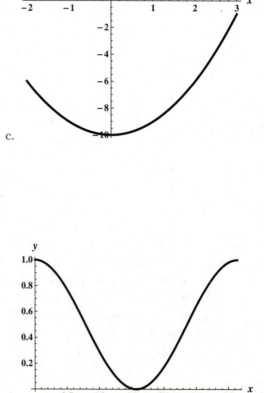

a. $f'(x) = 2x$, which is zero for $x = 0$.

b. We have that $f(-2) = -6$, $f(0) = -10$, and $f(3) = -1$, so the maximum value of f on this interval is -1 and the minimum is -10.

c.

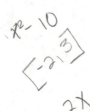

4.1.33

$cos^2(x)$

a. $f'(x) = -2\cos x \sin x$, which is zero for $x = 0$, $x = \pi/2$, and $x = \pi$. Since there are endpoints at $x = 0$ and $x = \pi$, only $(\pi/2, 1)$ is a critical point.

b. We have that $f(0) = 1$, $f(\pi/2) = 0$, and $f(\pi) = 1$, so the maximum value of f on this interval is 1 and the minimum is 0.

c.

4.1.35

a. $f'(x) = 3\cos 3x$, which is zero when $3x = \ldots - \pi/2, \pi/2, 3\pi/2, \ldots$, so when $x = \ldots - \pi/6, \pi/6, \pi/2, \ldots$. The only such values on the given interval are $x = -\pi/6$ and $x = \pi/6$.

b. We have $f(-\pi/4) = -\sqrt{2}/2 \approx -7.07$, $f(-\pi/6) = -1$, $f(\pi/6) = 1$, and $f(\pi/3) = 0$, so the absolute maximum of f is 1 and the absolute minimum is -1.

c.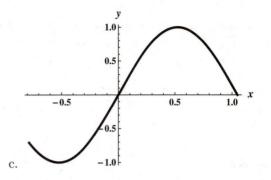

4.1.37

a.
$$f'(x) = \frac{x^2(4) - (4x - 3)(2x)}{x^4} = \frac{6x - 4x^2}{x^4}$$

so that $f'(x) = 0$ for $x = 0, 3/2$. Only $x = 3/2$ lies inside the given interval.

b. $f(1) = 1$, $f(4) = 13/16$, and $f(3/2) = 4/3$ so $4/3$ is the maximum value and $13/16$ the minimum value of f on $[1, 4]$.

c.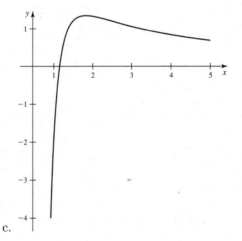

4.1.39

a.
$$f'(x) = \frac{\sqrt{4 - x^2} \cdot 1 - x \cdot \frac{1}{2}(4 - x^2)^{-1/2} \cdot (-2x)}{4 - x^2} =$$

$$\frac{\sqrt{4 - x^2} \cdot 1 - x \cdot \frac{1}{2}(4 - x^2)^{-1/2} \cdot (-2x)}{4 - x^2} \cdot \frac{\sqrt{4 - x^2}}{\sqrt{4 - x^2}} =$$

$$\frac{(4 - x^2) + x^2}{(4 - x^2)^{3/2}} = \frac{4}{(4 - x^2)^{3/2}}.$$

So there are no critical points.

b. There are no critical points, and the endpoints aren't included in the given interval. From this much information, it isn't possible to tell whether or not there are extrema, and if there are any, what they are. The graph pictured indicates that there aren't any extrema for this function on this interval.

c.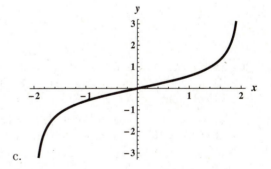

4.1.41 The stone will reach its maximum height when its velocity is zero, which occurs at the only critical point for this inverted parabola. We have that $v(t) = s'(t) = -32t + 64$, which is zero when $t = 2$. The height at this time is $s(2) = 256$, the maximum height.

4.1.43

a. Note that $P(n) = 50n - .5n^2 - 100$, so $P'(n) = 50 - n$, which is zero when $n = 50$. It is clear that this is a maximum, since the graph of P is an inverted parabola. The guide should take 50 people on the tour.

b. Given a domain of $[0, 45]$, since the only critical point is not in the domain, the maximum must occur at an endpoint. Since $P(0) = -100$, and $P(45) = \$1137.50$, he should take 45 people on the tour.

4.1.45

a. False. The derivative $f'(x) = \frac{1}{2\sqrt{x}}$ is never zero, and the function has no critical points.

b. False. For example, the function $f(x) = \begin{cases} \sin x & \text{if } -5 \le x \le 0, \\ -8 & \text{if } 0 < x \le 5. \end{cases}$ is not continuous on $[-5, 5]$, but has an absolute maximum of 1.

c. False. For example, the function $f(x) = (x-2)^3$ satisfies $f'(2) = 0$, but it has neither a maximum nor a minimum at $x = 2$.

d. True. This follows from the theorems in this section.

e. False. It could be the case that f doesn't exist at $x = 3$ either (which is a way of saying that 3 isn't in the domain of f.) It is good to remember that $x = c$ is the x-value of a critical point exactly when c is in the domain of f, and either $f'(c) = 0$ or $f'(c)$ doesn't exist.

4.1.47

c.

a. $f(x) = x^4 + 4x^3 - 8x^2$, so $f'(x) = 4x^3 + 12x^2 - 16x = 4x(x^2 + 3x - 4) = 4x(x+4)(x-1)$. Thus the critical points of $f(x)$ in $[-5, 2]$ are $x = -4, 0, 1$.

b. $f(-5) = -75$, $f(-4) = -128$, $f(0) = 0$, $f(1) = -3$, $f(2) = 16$, so the absolute minimum of $f(x)$ on $[-5, 2]$ is -128 at $x = -4$; its absolute maximum is 16 at $x = 2$.

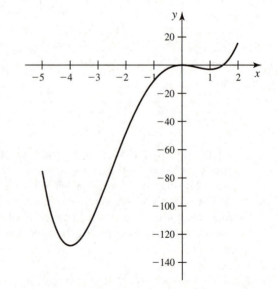

4.1.49

c.

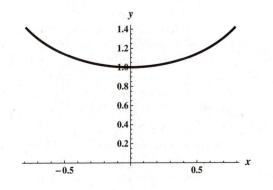

a. $f'(x) = \sec x \tan x$ which is zero when $\tan x = 0$ (since $\sec x$ is never zero.) So we are looking for where $\frac{\sin x}{\cos x} = 0$, which is when $\sin x = 0$, which is at $x = 0$

b. $f(-\pi/4) = \sqrt{2} = f(\pi/4)$ and $f(0) = 1$. So the absolute maximum for f is $\sqrt{2}$ and the absolute minimum is 1.

4.1.51

a.

$$f'(x) = \frac{\sqrt{x-4} \cdot 1 - x \cdot \frac{1}{2\sqrt{x-4}} \cdot 1}{x-4}$$

$$= \frac{\sqrt{x-4} \cdot 1 - x \cdot \frac{1}{2\sqrt{x-4}} \cdot 1}{x-4} \cdot \frac{2\sqrt{x-4}}{2\sqrt{x-4}}$$

$$= \frac{2x - 8 - x}{2(x-4)^{3/2}} = \frac{x-8}{2(x-4)^{3/2}}.$$

This expression is 0 when $x = 8$. So $(8, 4)$ is the only critical point.

b. $f(6) = 3\sqrt{2} = f(12)$, and $f(8) = 4$. Note that $3\sqrt{2} \approx 4.24 > 4$. So the absolute maximum of f on this interval is $3\sqrt{2}$, and the absolute minimum is 4.

c.

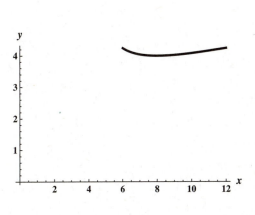

4.1.53 $f'(x) = \sqrt{x-a} + \frac{x}{2\sqrt{x-a}} = \frac{2x - 2a + x}{2\sqrt{x-a}} = \frac{3x - 2a}{2\sqrt{x-a}}$. This expression is zero when $x = \frac{2a}{3}$, however, that number is not in the domain of f if $a > 0$. However, if $a < 0$, then $\frac{2a}{3}$ is in the domain, and thus gives a critical point.

4.1.55 $f'(x) = x^4 - a^4$, which is zero when $x^4 = a^4$, or $|x| = a$. So there are critical points at $x = a$ and at $x = -a$.

4.1.57

a. $f'(\theta) = 2\cos\theta - \sin\theta$, which is zero when $\tan\theta = 2$. So one critical point occurs at $\theta = \tan^{-1}(2) \approx 1.107$. And since the tangent function is periodic with period π, there are will also be solutions at this number plus or minus integer multiples of π. On the given interval, these are located at approximately $1.107 - 2\pi \approx -5.176$, at $1.107 - \pi \approx -2.034$, and at $1.107 + \pi \approx 4.249$.

b. From the graph, it appears that there is a local minimum at about $\theta = -2.034$ and at $\theta = 4.249$, and there is a local maximum at about $\theta = -5.176$, and at about $\theta = 1.107$.

c. From the graph, it appears that the local minimum at about $\theta = -2.034$ is also an absolute minimum, as is the one at $\theta = 4.249$. The local maximum at about $\theta = -5.176$, and at about $\theta = 1.107$ are also absolute maximums. The value of the absolute maximum appears to be about 2.24 and the value of the absolute minimum appears to be about -2.24.

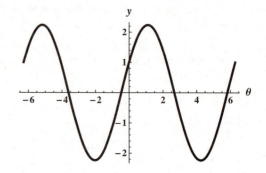

4.1.59

a.

$$f'(x) = (x-3)^{5/3} + (x+2) \cdot \frac{5}{3}(x-3)^{2/3}$$

$$= \frac{(x-3)^{2/3}}{3}(3x - 9 + 5x + 10)$$

$$= \frac{(x-3)^{2/3}}{3}(8x + 1).$$

This is zero when $x = 3$ and when $x = \frac{-1}{8}$. There are critical points at $x = \frac{-1}{8}$ and at $x = 3$.

b. From the graph, it appears that there is a local minimum of about -12.52 at $x = \frac{-1}{8}$.

c. The local minimum mentioned above is also an absolute minimum. The absolute maximum occurs at the left endpoint $x = -5$, where the value of f is about 51.23.

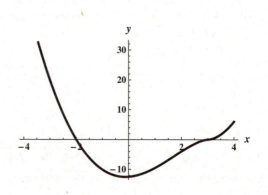

4.1.61

a.

$$h'(x) = \frac{(x^2 + 2x - 3)(-1) - (5 - x)(2x + 2)}{(x^2 + 2x - 3)^2} = \frac{x^2 - 10x - 7}{(x^2 + 2x - 3)^2}.$$

By the quadratic formula, the numerator is zero (making the quotient zero) when $x = \frac{10 \pm \sqrt{100 - 4(-7)}}{2} = 5 \pm \frac{1}{2}\sqrt{128} = 5 \pm 4\sqrt{2}$. Note that $5 + 4\sqrt{2}$ isn't in the domain, so the only critical point is at $x = 5 - 4\sqrt{2}$.

b. From the graph, it appears that the one critical point mentioned above yields a local maximum.

c. The function has no absolute maximum and no absolute minimum on the given interval.

4.1.63 Note that

$$g(x) = \begin{cases} 3 - x + 2x + 2 = x + 5 & \text{if } -2 \le x \le -1, \\ 3 - x - 2x - 2 = 1 - 3x & \text{if } -1 \le x \le 3. \end{cases}$$

There is an absolute maximum of 4 and an absolute minimum of -8. The absolute maximum occurs at $x = -1$, and the absolute minimum occurs at a $x = 3$.

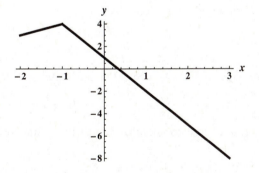

4.1.65

a. Since distance is rate times time, the time will be distance over rate. The swim distance is given by $\sqrt{2500 + x^2}$ meters, so the time for swimming is $\frac{\sqrt{2500 + x^2}}{2}$. For running, the distance is $50 - x$, so the time is $\frac{50-x}{4}$. Thus we have $T(x) = \dfrac{\sqrt{2500 + x^2}}{2} + \dfrac{50 - x}{4}$.

b.

d.

$$T'(x) = \frac{1}{2} \cdot \frac{1}{2} \left(x^2 + 2500\right)^{-1/2} \cdot 2x - \frac{1}{4}$$

$$= \frac{x}{2\sqrt{x^2 + 2500}} - \frac{1}{4}.$$

This expression is zero when $\frac{x^2}{x^2+2500} = \frac{1}{4}$, so when $4x^2 = x^2 + 2500$, which occurs when $x^2 = \frac{2500}{3}$. So $x = \sqrt{\left(\frac{2500}{3}\right)} \approx 28.868$.

c. $T(0) = 37.5$, $T(28.868) \approx 34.151$, and $T(50) = 25\sqrt{2} \approx 35.355$. The absolute minimum occurs at the only critical point. The minimal crossing time is approximately 34.151 seconds.

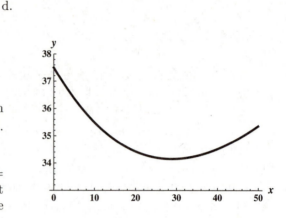

4.1.67

a. Note that since there is a local extreme value at 2 for f and since f is differentiable everywhere, we must have $f'(2) = 0$. $g(2) = 2f(2) + 1 = 1$. $h(2) = 2f(2) + 2 + 1 = 3$. $g'(2) = 2 \cdot f'(2) + f(2) = 0$. $h'(2) = 2f'(2) + f(2) + 1 = 1$.

b. h doesn't, since its derivative isn't zero at $x = 2$. However g might: for example, if $f(x) = (x - 2)^2$ then $g(x) = x(x - 2)^2 + 1$ has a local minimum at $x = 2$.

4.1.69

a. Because of the symmetry about the y-axis for an even function, a minimum at $x = c$ will correspond to a minimum at $x = -c$ as well.

b. Because of the symmetry about the origin, a minimum at $x = c$ will correspond to a maximum at $x = -c$. It is helpful to think about the symmetry about the origin as being the result of flipping about the y-xis and then flipping about the x-axis.

4.1.71

a. If $f(c)$ is a local maximum, then when x is near c but not equal to c, $f(c) \geq f(x)$, so $f(x) - f(c) \leq 0$.

b. When x is near to c but a little bigger than c, $x - c > 0$. So in this case, $\frac{f(x)-f(c)}{x-c} \leq 0$, since the numerator is negative (or 0) and the denominator is positive. Thus, $\lim\limits_{x \to c^+} \dfrac{f(x) - f(c)}{x - c} = f'(c) \leq 0$

c. When x is near to c but a little smaller than c, $x - c < 0$. So in this case, $\frac{f(x)-f(c)}{x-c} \geq 0$, since the numerator is negative (or 0) and the denominator is negative, making the quotient positive (or 0.) Thus, $\lim\limits_{x \to c^-} \dfrac{f(x) - f(c)}{x - c} = f'(c) \geq 0$.

d. From the above, we have that $f'(c) \leq 0$ and $f'(c) \geq 0$, so $f'(c) = 0$.

4.2   What Derivatives Tell Us

4.2.1 If f' is positive on an interval, f is increasing on that interval. If f' is negative on an interval, f is decreasing on that interval.

4.2.3

One such example is $f(x) = x^3$ at $x = 0$.

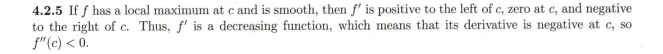

4.2.5 If f has a local maximum at c and is smooth, then f' is positive to the left of c, zero at c, and negative to the right of c. Thus, f' is a decreasing function, which means that its derivative is negative at c, so $f''(c) < 0$.

4.2.7 An inflection point is a point on the graph of a function where the concavity changes. Thus, if $(c, f(c))$ is an inflection point, either $f''(x) < 0$ for x a little less than c and $f''(x) > 0$ for x a little bigger than c, or vice versa.

4.2.9

Yes, for example, consider $f(x) = x^3$ on the interval $[1, 2]$. It is above the x axis, increasing, and concave up.

4.2.11

Such a function would be decreasing until $x = 2$, then increasing until $x = 5$, and then decreasing again after that.

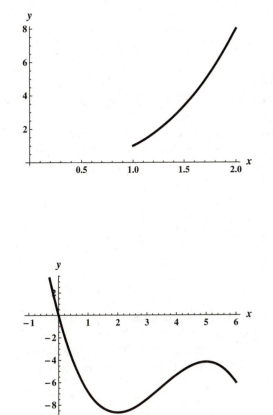

4.2.13

Such a function has extrema at 0, 2, and 4, where the y value is zero. The function should never go below the x axis.

4.2.15

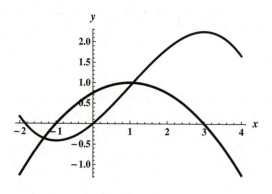

4.2.17

$f'(x) = -2x$, which is zero exactly when $x = 0$. On $(-\infty, 0)$ we note that $f' > 0$, so that f is increasing on this interval. On $(0, \infty)$, we note that $f' < 0$, so f is decreasing on this interval.

4.2.19

$f'(x) = 2(x-1)$, which is zero exactly when $x = 1$. On $(-\infty, 1)$ we note that $f' < 0$, so that f is decreasing on this interval. On $(1, \infty)$, we note that $f' > 0$, so f is increasing on this interval.

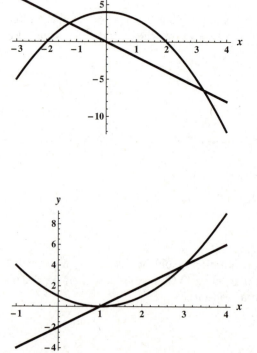

4.2.21

$f'(x) = 1 - 2x$, which is 0 when $x = 1/2$. On $(-\infty, 1/2)$ $f' > 0$ so f is increasing on this interval, while on $(1/2, \infty)$ $f' < 0$, so f is deceasing on this interval.

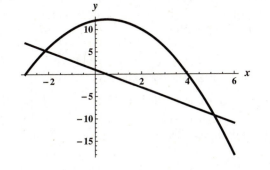

4.2.23 $f'(x) = -9\sin 3x$, which is 0 for $3x = -3\pi, -2\pi, -\pi, 0, \pi, 2\pi,$ and 3π, which corresponds to $x = -\pi, -2\pi/3, -\pi/3, 0, \pi/3, 2\pi/3$ and π. Note that $f'(-5\pi/6) = 9 > 0$, $f'(-\pi/2) = -9 < 0$, $f'(-\pi/6) = 9 > 0$, $f'(\pi/6) = -9 < 0$, $f'(\pi/2) = 9 > 0$, and $f'(5\pi/6) = -9 < 0$. Thus f is increasing on $(-\pi, -2\pi/3)$, on $(-\pi/3, 0)$, and on $(\pi/3, 2\pi/3)$, while f is decreasing on $(-2\pi/3, -\pi/3)$, on $(0, \pi/3)$, and on $(2\pi/3, \pi)$.

4.2.25 $f'(x) = \frac{4}{3}x^{1/3}$, which is 0 only for $x = 0$. Note that $f'(-1) = -4/3 < 0$ and $f'(1) = 4/3 > 0$, so f is decreasing on $(-\infty, 0)$ and increasing on $(0, \infty)$.

4.2.27 $f'(x) = -60x^4 + 300x^3 - 240x^2 = -60x^2(x^2 - 5x + 4) = -60x^2(x-4)(x-1)$. This is 0 for $x = 0$, $x = 1$, and $x = 4$. Note that $f'(-1) = -600 < 0$, $f'(1/2) \approx -26.25 < 0$, $f'(2) = 480 > 0$, and $f'(5) = -6000 < 0$. Thus f is increasing on $(1, 4)$ and is decreasing on $(-\infty, 1)$ and on $(4, \infty)$.

4.2.29

a. $f'(x) = 2x$, so $x = 0$ is the only critical point.

b. Note that $f' < 0$ for $x < 0$ and $f' > 0$ for $x > 0$, so f has a local minimum of $f(0) = 3$ at $x = 0$.

c. Note that $f(-3) = 12$, $f(0) = 3$ and $f(2) = 7$, so the absolute maximum is 12 and the absolute minimum is 3.

4.2.31

a. $f'(x) = x \cdot \frac{1}{2}(9 - x^2)^{-1/2} \cdot -2x + \sqrt{9 - x^2} \cdot 1 = \frac{9 - 2x^2}{\sqrt{9 - x^2}}$, which exists everywhere on $(-3, 3)$ and is zero only for $x = \pm\frac{3}{\sqrt{2}}$, so those are the only critical points.

b. Note that $f'(-2.5) < 0$ and $f'(0) > 0$, and $f'(2.5) < 0$, so f has a local minimum of $f(-3/\sqrt{2}) = -4.5$ and a local maximum of $f(3/\sqrt{2}) = 4.5$.

c. Note that $f(-3) = 0 = f(3)$. So the absolute maximum is 4.5 at $x = 3/\sqrt{2}$ and the absolute minimum is -4.5 at $x = -3/\sqrt{2}$.

4.2.33

a. $f'(x) = x^{2/3} + (x - 4) \cdot \frac{2}{3}x^{-1/3} = \frac{5x - 8}{3x^{1/3}}$, which is undefined at $x = 0$ and is 0 at $x = 8/5$. So these are the two critical points.

b. Note that $f'(-1) > 0$ and $f'(1) < 0$, and $f'(2) > 0$ so f has a local maximum at $x = 0$ of $f(0) = 0$ and a local minimum at $x = 8/5$ of $\frac{-48}{5 \cdot 5^{2/3}} \approx -3.28$.

c. Note that $f(-5) = -9\sqrt[3]{25} \approx -26.32$, $f(0) = 0$, and $f(5) = \sqrt[3]{25} \approx 2.92$, so the absolute maximum of f on $[-5, 5]$ is $\sqrt[3]{25}$ and the absolute minimum is $-9\sqrt[3]{25}$.

4.2.35 $f'(x) = -6x + 2$, so the only critical point is at $x = 1/3$. The domain of f is $(-\infty, \infty)$, and f is everywhere continuous. Thus $1/3$ is the only critical point in the domain. Since $f'(x) > 0$ for $x < 1/3$ and $f'(x) < 0$ for $x > 1/3$, there is a local maximum at $x = 1/3$. By Theorem 4.5, there is also an absolute maximum there, with $f(1/3) = -14/3$.

4.2.37 Note that A is continuous on $(0, \infty)$. $A'(r) = \dfrac{-24}{r^2} + 4\pi r = \dfrac{4\pi r^3 - 24}{r^2}$, which is 0 for $r = \sqrt[3]{6/\pi}$, so there is only one critical point on the stated interval. Note that $A' < 0$ on $(0, \sqrt[3]{6/\pi})$ and $A' > 0$ on $(\sqrt[3]{6/\pi}, \infty)$, so there is a local minimum of $A(\sqrt[3]{6/\pi}) = 6\sqrt[3]{36\pi}$. The local minimum mentioned above is an absolute minimum. There is no absolute maximum, since A is unbounded as $r \to \infty$.

4.2.39

The function sketched should be increasing and concave up everywhere.

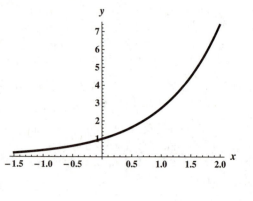

4.2.41

The function sketched should be decreasing everywhere, concave down for $x < 0$, and concave up for $x > 0$.

4.2.43 $f'(x) = 20x^3 - 60x^2$, and $f''(x) = 60x^2 - 120x = 60x(x - 2)$. This is 0 for $x = 0$ and for $x = 2$. Note that $f''(-1) > 0$, $f''(1) < 0$, and $f''(3) > 0$. So f is concave up on $(-\infty, 0)$, concave down on $(0, 2)$, and concave up on $(2, \infty)$. There are inflection points at $x = 0$ and $x = 2$.

4.2.45

$$f'(x) = \frac{(t+3) - (t-2)}{(t+3)^2} = \frac{5}{(t+3)^2}$$
$$f''(x) = \frac{-10}{(t+3)^3}$$

$f''(x) > 0$ for $x < -3$ and $f''(x) < 0$ for $x > -3$ so that f is concave up on $(-\infty, -3)$ and concave down on $(-3, \infty)$. Since neither $f'(x)$ nor $f''(x)$ is ever zero, there are no critical points and no inflection points.

4.2.47

$$f'(x) = \frac{2x(x^2+1) - 2x(x^2-1)}{(x^2+1)^2} = \frac{4x}{(x^2+1)^2}$$

$$f''(x) = \frac{(x^2+1)^2(4) - 16x^2(x^2+1)}{(x^2+1)^4} = \frac{4 - 12x^2}{(x^2+1)^3}$$

$f'(x) = 0$ only at $x = 0$. $f''(x) = 0$ for $x = \pm\sqrt{3}/3$, so there are points of inflection there. Since $f''(x) < 0$ on $(-\infty, -\sqrt{3}/3)$, $f''(x) > 0$ on $(-\sqrt{3}/3, \sqrt{3}/3)$, and $f''(x) < 0$ on $(\sqrt{3}/3, \infty)$, we see that f is concave up on $(-\sqrt{3}/3, \sqrt{3}/3)$ and concave down on $(-\infty, -\sqrt{3}/3) \cup (\sqrt{3}/3, \infty)$.

4.2.49 $g'(t) = 15t^4 - 120t^3 + 240t^2$, and $g''(t) = 60t^3 - 360t^2 + 480t = 60t(t-2)(t-4)$. Note that g'' is 0 for $t = 0, 2,$ and 4. Note also that $g'' < 0$ on $(-\infty, 0)$ and on $(2, 4)$, so g is concave down on those intervals, while $g'' > 0$ on $(0, 2)$ and on $(4, \infty)$, so g is concave up there. There are inflection points at $t = 0, 2,$ and 4.

4.2.51 $f'(x) = -2x$, so $x = 0$ is a critical point. $f''(x) = -2$, so $f''(0) = -2$ and the critical point yields a local maximum.

4.2.53 $f'(x) = 6x^2 - 6x = 6x(x-1)$, so $x = 0$ and $x = 1$ are critical points. $f''(x) = 12x - 6$, so $f''(1) = 6 > 0$, so the critical point at $x = 1$ yields a local minimum. Also, $f''(0) = -6 < 0$, so the critical point at 0 yields a local maximum.

4.2.55

$$f'(x) = -\frac{1}{x^2} + \frac{9}{x^4} = \frac{9 - x^2}{x^4}$$

$$f''(x) = \frac{x^4(-2x) - 4x^3(9 - x^2)}{x^8} = \frac{2(x^2 - 18)}{x^5}$$

There are critical points where $f'(x) = 0$, i.e. at $x = \pm 3$. At the critical points, we have

$$f''(-3) = \frac{2(9 - 18)}{(-3)^5} > 0, \qquad f''(3) = \frac{2(9 - 18)}{3^5} < 0$$

so that there is a local minimum at $x = -3$ and a local maximum at $x = 3$.

4.2.57

a. True. $f'(x) > 0$ implies that f is increasing, and $f''(x) < 0$ implies that f' is decreasing. So f is increasing, but at a decreasing rate.

b. False. In fact, if $f'(c)$ exists and isn't zero, then there isn't any kind of local extrema at $x = c$.

c. True. In fact, if two functions differ by a constant, then all of their derivatives are the same.

d. False. For example, consider $f(x) = x$ and $g(x) = x - 10$. Both are increasing, but $f(x)g(x) = x^2 - 10x$ is decreasing on $(-\infty, 5)$.

e. False. A continuous function with two local maxima must have a local minimum in between.

4.2.59

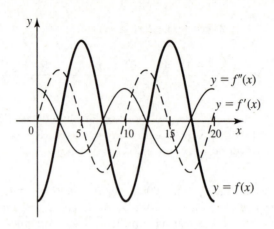

4.2.61 The graphs match as follows: (a) – (f) – (g);   (b) – (e) –(i);   (c) – (d) –(h). Note that (a) is always increasing, so its derivative must be always positive, and (f) switches from decreasing to increasing at 0, so its derivative must be negative for $x < 0$ and positive for $x > 0$. Note that (b) has three extrema where there are horizontal tangent lines, so its derivative must cross the x-axis three times, and (e) has two extrema, so its derivative must cross the x-axis two times.

4.2.63

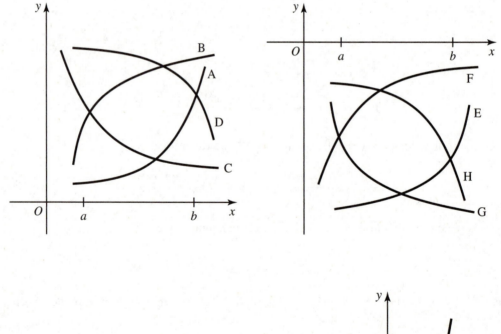

4.2.65

The graph sketched must have a flat tangent line at $x = -3/2$, $x = 0$, and $x = 1$, and must contain the points $(-2, 0)$, $(0, 0)$, and $(1, 0)$. The example to the right is only one possible such graph.

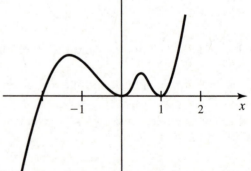

4.2.67

a. f is increasing on $(-2, 0)$ and on $(0, 2)$. It is decreasing on $(-3, -2)$.

b. There are critical points of f at $x = -2$ and at $x = 0$. There is a local minimum at $x = -2$ and no extrema at $x = 0$.

c. There are inflection points of f at $x = -1$ and at $x = 0$.

d. f is concave up on $(-3, -1)$ and on $(0, 2)$, while it is concave down on $(-1, 0)$.

  e.   f.

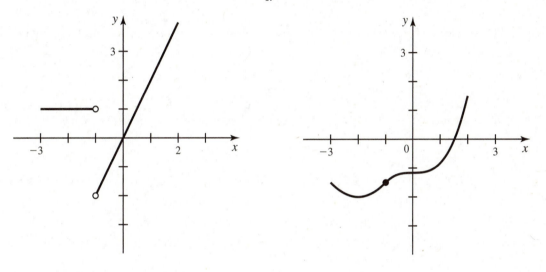

4.2.69 $f'(x) = x^3 - 5x^2 - 8x + 48 = (x-4)^2(x+3)$. (This can be obtained by using trial-and-error to determine that $x = 4$ is a root, and then using long division of polynomials to see that $f'(x) = (x - 4)(x^2 - x - 12)$.) Note that $f''(x) = 3x^2 - 10x - 8$, so $f''(-3) = 49 > 0$ and $f''(4) = 0$. So there is a local minimum at $x = -3$, but the test is inconclusive for $x = 4$. The first derivative test shows that there is neither a maximum nor a minimum at $x = 4$.

4.2.71 $f'(x) = 3x^2 + 4x + 4$, which is never 0. (Note that the discriminant $4^2 - (4)(3)(4) < 0$, so this quadratic has no real roots.) So there are no critical points.

4.2.73

a. $E = \frac{dD}{dp} \cdot \frac{p}{D} = -10\frac{p}{500-10p} = \frac{p}{p-50}$.

b. $E = \frac{12}{12-50} \cdot .045 = -1.42\%$.

c. If $D(p) = a - bp$, then $E(p) = -b \cdot \frac{p}{a-bp} = \frac{bp}{bp-a}$. So $E'(p) = \frac{(bp - a)(b) - bp(b)}{(bp - a)^2} = \frac{-ab}{(bp - a)^2}$, which is less than 0 for $a, b > 0$ and $p \neq a/b$.

d. If $D(p) = \frac{a}{p^b}$, then $E(p) = \frac{-ab}{p^{b+1}} \cdot \frac{p}{a/p^b} = -b$.

4.2.75

a. $\lim_{t\to\infty} \frac{300t^2}{t^2 + 30} \cdot \frac{1/t^2}{1/t^2} = \lim_{t\to\infty} \frac{300}{1 + (30/t^2)} = 300$.

b. Note that $P'(t) = \dfrac{(t^2 + 30)(600t) - 300t^2(2t)}{(t^2 + 30)^2} = \dfrac{18000t}{(t^2 + 30)^2}$. We want to maximize this, so we compute

its derivative $P''(t) = \dfrac{(t^2 + 30)^2 \cdot 18000 - 18000t \cdot 2(t^2 + 30) \cdot 2t}{(t^2 + 30)^4} = \dfrac{54000(10 - t^2)}{(t^2 + 30)^3}$. This is 0 for

$t = \sqrt{10}$, and an analysis of $P''(t)$ reveals that $P''(t) > 0$ for $t < \sqrt{10}$ and $P''(t) < 0$ for $t > \sqrt{10}$ so there is a local maximum for $P'(t)$ at $t = \sqrt{10}$.

c. Following the outline from the previous problem, we see that

$$P'(t) = \frac{2bKt}{(t^2 + b)^2}, \text{ and } P''(t) = \frac{2bK(b - 3t^2)}{(t^2 + b)^3}$$

$P''(t)$ is 0 for $t = \sqrt{b/3}$, and the first derivative test reveals that this is a local maximum.

4.2.77

a. $f'(x) = 3x^2 + 2ax + b$, and $f''(x) = 6x + 2a$, which is 0 only for $x = \frac{-a}{3}$. Note that the sign of $f''(x)$ is different for $x < -a/3$ and $x > -a/3$, so this does represent an inflection point.

b.

$$f(x^*) - f(x^* + x) = f(-a/3) - f(-a/3 + x) = (-a/3)^3 + a(-a/3)^2 + b(-a/3) + c -$$
$$((-a/3 + x)^3 + a(-a/3 + x)^2 + b(-a/3 + x) + c)$$
$$= (-a/3)^3 + a(-a/3)^2 + b(-a/3) + c - (-a/3)^3 - 3(-a/3)^2 x$$
$$- 3(-a/3)x^2 - x^3 - a(-a/3)^2 - 2a(-a/3)x - ax^2 - b(-a/3) - bx - c$$
$$= -x^3 + \left(\frac{a^2}{3} - b\right)x.$$

Also,

$$f(x^* - x) - f(x^*) = f(-a/3 - x) - f(-a/3)$$
$$= (-a/3)^3 + 3(-a/3)^2(-x) + 3(-a/3)(-x)^2 + (-x)^3 + a(-a/3)^2 + 2a(-a/3)(-x) +$$
$$a(-x)^2 + b(-a/3) + b(-x) + c - (-a/3)^3 - a(-a/3)^2 - b(-a/3) - c$$
$$= -x^3 + \left(\frac{a^2}{3} - b\right)x.$$

Thus the two expressions are the same for all x.

4.2.79

a. $f(-x) = \dfrac{1}{((-x)^2)^n + 1} = \dfrac{1}{x^{2n} + 1} = f(x)$, so f is even.

b. Note that $f(\pm 1) = \frac{1}{((-1)^2)^n + 1} = \frac{1}{2}$, for all n.

c.

$$f'(x) = -(x^{2n} + 1)^{-2}(2nx^{2n-1}) = -2nx^{2n-1}(x^{2n} + 1)^{-2}$$

$$f''(x) = -2nx^{2n-1} \cdot -2(x^{2n} + 1)^{-3} \cdot 2nx^{2n-1} + (x^{2n} + 1)^{-2} \cdot -2n(2n - 1)x^{2n-2}$$
$$= \frac{-2nx^{2n-2}((-2n - 1)x^{2n} + 2n - 1)}{(x^{2n} + 1)^3}$$

This is 0 when $x = 0$, and when $x = \pm \sqrt[2n]{\dfrac{2n-1}{2n+1}}$. An analysis of the sign of f'' shows that there is no sign change at $x = 0$.

d.

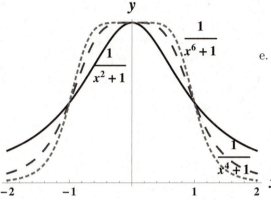

e. As n increases, the inflection points move further away from the y-axis, and closer to the point $(1, 1/2)$, although this movement is slight. The graphs become steeper and more "box-like."

4.2.81 $f'(x) = 4x^3 + 3ax^2 + 2bx + c$, and $f''(x) = 12x^2 + 6ax + 2b = 2(6x^2 + 3ax + b)$. Note that $f''(x) = 0$ exactly when $x = \dfrac{-3a \pm \sqrt{9a^2 - 24b}}{12}$. This represents no real solutions when $9a^2 - 24b < 0$, which occurs when $b > 3a^2/8$. When $b = 3a^2/8$, there is one root, but in this case the sign of f'' doesn't change at the double root $x = -a/4$, so there are no inflection points for f. In the case $b < 3a^2/8$, there are two roots of f'', both of which yield inflection points of f, as can be seen by the change in sign of f'' at its two roots.

4.3 Graphing Functions

4.3.1 Since the intervals of increase and decrease and the intervals of concavity must be subsets of the domain, it is helpful to know what the domain is at the outset.

4.3.3 No. Polynomials are continuous everywhere, so they have no vertical asymptotes. Also, polynomials in x always tend to $\pm\infty$ as $x \to \pm\infty$.

4.3.5 The maximum and minimum must occur at either an endpoint or a critical point. So to find the absolute maximum and minimum, it suffices to find all the critical points, and then compare the values of the function at those points and at the endpoints. The largest such value is the maximum and the smallest is the minimum.

4.3.7

The function sketched should be decreasing and concave down for $x < 3$ and decreasing and concave up for $x > 3$.

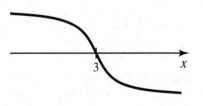

4.3.9 The domain of f is $(-\infty, \infty)$, and there is no symmetry. $f'(x) = x^2 - 4x - 5 = (x - 5)(x + 1)$. This is 0 when $x = -1, 5$. $f''(x) = 2x - 4$, which is 0 when $x = 2$. Note that $f'(-2) > 0$, $f'(0) < 0$, $f'(3) < 0$ and $f'(6) > 0$. So f is increasing on $(-\infty, -1)$ and on $(5, \infty)$. It is decreasing on $(-1, 5)$. There is a local maximum of $14/3$ at $x = -1$ and a local minimum of $-94/3$ at $x = 5$. Note also that $f''(x) < 0$ for $x < 2$ and $f''(x) > 0$ for $x > 2$, so there is an inflection point at $(2, -40/3)$, and f is concave down on $(-\infty, 2)$ and concave up on $(2, \infty)$. The y intercept is $f(0) = 2$.

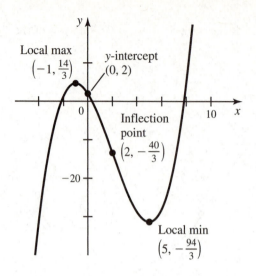

4.3.11 The domain of f is $(-\infty, \infty)$, and there is even symmetry, since $f(-x) = f(x)$. $f'(x) = 4x^3 - 12x = 4x(x^2 - 3)$. This is 0 when $x = \pm\sqrt{3}$ and when $x = 0$. $f''(x) = 12x^2 - 12 = 12(x^2 - 1)$, which is 0 when $x = \pm 1$. Note that $f'(-2) < 0$, $f'(-1) > 0$, $f'(1) < 0$, and $f'(2) > 0$. So f is decreasing on $(-\infty, -\sqrt{3})$ and on $(0, \sqrt{3})$. It is increasing on $(-\sqrt{3}, 0)$ and on $(\sqrt{3}, \infty)$.. There is a local maximum of 0 at $x = 0$ and local minimums of -9 at $x = \pm\sqrt{3}$. Note also that $f''(x) > 0$ for $x < -1$ and for $x > 1$ and $f''(x) < 0$ for $-1 < x < 1$, so there are inflection points at $x = \pm 1$. Also, f is concave down on $(-1, 1)$ and concave up on $(-\infty, -1)$ and on $(1, \infty)$. There is a y-intercept at $f(0) = 0$ and x-intercepts where $f(x) = x^4 - 6x^2 = x^2(x^2 - 6) = 0$, which is at $x = \pm\sqrt{6}$ and $x = 0$.

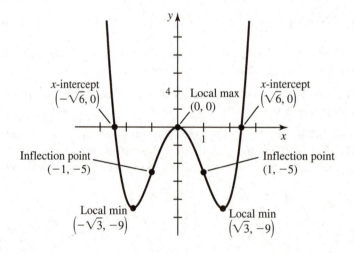

4.3.13 The domain of f is $(-\infty, \infty)$, and there is no symmetry. $f'(x) = 12x^3 + 12x^2 - 24x = 12x(x+2)(x-1)$. This is 0 when $x = -2$, when $x = 1$, and when $x = 0$. $f''(x) = 36x^2 + 24x - 24 = 12(3x^2 + 2x - 2)$, which is 0 when $x = \frac{-1 \pm \sqrt{7}}{3}$. These values are at approximately -1.21 and 0.55. Note that $f'(-3) < 0$, $f'(-1) > 0$, $f'(.5) < 0$, and $f'(2) > 0$. So f is decreasing on $(-\infty, -2)$ and on $(0, 1)$. It is increasing on $(-2, 0)$ and on $(1, \infty)$. There is a local maximum of 0 at $x = 0$ and a local minimum of -32 at $x = -2$ and a local minimum of -5 at $x = 1$. Let $r_1 < r_2$ be the two roots of $f''(x)$ mentioned above. Note that $f''(x) > 0$ for $x < r_1$ and for $x > r_2$ and $f''(x) < 0$ for $r_1 < x < r_2$, so there are inflection points at $x = r_1$ and at $x = r_2$. Also, f is concave down on (r_1, r_2) and concave up on $(-\infty, r_1)$ and on (r_2, ∞). There is a y-intercept at $f(0) = 0$ and x-intercepts where $f(x) = 3x^4 + 4x^3 - 12x^2 = x^2(3x^2 + 4x - 12) = 0$, which is at $x = \frac{-2 \pm 2\sqrt{10}}{3}$ and $x = 0$.

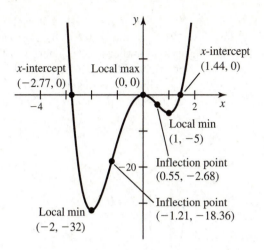

4.3.15 The domain of f is $(-\infty, 2) \cup (2, \infty)$, and there is no symmetry. Note that $\lim\limits_{x \to 2^+} f(x) = \infty$ and $\lim\limits_{x \to 2^-} f(x) = -\infty$, so there is a vertical asymptote at $x = 2$. There isn't a horizontal asymptote, since $\lim\limits_{x \to \pm\infty} f(x) = \pm\infty$. $f'(x) = \dfrac{(x-2)(2x) - x^2}{(x-2)^2} = \dfrac{x(x-4)}{(x-2)^2}$. This is 0 when $x = 4$ and when $x = 0$. $f''(x) = \dfrac{(x-2)^2(2x-4) - (x^2 - 4x)(2)(x-2)}{(x-2)^4} = \dfrac{8}{(x-2)^3}$. This is never 0. Note that $f'(-1) > 0$, $f'(1) < 0$, $f'(3) < 0$ and $f'(5) > 0$. So f is decreasing on $(0, 2)$ and on $(2, 4)$. It is increasing on $(-\infty, 0)$ and on $(4, \infty)$. There is a local maximum of 0 at $x = 0$ and a local minimum of 8 at $x = 4$. Note that $f''(x) > 0$ for $x > 2$ and $f''(x) < 0$ for $x < 2$, So f is concave up on $(2, \infty)$ and concave down on $(-\infty, 4)$. There are no inflection points, since the only change in concavity occurs at a vertical asymptote. The only intercept is $(0, 0)$.

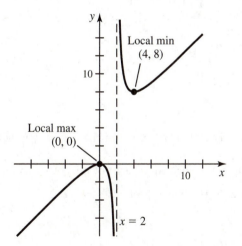

4.3.17 The domain of f is $(-\infty, -1) \cup (-1, 1) \cup (1, \infty)$, and there is no symmetry. Note that $\lim\limits_{x \to -1^+} f(x) = \infty$ and $\lim\limits_{x \to -1^-} f(x) = -\infty$, so there is a vertical asymptote at $x = -1$. Also, $\lim\limits_{x \to 1^+} f(x) = -\infty$ and $\lim\limits_{x \to 1^-} f(x) = \infty$, so there is a vertical asymptote at $x = 1$ Note that $\lim\limits_{x \to \pm\infty} \dfrac{3x - 5}{x^2 - 1} \cdot \dfrac{1/x^2}{1/x^2} = \lim\limits_{x \to \pm\infty} \dfrac{3/x}{1 - (1/x^2)} = 0$, so $y = 0$ is a horizontal asymptote.

$$f'(x) = \frac{(x^2 - 1)(3) - (3x - 5)(2x)}{(x^2 - 1)^2} = \frac{-3x^2 + 10x - 3}{(x^2 - 1)^2} = \frac{(-3x + 1)(x - 3)}{(x^2 - 1)^2}$$

$$f''(x) = \frac{(x^2 - 1)^2(-6x + 10) - (-3x^2 + 10x - 3)(2)(x^2 - 1)(2x)}{(x^2 - 1)^4} = \frac{2(3x^3 - 15x^2 + 9x - 5)}{(x^2 - 1)^3}.$$

$f'(x) = 0$ when $x = 3$ and when $x = 1/3$; $f''(x) = 0$ for $x \approx 4.4$. Let r_1 be this root of $f''(x)$. Note that $f'(-2) < 0$, $f'(-1/2) < 0$, $f'(1/2) > 0$, $f'(2) > 0$ and $f'(4) < 0$. So f is decreasing on $(-\infty, -1)$, on $(-1, 1/3)$ and on $(3, \infty)$. It is increasing on $(1/3, 1)$ and on $(1, 3)$. There is a local maximum of $1/2$ at $x = 3$ and a local minimum of $9/2$ at $x = 1/3$. Note that $f''(x) < 0$ for $x < -1$ and $f''(x) < 0$ for $1 < x < r_1$, while $f''(x) > 0$ for $-1 < x < 1$, and for $x > r_1$. Thus f is concave up on $(-1, 1)$ and on (r_1, ∞) and concave down on $(-\infty, -1)$ and on $(1, r_1)$. There is an inflection point at r_1. There is a y-intercept at $f(0) = 5$ and an x-intercept at $(5/3, 0)$.

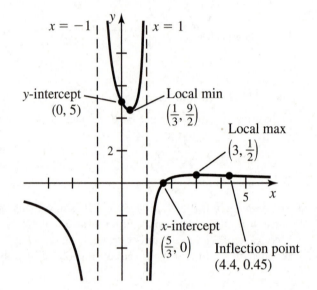

4.3.19 The domain of f is $(-\infty, -1/2) \cup (-1/2, \infty)$, and there is no symmetry. Since $\lim\limits_{x \to \pm\infty} \dfrac{x^2 + 12}{2x + 1} \cdot \dfrac{1/x}{1/x} = \lim\limits_{x \to \pm\infty} \dfrac{x + (12/x)}{2 + (1/x)} = \pm\infty$, there is no horizontal asymptote. Also, since $\lim_{x \to (-1/2)^-} f(x) = -\infty$ and $\lim_{x \to (-1/2)^+} f(x) = \infty$, there is a vertical asymptote at $x = -1/2$.

$$f'(x) = \frac{(2x + 1)(2x) - (x^2 + 12)(2)}{(2x + 1)^2} = \frac{2x^2 + 2x - 24}{(2x + 1)^2} = \frac{2(x + 4)(x - 3)}{(2x + 1)^2}$$

$$f''(x) = \frac{(2x + 1)^2(4x + 2) - (2x^2 + 2x - 24)(2)(2x + 1)(2)}{(2x + 1)^4} = \frac{98}{(2x + 1)^3}$$

$f'(x) = 0$ for $x = -4$ and $x = 3$, while $f''(x)$ is never zero. Note that $f'(x) > 0$ on $(-\infty, -4)$ and on $(3, \infty)$. So f is increasing on $(-\infty, -4)$ and on $(3, \infty)$. Also, $f'(x) < 0$ on $(-4, -1/2)$ and on $(-1/2, 3)$. So f is decreasing on those intervals. There is a local maximum of -4 at $x = -4$ and a local minimum of 3 at $x = 3$. Note also that $f''(x) < 0$ for $x < -1/2$, and $f''(x) > 0$ for $x > -1/2$, so f is concave down on $(-\infty, -1/2)$ and is concave up on $(-1/2, \infty)$. There are no inflection points since the only changes in concavity occur at the vertical asymptote. There are no x-intercepts since $x^2 + 12 > 0$ for all x, and the y-intercept is $f(0) = 12$.

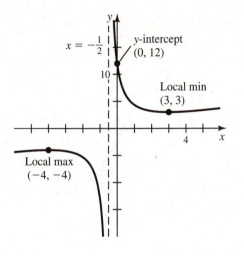

4.3.21 The domain of f is given to be $[-2\pi, 2\pi]$, and there is no symmetry, and no vertical asymptotes. There are no horizontal asymptotes to consider on this restricted domain. $f'(x) = 1 - 2\sin x$. This is 0 when $\sin x = 1/2$, which occurs on the given interval for $x = -11\pi/6, -7\pi/6, \pi/6$, and $5\pi/6$. $f''(x) = -2\cos x$, which is 0 for $x = -3\pi/2, -\pi/2, \pi/2$, and $3\pi/2$. Note that $f'(x) > 0$ on $(-2\pi, -11\pi/6)$, and on $(-7\pi/6, \pi/6)$, and on $(5\pi/6, 2\pi)$. So f is increasing on those intervals, while $f'(x) < 0$ on $(-11\pi/6, -7\pi/6))$ and on $(\pi/6, 5\pi/6)$, so f is decreasing there. f has local maximums at $x = -11\pi/6$ and at $x = \pi/6$ and local minimums at $x = -7\pi/6$ and at $x = 5\pi/6$. Note also that $f''(x) < 0$ on $(-2\pi, -3\pi/2)$ and on $(-\pi/2, \pi/2)$ and on $(3\pi/2, 2\pi)$, so f is concave down on those intervals, while $f''(x) > 0$ on $(-3\pi/2, -\pi/2)$ and on $(\pi/2, 3\pi/2)$, so f is concave up there and there are inflection points at $x = \pm 3\pi/2$ and $x = \pm \pi/2$. The y-intercept is $f(0) = 2$ and the x-intercept is at approximately -1.03.

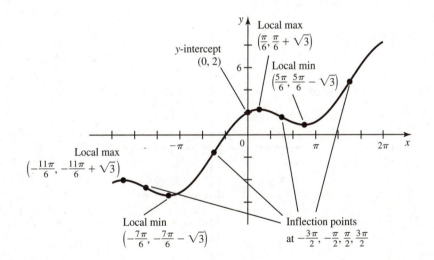

4.3.23 The domain of f is given to be $[0, 2\pi]$, so questions about symmetry and horizontal asymptotes aren't relevant. There are no vertical asymptotes. $f'(x) = \cos x - 1$. This is never 0 on $(0, 2\pi)$. $f''(x) = -\sin x$, which is 0 on the given interval only for $x = \pi$. Note that $f'(x) < 0$ on $(0, 2\pi)$, so f is decreasing on the given interval and there are no relative extrema. Note also that $f''(x) < 0$ on $(0, \pi)$ and $f''(x) > 0$ on $(\pi, 2\pi)$, so f is concave down on $(0, \pi)$ and is concave up on $(\pi, 2\pi)$, and there is an inflection point at $x = \pi$. The only intercept is the origin $(0, 0)$.

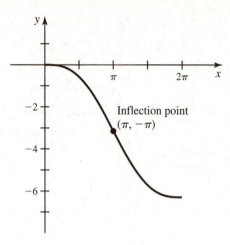

Inflection point
$(\pi, -\pi)$

4.3.25 $g(t)$ is symmetric around the y axis, and its domain is all t except for $t = 0$. As $t \to \infty$, $g(t) \to 0$, so that $y = 0$ is a horizontal asymptote. As $t \to 0$, $g(t) \to -\infty$ (from either side of 0), so that $t = 0$ is a vertical asymptote and $g(t)$ increases negatively without bound as t approaches zero from either direction.

$$g'(t) = \frac{-6}{t^3} + \frac{216}{t^5} = 6\left(\frac{36 - t^2}{t^5}\right)$$

$$g''(t) = 6\left(\frac{(-2t)t^5 - (36 - t^2)(5t^4)}{t^{10}}\right) = 18\left(\frac{t^2 - 60}{t^6}\right)$$

Thus g has critical points at $t = \pm 6$. At both critical points, $g''(t) < 0$, so these are both at maxima. Additionally, g has inflection points where $t^2 - 60 = 0$, or for $t = \pm 2\sqrt{15}$. $g''(t) > 0$ for $t \in (-\infty, -2\sqrt{15}) \cup (2\sqrt{15}, \infty)$ so g is concave upwards there; $g''(t) < 0$ for $t \in (-2\sqrt{15}, 2\sqrt{15})$ (except of course for $t = 0$, where g is undefined), so g is concave downwards there.

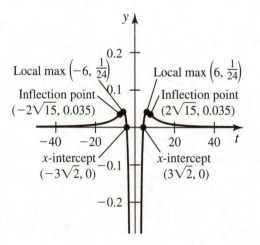

Local max $\left(-6, \frac{1}{24}\right)$ Local max $\left(6, \frac{1}{24}\right)$

Inflection point   Inflection point
$(-2\sqrt{15}, 0.035)$ $(2\sqrt{15}, 0.035)$

x-intercept x-intercept
$(-3\sqrt{2}, 0)$ $(3\sqrt{2}, 0)$

4.3.27 $f(x)$ has no symmetries. Its domain is $x \neq 1$.

$$\lim_{x \to \infty} f(x) = \lim_{x \to \infty} \frac{\sqrt{1 + 2/x^2}}{1 - 1/x} = 1$$

$$\lim_{x \to -\infty} f(x) = \lim_{x \to \infty} -\frac{\sqrt{1 + 2/x^2}}{1 - 1/x} = -1$$

so that f has horizontal asymptotes at $y = 1$ and $y = -1$. f has a vertical asymptote at $x = 1$ since the denominator vanishes there, and $\lim\limits_{x \to 1^-} f(x) = -\infty$ while $\lim\limits_{x \to 1^+} f(x) = \infty$.

$$f'(x) = \frac{(1/2\sqrt{x^2 + 2})(x - 1)(2x) - \sqrt{x^2 + 2}}{(x - 1)^2} = -\frac{x + 2}{\sqrt{x^2 + 2}(x - 1)^2}$$

$$f''(x) = -\frac{(x - 1)^2\sqrt{x^2 + 2} - (x + 2)(2(x - 1)\sqrt{x^2 + 2} + (x - 1)^2 x(x^2 + 2)^{-1/2})}{(x - 1)^4(x^2 + 2)}$$

$$= \frac{2(x^3 + 3x^2 + 5)}{(x^2 + 2)^{3/2}(x - 1)^3}$$

Thus $f''(x) = 0$ when $x^3 + 3x^2 + 5 = 0$. This cubic cannot be factored; its real root is approximately -3.426. Further calculations with $f''(x)$ reveal that $f(x)$ is concave upwards for $x > 1$ and for $x < -3.426$; $f(x)$ is concave downwards for $-3.426 < x < 1$. Since $f'(-2) = 0$ and $f''(-2) < 0$, there is a local maximum at $x = -2$.

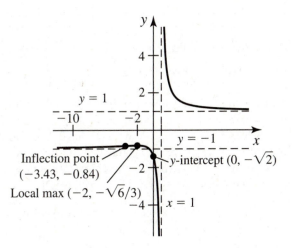

4.3.29 Note that the domain is given as $[-\pi, \pi]$.

$$f'(x) = \cos x + 2\sin 2x = \cos x + 4\sin x \cos x = \cos x(1 + 4\sin x)$$
$$f''(x) = -\sin x(1 + 4\sin x) + \cos x(4\cos x) = 4(\cos^2 x - \sin^2 x) - \sin x$$

The critical points are where $f'(x) = 0$, so $x = \pm\pi/2$ or $x = \arcsin(-1/4) \approx -.253, -2.889$. $f''(\pi/2) = -5$ and $f''(-\pi/2) = -3$, so there are maxima at both of these points. $f''(-.253) \approx 3.75$ and $f''(-2.889) \approx 3.75$, so there are minima at both of these points. Solving $f''(x) = 0$ using technology gives $x \approx -2.259, -0.883, 0.704$, and 2.437, so we see that there are points of inflection at these values. Looking at values of $f''(x)$ away from the points of inflection shows that f is concave upwards on $(-\pi, -2.259) \cup (-.883, .704) \cup (2.437, \pi)$ and concave downwards on $(-2.259, -.883) \cup (.704, 2.437)$. Finally, to find the x-intercepts, we must solve $f(x) = 0$, so $\sin x - \cos 2x = 0$. Substituting, we get $2\sin^2 x + \sin x - 1 = 0$, so that the x-intercepts occur when $\sin x = 1/2$ or $\sin x = -1$. This happens for $x = -\pi/2$, $x = \pi/6$, and $x = 5\pi/6$.

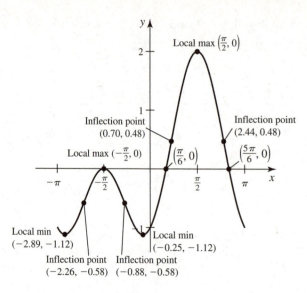

4.3.31 The domain of f is given to be $[-2\pi, 2\pi]$. There are no vertical asymptotes. Note that $f(-x) = \dfrac{(-x)(\sin(-x))}{((-x)^2 + 1)} = \dfrac{x \sin x}{x^2 + 1} = f(x)$, f has even symmetry. Questions about horizontal asymptotes aren't relevant since the given domain is an interval with finite length. $f'(x) = \dfrac{(x^2 + 1)(x \cos x + \sin x) - x \sin x \cdot (2x)}{(x^2 + 1)^2}$, which can be simplified to $\dfrac{x(x^2 + 1)\cos x + (1 - x^2)\sin x}{(x^2 + 1)^2}$, and with the aid of a computer algebra system, the roots of this expression can be found to be approximately ± 4.51 and ± 1.36, as well as $x = 0$. We will call the non-zero roots $\pm r_1$ and $\pm r_2$ where $0 < r_1 < r_2$. Note that $f'(x) < 0$ on $(-2\pi, -r_2)$ and on $(-r_1, 0)$ and (r_1, r_2), so f is decreasing there, while $f'(x) > 0$ on $(-r_2, -r_1)$, on $(0, r_1)$, and on $(r_2, 2\pi)$, so f is increasing on these intervals. There are local maxima at $x = \pm r_1$ and local minima at $x = 0$ and at $x = \pm r_2$. $f''(x)$ has numerator $(x^2 + 1)^2((x^3 + x)(-\sin x) + \cos x(3x^2 + 1) + (1 - x^2)(\cos x) + \sin x(-2x)) - ((x^3 + x)\cos x + (1 - x^2)\sin x)(4x)(x^2 + 1)$ and denominator $(x^2 + 1)^4$. This simplifies to

$$f''(x) = \frac{(-x^5 - 7x)\sin x + (-2x^4 + 2)\cos x}{(x^2 + 1)^3},$$

which is 0 at approximately ± 5.96 and ± 2.56 and $\pm.49$. We will call these 6 roots $\pm r_3$, $\pm r_4$ and $\pm r_5$ where $0 < r_3 < r_4 < r_5$. Note that $f''(x) < 0$ on $(-2\pi, -r_5)$ and on $(-r_4, -r_3)$, and on (r_3, r_4), and on $(r_5, 2\pi)$, so f is concave down on these intervals, while $f''(x) > 0$ on $(-r_5, -r_4)$, and on $(-r_3, r_3)$, and on (r_4, r_5), so f is concave up on these intervals. There are points of inflection at each of $\pm r_3$, $\pm r_4$, and $\pm r_5$. There is an x-intercept at $(0, 0)$, which is also the y-intercept, as well as x-intercepts at $\pm 2\pi$.

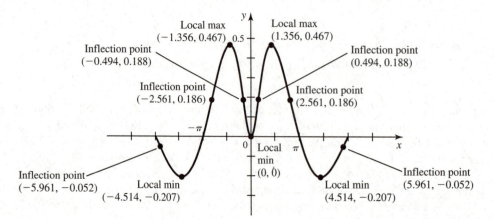

4.3.33

a. False. Maximums and minimums can also occur at points where $f'(x)$ doesn't exist. Also, it is possible to have a zero of f' which doesn't lead to an extreme point.

b. False. Inflection points can also occur at points where $f''(x)$ doesn't exist, and a zero of f'' might not lead to an inflection point.

c. False. For example, $f(x) = \dfrac{(x^2 - 9)(x^2 - 16)}{(x + 3)(x - 4)}$ doesn't have a vertical asymptote at $x = -3$ or $x = 4$.

d. True. The limit of a rational function, as $x \to \infty$ is a finite number only in the case that the degree of the denominator is greater than or equal to that of the numerator. If they both have the same degree, the limit is the ratio of the leading coefficients, and this is also true of the limit as $x \to -\infty$. In the case where the denominator has greater degree than the numerator, the limit as 0 as $x \to -\infty$ and is $x \to \infty$.

4.3.35 $f'(x)$ is 0 on the interior of the given interval at $x = \pm 3\pi/2$, $x = \pm\pi$, $x = \pm\pi/2$, and at $x = 0$. $f(x) > 0$ on $(-2\pi, -3\pi/2)$, and on $(-\pi, -\pi/2)$, and on $(0, \pi/2)$, and on $(\pi, 3\pi/2)$, so f is increasing on those intervals. $f'(x) < 0$ on $(-3\pi/2, -\pi)$, and on $(-\pi/2, 0)$, and on $(\pi/2, \pi)$, and on $(3\pi/2, 2\pi)$, so f is decreasing on those intervals. There are local maxima at $x = \pm 3\pi/2$ and $x = \pm\pi/2$, and local minima at $x = 0$ and at $x = \pm\pi$. An example of such a function is sketched.

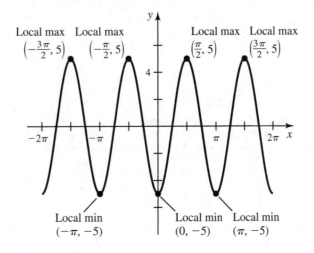

4.3.37 $f'(x)$ is 0 at $x = -2$, and doesn't exist at $x = 0$ and $x = 6$. $f''(x) > 0$ on $(-\infty, -2)$, and on $(6, \infty)$, so f is increasing on those intervals. $f'(x) < 0$ on $(-2, 0)$, and on $(0, 6)$, so f is decreasing on those intervals. There is a local maximum at $x = -2$. An example of such a function is sketched.

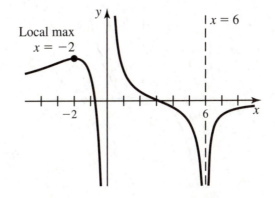

4.3.39 $f'(x)$ is 0 at $x = 1$ and $x = 3$. $f'(x) > 0$ on $(0, 1)$, and on $(3, 4)$, so f is increasing on those intervals. $f'(x) < 0$ on $(1, 3)$, so f is decreasing on that interval. There is a local maximum at $x = 1$ and a local minimum at $x = 3$. $f''(x)$ changes sign at $x = 2$ from negative to positive, so $x = 2$ is an inflection point where the concavity of f changes from down to up. An example of such a function is sketched.

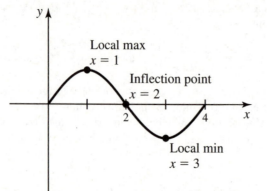

4.3.41 The domain of f is $(-\infty, \infty)$ and there is no symmetry. There are no asymptotes since f is a polynomial. $f'(x) = 3x^2 - 12x - 135 = 3(x - 9)(x + 5)$, which is 0 for $x = 9$ and $x = -5$. $f'(x) > 0$ on $(-\infty, -5)$, and on $(9, \infty)$, so f is increasing on those intervals. $f'(x) < 0$ on $(-5, 9)$, so f is decreasing on that interval. There is a local maximum at $x = -5$ and a local minimum at $x = 9$. $f''(x) = 6x - 12$, which is 0 for $x = 2$. $f''(x) > 0$ on $(2, \infty)$, so f is concave up on that interval. $f''(x) < 0$ on $(-\infty, 2)$, so f is concave down on that interval. There is a point of inflection at $x = 2$. The y-intercept is 0 and the x-intercepts are -9 and 15.

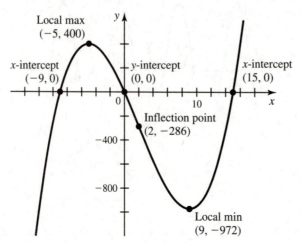

4.3.43

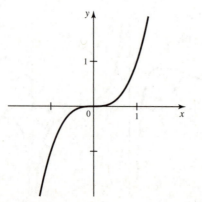

4.3.45

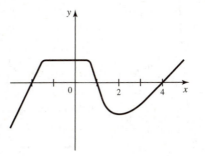

4.3.47 The domain of f is $(-\infty, -2] \cup (2, \infty)$ and there is no symmetry. There is a vertical asymptote at $x = 2$ since $\displaystyle\lim_{x \to 2^+} \frac{-x\sqrt{x^2-4}}{x-2} = -\infty$. There are no horizontal asymptotes. Now,

$$f'(x) = \frac{(x-2)(-x^2(x^2-4)^{-1/2} + (x^2-4)^{1/2}(-1)) - ((-x)(x^2-4)^{1/2})}{(x-2)^2}.$$

When simplified, this can be written as $\dfrac{-x^2 + 2x + 4}{(x-2)\sqrt{x^2-4}}$, and this quantity is 0 on the given domain only for $x = 1 + \sqrt{5} \approx 3.236$. $f'(x) > 0$ on $(-\infty, -2)$, and on $(2, 1+\sqrt{5})$, so f is increasing on those intervals. $f'(x) < 0$ on $(1 + \sqrt{5}, \infty)$, so f is decreasing on that interval. There is a local maximum at $x = 1 + \sqrt{5}$. $f''(x)$ simplifies to be $\dfrac{-4(x+4)}{(x-2)^2(x+2)\sqrt{x^2-4}}$, which is 0 at $x = -4$ $f''(x) > 0$ on $(-4, -2)$, so f is concave up on that interval. $f''(x) < 0$ on $(-\infty, -4)$ and on $(1 + \sqrt{5}, \infty)$, so f is concave down on those intervals. There is a point of inflection at $x = -4$.

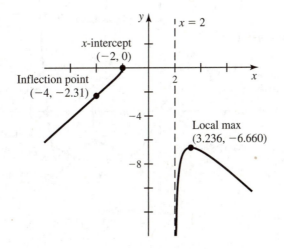

4.3.49 The domain of f is $(-\infty, \infty)$ and there is no symmetry. There are no asymptotes since f is a polynomial. $f'(x) = 12x^3 - 132x^2 + 120x = 12x(x-10)(x-1)$, which is 0 for $x = 0$, $x = 1$, and $x = 10$. $f'(x) > 0$ on $(0, 1)$, and on $(10, \infty)$, so f is increasing on those intervals. $f'(x) < 0$ on $(-\infty, 0)$ and on $(1, 10)$, so f is decreasing on those intervals. There is a local maximum at $x = 1$, and local minima at $x = 0$ and at $x = 10$. $f''(x) = 36x^2 - 264x + 120 = 12(3x^2 - 22x + 10)$. This is 0 at approximately $x = .487$ and $x = 6.846$. Let these two roots be r_1 and r_2 with $r_1 < r_2$. $f''(x) > 0$ on $(-\infty, r_1)$ and on (r_2, ∞), so f is concave up on those intervals. $f''(x) < 0$ on (r_1, r_2), so f is concave down on that interval. There are points of inflection at $x = r_1$ and $x = r_2$.

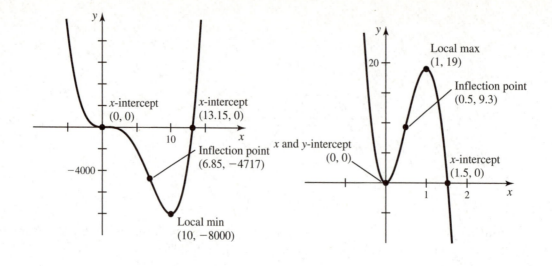

4.3.51 The domain of f is $(-\infty, \infty)$ and there is no symmetry. There are no asymptotes since f is a polynomial. $f'(x) = 60x^5 - 180x^4 - 300x^3 + 900x^2 + 240x - 720 = 60(x+2)(x+1)(x-1)(x-2)(x-3)$, which is 0 for $x = -2$, $x = -1$, $x = 2$, and $x = 3$. $f'(x) > 0$ on $(-2, -1)$ and on $(1, 2)$, and on $(3, \infty)$, so f is increasing on those intervals. $f'(x) < 0$ on $(-\infty, -2)$ and on $(-1, 2)$, and on $(2, 3)$, so f is decreasing on those intervals. There are local minima at $x = -2$, $x = 1$, and $x = 3$, and local maxima at $x = -1$ and $x = 2$. $f''(x) = 300x^4 - 720x^3 - 900x^2 + 1800x + 240 = 60(5x^4 - 12x^3 - 15x^2 + 30x + 4)$. Using a computer algebra system, we find that this has 4 real roots, which we will call $r_1 < r_2 < r_3 < r_4$. They values of these roots are approximately -1.61, $-.13$, 1.5, and 2.63. $f''(x) > 0$ on $(-\infty, r_1)$, and on (r_2, r_3) and on (r_4, ∞), so f is concave up on those intervals. $f''(x) < 0$ on (r_1, r_2) and on (r_3, r_4), so f is concave down on those intervals. There are points of inflection at each r_i for i from 1 to 4.

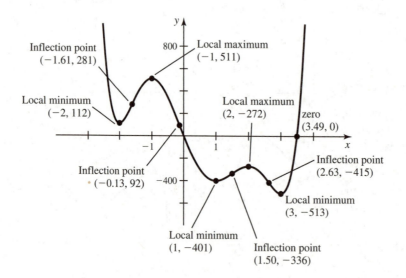

4.3.53 The domain of f is $(-\infty, \infty)$ and there is odd symmetry, since $f(-x) = -f(x)$. There are no vertical asymptotes, but $y = 0$ is a horizontal asymptote, since $\lim\limits_{x\to\infty} \dfrac{x\sqrt{|x^2-1|}}{x^4+1} \cdot \dfrac{1/x^4}{1/x^4} = \lim\limits_{x\to\infty} \dfrac{\sqrt{(1/x^4)-(1/x^6)}}{1+(1/x^4)} = 0$. Note that

$$f(x) = \begin{cases} \frac{x\sqrt{x^2-1}}{x^4+1} & \text{if } |x| \geq 1; \\ \frac{x\sqrt{1-x^2}}{x^4+1} & \text{if } |x| < 1. \end{cases}$$

Differentiating each part of the above and simplifying yields

$$f'(x) = \begin{cases} \frac{-2x^6+3x^4+2x^2-1}{(x^4+1)^2\sqrt{x^2-1}} & \text{if } |x| > 1; \\ \frac{2x^6-3x^4-2x^2+1}{(x^4+1)^2\sqrt{1-x^2}} & \text{if } |x| < 1. \end{cases}$$

The roots of this expression (on the respective domains) are approximately -1.37, $-.6$, $.6$, and 1.37. Also, this derivative doesn't exist at $x = \pm 1$. Let the roots of f' be $\pm r_1$ and $\pm r_2$ where $0 < r_1 < r_2$. An analysis of the sign of f' shows that f is increasing on $(-r_2, -1)$, on $(-r_1, r_1)$, and on $(1, r_2)$, while f is decreasing on $(-\infty, -r_2)$, on $(-1, -r_1)$, on $(r_1, 1)$, and on (r_2, ∞), so there are local maxima at $x = -1$, $x = r_1$, and $x = r_2$, and local minima at $x = -r_1$, $x = -r_2$, and $x = 1$. An analysis without computer of $f''(x)$ is not for the fainthearted. In the case $|x| > 1$ the second derivative is given by $\dfrac{x\left(6x^{10} - 19x^8 - 12x^6 + 42x^4 - 18x^2 - 3\right)}{\left(x^2 - 1\right)^{3/2}\left(x^4 + 1\right)^3}$, and for the case $|x| < 1$ we have $\dfrac{x\left(6x^{10} - 19x^8 - 12x^6 + 42x^4 - 18x^2 - 3\right)}{\left(1 - x^2\right)^{3/2}\left(x^4 + 1\right)^3}$. There is a root of approximately ± 1.79, and 0 is a root as well. Let the non-zero roots be $\pm r_3$ where $r_3 > 0$. An analysis of the sign of f'' reveals that f is concave down on $(-\infty, -r_3)$ and on $(0, 1)$ and on $(1, r_3)$, while it is concave up on $(-r_3, -1)$ and on $(-1, 0)$, and on (r_3, ∞). There are inflection points at $\pm r_3$ and at 0. The x-intercepts are ± 1 and 0.

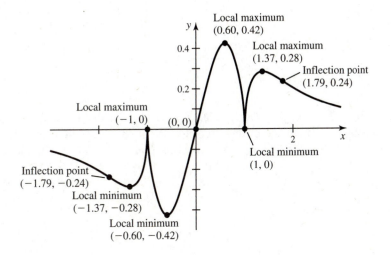

4.3.55

a. f has even symmetry, so we will analyze the function on $(0, 2\pi]$, and use the symmetry to graph the function over $[-2\pi, 0)$. $f'(x) = \dfrac{x^2(-3\cos^2 x(-\sin x)) - (1 - \cos^3 x)(2x)}{x^4} = \dfrac{3x\sin x\cos^2 x - 2(1 - \cos^3 x)}{x^3}$. This has no roots on $(0, 2\pi)$, and in fact is always negative, so f is decreasing on $(0, 2\pi)$. $f''(x)$ when simplified is given by $\dfrac{3\left(\left(x^2 - 2\right)\cos^3(x) - 2x^2\sin^2(x)\cos(x) - 4x\sin(x)\cos^2(x) + 2\right)}{x^4}$. The roots of f'' on $(0, 2\pi)$ are $r_1 \approx .89$, $r_2 \approx 2.47$, $r_3 \approx 3.48$, $r_4 \approx 4.76$, and $r_5 \approx 5.5$. An analysis of the sign of f'' reveals that there is a change in concavity at each of these roots, starting with concavity downward on $(-r_1, r_1)$. So each of $\pm r_i$ is an inflection point.

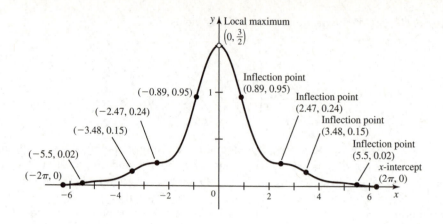

b. f has even symmetry, so we will analyze the function on $(0, 2\pi]$, and use the symmetry to graph the function over $[-2\pi, 0)$. $f'(x) = \dfrac{x^2(-5\cos^4 x(-\sin x) - (1 - \cos^5 x)(2x))}{x^4} = \dfrac{5x\cos^4 x \sin x - 2(1 - \cos^5 x)}{x^3}$. This has roots on $(0, 2\pi)$ of $r_1 \approx 2.41$ and $r_2 \approx 2.83$. An analysis of the sign of f' shows that f is decreasing on $(0, r_1)$, increasing on (r_1, r_2), decreasing on $(r_2, 2\pi)$, so there is a local minimum at r_1 and a local maximum at r_2. $f''(x)$ when simplified is given by

$$\frac{(5x^2 - 6)\cos^5(x) - 20x^2 \sin^2(x)\cos^3(x) - 20x \sin(x)\cos^4(x) + 6}{x^4}.$$

The roots of f'' on $(0, 2\pi)$ are $r_3 \approx .63$, $r_4 \approx 2.62$, $r_5 \approx 3.45$, $r_6 \approx 4.96$, and $r_7 \approx 5.74$. An analysis of the sign of f'' reveals that there is a change in concavity at each of these roots, starting with concavity downward on $(-r_3, r_3)$. So each of $\pm r_i$ is an inflection point for $i = 3, 4, 5, 6$ and 7.

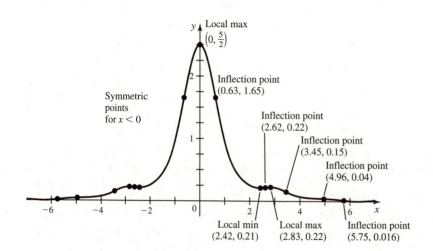

4.3.57

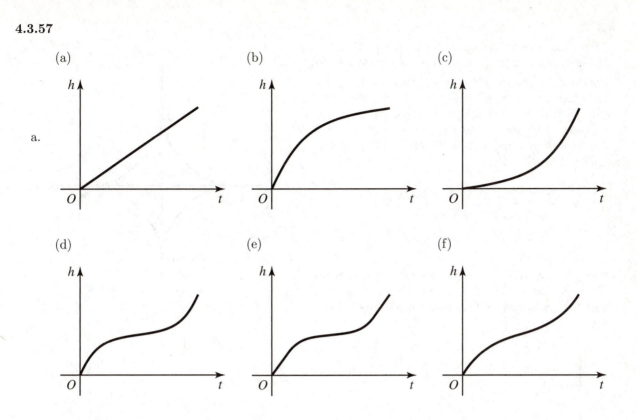

(a) (b) (c)

(d) (e) (f)

a.

b. The water is being poured in at a constant rate, so the depth is always increasing, so $y = h(t)$ is an increasing function.

c. (a) No concavity

(b) Always concave down.

(c) Always concave up.

(d) Concave down for for the first half and concave up for the second half.

(e) At the beginning, in the middle, and at the end, there is no concavity. In the lower middle it is concave down and in the upper middle it will be concave up.

(f) This is concave down for the first half, and concave up for the second half.

d. (a) $h'(t)$ is constant, so there is no local max/min.

(b) $h'(t)$ is maximal at $t = 0$.

(c) $h'(t)$ is maximal at $t = 10$.

(d) $h'(t)$ is maximal at $t = 0$ and $t = 10$.

(e) $h'(t)$ is maximal on the first and last straight parts of $h(t)$.

(f) $h'(t)$ is maximal at $t = 0$ and $t = 10$.

4.3.59

If $f''(x) > 0$ on $(-\infty, 0)$ and on $(0, \infty)$, then $f'(x)$ is increasing on both of those intervals. But if there is a local max at 0, the function f must be switching from increasing to decreasing there. This means that f' must be switching from positive to negative. But if f' is switching from positive to negative, but increasing, there must be a singular point at $x = 0$.

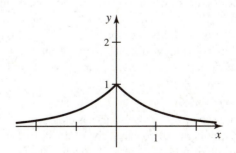

4.3.61

The equation is valid on only for $|x| \leq 1$ and $|y| \leq 1$. Using implicit differentiation, we have $(2/3)x^{-1/3} + (2/3)y^{-1/3}y' = 0$, so $y' = \frac{-y^{1/3}}{x^{1/3}}$. This is 0 for $y = 0$ (in which case $x = \pm 1$) and doesn't exist for $x = 0$ (in which case $x = \pm 1$.) In the first quadrant the curve is decreasing, in the 2nd it is increasing, in the 3rd it is decreasing, and in the 4th it is increasing. Differentiating y' yields

$$y'' = \frac{x^{1/3}(-1/3)y^{-2/3}y' + y^{1/3}(1/3)(x^{-2/3})}{x^{2/3}}$$
$$= \frac{y^{2/3} + x^{2/3}}{3x^{4/3}y^{1/3}},$$

which is positive when y is positive and negative when y is negative, so the curve is concave up in the first and 2nd quadrants, and concave down in the 3rd and 4th.

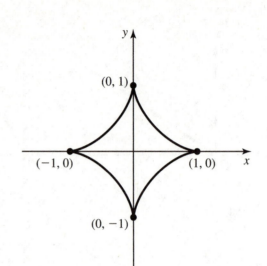

4.3.63

First note that the expression is symmetric when x and y are switched, so the curve should be symmetric about the line $y = x$. Also, if $y = x$, then $2x^3 = 3x^2$, so either $x = 0$ or $x = 3/2$, so this is where the curve intersects the line $y = x$. Differentiating implicitly yields $3x^2 + 3y^2y' = 3xy' + 3y$, so $y' = \frac{y-x^2}{y^2-x}$. This is 0 when $y = x^2$, but this occurs on the curve when $x^3 + x^6 = 3x^3$, which yields $x = 0$ (and $y = 0$), or $x^3 = 2$, so $x = \sqrt[3]{2} \approx 1.26$. Note also that the derivative doesn't exist when $x = y^2$, which again yields $(0,0)$ and $y^6 + y^3 = 3y^3$, or $y = \sqrt[3]{2}$. So there should be a flat tangent line at approximately $(1.26, 1.59)$ and a vertical tangent line at about $(1.59, 1.26)$. Differentiating again and solving for y'' yields $y''(x) = \frac{2xy\left(x^3 - 3xy + y^3 + 1\right)}{(x - y^2)^3} = \frac{2xy}{(x - y^2)^3}$. In the first quadrant, when $x > y^2$, the curve is concave up, when $x < y^2$, the curve is concave down. In both the 2nd and 4th quadrants, the curve is concave up.

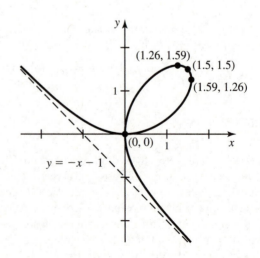

4.3.65

Note that the curve is symmetric about both the
x-axis and the y-axis, so we can just consider the
first quadrant, and obtain the rest by reflection.
Differentiating implicitly and solving for y' yields
$y'(x) = \dfrac{2x^3 - 5x}{2y\,(y^2 - 2)}$. The numerator is negative on
$(0, \sqrt{5/2})$ and positive on $(\sqrt{5/2}, \infty)$, while the denominator is negative for $0 < y < \sqrt{2}$ and positive for $y > \sqrt{2}$. Thus the relation is increasing
in the rectangle $(0, \sqrt{5/2}) \times (0, \sqrt{2})$ and in the region $(\sqrt{5/2}, \infty) \times (\sqrt{2}, \infty)$, while it is decreasing in
the other regions in the first quadrant. There are
vertical tangent lines when $y = \sqrt{2}$. When $y = 2$
and $x = 0$ there is a horizontal tangent line. Note
that when $x = 0$, we have $y = 0$ or $y = \pm 2$, while
if $y = 0$, we have $x = 0$ or $x = \pm\sqrt{5}$. Also, if
$y = \sqrt{2}$ then $x = 1$ or $x = 2$. So some sample points
to plot are $(0,0)$, $(\pm\sqrt{5}, 0)$, $(0, \pm 2)$, $(\pm 1, \pm\sqrt{2})$, and
$(\pm\sqrt{5}, \pm\sqrt{2})$. Also, when $1 < x < 2$, there are no
corresponding y values on the curve.

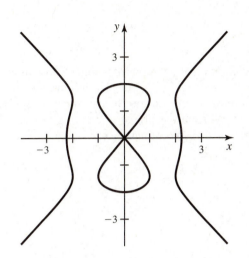

4.3.67

Note that the curve requires $-1 \le x < 1$. Note also
that the curve is symmetric about both the x-axis and
the y-axis, so we can just consider the first quadrant,
and obtain the rest by reflection. Differentiating implicitly yields $4x^3 - 2x + 2yy' = 0$, so $y' = \dfrac{x - 2x^3}{y}$.
This is 0 in the first quadrant for $x = \sqrt{2}/2$. Note also
that there is a vertical tangent line at the point $(1, 0)$.
The derivative is positive on $(0, \sqrt{2}/2)$ and negative
on $(\sqrt{2}/2, 1)$, so in the first quadrant the curve is increasing on that first interval and decreasing on the
second. Differentiating again and solving for y'' (and
rewriting) yields $y''(x) = \dfrac{x^4(2x^2 - 3)}{y}$, which is negative in the first quadrant for $0 < x < 1$, so this curve
is concave down in the first quadrant. The rest of the
curve can be found by reflection.

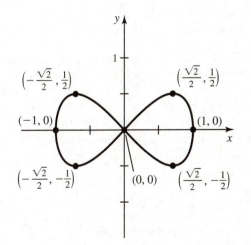

4.3.69

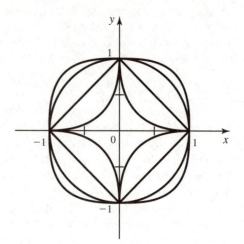

The n increases, the curves retain their symmetry, but move "outward." That is, the curves enclose a greater area. It appears that the figures approach the 2×2 square centered at the origin with sides parallel to the coordinate axes.

4.4   Optimization Problems

4.4.1 ... objective... constraints ....

4.4.3 The constraint is $x + y = 10$, so we can express $y = 10 - x$ or $x = 10 - y$. Therefore the objective function can be expressed $Q = x^2(10 - x)$ or $Q = (10 - y)^2 y$.

4.4.5 Let x and y be the dimensions of the rectangle. The perimeter is $2x + 2y$, so the constraint is $2x + 2y = 10$, which gives $y = 5 - x$. The objective function to be maximized is the area of the rectangle, $A = xy$. Thus we have $A = xy = x(5 - x) = 5x - x^2$. We have $x, y \geq 0$, which also implies $x \leq 5$ (otherwise $y < 0$). Therefore we need to maximize $A(x) = 5x - x^2$ for $0 \leq x \leq 5$. The critical points of the objective function satisfy $A'(x) = 5 - 2x = 0$, which has the solution $x = 5/2$. To find the absolute maximum of A, we check the endpoints of $[0, 5]$ and the critical point $x = 5/2$. Because $A(0) = A(5) = 0$ and $A(5/2) = 25/4$, the absolute maximum occurs when $x = y = 5/2$, so width = length = $5/2$ m.

4.4.7 Let x and y be the two non-negative numbers. The constraint is $x + y = 23$, which gives $y = 23 - x$. The objective function to be maximized is the product of the numbers, $P = xy$. Using $y = 23 - x$, we have $P = xy = x(23 - x) = 23x - x^2$. Now x must be at least 0, and cannot exceed 23 (otherwise $y < 0$). Therefore we need to maximize $P(x) = 23x - x^2$ for $0 \leq x \leq 23$. The critical points of the objective function satisfy $P'(x) = 23 - 2x = 0$, which has the solution $x = 23/2$. To find the absolute maximum of P, we check the endpoints of $[0, 23]$ and the critical point $x = 23/2$. Because $P(0) = P(23) = 0$ and $P(23/2) = (23/2)^2$, the absolute maximum occurs when $x = y = 23/2$.

4.4.9 Let x and y be the two positive numbers. The constraint is $xy = 50$, which gives $y = 50/x$. The objective function to be minimized is the sum of the numbers, $S = x + y$. Using $y = 50/x$, we have $S = x + y = x + \frac{50}{x}$. Now x can be any positive number, so we need to maximize $S(x) = x + 50/x$ on the interval $(0, \infty)$. The critical points of the objective function satisfy $S'(x) = 1 - \frac{50}{x^2} = 0$, which has the solution $x = \sqrt{50} = 5\sqrt{2}$. By the First (or Second) Derivative Test, this critical point corresponds to a local minimum, and by Theorem 4.5, this solitary local minimum is also the absolute minimum on the interval $(0, \infty)$. Therefore the numbers with minimum sum are $x = 5\sqrt{2}$ and $y = 50/\sqrt{50} = \sqrt{50} = 5\sqrt{2}$, so $x = y = 5\sqrt{2}$.

4.4.11 Let x be the length of the sides of the base of the box and y be the height of the box. The volume is $x \cdot x \cdot y = 100$, so the constraint is $x^2 y = 100$, which gives $y = 100/x^2$. The objective function to be minimized is the surface area S of the box, which consists of $2x^2$ (for the top and base) + $4xy$ (for the 4 sides); therefore $S = 2x^2 + 4xy$. Using $y = 100x^2$, we have $S = 2x^2 + 4xy = 2x^2 + 4x \cdot \frac{100}{x^2} = 2x^2 + \frac{400}{x}$. The base side length can be any $x > 0$, so we need to maximize $S(x) = 2x^2 + 400/x$ on the interval $(0, \infty)$. The critical points of the objective function satisfy $S'(x) = 4x - \frac{400}{x^2} = 0$; clearing denominators gives $4x^3 = 400$ so $x = \sqrt[3]{100}$. By the First (or Second) Derivative Test, this critical point corresponds to a local minimum, and by Theorem 4.5, this solitary local minimum is also the absolute minimum on the interval $(0, \infty)$. Therefore the dimensions of the box with minimum surface area are $x = \sqrt[3]{100}$ and $y = 100/\sqrt[3]{100}^2 = \sqrt[3]{100}$, so length = width = height = $\sqrt[3]{100}$ m.

4.4.13 Let x be the length of the sides of the base of the box and y be the height of the box. The volume of the box is $x \cdot x \cdot y = x^2y$, so the constraint is $x^2y = 16$, which gives $y = 16/x^2$. Let c be the cost per square foot of the material used to make the sides. Then the cost to make the base is $2cx^2$, the cost to make the 4 sides is $4cxy$, and the cost to make the top is $\frac{1}{2}cx^2$. The objective function to be minimized is the total cost, which is $C = 2cx^2 + 4cxy + \frac{1}{2}cx^2 = \frac{5}{2}cx^2 + 4cx \cdot \frac{16}{x^2} = c\left(\frac{5x^2}{2} + \frac{64}{x}\right)$. The base side length can be any $x > 0$, so we need to maximize $C(x) = c(5x^2/2 + 64/x)$ on the interval $(0, \infty)$. The critical points of the objective function satisfy $5x - \frac{64}{x^2} = 0$, which gives $x^3 = 64/5$ or $x = 4/\sqrt[3]{5}$. By the First (or Second) Derivative Test, this critical point corresponds to a local minimum, and by Theorem 4.5, this solitary local minimum is also the absolute minimum on the interval $(0, \infty)$. Therefore the box with minimum cost has base $4/\sqrt[3]{5}$ ft by $4/\sqrt[3]{5}$ ft and height $y = 16/(4/\sqrt[3]{5})^2 = 5^{2/3}$ ft.

4.4.15

a. Let x be the distance from the point on the shoreline nearest to the boat to the point where the woman lands on shore; then the remaining distance she must travel on shore is $6 - x$. By the Pythagorean theorem, the distance the woman must row is $\sqrt{x^2 + 16}$. So the time for the rowing leg is $\frac{\text{distance}}{\text{rate}} = \frac{\sqrt{x^2+16}}{2}$ and the time for the walking leg is $\frac{\text{distance}}{\text{rate}} = \frac{6-x}{3}$. The total travel time for the trip is the objective function $T(x) = \frac{\sqrt{x^2 + 16}}{2} + \frac{6 - x}{3}$. We wish to minimize this function for $0 \le x \le 6$. The critical points of the objective function satisfy $T'(x) = \frac{x}{2\sqrt{x^2 + 16}} - \frac{1}{3} = 0$ which when simplified gives $5x^2 = 64$ and so $x = 8/\sqrt{5}$ is the only critical point in $(0, 6)$. From the First Derivative Test we see that T has a local minimum at this point, so $x = 8/\sqrt{5}$ must give the minimum value of T on $[0, 6]$.

b. Let $v > 0$ be the woman's rowing speed. Then the total travel time is now given by $T(x) = \frac{\sqrt{x^2 + 16}}{v} + \frac{6 - x}{3}$. The derivative of the objective function is $T'(x) = \frac{x}{v\sqrt{x^2 + 16}} - \frac{1}{3}$. If we try to solve the equation $T'(x) = 0$ as in part (a) above, we see that there is at most one solution $x > 0$. Therefore there can be at most one critical point of T in the interval $(0, 6)$. Observe also that $T'(0) = -1/3 < 0$ so the absolute minimum of T on $[0, 6]$ cannot occur at $x = 0$. So one of two things must happen: there is a unique critical point for T in $(0, 6)$ which is the absolute minimum for T on $[0, 6]$, and then $T'(6) > 0$; or, T is decreasing on $[0, 6]$, and then $T'(6) \le 0$ (the quickest way to the restaurant is to row directly in this case). The condition $T'(6) \le 0$ is equivalent to $\frac{6}{\sqrt{6^2+16}} \le \frac{v}{3}$ which gives $v \ge 9/\sqrt{13}$ mi/hr.

4.4.17 Let L be the ladder length and x be the distance between the foot of the ladder and the fence. The Pythagorean theorem gives the relationship $L^2 = (x+5)^2 + b^2$, where b is the height of the top of the ladder. We see that $b/(x + 5) = 8/x$ by similar triangles, which gives $b = 8(x + 5)/x$. Substituting in the expression for L^2 above gives $L^2 = (x + 5)^2 + 64\frac{(x+5)^2}{x^2} = (x + 5)^2 \left(1 + \frac{64}{x^2}\right)$. It suffices to minimize L^2 instead of L. However in this case x and b must satisfy $x, b \le 20$. Solving $20 = 8(x + 5)/x$ for x gives $x = 10/3$, so the condition $b \le 20$ corresponds to $x \ge 10/3$, and we see that we must minimize L^2 for $10/3 \le x \le 20$. We have $\frac{d}{dx}L^2 = (x + 5)^2 \left(-\frac{128}{x^3}\right) + 2(x + 5)\left(1 + \frac{64}{x^2}\right) = \frac{2(x + 5)(x^3 - 320)}{x^3}$. Because $x > 0$, the only critical point is $x = \sqrt[3]{320} \approx 6.84$. By the First Derivative Test, this critical point corresponds to a local minimum, and by Theorem 4.5, this solitary local minimum is also the absolute minimum on the interval $[10/3, 20]$. Substituting $x \approx 6.84$ in the expression for L^2 we find the length of the shortest ladder $L \approx 18.22$ ft.

4.4.19 Let the coordinates of the base of the rectangle be $(x, 0)$ and $(-x, 0)$ where $0 \leq x \leq 5$. Then the width of the rectangle is $2x$ and the height is $\sqrt{25 - x^2}$, so the area A is given by $A(x) = 2x\sqrt{25 - x^2}$. The critical points of this function satisfy $A'(x) = 2\sqrt{25 - x^2} + \dfrac{2x \cdot (-x)}{\sqrt{25 - x^2}} = \dfrac{2(25 - 2x^2)}{\sqrt{25 - x^2}} = 0$ which has unique solution $x = 5/\sqrt{2}$ in $(0, 5)$. We have $A(0) = A(5) = 0$, so the rectangle of maximum area has width $2x = 10/\sqrt{2}$ cm, height $y = \sqrt{25 - (25/2)} = 5/\sqrt{2}$ cm.

4.4.21 If we remove a sector of angle θ from a circle of radius 20, the remaining circumference is $2\pi \cdot 20 - \theta \cdot 20 = 20(2\pi - \theta)$. This length will be the circumference of the cone that is formed, so the base of the cone has radius $r = 20(2\pi - \theta)/2\pi$. The radius ranges from 0 to 20, and all possible cones formed have side length 20. The height h of the cone is given by the Pythagorean theorem: $h^2 + r^2 = 20^2$, so $h = \sqrt{400 - r^2}$. The objective function to be maximized is the volume of the cone given by $V = \frac{\pi}{3}r^2h = \frac{\pi}{3}r^2\sqrt{400 - r^2}$. The critical points of this function satisfy $V'(r) = \dfrac{\pi}{3}\left(2r\sqrt{400 - r^2} + \dfrac{r^2 \cdot (-r)}{\sqrt{400 - r^2}}\right) = \dfrac{\pi}{3} \cdot \dfrac{r(800 - 3r^2)}{\sqrt{400 - r^2}} = 0$, which has unique solution $r = \sqrt{800}/\sqrt{3} = 20\sqrt{6}/3$ in $(0, 20)$. Since $V(0) = V(20) = 0$, the maximum volume must occur when $r = 20\sqrt{6}/3$. This value of r corresponds to the angle $\theta = 2\pi\left(1 - \frac{\sqrt{6}}{3}\right)$.

4.4.23 Let x and y be the dimensions of the flower garden; the area of the flower garden is 30, so we have the constraint $xy = 30$ which gives $y = 30/x$. The dimensions of the garden and borders are $x + 4$ and $y + 2$, so the objective function to be minimized for $x > 0$ is $A = (x + 4)(y + 2) = (x + 4)\left(\frac{30}{x} + 2\right) = 2x + \frac{120}{x} + 38$. The critical points of $A(x)$ satisfy $A'(x) = 2 - \frac{120}{x^2} = 0$, which has unique solution $x = \sqrt{60} = 2\sqrt{15}$. By the First (or Second) Derivative test, this critical point gives a local minimum, which by Theorem 4.5 must be the absolute minimum of A over $(0, \infty)$. The corresponding value of y is $30/2\sqrt{15} = \sqrt{15}$, so the dimensions are $\sqrt{15}$ by $2\sqrt{15}$ m.

4.4.25 The radius r and height h of the barrel satisfy the constraint $r^2 + h^2 = d^2$, which we can rewrite as $r^2 = d^2 - h^2$. The volume of the barrel is given by $V = \pi r^2 h = \pi(d^2 - h^2)h = \pi(d^2 h - h^3)$. The height h must satisfy $0 \leq h \leq d$, so we need to maximize $V(h)$ on the interval $[0, d]$. The critical points of V satisfy $V'(h) = \pi(d^2 - 3h^2) = 0$. The only critical point in $(0, d)$ is $h = d/\sqrt{3}$, which gives the maximum volume since at the endpoints $V(0) = V(d) = 0$. The corresponding r value satisfies $r^2 = d^2 - d^2/3 = 2d^2/3$, so $r = \sqrt{2}d/\sqrt{3}$ and we see that the ratio r/h that maximizes the volume is $\sqrt{2}$.

4.4.27 Let h be the height of the cylindrical tower and r the radius of the dome. The cylinder has volume $\pi r^2 h$, and the hemispherical dome has volume $2\pi r^3/3$ (half the volume of a sphere of radius r). The total volume is 750, so we have the constraint $\pi r^2 h + \frac{2\pi r^3}{3} = 750$ which gives $h = \frac{750}{\pi r^2} - \frac{2r}{3}$. We must have $h \geq 0$, which is equivalent to $r \leq \sqrt[3]{1125/\pi}$. The objective function to be maximized is the cost of the metal to make the silo, which is proportional to the surface area of the cylinder $(= 2\pi r h)$ plus 1.5 times the surface area of the hemisphere $(= 2\pi r^2)$. So we can take as objective function $C = 2\pi r h + 1.5 \cdot 2\pi r^2 = 2\pi r\left(\frac{750}{\pi r^2} - \frac{2r}{3}\right) + 3\pi r^2 = \frac{1500}{r} + \frac{5}{3}\pi r^2$. The critical points of $C(r)$ satisfy $C'(r) = -\frac{1500}{r^2} + \frac{10}{3}\pi r = 0$ which gives $\pi r^3 = 450$ and hence $r = \sqrt[3]{450/\pi}$. The corresponding value of h is $h = \dfrac{750}{\pi r^2} - \dfrac{2r}{3} = \dfrac{750r}{\pi r^3} - \dfrac{2r}{3} = \left(\dfrac{750}{450} - \dfrac{2}{3}\right)r = r$. By the First (or Second) Derivative Test, this critical point corresponds to a local minimum, and by Theorem 4.5, this solitary local minimum is also the absolute minimum on the interval $[0, \sqrt[3]{1125/\pi}]$. Therefore the dimensions that minimize the cost are $r = h = \sqrt[3]{450/\pi}$ m.

4.4.29 Let x be the distance between the point and the weaker light source; then $12 - x$ is the distance to the stronger light source. The intensity is proportional to $I(x) = \dfrac{1}{x^2} + \dfrac{2}{(12 - x)^2}$, so we can take this as our objective function to be minimized for $0 < x < 12$. The critical points of $I(x)$ satisfy $I'(x) = -\dfrac{2}{x^3} + \dfrac{4}{(12 - x)^3} = 0$ which gives $\left(\frac{12-x}{x}\right)^3 = 2$, or $\frac{12-x}{x} = \sqrt[3]{2}$, or $x = \frac{12}{\sqrt[3]{2}+1} \approx 5.31$. By the First (or Second) Derivative Test, this critical point corresponds to a local minimum, and by Theorem 4.5, this solitary local minimum is also the absolute minimum on the interval $(0, 12)$. Therefore the intensity is weakest at the point $12/(\sqrt[3]{2} + 1) \approx 5.31$ m from the weaker source.

4.4.31 Let x be the distance from the point on shore nearest the island to the point where the underwater cable meets the shore, and let y be the be the length of the underwater cable. By the Pythagorean theorem, $y = \sqrt{x^2 + 3.5^2}$. The objective function to be minimized is the cost given by $C(x) = 2400\sqrt{x^2 + 3.5^2} + 1200 \cdot (8 - x) = 2400\sqrt{x^2 + 3.5^2} - 1200x + 9600$. We wish to minimize this function for $0 \le x \le 8$. The critical points of $C(x)$ satisfy $C'(x) = \dfrac{2400x}{\sqrt{x^2 + 3.5^2}} - 1200 = 1200\left(\dfrac{2x}{\sqrt{x^2 + 3.5^2}} - 1\right) = 0$, which we solve to obtain $x = 7\sqrt{3}/6$. By the First Derivative Test, this critical point corresponds to a local minimum, and by Theorem 4.5, this solitary local minimum is also the absolute minimum on the interval $[0, 8]$. Therefore the optimal point on shore has distance $x = 7\sqrt{3}/6$ mi from the point on shore nearest the island, in the direction of the power station.

4.4.33

a. Using the Pythagorean theorem, we find that the height of this triangle is 2. Let x be the distance from the point P to the base of the triangle; then the distance from P to the top vertex is $2 - x$ and the distance to each of the base vertices is $\sqrt{x^2 + 4}$, again by the Pythagorean theorem. Therefore the sum of the distances to the three vertices is given by $S(x) = 2\sqrt{x^2 + 4} + 2 - x$. We wish to minimize this function for $0 \le x \le 2$. The critical points of $S(x)$ satisfy $S'(x) = \frac{2x}{\sqrt{x^2+4}} - 1 = 0$, which has unique solution $x = 2/\sqrt{3}$ in $(0, 2)$. By the First Derivative Test, this critical point corresponds to a local minimum, and by Theorem 4.5, this solitary local minimum is also the absolute minimum on the interval $[0, 2]$. Therefore the optimal location for P is $2/\sqrt{3}$ units above the base.

b. In this case the objective function to be minimized is $S(x) = 2\sqrt{x^2 + 4} + h - x$ where $0 \le x \le h$. Exactly as above, we find that the only critical point $x > 0$ is $x = 2/\sqrt{3}$. This will give the absolute minimum on $[0, h]$ as long as $h \ge 2/\sqrt{3}$. When $h < 2/\sqrt{3}$, $S(x)$ is decreasing on $[0, h]$ and the minimum occurs at the endpoint $x = h$.

4.4.35 The critical points of the function $a(\theta)$ satisfy

$$a'(\theta) = \omega^2 r\left(-\sin\theta - \frac{2r\sin 2\theta}{L}\right) = -\omega^2 r \sin\theta\left(1 + \frac{4r\cos\theta}{L}\right) = 0,$$

using the identity $\sin 2\theta = 2\sin\theta\cos\theta$. There are two cases to consider separately: (a) $0 < L < 4r$ and (b) $L \ge 4r$. In case (a) the critical points in $[0, 2\pi]$ are $\theta = 0, \pi, 2\pi$ and also $\theta = \cos^{-1}(-L/(4r))$ and $2\pi - \cos^{-1}(-L/(4r))$. Comparing the values of $a(\theta)$ at these points shows that the maximum acceleration occurs at $\theta = 0$ and 2π and the minimum occurs at $\theta = \cos^{-1}(-L/(4r))$ and $2\pi - \cos^{-1}(-L/(4r))$. (There is a local maximum at $\theta = \pi$.) In case (b) the only critical points are $\theta = 0, \pi$ and 2π, and comparing the values of $a(\theta)$ at these points shows that the maximum acceleration occurs at $\theta = 0$ and 2π as in case (a), whereas the minimum occurs at $\theta = \pi$ in this case.

4.4.37

a. Let r and h be the radius and height of the can. The volume of the can is $V = \pi r^2 h$, which gives the constraint $\pi r^2 h = 354$ or $h = 354/(\pi r^2)$. The objective function to be minimized is the surface area, which consists of $2\pi r^2$ (for the top and bottom of the can) and $2\pi rh$ (for the side of the can). Therefore the objective function to be minimized is $A = 2\pi r^2 + 2\pi rh = 2\pi\left(r^2 + r\left(\frac{354}{\pi r^2}\right)\right) = 2\pi\left(r^2 + \frac{354}{\pi r}\right)$. We need to minimize $A(r)$ for $r > 0$. The critical points of $A(r)$ satisfy $A'(r) = 2\pi\left(2r - \frac{354}{\pi r^2}\right) = 0$, which gives $r = \sqrt[3]{(177/\pi)} \approx 3.83$ cm. The corresponding value of h is $h = \frac{354}{\pi r^2} = \frac{354r}{\pi r^3} = 2r \cdot \frac{177}{\pi r^3} = 2r$, so $h = 2\sqrt[3]{(177/\pi)} \approx 7.67$ cm. By the First (or Second) Derivative Test, this critical point corresponds to a local minimum, and by Theorem 4.5, this solitary local minimum is also the absolute minimum on the interval $(0, \infty)$.

b. We modify the objective function in part (a) above to account for the fact that the top and bottom of the can have double thickness: $A = 4\pi r^2 + 2\pi rh = 2\pi\left(2r^2 + r\left(\frac{354}{\pi r^2}\right)\right) = 4\pi\left(r^2 + \frac{177}{\pi r}\right)$. We need to minimize $A(r)$ for $r > 0$. The critical points of $A(r)$ satisfy $A'(r) = 4\pi\left(2r - \frac{177}{\pi r^2}\right) = 0$, which gives $r = \sqrt[3]{(177/2\pi)} \approx 3.04$ cm. The corresponding value of h is $h = \frac{354}{\pi r^2} = \frac{354r}{\pi r^3} = 4r \cdot \frac{177}{2\pi r^3} = 4r$, so $h = 4\sqrt[3]{(177/2\pi)} = 2\sqrt[3]{(708/\pi)} \approx 12.17$ cm. These dimensions are closer to those of a real soda can.

4.4.39 The viewing angle θ is given by $\theta = \cot^{-1}\left(\frac{x}{10}\right) - \cot^{-1}\left(\frac{x}{3}\right)$, and we wish to maximize this function for $x > 0$. The critical points satisfy $\theta'(x) = -\dfrac{1}{1 + \left(\frac{x}{10}\right)^2} \cdot \dfrac{1}{10} - (-)\dfrac{1}{1 + \left(\frac{x}{3}\right)^2} \cdot \dfrac{1}{3} = \dfrac{3}{x^2 + 3^2} - \dfrac{10}{x^2 + 10^2} = 0$ which simplifies to $3(x^2 + 100) = 10(x^2 + 9)$ or $x^2 = 30$. Therefore $x = \sqrt{30} \approx 5.5$ ft is the only critical point in $(0, \infty)$. By the First (or Second) Derivative Test, this critical point corresponds to a local maximum, and by Theorem 4.5, this solitary local maximum must be the absolute maximum on the interval $(0, \infty)$.

4.4.41 Let the radius of the ferris wheel have length r, and let α be the angle the specific seat on the ferris wheel makes with the center of the wheel (see figure). This point has coordinates $(r\cos\alpha, r + r\sin\alpha)$ so the distance from the seat to the base of the wheel is

$$d = \sqrt{r^2\cos^2\alpha + r^2(1 + \sin\alpha)^2} = \sqrt{2}\,r\sqrt{1 + \sin\alpha}.$$

Therefore the observer's angle satisfies $\tan\theta = \frac{r\sqrt{2}}{20}\sqrt{1 + \sin\alpha}$. Think of θ and α as functions of time t and differentiate: $\sec^2\theta \dfrac{d\theta}{dt} = \dfrac{r\sqrt{2}}{20} \cdot \dfrac{\cos\alpha}{2\sqrt{1 + \sin\alpha}} \dfrac{d\alpha}{dt} = \dfrac{\pi r\sqrt{2}}{40} \cdot \dfrac{\cos\alpha}{\sqrt{1 + \sin\alpha}}$. Therefore

$$\frac{d\theta}{dt} = \frac{\pi r\sqrt{2}}{40}\frac{\cos^2\theta\cos\alpha}{\sqrt{1 + \sin\alpha}}.$$

Observe that $\left|\dfrac{d\theta}{dt}\right| = \dfrac{\pi r\sqrt{2}}{40}\dfrac{\cos^2\theta|\cos\alpha|}{\sqrt{1 + \sin\alpha}}\dfrac{\sqrt{1 - \sin\alpha}}{\sqrt{1 - \sin\alpha}}$, which can be written as $\frac{\pi r\sqrt{2}}{40}\cos^2\theta\sqrt{1 - \sin\alpha}$. When the seat on the ferris wheel is at its lowest point we have $\theta = 0$ and $\alpha = -\pi/2$, which gives $\cos^2\theta = 1$ and $\sqrt{1 - \sin\alpha} = \sqrt{2}$. At any other point on the wheel we have $\cos^2\theta \le 1$ and $\sqrt{1 - \sin\alpha} < \sqrt{2}$, so θ is changing most rapidly when the seat is at its lowest point.

4.4.43 Let r and h be the radius and height of the cylinder. The distance d from the centroid of the cylinder (the midpoint of the cylinder's axis of rotation) to any point on the top or bottom edge satisfies $d^2 = r^2 + \left(\frac{h}{2}\right)^2$ so the constraint is $r^2 + (h/2)^2 = R^2$. The volume of the cylinder is given by $V = \pi r^2 h = \pi\left(R^2 - \left(\frac{h}{2}\right)^2\right)h = \pi\left(R^2 h - \frac{h^3}{4}\right)$. Since $r, h \ge 0$ we must have $0 \le h \le 2R$. We wish to maximize $V(h)$ on this interval. The critical points of $V(h)$ satisfy $V'(h) = \pi\left(R^2 - \frac{3h^2}{4}\right) = 0$ which gives $h = 2R/\sqrt{3}$, and from the constraint we obtain $r = \sqrt{2}R/\sqrt{3}$. The volume $V(h) = 0$ at the endpoints $h = 0$ and $h = 2R$, so the maximum volume must occur at this critical point.

4.4.45

a. Let r and h be the radius and height of the inscribed cylinder. The region that lies above the cylinder inside the cone is a cone with radius r and height $H - h$; by similar triangles we have $\frac{H - h}{r} = \frac{H}{R}$ so $h = \frac{H}{R}(R - r)$. The volume of the cylinder is $V = \pi r^2 h = \frac{\pi H}{R}\left(Rr^2 - r^3\right)$, which we must maximize over $0 \le r \le R$. The critical points of $V(r)$ satisfy $V'(r) = \frac{\pi H}{R}\left(2Rr - 3r^2\right) = 0$, which has unique solution $r = 2R/3$ in $(0, R)$. Since $V(r) = 0$ at the endpoints $r = 0$ and $r = R$, the cylinder with maximum volume has radius $r = 2R/3$, height $h = H/3$ and volume $V = \pi r^2 h = \frac{4\pi}{27}R^2 H = \frac{4}{9} \cdot \frac{\pi}{3}R^2 H$, i.e. $4/9$ the volume of the cone.

b. The lateral surface area of the cylinder is $A = 2\pi r h = 2\pi r \cdot \frac{H}{R}(R - r) = \frac{2\pi H}{R}r(R - r)$. This function takes its maximum over $0 \le r \le R$ at $r = R/2$, so the cylinder with maximum lateral surface area has dimensions $r = R/2$ and $h = H/2$.

4.4.47 Let R and H be the radius and height of the larger cone and let r and h be the radius and height of the smaller inscribed cone. The region that lies above the smaller cone inside the larger cone is a cone with radius r and height $H - h$; by similar triangles we have $\frac{H - h}{r} = \frac{H}{R}$ so $h = \frac{H}{R}(R - r)$. The volume of the smaller cone is $V = \frac{\pi}{3}r^2 h = \frac{\pi H}{3R}\left(Rr^2 - r^3\right)$, which we must maximize over $0 \le r \le R$. The critical points of $V(r)$ satisfy $V'(r) = \frac{\pi H}{3R}\left(2Rr - 3r^2\right) = 0$ which has unique solution $r = 2R/3$ in $(0, R)$. Since $V(r) = 0$ at the endpoints $r = 0$ and $r = R$, the smaller cone with maximum volume has radius $r = 2R/3$ and height $h = H/3$, so the optimal ratio of the heights is 3:1.

4.4.49

Following the hint, place two points P and Q above the midpoint of the base of the square, at distances x and y to the sides (see figure), where $0 \le x, y \le 1/2$. Then join the bottom vertices of the square to P, the upper vertices to Q and join P to Q. This road system has total length $L = 2\sqrt{x^2 + \frac{1}{4}} + 2\sqrt{y^2 + \frac{1}{4}} + (1 - x - y) = 1 + \left(\sqrt{4x^2 + 1} - x\right) + \left(\sqrt{4y^2 + 1} - y\right)$. We can minimize the contributions from x and y separately; the critical points of the function $f(x) = \sqrt{4x^2 + 1} - x$ satisfy $f'(x) = \frac{4x}{\sqrt{4x^2+1}} - 1 = 0$ which gives $\sqrt{4x^2 + 1} = 4x$, so $12x^2 = 1$ and $x = 1/(2\sqrt{3})$. By the First (or Second) Derivative Test, this critical point corresponds to a local minimum, and by Theorem 4.5, this solitary local minimum is also the absolute minimum on the interval $[0, 1/2]$. The minimum value of $f(x)$ on this interval is $f(1/(2\sqrt{3})) = \sqrt{3}/2$, so the shortest road system has length $L = 1 + 2 \cdot \frac{\sqrt{3}}{2} = 1 + \sqrt{3}$.

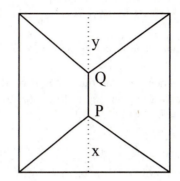

4.4.51 Let x be the distance between the point on the track nearest your initial position to the point where you catch the train. If you just catch the back of the train, then the train will have travelled $x + 1/3$ miles, which will require time $T = \dfrac{\text{distance}}{\text{rate}} = \dfrac{x + \frac{1}{3}}{20}$. The distance you must run is $\sqrt{x^2 + 1/(16)}$, so your running speed must be $v = \dfrac{\text{distance}}{\text{time}} = \dfrac{20\sqrt{x^2 + \frac{1}{16}}}{x + \frac{1}{3}}$. We wish to minimize this function for $x \ge 0$. The derivative of $v(x)$ can be written $v'(x) = \left(\dfrac{x}{x^2 + \frac{1}{16}} - \dfrac{1}{x + \frac{1}{3}}\right) v(x)$, so the critical points of $v(x)$ satisfy $\dfrac{x}{x^2 + \frac{1}{16}} = \dfrac{1}{x + \frac{1}{3}}$ so $x\left(x + \dfrac{1}{3}\right) = x^2 + \dfrac{1}{16}$ which gives $x = 3/16$ mi. By the First (or Second) Derivative Test, this critical point corresponds to a local minimum, and by Theorem 4.5, this solitary local minimum is also the absolute minimum on the interval $[0, \infty)$. The minimum running speed is $v\left(\dfrac{3}{16}\right) = \dfrac{20\sqrt{\left(\frac{3}{16}\right)^2 + \frac{1}{16}}}{\frac{3}{16} + \frac{1}{3}} = \dfrac{60\sqrt{9 + 16}}{9 + 16} = 12$ mph.

4.4.53

a. A point on the line $y = 3x + 4$ has the form $(x, 3x + 4)$, which has distance L to the origin given by $L^2 = x^2 + (3x + 4)^2 = 10x^2 + 24x + 16$. Since L is positive, it suffices to minimize L^2. The quadratic function $10x^2 + 24x + 16$ takes its minimum at $x = -24/20 = -6/5$, and the corresponding value of $y = 2/5$. Therefore the point closest to the origin on this line is $(-6/5, 2/5)$.

b. A point on the parabola $y = 1 - x^2$ has the form $(x, 1 - x^2)$, which has distance L to the point $(1,1)$ given by $L^2 = (x - 1)^2 + \left(1 - (1 - x^2)\right)^2 = x^4 + x^2 - 2x + 1$. Since L is positive, it suffices to minimize L^2. The critical points of L^2 satisfy $\frac{dL^2}{dx} = 4x^3 + 2x - 2 = 2(2x^3 + x - 1) = 0$ This cubic equation has a unique root $x \approx 0.59$, so the point closest to $(1, 1)$ on this parabola is approximately $(0.59, 0.65)$.

c. A point on the curve $y = \sqrt{x}$ has the form $(x, \sqrt{x})$, which has distance L to the point $(p, 0)$ given by $L^2 = (x - p)^2 + \sqrt{x}^2 = x^2 + (1 - 2p)x + p^2$. Since L is positive, it suffices to minimize L^2 for $x \ge 0$. This quadratic function takes its minimum at $x = -(1 - 2p)/2 = p - 1/2$, so in case (i) the minimum occurs at the point $(p - 1/2, \sqrt{p - 1/2})$ and in case (ii) there are no critical points for $x > 0$, the function L^2 is increasing on $[0, \infty)$ so the minimum occurs at $(0, 0)$.

4.4.55

a. We find $g(0) = 0$, $g(40) = 30$ and $g(60) = 25$ miles per gallon. The value at $v = 0$ is reasonable since when a car first starts moving it needs a lot of power from its engine, so the gas mileage is very low. The decline from 30 to 25 mi/gal as v increases from 40 mi/hr to 60 mi/hr reflects the fact that gas mileage tends to decrease at speeds over 55 mi/hr.

b. The quadratic function $g(v) = (85v - v^2)/60$ takes its maximum value at $v = 85/2 = 42.5$ mi/hr.

c. At speed v the amount of gas needed to drive one mile is $1/g(v)$ and the time it takes is $1/v$. Hence the cost of gas for one mile is $p/g(v)$ and the cost for the driver is w/v, and so the cost for L miles is $C(v) = Lp/g(v) + Lw/v$.

d. We have $C(v) = 400\left(\frac{4}{g(v)} + \frac{20}{v}\right) = 1600\left(\frac{1}{g(v)} + \frac{5}{v}\right)$. The critical points of $C(v)$ satisfy $\frac{g'(v)}{g(v)^2} + \frac{5}{v^2} = 0$, which simplifies to $v^2 g'(v) + 5g(v)^2 = 0$. Substituting the formula for $g(v)$ above and using $g'(v) = (85 - 2v)/60$, we can factor out v^2 and reduce to the quadratic equation $v^2 - 194v + 8245 = 0$, which has roots $v \approx 62.9, 131.1$. The First (or Second) Derivative Test shows that $C(v)$ has a local minimum at $v \approx 62.9$, which is the unique critical point for $0 \le v \le 131$. Therefore the cost is minimized at this value of v.

e. Since L is a constant factor in the cost function $C(v)$, changing L will not change the critical points of $C(v)$.

f. The critical points of $C(v)$ now satisfy the equation $\dfrac{4.2g'(v)}{g(v)^2} + \dfrac{20}{v^2} = 0$, which simplifies to $4.2v^2 g'(v) + 20g(v)^2 = 0$. As above, substituting the formula for $g(v)$ above and using $g'(v) = (85 - 2v)/60$, we can factor out v^2 and reduce to the quadratic equation $v^2 - 195.2v + 8296 = 0$, which has roots $v \approx 62.5, 132.7$. As in part (d), the minimum cost occurs for $v \approx 62.5$, slightly less than the speed in part (d).

g. The critical points of $C(v)$ now satisfy the equation $\dfrac{4g'(v)}{g(v)^2} + \dfrac{15}{v^2} = 0$, which simplifies to $4v^2 g'(v) + 15g(v)^2 = 0$. As above, substituting the formula for $g(v)$ above and using $g'(v) = (85 - 2v)/60$, we can factor out v^2 and reduce to the quadratic equation $v^2 - 202v + 8585 = 0$, which has roots $v \approx 60.8, 141.2$. As in part (d), the minimum cost occurs for $v \approx 60.8$, less than the speed in part (d).

4.4.57

a. Let x, $d - x$ be the distances from the point where the rope meets the ground to the poles of height m, n respectively. Then the rope has length $L(x) = \sqrt{x^2 + m^2} + \sqrt{(d-x)^2 + n^2}$. We wish to minimize this function for $0 \le x \le d$. The critical points of $L(x)$ satisfy $L'(x) = \dfrac{x}{\sqrt{x^2 + m^2}} - \dfrac{d - x}{\sqrt{(d-x)^2 + n^2}} = 0$, which is equivalent to $\dfrac{x}{\sqrt{x^2 + m^2}} = \dfrac{d - x}{\sqrt{(d-x)^2 + n^2}}$, or in terms of the angles θ_1 and θ_2 in the figure, $\sec\theta_1 = \sec\theta_2$ and therefore $\theta_1 = \theta_2$. Observe that $L'(0) < 0$ and $L'(d) > 0$, so the minimum value of $L(x)$ must occur at some $x \in (0, d)$. There must be exactly one critical point, because as x ranges from 0 to d, θ_1 decreases and θ_2 increases, and so $\theta_1 = \theta_2$ can occur for at most one value of x.

b. Because the speed of light is constant, travel time is minimized when distance is minimized, which we saw in part (a) occurs when $\theta_1 = \theta_2$.

4.4.59 Let the angle of the cuts with the horizontal be ϕ_1 and ϕ_2, where $\phi_1 + \phi_2 = \theta$. The volume of the notch is proportional to $\tan\phi_1 + \tan\phi_2 = \tan\phi_1 + \tan(\theta - \phi_1)$, so it suffices to minimize the objective function $V(\phi_1) = \tan\phi_1 + \tan(\theta - \phi_1)$ for $0 \le \phi_1 \le \theta$. The critical points of $V(\phi_1)$ satisfy $\sec^2\phi_1 - \sec^2(\theta - \phi_1) = 0$, which is equivalent to the condition $\cos\phi_1 = \cos(\theta - \phi_1)$. This is satisfied if $\phi_1 = \theta - \phi_1$ which gives $\phi_1 = \theta/2$. There are no other solutions in $(0, \theta)$, since $\cos\phi_1$ is decreasing and $\cos(\theta - \phi_1)$ is increasing on $(0, \theta)$ and therefore can intersect at most once. So the only critical point occurs when $\phi_1 = \phi_2 = \theta/2$, and the First Derivative Test shows that this critical point is a local minimum; by Theorem 4.5, this must be the absolute minimum on $[0, \theta]$.

4.5 Linear Approximation and Differentials

4.5.1

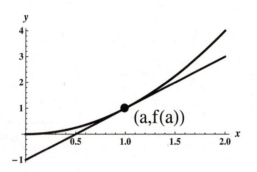

4.5.3 If f is differentiable at the point, then near that point, f is approximately linear, so the function nearly coincides with the tangent line at that point.

4.5.5 The relationship is given by $dy = f'(x)dx$, which is the linear approximation of the change Δy in $y = f(x)$ corresponding to a change dx in x.

4.5.7

$12 - x^2; \ a = 2; \ f(2.1) \quad -2x$

a. Note that $f(a) = f(2) = 8$ and $f'(a) = -2a = -4$, so the linear approximation has equation

$$y = L(x) = f(a) + f'(a)(x - a)$$
$$= 8 + (-4)(x - 2) = -4x + 16.$$

b.

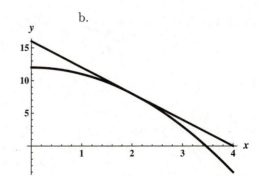

c. We have $f(2.1) \approx L(2.1) = 7.6$.

d. The percentage error is $100 \cdot \frac{|7.6 - 7.59|}{7.59} \approx 0.13\%$.

4.5.9

a. $f(0) = 1$ and $f'(x) = -1/(1+x)^2$ so that $f'(0) = -1$. Thus the linear approximation has the equation $y = L(x) = f(0) + f'(0)(x - 0) = 1 - x$.

c. $f(-0.1) \approx L(-0.1) = 1 - (-0.1) = 1.1$.

d. Noting that $f(-0.1) = 1/.9 = 10/9$, the percentage error is

$$100 \cdot \frac{|f(-0.1) - L(-0.1)|}{f(-0.1)} = 100 \cdot \frac{10/9 - 11/10}{10/9} = 1\%$$

b.

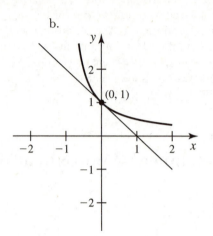

4.5.11

a. Note that $f(a) = f(0) = \cos 0 = 1$ and $f'(a) = -\sin a = 0$, so the linear approximation has equation

$$y = L(x) = f(a) + f'(a)(x - a) = 1.$$

c. We have $f(-0.01) \approx L(-0.01) = 1$.

d. The percentage error is $100 \cdot \frac{|1 - \cos(-0.01)|}{\cos(-0.01)} \approx 0.005\%$.

b.

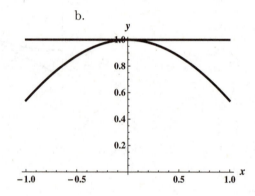

4.5.13 Let $f(x) = 1/x$, $a = 200$. Then $f(a) = 0.005$ and $f'(a) = -1/a^2 = -0.000025$, so the linear approximation to f near $a = 200$ is $L(x) = f(a) + f'(a)(x - a) = 0.005 - 0.000025(x - 200)$. Therefore $\frac{1}{203} = f(203) \approx L(203) = .004925$.

4.5.15 Let $f(x) = \sqrt{x}$, $a = 144$. Then $f(a) = 12$ and $f'(a) = 1/(2\sqrt{a}) = 1/24$, so the linear approximation to f near $a = 144$ is $L(x) = f(a) + f'(a)(x - a) = 12 + \frac{1}{24}(x - 144)$. Therefore $\sqrt{146} = f(146) \approx L(146) = 12\frac{1}{12}$.

4.5.17 Let $f(x) = 1/(1 + x)$ with $a = 0$. Then $f'(x) = -1/(1+x)^2$ so that $f'(0) = -1, f(0) = 1$. Then $f(x) \approx L(x) = f(0) + f'(0)(x - 0) = 1 - x$. Thus

$$\frac{1}{1.05} = f(.05) \approx L(.05) = .95$$

4.5.19 Let $f(x) = \sin(x + \pi/4)$ with $a = 0$. Then $f'(x) = \cos(x + \pi/4)$, so that $f'(0) = f(0) = \sqrt{2}/2$. Then $f(x) \approx L(x) = f(0) + f'(0)(x - 0) = (x + 1)\sqrt{2}/2$. Thus

$$\sin\left(\frac{\pi}{4} + 0.1\right) = f(0.1) \approx L(0.1) = \frac{1.1\sqrt{2}}{2} \approx .7778$$

4.5.21 Let $f(x) = 1/\sqrt[3]{x}$, $a = 512$. Then $f(a) = 1/8$ and $f'(a) = -1/(3a^{4/3}) = -1/12{,}288$, so the linear approximation to f near $a = 512$ is $L(x) = f(a) + f'(a)(x-a) = \frac{1}{8} - \frac{1}{12{,}288}(x-512)$. Therefore $\frac{1}{\sqrt[3]{510}} = f(510) \approx L(510) = \frac{769}{6144} \approx 0.1252$.

4.5.23 Note that $V'(r) = 4\pi r^2$, so $\Delta V \approx V'(a)\Delta r = 4\pi a^2 \Delta r$. Substituting $a = 5$ and $\Delta r = 0.1$ gives $\Delta V \approx 4\pi \cdot 25 \cdot 0.1 = 10\pi \approx 31.42$ ft^3.

4.5.25 Note that $S'(r) = \pi\sqrt{r^2+h^2} + \pi r \cdot \dfrac{r}{\sqrt{r^2+h^2}} = \pi\dfrac{2r^2+h^2}{\sqrt{r^2+h^2}}$, so $\Delta S \approx S'(a)\Delta r = \pi\dfrac{2a^2+h^2}{\sqrt{a^2+h^2}}\Delta r$. Substituting $h = 6$, $a = 10$ and $\Delta r = -0.1$ gives $\Delta S \approx \pi\dfrac{236}{\sqrt{136}}(-0.1) = \dfrac{-59\pi}{5\sqrt{34}} \approx -6.36$ m^2.

4.5.27 We have $f'(x) = 2$, so $dy = 2\,dx$.

4.5.29 We have $f'(x) = -3/x^4$, so $dy = -\frac{3}{x^4}\,dx$.

4.5.31 We have $f'(x) = a\sin x$, so $dy = a\sin x\,dx$.

4.5.33 We have $f'(x) = 9x^2 - 4$, so $dy = (9x^2 - 4)\,dx$.

4.5.35

   a. True. Note that $f(0) = 0$ and $f'(0) = 0$, so the linear approximation at 0 is in fact $L(x) = 0$.

   b. False. The function $f(x) = |x|$ is not differentiable at $x = 0$, so there is no good linear approximation at 0.

   c. True. For linear functions, the linear approximation at any point and the function are equal.

4.5.37

   a. Note that $f(a) = f(0) = 1$ and $f'(a) = -1/(1+a)^2 = -1$, so the linear approximation has equation $y = L(x) = f(a) + f'(a)(x-a) = 1 - x$.

   b.

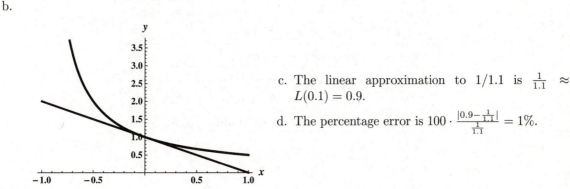

   c. The linear approximation to $1/1.1$ is $\frac{1}{1.1} \approx L(0.1) = 0.9$.

   d. The percentage error is $100 \cdot \dfrac{|0.9 - \frac{1}{1.1}|}{\frac{1}{1.1}} = 1\%$.

4.5.39

a. $f(0) = 4$. $f'(x) = \dfrac{1}{3}(x + 64)^{-2/3}$, so $f'(0) = \frac{1}{48}$. Thus $L(x) = f(0) + f'(0)(x - 0) = 4 + \frac{x}{48}$

b.

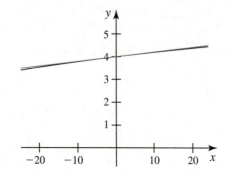

c. $\sqrt[3]{62.5} = f(-1.5) \approx L(-1.5) = 4 - \frac{1}{32} = 3.96875$.

d. $f(-1.5) = \sqrt[3]{62.5} \approx 3.9685$, so the percent error is

$$100 \cdot \frac{3.96875 - 3.9685}{3.9685} \approx 0.006\%$$

4.5.41 Note that $f(a) = f(8) = 2$ and $f'(a) = (1/3)a^{-2/3} = 1/12$, so the linear approximation has equation $y = L(x) = f(a) + f'(a)(x - a) = 2 + \frac{1}{12}(x - 8) = \frac{x}{12} + \frac{4}{3}$.

x	Linear approx	Exact value	Percent error
8.1	$2.008\overline{3}$	2.00829885	1.717×10^{-3}
8.01	$2.0008\overline{3}$	2.000832986	1.734×10^{-5}
8.001	$2.00008\overline{3}$	2.00008333	1.736×10^{-7}
8.0001	$2.0000083\overline{3}$	2.00000833	1.735×10^{-9}
7.9999	$1.999991\overline{6}$	1.999991667	1.738×10^{-9}
7.999	$1.99991\overline{6}$	1.999916663	1.738×10^{-7}
7.99	$1.9991\overline{6}$	1.999166319	1.736×10^{-5}
7.9	$1.991\overline{6}$	1.991631701	1.736×10^{-3}

The percentage errors become extremely small as x approaches 8 In fact, each time we decrease Δx by a factor of 10, the percentage error decreases by a factor of 100.

4.5.43

a. The linear approximation near $x = 1$ is more accurate for f because the rate at which f' is changing at 1 is smaller than the rate at which g' is changing at 1. The graph of f bends away from the linear function more slowly than the graph of g.

b. The larger the value of $|f''(a)|$, the greater the deviation of the curve $y = f(x)$ from the tangent line at points near $x = a$.

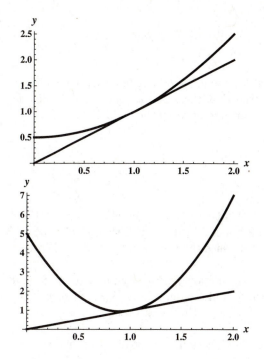

4.6 Mean Value Theorem

4.6.1

If f is a continuous function on the closed interval $[a, b]$ and is differentiable on (a, b) and the slope of the secant line that joins $(a, f(a))$ and $(b, f(b))$ is zero, then there is at least one value c in (a, b) at which the slope of the line tangent to f at $(c, f(c))$ is also zero.

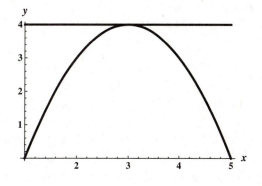

4.6.3 The function $f(x) = |x|$ is not differentiable at 0.

4.6.5

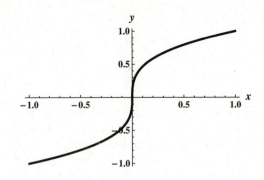

We seek a function over an interval for which it
isn't true that there is a tangent line parallel to
the secant line between the endpoints.

4.6.7 The function f is differentiable on $[0,1]$ and $f(0) = f(1) = 0$, so Rolle's theorem applies. We wish
to find a point x in $(0,1)$ such that $f'(x) = 0$; we have $f'(x) = (x-1)^2 + 2x(x-1) = (x-1)(3x-1)$, so
$x = 1/3$ satisfies the conclusion of Rolle's theorem.

4.6.9 The function f is differentiable on $[\pi/8, 3\pi/8]$ and $f(\pi/8) = f(3\pi/8) = 0$, so Rolle's theorem applies.
We wish to find a point x in $(\pi/8, 3\pi/8)$ such that $f'(x) = 0$; we have $f'(x) = -4\sin 4x$, so $x = \pi/4$ satisfies
the conclusion of Rolle's theorem.

4.6.11 The function f is not differentiable at $x = 0$, so Rolle's theorem does not apply.

4.6.13 The average rate of change of the temperature from 3.2 km to 6.1 km is $\frac{-10.3-8.0}{6.1-3.2} \approx -6.3°$/km.
Based on this, we cannot conclude that the lapse rate exceeds the critical value of 7°/ km.

4.6.15

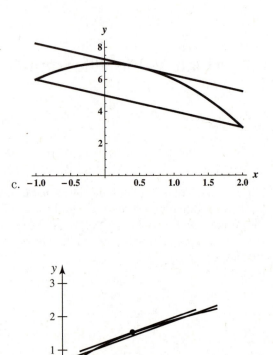

a. The function f is differentiable on $[-1,2]$ so the
mean value theorem applies.

b. The average rate of change of f on $[-1,2]$ is
$\frac{f(2)-f(-1)}{2-(-1)} = \frac{3-6}{3} = -1$. We wish to find a point
c in $(-1,2)$ such that $f'(c) = -1$, or equivalently
$-2c = -1$ which gives $c = 1/2$.

c.

4.6.17

a. $\sqrt{x}$ is continuous on $[1,4]$ and differentiable on
$(1,4)$, so the Mean Value Theorem applies.

b. The average rate of change of $f(x)$ on $[1,4]$ is

$$\frac{f(4) - f(1)}{4 - 1} = \frac{1}{3}$$

and $f'(x) = \frac{1}{2\sqrt{x}}$. The required x is therefore such
that $2\sqrt{x} = 3$, so $x = \frac{9}{4}$.

c.

4.6.19

a. $x^{-1/3}$ is continuous on $[1/8, 8]$ and is differentiable on $(1/8, 8)$. Thus the Mean Value Theorem applies.

b. The average rate of change of $f(x)$ on $[1/8, 8]$ is

$$\frac{f(8) - f(1/8)}{8 - 1/8} = \frac{1/2 - 2}{63/8} = -\frac{4}{21}$$

and $f'(x) = -1/(3x^{4/3})$, so we want x such that $-1/(3x^{4/3}) = -4/21$, or $21 = 12x^{4/3}$. Thus $x = \left(\frac{7}{4}\right)^{3/4} \approx 1.52$.

c.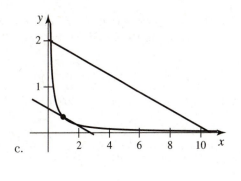

4.6.21

a. The mean value theorem does not apply since the function f is not differentiable at $x = 0$.

b. Even though the mean value theorem doesn't apply, it still happens to be the case that there is a number c between -8 and 8 where the tangent line has slope $\frac{f(8) - f(-8)}{8 - (-8)} = \frac{1}{2}$. This occurs where $\frac{2}{3}c^{-2/3} = 1/2$, which is $c = \frac{8}{9} \cdot \sqrt{3}$.

c.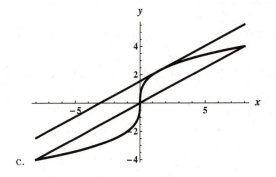

4.6.23

a. False. The function f is not differentiable at $x = 0$.

b. True. If $f(x) - g(x) = c$ is constant, then $f'(x) - g'(x) = 0$.

c. False. If $f'(x) = 0$ then we can conclude that $f(x) = c$ for some constant.

4.6.25 $h'(x) = p'(x) = f'(x)$ since all three functions differ by a constant. $g'(x) \neq f'(x)$ since the functions differ by a multiple, not a constant.

4.6.27 The secant line between the endpoints has slope $\dfrac{f(4) - f(-4)}{4 - (-4)} = \dfrac{4 - 1}{8} = \dfrac{3}{8}$. The slope of the tangent line to the graph appears to have this value at approximately -2.5 and at about 2.6. These are eyeballed estimates, so your personal estimate may differ.

4.6.29 The average speed of the car over the 28 minute period ($= 28/60$ hr) is $\frac{30-0}{28/60} \approx 64$mi/hr, so the officer can conclude by the mean value theorem that at some point the car exceeded the speed limit.

4.6.31 The runner's average speed is $6.2/(32/60) \approx 11.6$ mi /hr. By the mean value theorem, the runner's speed was 11.6 mi/hr at least once. By the intermediate value theorem, all speeds between 0 and 11.6 mi/hr were reached. Because the initial and final speed was 0 mi/hr, the speed of 11 mi/hr was reached at least twice.

4.6.33 Observe that

$$\frac{f(b) - f(a)}{b - a} = \frac{A(b^2 - a^2) + B(b - a)}{b - a} = A(a + b) + B$$

and $f'(c) = 2Ac + B$, so the point c that satisfies the conclusion of the mean value theorem is $c = (a + b)/2$.

4.6.35 Note that $f'(x) = 2\tan x \sec^2 x$ and $g'(x) = 2\sec x \sec x \tan x = 2\tan x \sec^2 x$, so $f'(x) = g'(x)$. This implies that $f - g$ is a constant, which also follows from the trigonometric identity $\sec^2 x = \tan^2 x + 1$.

4.6.37 Bolt's average speed during the race was $\frac{100}{9.58}$ m/s $= \frac{100}{9.58} \cdot \frac{3600}{1000}$ km/hr ≈ 37.58 km/hr, so by the mean value theorem he must have exceeded 37 km/hr during the race.

4.6.39

a. If $g(x) = x$ then $g'(x) = 1$ and hence $\dfrac{f(b) - f(a)}{g(b) - g(a)} = \dfrac{f(b) - f(a)}{b - a} = \dfrac{f'(c)}{g'(c)} = f'(c)$.

b. We have $\dfrac{f(b) - f(a)}{g(b) - g(a)} = \dfrac{0 - (-1)}{6 - 2} = \dfrac{1}{4}$; $\dfrac{f'(c)}{g'(c)} = \dfrac{2c}{4} = \dfrac{c}{2}$; so $c = 1/2$.

4.7 L'Hôpital's Rule

4.7.1 If $\lim\limits_{x \to a} f(x) = 0$ and $\lim\limits_{x \to a} g(x) = 0$, then we say $\lim\limits_{x \to a} f(x)/g(x)$ is of indeterminate form $0/0$.

4.7.3 Take the limit of the quotient of the derivatives of the numerator and denominator.

4.7.5 If $\lim\limits_{x \to a} f(x)g(x)$ has the form $0 \cdot \infty$, then $\lim\limits_{x \to a} \dfrac{f(x)}{1/g(x)}$ has the indeterminate form $0/0$ or ∞/∞.

4.7.7

$$\lim_{x \to 1^-} (x - 1)\tan(\pi x/2) = \lim_{x \to 1^-} \frac{(x-1)\sin(\pi x/2)}{\cos(\pi x/2)}$$

which is of the form $0/0$.

4.7.9 L'Hôpital's rule gives $\lim\limits_{x \to 2} \dfrac{x^2 - 2x}{8 - 6x + x^2} = \lim\limits_{x \to 2} \dfrac{2x - 2}{-6 + 2x} = \dfrac{2}{-2} = -1$.

4.7.11 L'Hôpital's rule gives $\lim\limits_{x \to 0} \dfrac{3\sin 4x}{5x} = \lim\limits_{x \to 0} \dfrac{12\cos 4x}{5} = \dfrac{12}{5}$.

4.7.13 L'Hôpital's rule gives $\lim\limits_{u \to \pi/4} \dfrac{\tan u - \cot u}{u - \pi/4} = \lim\limits_{u \to \pi/4} \dfrac{\sec^2 u + \csc^2 u}{1} = 2 + 2 = 4$.

4.7.15 Apply l'Hôpital's rule twice: $\lim\limits_{x \to 0} \dfrac{1 - \cos 3x}{8x^2} = \lim\limits_{x \to 0} \dfrac{3\sin 3x}{16x} = \lim\limits_{x \to 0} \dfrac{9\cos 3x}{16} = \dfrac{9}{16}$.

4.7.17 Apply l'Hôpital's rule twice:

$$\lim_{x \to -1} \frac{x^3 - x^2 - 5x - 3}{x^4 + 2x^3 - x^2 - 4x - 2} = \lim_{x \to -1} \frac{3x^2 - 2x - 5}{4x^3 + 6x^2 - 2x - 4}$$
$$= \lim_{x \to -1} \frac{6x - 2}{12x^2 + 12x - 2} = 4$$

4.7.19 L'Hôpital's rule gives $\lim\limits_{v \to 3} \dfrac{v - 1 - \sqrt{v^2 - 5}}{v - 3} = \lim\limits_{v \to 3} \dfrac{1 - \dfrac{v}{\sqrt{v^2 - 5}}}{1} = -\dfrac{1}{2}$.

4.7.21 L'Hôpital's rule gives $\lim\limits_{h \to 0} \dfrac{\sin(x + h) - \sin x}{h} = \lim\limits_{h \to 0} \dfrac{\cos(x + h)}{1} = \cos x$. (Notice that this limit is the definition of the derivative of $\sin x$.)

4.7.23 Apply l'Hôpital's rule three times:

$$\lim_{x\to\infty}\frac{3x^4-x^2}{6x^4+12}=\lim_{x\to\infty}\frac{12x^3-2x}{24x^3}=\lim_{x\to\infty}\frac{36x^2-2}{72x^2}=\lim_{x\to\infty}\frac{72x}{144x}=\frac{1}{2}.$$

4.7.25 The degree of the numerator is less than that of the denominator, so the limit is zero:

$$\lim_{x\to\infty}\frac{8-4x^2}{3x^3+x-1}=\lim_{x\to\infty}\frac{8/x^2-4}{3x+1/x-1/x^2}=0$$

4.7.27 Observe that $\displaystyle\lim_{x\to0}x\csc x=\lim_{x\to0}\frac{x}{\sin x}=\lim_{x\to0}\frac{1}{\cos x}=1$, by l'Hôpital's rule.

4.7.29 Observe that $\displaystyle\lim_{x\to(\pi/2)^-}\left(\frac{\pi}{2}-x\right)\sec x=\lim_{x\to(\pi/2)^-}\frac{\pi/2-x}{\cos x}=\lim_{x\to(\pi/2)^-}\frac{-1}{-\sin x}=1$ by l'Hôpital's rule.

4.7.31 Observe that $\displaystyle\lim_{x\to0^+}\left(\cot x-\frac{1}{x}\right)=\lim_{x\to0^+}\left(\frac{\cos x}{\sin x}-\frac{1}{x}\right)=\lim_{x\to0^+}\frac{x\cos x-\sin x}{x\sin x}$. Apply l'Hôpital's rule twice:

$$\lim_{x\to0^+}\frac{x\cos x-\sin x}{x\sin x}=\lim_{x\to0^+}\frac{\cos x-x\sin x-\cos x}{\sin x+x\cos x}=-\lim_{x\to0^+}\frac{x\sin x}{\sin x+x\cos x}$$

$$=-\lim_{x\to0^+}\frac{\sin x+x\cos x}{\cos x+\cos x-x\sin x}=-\frac{0}{2}=0.$$

4.7.33 Observe that

$$\lim_{\theta\to(\pi/2)^-}(\tan\theta-\sec\theta)=\lim_{\theta\to(\pi/2)^-}\left(\frac{\sin\theta}{\cos\theta}-\frac{1}{\cos\theta}\right)=\lim_{\theta\to(\pi/2)^-}\frac{\sin\theta-1}{\cos\theta},$$

and by l'Hôpital's rule

$$\lim_{\theta\to(\pi/2)^-}\frac{\sin\theta-1}{\cos\theta}=\lim_{\theta\to(\pi/2)^-}\frac{\cos\theta}{-\sin\theta}=\frac{0}{-1}=0.$$

4.7.35

a. False; $\displaystyle\lim_{x\to2}x^2-1=3$, so l'Hôpital's rule does not apply. In fact, $\displaystyle\lim_{x\to2}\frac{x-2}{x^2-1}=\frac{0}{3}=0$.

b. False; l'Hôpital's rule does not say $\displaystyle\lim_{x\to a}f(x)g(x)=\lim_{x\to a}f'(x)\lim_{x\to a}g'(x)$. In fact, $\displaystyle\lim_{x\to0}x\sin x=0\cdot0=0$.

c. False.

$$\lim_{x\to2}\frac{x^3-2x^2+x-2}{x^2-1}=\frac{0}{3}=0$$

$$\lim_{x\to2}\frac{3x^2-4x+1}{2x}=\frac{5}{4}$$

so the limits are not equal.

d. True. We must use L'Hôpital's rule to evaluate the left-hand limit since it has the form $0/0$.

$$\lim_{x\to2}\frac{x^3-2x^2+x-2}{x^2-4}=\lim_{x\to2}\frac{3x^2-4x+1}{2x}=\frac{5}{4}$$

and this is exactly the limit on the right-hand side.

4.7.37 Observe that $\lim\limits_{x\to\infty} \dfrac{2x^3 - x^2 + 1}{5x^3 + 2x} = \dfrac{2}{5}$ by Theorem 2.7. We can also use l'Hôpital's rule:

$$\lim_{x\to\infty} \frac{2x^3 - x^2 + 1}{5x^3 + 2x} = \lim_{x\to\infty} \frac{6x^2 - 2x}{15x^2 + 2} = \lim_{x\to\infty} \frac{12x - 2}{30x} = \lim_{x\to\infty} \frac{12}{30} = \frac{2}{5}.$$

4.7.39 By l'Hôpital's rule, $\lim\limits_{x\to 6} \dfrac{(5x+2)^{1/5} - 2}{x^{-1} - 6^{-1}} = \lim\limits_{x\to 6} \dfrac{(5x+2)^{-4/5}}{-x^{-2}} = -\dfrac{9}{4}$.

4.7.41 Observe that $(\sqrt{x-2} - \sqrt{x-4}) \cdot \dfrac{\sqrt{x-2} + \sqrt{x-4}}{\sqrt{x-2} + \sqrt{x-4}} = \dfrac{x - 2 - (x-4)}{\sqrt{x-2} + \sqrt{x-4}} = \dfrac{2}{\sqrt{x-2} + \sqrt{x-4}}$, so

$\lim\limits_{x\to\infty} \sqrt{x-2} - \sqrt{x-4} = \lim\limits_{x\to\infty} \dfrac{2}{\sqrt{x-2} + \sqrt{x-4}} = 0$.

4.7.43 Make the substitution $t = 1/x$; then $\lim\limits_{x\to\infty} x^3 \left(\dfrac{1}{x} - \sin\dfrac{1}{x} \right) = \lim\limits_{t\to 0+} \dfrac{t - \sin t}{t^3} = \lim\limits_{t\to 0+} \dfrac{1 - \cos t}{3t^2} =$

$\lim\limits_{t\to 0+} \dfrac{\sin t}{6t} = \lim\limits_{t\to 0+} \dfrac{\cos t}{6} = \dfrac{1}{6}$, using l'Hôpital's rule.

4.7.45 Observe that $\lim\limits_{x\to 1+} \left(\dfrac{1}{x-1} - \dfrac{1}{\sqrt{x-1}} \right) = \lim\limits_{x\to 1+} \dfrac{1 - \sqrt{x-1}}{x-1} = \dfrac{1}{0} = \infty$.

4.7.47 $\lim\limits_{\theta\to\infty} \dfrac{\sin 2\theta - \theta^3}{3\theta^3} = \lim\limits_{\theta\to\infty} \dfrac{(\sin 2\theta)/\theta^3 - 1}{3} = -\dfrac{1}{3}$.

4.7.49 L'Hôpital's rule gives $\lim\limits_{x\to\infty} \dfrac{\sqrt{ax+b}}{\sqrt{cx+d}} = \lim\limits_{x\to\infty} \dfrac{a}{\sqrt{ax+b}} \cdot \dfrac{\sqrt{cx+d}}{c} = \dfrac{a}{c} \lim\limits_{x\to\infty} \dfrac{\sqrt{cx+d}}{\sqrt{ax+b}}$, which is the same

form as the original limit, so l'Hôpital's rule fails in this case. We can evaluate this limit as follows: first

observe that $\lim\limits_{x\to\infty} \dfrac{ax+b}{cx+d} = \dfrac{a}{c}$ by l'Hôpital's rule; therefore $\lim\limits_{x\to\infty} \dfrac{\sqrt{ax+b}}{\sqrt{cx+d}} = \lim\limits_{x\to\infty} \sqrt{\dfrac{ax+b}{cx+d}} = \sqrt{\dfrac{a}{c}}$.

4.7.51 The triangle ABP has base $1 - \cos\theta$ and height $\sin\theta$, so its area is $f(\theta) = \frac{1}{2}\sin\theta(1 - \cos\theta)$. The sector OBP has area $\theta/2$, and the triangle OBP has base 1 and height $\sin\theta$; therefore $g(\theta) = \frac{1}{2}(\theta - \sin\theta)$. We have $\lim\limits_{\theta\to 0} \dfrac{g(\theta)}{f(\theta)} = \lim\limits_{\theta\to 0} \dfrac{\theta - \sin\theta}{\sin\theta(1 - \cos\theta)} = \lim\limits_{\theta\to 0} \dfrac{\theta - \sin\theta}{\sin\theta - (1/2)\sin 2\theta}$. Three applications of l'Hôpital's rule gives $\lim\limits_{\theta\to 0} \dfrac{g(\theta)}{f(\theta)} = \lim\limits_{\theta\to 0} \dfrac{1 - \cos\theta}{\cos\theta - \cos 2\theta} = \lim\limits_{\theta\to 0} \dfrac{\sin\theta}{2\sin 2\theta - \sin\theta} = \lim\limits_{\theta\to 0} \dfrac{\cos\theta}{4\cos 2\theta - \cos\theta} = \dfrac{1}{3}$.

4.8   Antiderivatives

4.8.1 Derivative, antiderivative.

4.8.3 $x + C$, where C is any constant.

4.8.5 $\dfrac{x^{p+1}}{p+1} + C$, where C is any real number and $p \neq -1$.

4.8.7 Observe that $F(-1) = 4 + C = 4$, so $C = 0$.

4.8.9 The antiderivatives of $5x^4$ are $x^5 + C$. Check: $\frac{d}{dx}(x^5 + C) = 5x^4$.

4.8.11 The antiderivatives of $\sin 2x$ are $-(1/2)\cos 2x + C$. Check: $\frac{d}{dx}(-\frac{1}{2}\cos 2x + C) = \sin 2x$.

4.8.13 The antiderivatives of $3\sec^2 x$ are $3\tan x + C$. Check: $\frac{d}{dx}(3\tan x + C) = 3\sec^2 x$.

4.8.15 The antiderivatives of $-2/y^3 = -2y^{-3}$ are $y^{-2} + C$. Check: $\frac{d}{dy}(y^{-2} + C) = -2y^{-3}$.

4.8.17 $\int (3x^5 - 5x^9)\,dx = 3 \cdot \frac{x^6}{6} - 5 \cdot \frac{x^{10}}{10} + C = \frac{1}{2}x^6 - \frac{1}{2}x^{10} + C$. Check: $\frac{d}{dx}(\frac{1}{2}x^6 - \frac{1}{2}x^{10} + C) = 3x^5 - 5x^9$.

4.8.19 $\int \left(4\sqrt{x} - \frac{4}{\sqrt{x}}\right) dx = \int (4x^{1/2} - 4x^{-1/2})\,dx = 4 \cdot \frac{x^{3/2}}{3/2} - 4 \cdot \frac{x^{1/2}}{1/2} + C = \frac{8}{3}x^{3/2} - 8x^{1/2} + C$. Check: $\frac{d}{dx}(\frac{8}{3}x^{3/2} - 8x^{1/2} + C) = 4\sqrt{x} - \frac{4}{\sqrt{x}}$.

4.8.21 Note that $\frac{d}{ds}(5s+3)^3 = 15(5s+3)^2$; therefore $\int (5s+3)^2\,ds = \frac{(5s+3)^3}{15} + C$.

4.8.23 $\int (3x^{1/3} + 4x^{-1/3} + 6)\,dx = 3 \cdot \frac{3}{4}x^{4/3} + 4 \cdot \frac{3}{2}x^{2/3} + 6x + C = \frac{9}{4}x^{4/3} + 6x^{2/3} + 6x + C$. Check: $\frac{d}{dx}(\frac{9}{4}x^{4/3} + 6x^{2/3} + 6x + C) = 3x^{1/3} + 4x^{-1/3} + 6$.

4.8.25 Using Table 4.5 (formulas 1 and 2), $\int (\sin 2y + \cos 3y)\,dy = -\frac{1}{2}\cos 2y + \frac{1}{3}\sin 3y + C$. Check: $\frac{d}{dy}(\frac{-1}{2}\cos 2y + \frac{1}{3}\sin 3y + C) = \sin 2y + \cos 3y$.

4.8.27 Using Table 4.5 (formula 3), $\int (\sec^2 x - 1)\,dx = \tan x - x + C$. Check: $\frac{d}{dx}(\tan x - x + C) = \sec^2 x - 1$.

4.8.29 Using Table 4.5 (formulas 3 and 5), $\int (\sec^2 \theta + \sec\theta \tan\theta)\,d\theta = \tan\theta + \sec\theta + C$. Check: $\frac{d}{d\theta}(\tan\theta + \sec\theta + C) = \sec^2 \theta + \sec\theta \tan\theta$.

4.8.31 We have $F(x) = \int (x^5 - 2x^{-2} + 1)\,dx = \frac{x^6}{6} + 2x^{-1} + x + C$; substituting $F(1) = 0$ gives $\frac{1}{6} + 2 + 1 + C = 0$, so $C = -\frac{19}{6}$, and thus $F(x) = \frac{x^6}{6} + \frac{2}{x} + x - \frac{19}{6}$.

4.8.33 We have $F(v) = \int \sec v \tan v\,dv = \sec v + C$; substituting $F(0) = 2$ gives $\sec 0 + C = 1 + C = 2$, so $C = 1$, and thus $F(v) = \sec v + 1$.

4.8.35 We have $f(x) = \int (2x - 3)\,dx = x^2 - 3x + C$; substituting $f(0) = 4$ gives $C = 4$, so $f(x) = x^2 - 3x + 4$.

4.8.37 We have $g(x) = \int 7x \left(x^6 - \frac{1}{7}\right) dx = \int (7x^7 - x)\,dx = \frac{7}{8}x^8 - \frac{x^2}{2} + C$; substituting $g(1) = 2$ gives $\frac{7}{8} - \frac{1}{2} + C = 2$, so $C = \frac{13}{8}$, and thus $g(x) = \frac{7}{8}x^8 - \frac{x^2}{2} + \frac{13}{8}$.

4.8.39 We have $f(u) = \int 4(\cos u - \sin 2u)\,du = 4(\sin u + \frac{1}{2}\cos 2u) + C = 4\sin u + 2\cos 2u + C$; substituting $f(\pi/6) = 0$ gives $4\sin\frac{\pi}{6} + 2\cos\frac{\pi}{3} + C = 2 + 1 + C = 0$, so $C = -3$, and thus $f(u) = 4\sin u + 2\cos 2u - 3$.

4.8.41

We have $f(x) = \int (2x - 5)\,dx = x^2 - 5x + C$; substituting $f(0) = 4$ gives $C = 4$, so $f(x) = x^2 - 5x + 4$.

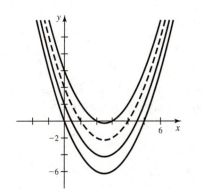

4.8.43

We have $f(x) = \int (3x + \sin \pi x)\,dx = \dfrac{3x^2}{2} - \dfrac{\cos \pi x}{\pi} + C$; substituting $f(2) = 3$ gives $6 - \dfrac{\cos 2\pi}{\pi} + C = 6 - \dfrac{1}{\pi} + C = 3$, so $C = \dfrac{1}{\pi} - 3$, and thus $f(x) = \dfrac{3x^2}{2} - \dfrac{\cos \pi x}{\pi} + \dfrac{1 - 3\pi}{\pi}$.

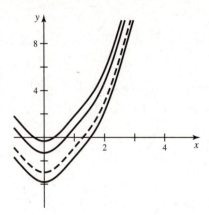

4.8.45

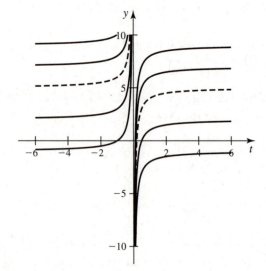

We have $f(t) = \int \frac{1}{t^2}\,dt = -\frac{1}{t} + C$. $f(1) = 4$ gives $4 = -1 + C$ so $C = 5$. Thus $f(t) = 5 - \frac{1}{t}$.

4.8.47

We have $s(t) = \int (2t + 4)\,dt = t^2 + 4t + C$; substituting $s(0) = 0$ gives $C = 0$, so $s(t) = t^2 + 4t$.

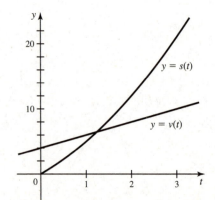

4.8.49

We have $s(t) = \int 2\sqrt{t}\,dt = 2 \cdot \frac{2}{3}t^{3/2} + C = \frac{4}{3}t^{3/2} + C$; substituting $s(0) = 1$ gives $C = 1$, so $s(t) = \frac{4}{3}t^{3/2} + 1$.

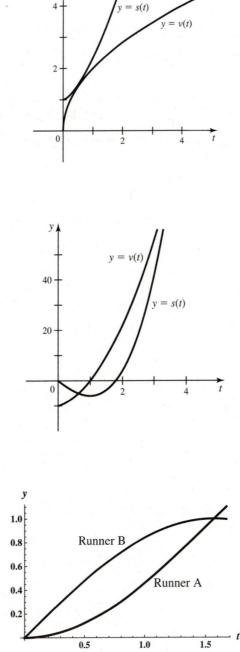

4.8.51

We have $s(t) = \int (6t^2 + 4t - 10)\,dt = 2t^3 + 2t^2 - 10t + C$; substituting $s(0) = 0$ gives $C = 0$, so $s(t) = 2t^3 + 2t^2 - 10t$.

4.8.53

Runner A has position function $s(t) = \int \sin t\,dt = -\cos t + C$; the initial condition $s(0) = 0$ gives $C = 1$, so $s(t) = 1 - \cos t$. Runner B has position function $S(t) = \int \cos t\,dt = \sin t + C$; the initial condition $S(0) = 0$ gives $C = 0$, so $S(t) = \sin t$. The smallest $t > 0$ where $s(t) = S(t)$ is $t = \pi/2$ s.

4.8.55

a. We have $v(t) = -9.8t + v_0$ and $v_0 = 30$, so $v(t) = -9.8t + 30$.

b. The height of the softball above ground is given by $s(t) = \int (-9.8t + 30)\,dt = -4.9t^2 + 30t + s_0 = -4.9t^2 + 30t$.

c. The ball reaches its maximum height when $v(t) = -9.8t + 30 = 0$, which gives $t = 30/9.8 \approx 3.06$ s; the maximum height is $s(30/9.8) \approx 45.92$ m.

   d. The ball strikes the ground when $s(t) = 0$ (and $t > 0$), which gives $t(30 - 4.9t) = 0$, so $t = 30/4.9 \approx$ 6.12 s.

4.8.57

   a. We have $v(t) = -9.8t + v_0$ and $v_0 = 10$, so $v(t) = -9.8t + 10$.

   b. The height of the payload above ground is given by $s(t) = \int(-9.8t + 10)\,dt = -4.9t^2 + 10t + s_0 = -4.9t^2 + 10t + 400$.

   c. The payload reaches its maximum height when $v(t) = -9.8t + 10 = 0$, which gives $t = 10/9.8 \approx 1.02$ s; the maximum height is $s(10/9.8) \approx 405.10$ m.

   d. The payload strikes the ground when $s(t) = 0$ (and $t > 0$), which gives $-4.9t^2 + 10t + 400 = 0$, so $t \approx 10.11$ s.

4.8.59

   a. True, because $F'(x) = G'(x)$.

   b. False; f is the derivative of F.

   c. True; $\int f(x)\,dx$ is the most general antiderivative of $f(x)$,which is $F(x) + C$.

   d. False; a function cannot have more than one derivative.

   e. False; one can only conclude that $F(x)$ and $G(x)$ differ by a constant.

4.8.61

$$\int \frac{\sqrt{2x} + \sqrt[3]{8x}}{x}\,dx = \int \sqrt{2}x^{-1/2} + 2x^{-2/3}\,dx = 2\sqrt{2x} + 6\sqrt[3]{x} + C$$

Check:

$$\frac{d}{dx}(2\sqrt{2x} + 6\sqrt[3]{x} + C) = \frac{d}{dx}(2\sqrt{2}x^{1/2} + 6x^{1/3} + C) = \sqrt{2}x^{-1/2} + 2x^{-2/3} = \frac{\sqrt{2x} + \sqrt[3]{8x}}{x}$$

4.8.63 $\int(\csc^2\theta + 2\theta^2 - 3\theta)\,d\theta = -\cot\theta + \frac{2}{3}\theta^3 - \frac{3}{2}\theta^2 + C$. Check: $\frac{d}{d\theta}(-\cot\theta + \frac{2}{3}\theta^3 - \frac{3}{2}\theta^2 + C) = \csc^2\theta + 2\theta^2 - 3\theta$.

4.8.65

$$\int \frac{1 + \sqrt{x}}{x^2}\,dx = \int (x^{-2} + x^{-3/2})\,dx = -x^{-1} - 2x^{-1/2} + C$$

Check:

$$\frac{d}{dx}(-x^{-1} - 2x^{-1/2} + C) = x^{-2} + x^{-3/2} = \frac{1 + \sqrt{x}}{x^2}$$

4.8.67 $\int \sqrt{x}(2x^6 - 4\sqrt[3]{x})\,dx = \int(2x^{13/2} - 4x^{5/6})\,dx = 2 \cdot \frac{2}{15}x^{15/2} - 4 \cdot \frac{6}{11}x^{11/6} + C = \frac{4}{15}x^{15/2} - \frac{24}{11}x^{11/6} + C$. Check: $\frac{d}{dx}(\frac{4}{15}x^{15/2} - \frac{24}{11}x^{11/6} + C) = 2x^{13/2} - 4x^{5/6} = \sqrt{x}(2x^6 - 4\sqrt[3]{x})$.

4.8.69 We have $F'(x) = \int \cos x\,dx = \sin x + C$; $F'(0) = 3$ so $C = 3$. Then $F(x) = \int(\sin x + 3)\,dx = -\cos x + 3x + C$; $F(\pi) = 4$ gives $1 + 3\pi + C = 4$ so $C = 3 - 3\pi$ and $F(x) = -\cos x + 3x + 3 - 3\pi$.

4.8.71 We have $F''(x) = \int(672x^5 + 24x)\,dx = 672 \cdot \frac{x^6}{6} + 24 \cdot \frac{x^2}{2} + C$; $F''(0) = 0$ so $C = 0$. Next $F'(x) = \int(112x^6 + 12x^2)\,dx = 112 \cdot \frac{x^7}{7} + 12 \cdot \frac{x^3}{3} + C$; $F'(0) = 2$ so $C = 2$. Finally $F(x) = \int(16x^7 + 4x^3 + 2)\,dx = 16 \cdot \frac{x^8}{8} + 4 \cdot \frac{x^4}{4} + 2x + C$; $F(0) = 1$ so $C = 1$ and $F(x) = 2x^8 + x^4 + 2x + 1$.

4.8.73

a. We have $Q(t) = \int 0.1(100 - t^2)\,dt = 0.1\left(100t - \frac{t^3}{3}\right) + C$; $Q(0) = 0$, so $C = 0$ and $Q(t) = 10t - t^3/30$ gal.

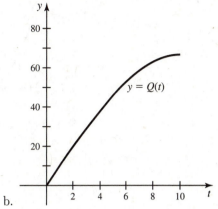

b.

c. $Q(10) = 200/3 \approx 67$ gal.

4.8.75

$$\int \sin^2 x\,dx = \frac{1}{2}\int (1 - \cos 2x)\,dx = \frac{1}{2}\left(x - \frac{\sin 2x}{2}\right) + C = \frac{x}{2} - \frac{\sin 2x}{4} + C$$

$$\int \cos^2 x\,dx = \frac{1}{2}\int (1 + \cos 2x)\,dx = \frac{1}{2}\left(x + \frac{\sin 2x}{2}\right) + C = \frac{x}{2} + \frac{\sin 2x}{4} + C.$$

4.8.77 Check that $\dfrac{d}{dx}(\sqrt{x^2 + 1}) = \dfrac{2x}{2\sqrt{x^2 + 1}} = \dfrac{x}{\sqrt{x^2 + 1}}$.

4.8.79 Check that $\dfrac{d}{dx}\left(-\dfrac{1}{2(x^2 - 1)}\right) = -\dfrac{1}{2}\dfrac{d}{dx}(x^2 - 1)^{-1} = -\dfrac{1}{2}(-1)(x^2 - 1)^{-2}(2x) = \dfrac{x}{(x^2 - 1)^2}$.

4.9 Chapter Four Review

4.9.1

a. False. The point $(c, f(c))$ is a critical point for f, but is not necessarily a local maximum or minimum. Example: $f(x) = x^3$ at $c = 0$.

b. False. The fact that $f''(c) = 0$ does not necessarily imply that f changes concavity at c. Example: $f(x) = x^4$ at $c = 0$.

c. True. Both are antiderivatives of $2x$.

d. True. The function has a maximum on the closed interval determined by the two local minima, and the only way the maximum can occur at the endpoints is if the function is constant, in which case every point is a local max and min.

4.9.3

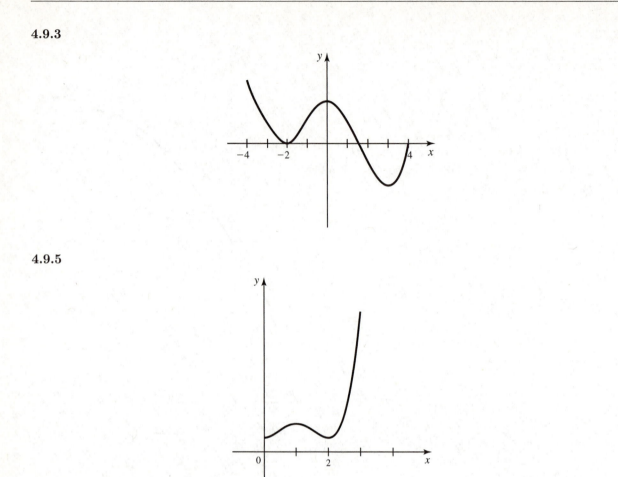

4.9.5

4.9.7 The critical points satisfy $f'(x) = 6x^2 - 6x - 36 = 6(x-3)(x+2) = 0$, so the critical points are $x = 3, -2$. Since $\lim\limits_{x \to \infty} f(x) = \infty$ and $\lim\limits_{x \to -\infty} f(x) = -\infty$, this function has no absolute max or min on $(-\infty, \infty)$.

4.9.9

$$f'(x) = \frac{(x+1)(2x) - (x^2+8)(1)}{(x+1)^2}$$
$$= \frac{x^2 + 2x - 8}{(x+1)^2}$$

The critical points are the points where $f'(x) = 0$ or $f'(x)$ is undefined; this happens when $x^2 + 2x - 8 = (x+4)(x-2) = 0$ or when $(x+1)^2 = 0$, so the critical points are at $x = -4, 2$ and at $x = -1$. On the given interval, $[-5, 5]$, the function is unbounded since it has a vertical asymptote at $x = -1$. Since $\lim\limits_{x \to -1^-} f(x) = -\infty$ and $\lim\limits_{x \to -1^+} f(x) = \infty$, it has no absolute minimum or maximum.

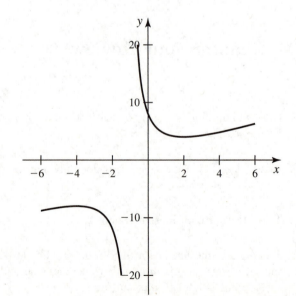

4.9.11

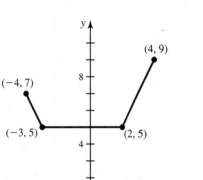

All points in the interval $[-3, 2]$ are critical points. The absolute max occurs at $(4, 9)$; there are no local maxima. All points $(x, 5)$ for x in the interval $[-3, 2]$ are absolute and local minima.

4.9.13 The derivatives of f are $f'(x) = 2x^3 - 6x + 4$, $f''(x) = 6x^2 - 6$. Observe that $f'(x) = 2(x-1)^2(x+2)$, so we have critical points $x = 1, -2$. Solving $f''(x) = 6(x^2 - 1) = 0$ gives possible inflection points at $x = \pm 1$. Testing the sign of $f'(x)$ shows that f is decreasing on the interval $(-\infty, -2)$ and is increasing on $(-2, \infty)$. The First Derivative Test shows that a local minimum occurs at $x = -2$, and that the critical point $x = 1$ is neither a local max or min. Testing the sign of $f''(x)$ shows that f is concave down on the interval $(-1, 1)$ and is concave up on the intervals $(-\infty, -1)$ and $(1, \infty)$. Therefore inflection points occur at $x = \pm 1$. Using a numerical solver, we see that the graph has x-intercepts at $x \approx -2.92, -0.22$. We also observe that $\lim\limits_{x \to \pm\infty} f(x) = \infty$, so f has no absolute maximum and an absolute minimum at $x = -2$.

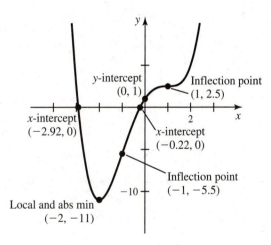

4.9.15 The derivatives of f are $f'(x) = -4\pi \sin[\pi(x - 1)]$, $f''(x) = -4\pi^2 \cos[\pi(x - 1)]$. Solving $f'(x) = 0$ gives critical point $x = 1$, and solving $f''(x) = 0$ gives possible inflection points at $x = 1/2, 3/2$. Testing the sign of $f'(x)$ shows that f is decreasing on the interval $(1, 2)$ and increasing on $(0, 1)$. The First Derivative Test shows that a local maximum occurs at $x = 1$. By Theorem 4.5, this solitary local maximum must be the absolute maximum for f on the interval $[0, 2]$. Testing the sign of $f''(x)$ shows that f is concave down on the interval $(1/2, 3/2)$ and concave up on the intervals $(0, 1/2)$ and $(3/2, 2)$. Therefore inflection points occur at $x = 1/2, 3/2$. These points are also the x-intercepts of the graph. We also observe that $f(0) = f(2) = -4$, so f takes its absolute minimum at these points.

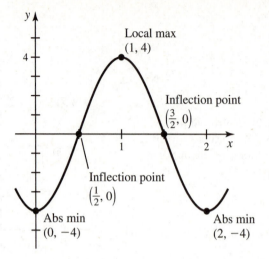

4.9.17 Note that the domain of f is the interval $(0, \infty)$. The derivatives of f are $f'(x) = \frac{1}{3}x^{-2/3} - \frac{1}{2}x^{-1/2}$, $\quad f''(x) = -\frac{2}{9}x^{-5/3} + \frac{1}{4}x^{-3/2}$. Solving $f'(x) = 0$ gives critical point $x = (2/3)^6$, and solving $f''(x) = 0$ gives a possible inflection point at $x = (8/9)^6$. Testing the sign of $f'(x)$ shows that f is increasing on the interval $(0, (2/3)^6)$ and decreasing on the interval $((2/3)^6, \infty)$. The First Derivative Test shows that a local maximum occurs at $x = (2/3)^6$ and By Theorem 4.5 this solitary local maximum must be the absolute maximum of f on the interval $(0, \infty)$. Testing the sign of $f''(x)$ shows that f is concave down on the interval $(0, (8/9)^6)$ and concave up on the interval $((8/9)^6, \infty)$. Therefore an inflection point occurs at $x = (8/9)^6$. Using a numerical solver, we find that the graph has x-intercept at $x \approx 23.77$. Since $\lim_{x \to \infty} f(x) = -\infty$, f has no absolute minimum on the interval $(0, \infty)$.

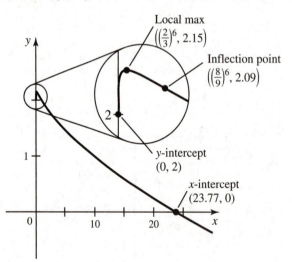

4.9.19 The derivatives of f are $f'(x) = \frac{2}{3}x^{-1/3} + \frac{1}{3}(x+2)^{-2/3}$, $\quad f''(x) = -\frac{2}{9}\left(x^{-4/3} + (x+2)^{-5/3}\right)$. Solving $f'(x) = 0$ gives critical points $x \approx -2.57, -1.56$; we also have critical points at $x = -2, 0$ since $f'(x)$ is undefined at these points. Solving $f''(x) = 0$ numerically gives a possible inflection point at $x \approx -6.43$. We also have possible inflection points at $x = -2, 0$ since $f''(x)$ is undefined at these points. Testing the sign of $f'(x)$ shows that f is decreasing on the intervals $(-\infty, -2.57)$ and $(-1.56, 0)$ and increasing on $(-2.57, -1.56)$ and $(0, \infty)$. The First Derivative Test shows that local mins occur at $x \approx -2.57$ and $x = 0$ and a local max occurs at $x \approx -1.56$. Testing the sign of $f''(x)$ shows that f is concave down on the intervals $(-\infty, -6.43)$, $(-2, 0)$ and $(0, \infty)$ and concave up on the interval $(-6.43, -2)$. Therefore inflection points occur at $x \approx -6.43$ and $x = -2$. Since $\lim_{x \to \pm\infty} f(x) = \infty$, f has no absolute maximum. The absolute minimum occurs at $x \approx -2.57$.

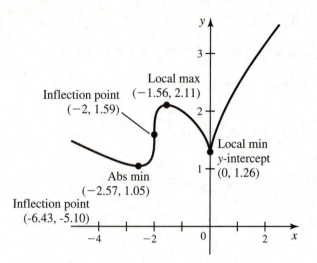

4.9.21 The objective function to be maximized is the volume of the cone, given by $V = \pi r^2 h/3$. By the Pythagorean theorem, r and h satisfy the constraint $h^2 + r^2 = 16$, which gives $r^2 = 16 - h^2$. Therefore $V(h) = \frac{\pi}{3} h(16 - h^2) = \frac{\pi}{3}(16h - h^3)$. We must maximize this function for $0 \le h \le 4$. The critical points of $V(h)$ satisfy $V'(h) = \frac{\pi}{3}(16 - 3h^2) = 0$, which has unique solution $h = 4/\sqrt{3} = 4\sqrt{3}/3$ in $(0,4)$. Since $V(0) = V(4) = 0$, $h = 4\sqrt{3}/3$ gives the maximum value of $V(h)$ on $[0,4]$. The corresponding value of r satisfies $r^2 = 16 - \frac{16}{3} = \frac{32}{3}$, so $r = \frac{4\sqrt{2}}{\sqrt{3}} = \frac{4\sqrt{6}}{3}$.

4.9.23

a. The constraint $a + b = 23$ gives $b = 23 - a$, so the objective function to be minimized/maximized is $Q(a) = a^2 + b^2 = a^2 + (23 - a)^2 = 2a^2 - 46a + 529$ We have $0 \le a$, and $a \le 23$ (otherwise $b < 0$). Therefore we must maximize/minimize $Q(a)$ on the closed interval $[0, 23]$. The critical points of $Q(a)$ satisfy $Q'(a) = 4a - 46 = 0$, which gives $a = 23/2$ as the only critical point. Note that $Q(0) = Q(23) = 529$ and $Q(23/2) = 529/2$, so the minimum occurs when $a = b = 23/2$.

b. Using the result above, the maximum occurs when $a = 0, b = 23$ or $a = 23, b = 0$.

4.9.25

a. The average rate of change of $P(t)$ on the interval $[0,8]$ is $\frac{P(8) - P(0)}{8 - 0} = \frac{800/9 - 0}{8} = \frac{100}{9}$ cells/week.

b. We solve $P'(t) = \frac{100}{(t+1)^2} = \frac{100}{9}$ which gives $(t + 1)^2 = 9$, so $t = 2$ weeks.

4.9.27 L'Hôpital's rule gives $\lim\limits_{t \to 0} \dfrac{1 - \cos 6t}{2t} = \lim\limits_{t \to 0} \dfrac{6\sin 6t}{2} = 0$.

4.9.29 Observe that $\lim\limits_{\theta \to 0} \dfrac{3\sin^2 2\theta}{\theta^2} = 3\left(\lim\limits_{\theta \to 0} \dfrac{\sin 2\theta}{\theta}\right)^2$; l'Hôpital's rule gives $\lim\limits_{\theta \to 0} \dfrac{\sin 2\theta}{\theta} = \lim\limits_{\theta \to 0} \dfrac{2\cos 2\theta}{1} = 2$, so $\lim\limits_{\theta \to 0} \dfrac{3\sin^2 2\theta}{\theta^2} = 3 \cdot 2^2 = 12$.

4.9.31 Observe that $2\theta \cot 3\theta = 2\cos 3\theta \cdot \dfrac{\theta}{\sin 3\theta}$; $\lim\limits_{\theta \to 0} \dfrac{\theta}{\sin 3\theta} = \lim\limits_{\theta \to 0} \dfrac{1}{3\cos 3\theta} = \dfrac{1}{3}$ by l'Hôpital's rule, so $\lim\limits_{\theta \to 0} 2\theta \cot 3\theta = 2 \cdot 1 \cdot \dfrac{1}{3} = \dfrac{2}{3}$.

4.9.33 Apply l'Hôpital's rule twice:

$$\lim_{x \to 1} \frac{x^4 - x^3 - 3x^2 + 5x - 2}{x^3 + x^2 - 5x + 3} = \lim_{x \to 1} \frac{4x^3 - 3x^2 - 6x + 5}{3x^2 + 2x - 5} = \lim_{x \to 1} \frac{12x^2 - 6x - 6}{6x + 2} = 0.$$

4.9.35 $\int \left(\frac{1}{x^2} - \frac{2}{x^{5/2}} \right) dx = \int (x^{-2} - 2x^{-5/2}) \, dx = -x^{-1} - 2 \cdot \left(-\frac{2}{3} \right) x^{-3/2} = -\frac{1}{x} + \frac{4}{3} x^{-3/2} + C.$

4.9.37 Using Table 4.5 (formula 1), $\int (1 + \cos 3\theta) \, d\theta = \theta + \frac{\sin 3\theta}{3} + C.$

4.9.39 Using Table 4.5 (formula 5), $\int \sec 2x \tan 2x \, dx = \frac{1}{2} \sec 2x + C.$

4.9.41

$$\int (4x^{1/3} - 7x^{2/5} + 10x^{3/7}) \, dx = 3x^{4/3} - 5x^{7/5} + 7x^{10/7}$$

4.9.43 $\int \left(\sqrt[4]{x^3} + \sqrt{x^5} \right) dx = \int (x^{3/4} + x^{5/2}) \, dx = \frac{4}{7} x^{7/4} + \frac{2}{7} x^{7/2} + C.$

4.9.45 We have $f(t) = \int (\sin t + 2t) \, dt = -\cos t + t^2 + C$; $f(0) = -1 + C = 5$, so $C = 6$ and $f(t) = -\cos t + t^2 + 6.$

4.9.47 Using the identity $\sin^2 x = (1 - \cos 2x)/2$ and Table 4.5 (formula 1), we see that $h(x) = \frac{1}{2} \int (1 - \cos 2x) \, dx = \frac{1}{2} \left(x - \frac{\sin 2x}{2} \right) + C = \frac{x}{2} - \frac{\sin 2x}{4} + C.$ We have $h(1) = \frac{1}{2} - \frac{\sin 2}{4} + C = 1$, so $C = \frac{1}{2} + \frac{\sin 2}{4}$ and so $h(x) = \frac{x}{2} - \frac{\sin 2x}{4} + \frac{1}{2} + \frac{\sin 2}{4}.$

4.9.49 The velocity of the rocket is given by $v(t) = -9.8t + v_0 = -9.8t + 120$, and the position function of the rocket is $s(t) = \int (-9.8t + 120) \, dt = -4.9t^2 + 120t + s_0 = -4.9t^2 + 120t + 125$. The rocket reaches its maximum height when $v(t) = 0$, which occurs at $t = 120/9.8 \approx 12.24$ s; the maximum height is $s(120/9.8) \approx 859.69$ m. The rocket hits the ground when $s(t) = 0$; solving this quadratic equation gives $t \approx 25.49$ s.

4.9.51 Observe that $\lim\limits_{x \to \infty} \dfrac{2x^5 - x + 1}{5x^6 + x} = 0$ by Theorem 2.7. We can also use l'Hôpital's rule:

$$\lim_{x \to \infty} \frac{2x^5 - x + 1}{5x^6 + x} = \lim_{x \to \infty} \frac{10x^4 - 1}{30x^5 + 1} = \lim_{x \to \infty} \frac{40x^3}{150x^4} = \lim_{x \to \infty} \frac{4}{15x} = 0.$$

4.9.53

a. Make the change of variables $y = x^n$ and apply l'Hôpital's rule twice: $\lim\limits_{x \to 0} \dfrac{1 - \cos x^n}{x^{2n}} = \lim\limits_{y \to 0} \dfrac{1 - \cos y}{y^2} = \lim\limits_{y \to 0} \dfrac{\cos y}{2} = \dfrac{1}{2}.$

b. Apply l'Hôpital's rule: $\lim\limits_{x \to 0} \dfrac{1 - \cos^n x}{x^2} = \lim\limits_{x \to 0} \dfrac{n \cos^{n-1} x \sin x}{2x} = \dfrac{n}{2} \left(\lim\limits_{x \to 0} \cos^{n-1} x \right) \left(\lim\limits_{x \to 0} \dfrac{\sin x}{x} \right) = \dfrac{n}{2} \cdot 1 \cdot 1 = \dfrac{n}{2}.$

Chapter 5

Integration

5.1 Approximating Areas under Curves

5.1.1

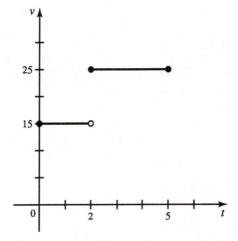

In the first 2 seconds, the object moves $15 \cdot 2 = 30$ meters. In the next three seconds, the object moves $25 \cdot 3 = 75$ meters, so the total displacement is $75 + 30 = 105$ meters.

5.1.3 Subdivide the interval from 0 to $\pi/2$ into subintervals. On each subinterval, pick a sample point (like the left endpoint, or the right endpoint, or the midpoint, for example) and call the first sample point x_1 and the second sample point x_2 and so on. For each sample point x_i, calculate the area of the rectangle which lies over the subinterval and has height $f(x_i) = \cos(x_i)$. Do this for each subinterval, and add the areas of the corresponding rectangles together. This will give an approximation to the area under the curve, with generally a better approximation occurring as n increases – where n is the number of subintervals used.

5.1.5 Since the interval $[1,3]$ has length 2, if we subdivide it into 4 subintervals, each will have length $\Delta x = \frac{2}{4} = \frac{1}{2}$. The grid points will be $x_0 = 1$, $x_1 = 1 + \Delta x = 1.5$, $x_2 = 1 + 2\Delta x = 2$, $x_3 = 1 + 3\Delta x = 2.5$, and $x_4 = 1 + 4\Delta x = 3$. If we use the left-hand side of each subinterval, we will use 1, 1.5, 2, and 2.5. If we use the right-hand side of each subinterval, we will use 1.5, 2, 2.5, and 3. If we use the midpoint of each subinerval, we will use 1.25, 1.75, 2.25, and 2.75.

5.1.7 It is an underestimate. If we use the right-hand side of each subinterval to determine the height of the rectangles, the height of each rectangle will be the minimum of f over the subinterval, so the sum of the areas of the rectangles will be less than the corresponding area under the curve.

169

5.1.9

a. On the first subinterval, the midpoint is .5, and $v(.5) = 1.75$. On the 2nd subinterval, the midpoint is 1.5 and $v(1.5) = 7.75$. Continuing in this manner, we obtain the estimate to the displacement of

$$v(.5) \cdot 1 + v(1.5) \cdot 1 + v(2.5) \cdot 1 + v(3.5) \cdot 1 = 1.75 + 7.75 + 19.75 + 37.75 = 67.$$

b. This time the midpoints are at .25, .75, 1.25 . . . . Each subinterval has length $\frac{1}{2}$. Thus, the estimate is given by

$$v(.25) \cdot .5 + v(.75) \cdot .5 + v(1.25) \cdot .5 + v(1.75) \cdot .5 + v(2.25) \cdot .5 + v(2.75) \cdot .5 + v(3.25) \cdot .5 + v(3.75) \cdot .5$$
$$= .5(1.1875 + 2.6875 + 5.6875 + 10.1875 + 16.1875 + 23.6875 + 32.6875 + 43.1875)$$
$$= .5(135.5) = 67.75.$$

5.1.11

The left-hand grid points are 0, 2, 4, and 6. The length of each subinterval is 2. So the left Riemann sum is given by

$$v(0) \cdot 2 + v(2) \cdot 2 + v(4) \cdot 2 + v(6) \cdot 2$$
$$= 2 \cdot \left(\frac{1}{1} + \frac{1}{5} + \frac{1}{9} + \frac{1}{13} \right) \approx 2.776.$$

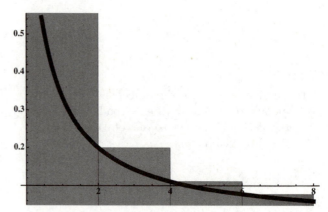

5.1.13

The left-hand grid points are 0, 3, 6, 9, and 12. The length of each subinterval is 3. So the left Riemann sum is given by

$$\sum_{i=1}^{5} v(3(i-1)) \cdot 3$$
$$= 12 + 24 + 12\sqrt{7} + 12\sqrt{10} + 12\sqrt{13}$$
$$\approx 148.96.$$

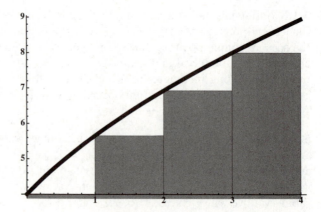

5.1.15 The left Riemann sum is given by $f(1) + f(2) + f(3) + f(4) + f(5) = 2 + 3 + 4 + 5 + 6 = 20$. The right Riemann sum is given by $f(2) + f(3) + f(4) + f(5) + f(6) = 3 + 4 + 5 + 6 + 7 = 25$.

5.1.17

a.

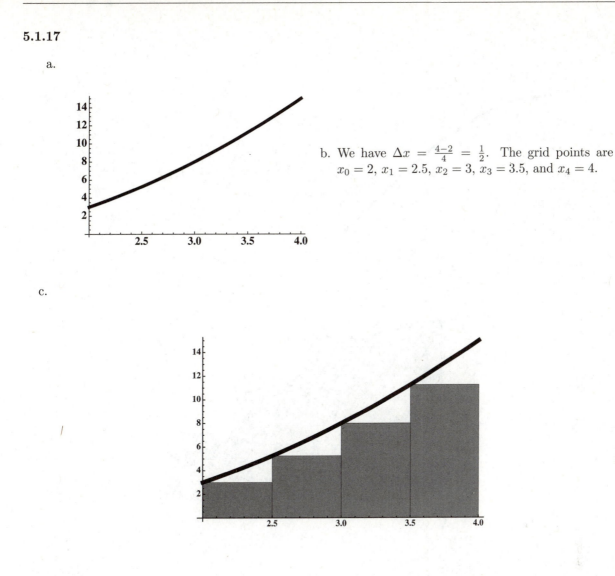

b. We have $\Delta x = \frac{4-2}{4} = \frac{1}{2}$. The grid points are $x_0 = 2$, $x_1 = 2.5$, $x_2 = 3$, $x_3 = 3.5$, and $x_4 = 4$.

c.

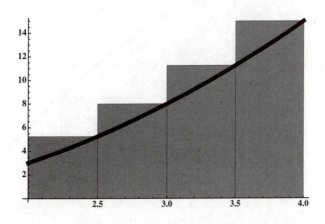

d. The left Riemann sum is $(3 + 5.25 + 8 + 11.25)(.5) = 13.75$, which is an underestimate of the area under the curve. The right Riemann sum is $(5.25 + 8 + 11.25 + 15)(.5) = 19.75$ which is an overestimate of the area under the curve.

5.1.19

a.

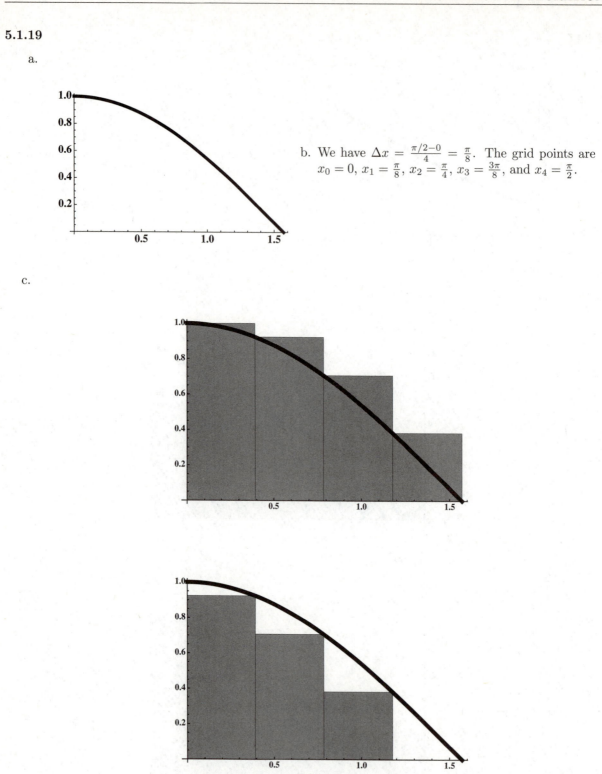

b. We have $\Delta x = \frac{\pi/2 - 0}{4} = \frac{\pi}{8}$. The grid points are $x_0 = 0$, $x_1 = \frac{\pi}{8}$, $x_2 = \frac{\pi}{4}$, $x_3 = \frac{3\pi}{8}$, and $x_4 = \frac{\pi}{2}$.

c.

d. The left Riemann sum is $\frac{\pi}{8} \cdot (1 + \cos(\pi/8) + \cos(\pi/4) + \cos(3\pi/8)) \approx 1.18$, which is an overestimate of the area under the curve. The right Riemann sum is $\frac{\pi}{8} \cdot (\cos(\pi/8) + \cos(\pi/4) + \cos(3\pi/8) + 0) \approx 0.79$ which is an unerestimate of the area under the curve.

5.1.21 We have $\Delta x = 2$, so the midpoints are 1, 3, 5, 7, and 9. So the midpoint Riemann sum is $2(f(1) + f(3) + f(5) + f(7) + f(9)) = 670$.

5.1.23

a.

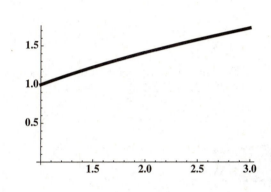

b. We have $\Delta x = \frac{3-1}{4} = \frac{1}{2}$. So the midpoints are 1.25, 1.75, 2.25, and 2.75.

c.

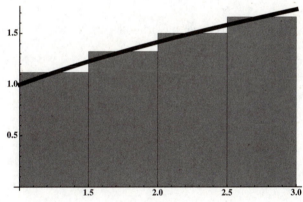

d. The midpoint Riemann sum is $.5(\sqrt{1.25} + \sqrt{1.75} + \sqrt{2.25} + \sqrt{2.75}) \approx 2.8$.

5.1.25

a.

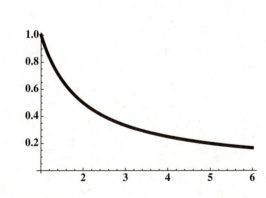

b. We have $\Delta x = \frac{6-1}{5} = 1$. So the midpoints are 1.5, 2.5, 3.5, 4.5, and 5.5.

c.

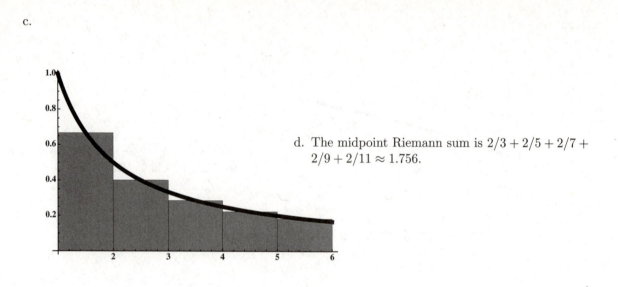

d. The midpoint Riemann sum is $2/3 + 2/5 + 2/7 + 2/9 + 2/11 \approx 1.756$.

5.1.27 Note that $\Delta x = \frac{2-0}{4} = .5$. So the left Riemann sum is given by $.5(5 + 3 + 2 + 1) = 5.5$ and the right Riemann sum is given by $.5(3 + 2 + 1 + 1) = 3.5$.

5.1.29

a.

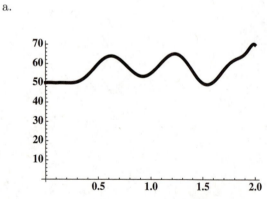

b. With $n = 2$, we have $\Delta x = \frac{2-0}{2} = 1$, so the midpoints are .5 and 1.5. So the midpoint Riemann sum is $60 + 50 = 110$. For $n = 4$, we have $\Delta x = \frac{2-0}{4} = \frac{1}{2}$, so the midpoints are .25, .75, 1.25, and 1.75. The midpoint Riemann sum in this case is $(.5)(50 + 60 + 65 + 60) = \frac{235}{2} = 117.5$.

5.1.31

a. $\displaystyle\sum_{i=1}^{5} i$.

b. $\displaystyle\sum_{i=1}^{6} i + 3$.

c. $\displaystyle\sum_{i=1}^{4} i^2$.

d. $\displaystyle\sum_{i=1}^{4} \frac{1}{i}$.

5.1.33

a. $\displaystyle\sum_{k=1}^{10} k = 1+2+3+4+5+6+7+8+9+10 = 55.$

b. $\displaystyle\sum_{k=1}^{6}(2k+1) = 3+5+7+9+11+13 = 48.$

c. $\displaystyle\sum_{k=1}^{4} k^2 = 1+4+9+16 = 30.$ d. $\displaystyle\sum_{n=1}^{5}(1+n^2) = 2+5+10+17+26 = 60.$

e. $\displaystyle\sum_{m=1}^{3}\frac{2m+2}{3} = \frac{4}{3}+\frac{6}{3}+\frac{8}{3} = 6.$ f. $\displaystyle\sum_{j=1}^{3}(3j-4) = -1+2+5 = 6.$

g. $\displaystyle\sum_{p=1}^{5}(2p+p^2) = 3+8+15+24+35 = 85.$ h. $\displaystyle\sum_{n=0}^{4}\sin\frac{n\pi}{2} = 0+1+0+-1+0 = 0.$

5.1.35   Note that $\Delta x = \frac{1}{10}$, and $x_i = a+i\Delta x = \frac{i}{10}$.

a. The left Riemann sum is given by $\displaystyle\sum_{i=0}^{39}\sqrt{i/10}\cdot(1/10) \approx 5.22697.$ The right Riemann sum is given by

$$\sum_{i=1}^{40}\sqrt{i/10}\cdot(1/10) \approx 5.42697.$$

The midpoint Riemann sum is given by

$$\sum_{i=0}^{39}\sqrt{(1/20)+(i/10)}\cdot(1/10) \approx 5.33515.$$

b. It appears that the actual area is about $5+1/3$.

5.1.37   Note that $\Delta x = \frac{1}{15}$, and $x_i = 2+i\Delta x = 2+(i/15)$. So $f(x_i) = (2+(i/15))^2 - 1$.

a. The left Riemann sum is given by $\displaystyle\sum_{i=0}^{74}[(2+(i/15))^2 - 1]\cdot(1/15) \approx 105.17.$ The right Riemann sum is

given by $\displaystyle\sum_{i=1}^{75}[(2+(i/15))^2 - 1]\cdot(1/15) \approx 108.17.$ The midpoint Riemann sum is given by $\displaystyle\sum_{i=0}^{74}[((61/30) +$

$(i/15))^2 - 1]\cdot(1/15) \approx 106.665.$

b. It appears that the actual area is about $106+2/3$.

5.1.39 The right Riemann sum is given by $A_n = \displaystyle\sum_{i=1}^{n} f(x_i)\Delta x = \sum_{i=1}^{n}\left(4 - \left(-2+\frac{4i}{n}\right)^2\right)\cdot\frac{4}{n}.$

n	A_n
10	10.56
30	10.6548
60	10.6637
80	10.665

It appears that A_n is approaching $10\frac{2}{3}$.

5.1.41 The right Riemann sum is given by $A_n = \sum_{i=1}^{n} f(x_i)\Delta x = \sum_{i=1}^{n} \left(2 - 2\sin\left(\frac{-\pi}{2} + \frac{\pi i}{n}\right)\right) \cdot \frac{\pi}{n}$.

n	A_n
10	5.6549
30	6.0737
60	6.1785
80	6.2046

It appears that A_n is approaching 2π.

5.1.43

a. True. Because the curve is a straight line, the region under the curve and over each subinterval is a trapezoid. The formula for the area of such a trapezoid over $[x_i, x_{i+1}]$ is $\frac{f(x_i) + f(x_{i+1})}{2} \cdot \Delta x = \frac{2x_i + 5 + 2x_{i+1} + 5}{2} \cdot \Delta x = (x_i + x_{i+1} + 5) \cdot \Delta x$ and the area given by using the midpoint formula is $f\left(\frac{x_i + x_{i+1}}{2}\right)\Delta x = (x_i + x_{i+1} + 5)\Delta x$. So the area under the curve is exactly given by the midpoint Riemann sum. Note that this holds for any straight line.

b. False. The left Riemann sum will underestimate the area under an increasing function.

c. True. The value of f at the midpoint will always be between the value of f at the endpoints, if f is monotonic increasing or monotonic decreasing.

5.1.45

We have $\Delta x = \frac{3-1}{5} = \frac{2}{5}$. The sum of the areas of the rectangles is

$$\left(\frac{10}{11} + \frac{2}{3} + \frac{1}{2} + \frac{10}{23} + \frac{1}{3}\right) \cdot \frac{2}{5} \approx 1.13755.$$

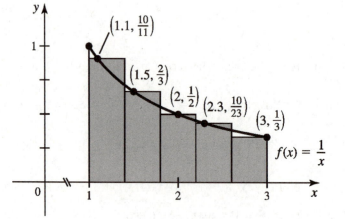

5.1.47 $\sum_{i=1}^{50} \left(\frac{2i}{25} + 1\right) \cdot \frac{2}{25} = 12.16$.

5.1.49 $\sum_{i=0}^{31} \left(3 + \frac{1}{8} + \frac{i}{4}\right)^3 \cdot \frac{1}{4} \approx 3639.13$.

5.1.51 This is the right Riemann sum for f on the interval $[1, 5]$ for $n = 4$.

5.1.53 This is the right Riemann sum for f on the interval $[1.5, 5.5]$ for $n = 4$.

5.1.55 For all of the calculations below, we have $\Delta x = \frac{1}{2}$, and grid points $x_0 = 0$, $x_1 = .5$, $x_2 = 1$, $x_3 = 2.5$, and $x_4 = 2$.

a.

The left Riemann sum is given by

$$\frac{1}{2}\left(f(0) + f(.5) + f(1) + f(1.5)\right) =$$

$$\frac{1}{2}\left(2 + 2.25 + 3 + 4.25\right) = 5.75.$$

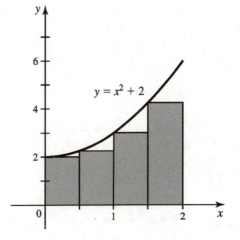

b.

The midpoint Riemann sum is given by

$$\frac{1}{2}\left(f(.25) + f(.75) + f(1.25) + f(1.75)\right) =$$

$$\frac{1}{2}\left(2.0625 + 2.5625 + 3.5625 + 5.0625\right) = 6.625.$$

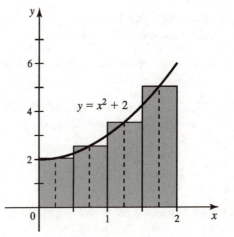

c.

The right Riemann sum is given by

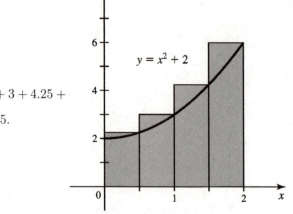

$$\frac{1}{2}\left(f(.5) + f(1) + f(1.5) + f(2)\right) = \frac{1}{2}\left(2.25 + 3 + 4.25 + \right.$$

$$= 7.75.$$

5.1.57 We have $\Delta x = \frac{7-1}{6} = 1$. The left Riemann sum is given by $1(10 + 9 + 7 + 5 + 2 + 1) = 34$ and the right Riemann sum is given by $1(9 + 7 + 5 + 2 + 1 + 0) = 24$.

5.1.59

a. During the first second, the velocity steadily increases from 0 to 20, then it remains constant until $t = 3$. From $t = 3$ until $t = 5$ it steadily decreases, and then remains constant until $t = 6$.

b. Between $t = 0$ and $t = 2$ the area under the curve is $\frac{1}{2} \cdot 1 \cdot 20 + 1 \cdot 20 = 30$.

c. Between $t = 2$ and $t = 5$ the displacement is the sum of the area of a rectangle with area 20 and a trapezoid with area 30, so the displacement is 50 meters.

d. Between $t = 0$ and $t = 5$ the displacement is 80. Between $t = 5$ and any time $t > 5$ the displacement is $10(t - 5)$ so the displacement between $t = 0$ and $t > 5$ is $80 + 10(t - 5)$.

5.1.61

a. Between 0 and 5, the area under the curve is given by the area of a square of area 4 and the area of a trapezoid of area 10.5, so the total area is 14.5.

b. Between 5 and 10, the area under the curve is given by the area of a trapezoid of area 5.5 and the area of a rectangle of area $4 \cdot 6 = 24$, so the total area is 29.5.

c. The mass of the entire rod would be the total area under the curve from 0 to 10, which would be $14.5 + 29.5 = 44$.

d. At $x = \frac{19}{3}$ there is a mass of 22 on each side. Note that from 0 to 6 the mass is 20, so the center of mass is a little greater than 6.

5.1.63 If $0 \leq t \leq 2$, the displacement is $30t$. If $2 \leq t \leq 2.5$, the displacement is $60 + 50(t - 2)$. If $2.5 \leq t \leq 3$, the displacement is $85 + 44(t - 2.5)$. Thus, $d(t) = \begin{cases} 30t & \text{if } 0 \leq t \leq 2, \\ 50t - 40 & \text{if } 2 \leq t \leq 2.5, \\ 44t - 25 & \text{if } 2.5 \leq t \leq 3. \end{cases}$

5.1.65 Using the left Riemann sum $\sum_{i=0}^{n-1} \left| \left(-1 + \frac{2i}{n} \right) \left(\left(-1 + \frac{2i}{n} \right)^2 - 1 \right) \right| \cdot \frac{2}{n}$, we have

n	16	32	64
A_n	.492188	.498047	.499512

It appears that the areas are approaching .5.

5.1.67 Using the left Riemann sum $\sum_{i=0}^{n-1} \left| 1 - \left(-1 + \frac{3i}{n} \right)^3 \right| \cdot \frac{3}{n}$, we have

n	16	32	64
A_n	4.33054	4.52814	4.63592

It appears that the areas are approaching 4.75.

5.1.69 The midpoint Riemann sum gives

$$\sum_{i=0}^{n-1} f\left(a + \frac{b-a}{2n} + \frac{i(b-a)}{n} \right) \cdot \frac{b-a}{n} = \sum_{i=0}^{n-1} \left(m\left(a + \frac{b-a}{2n} + \frac{i(b-a)}{n} \right) + c \right) \cdot \frac{b-a}{n} =$$

$$m \cdot a \cdot n \cdot \frac{b-a}{n} + \frac{m(b-a)^2 n}{2n^2} + \frac{(n-1)(n)}{2} \cdot \frac{m(b-a)^2}{n^2} + \frac{cn(b-a)}{n} =$$

$$m \cdot a \cdot (b-a) + \frac{m(b-a)^2}{2n} + \frac{m(b-a)^2}{2} - \frac{m(b-a)^2}{2n} + c(b-a) =$$

$$m \cdot a \cdot (b-a) + \frac{m(b-a)^2}{2} + c(b-a) = (b-a) \cdot \left(\frac{m(a+b)}{2} + c \right).$$

This proves that the midpoint Riemann sum is independent of n. Since the region in question is a trapezoid, we know that the exact area is given by the width of the subinterval times the average value at the endpoints, which is $(b-a)\left(\frac{f(a)+f(b)}{2} \right) = (b-a)\left(\frac{ma+c+mb+c}{2} \right) = (b-a)\left(\frac{m(a+b)}{2} + c \right).$

5.2 Definite Integrals

5.2.1 The net area is the difference between the area above the x-axis and below the curve, and below the x-axis and above the curve.

5.2.3 When the function is strictly above the x-axis, the net area is equal to the area. The net area differs from the area when the function dips below the x-axis so that the area below the x-axis and above the curve is nonzero.

5.2.5 Since each of the functions $\sin x$ and $\cos x$ have the same amount of area above the x-axis as below between 0 and 2π, these both have value 0.

5.2.7 Because a region "from $x = a$ to $x = a$" has no width, its area is zero. This is akin to asking for the area of a one-dimensional object.

5.2.9 This integral represents the area under $y = x$ between $x = 0$ and $x = a$, which is a right triangle. The length of the base of the triangle is a and the height is a, so the area is $\frac{1}{2} \cdot a^2$, so $\int_0^a x \, dx = \frac{a^2}{2}$.

5.2.11 We have $\Delta x = \frac{4}{4} = 1$.

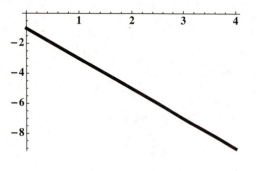

The left Riemann sum is

$$1(f(0)+f(1)+f(2)+f(3)) = -1+-3+-5+-7 = -16.$$

The right Riemann sum is

$$1(f(1) + f(2) + f(3) + f(4))$$
$$= -3 + -5 + -7 + -9 = -24.$$

The midpoint Riemann sum is

$$1(f(.5) + f(1.5) + f(2.5) + f(3.5)) =$$
$$-2 + -4 + -6 + -8 = -20.$$

5.2.13 We have $\Delta x = \frac{\pi/2}{4} = \frac{\pi}{8}$. The left Riemann sum is

$$\frac{\pi}{8}(f(\pi/2) + f(5\pi/8) + f(6\pi/8) + f(7\pi/8)) = 0 - \sqrt{2}/2 - 1 - \sqrt{2}/2 = \frac{\pi}{8} \cdot (-1 - \sqrt{2}) \approx -.948059.$$

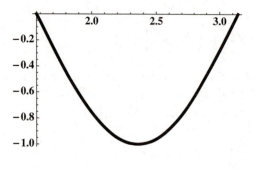

The right Riemann sum is

$$\frac{\pi}{8}(f(5\pi/8) + f(6\pi/8) + f(7\pi/8) + f(\pi)) =$$
$$-\sqrt{2}/2 - 1 - \sqrt{2}/2 - 0 = \frac{\pi}{8} \cdot (-1 - \sqrt{2}) \approx -.948059.$$

The midpoint Riemann sum is

$$\frac{\pi}{8}(f(9\pi/16)+f(11\pi/16)+f(13\pi/16)+f(15\pi/16)) \approx$$
$$\frac{\pi}{8} \cdot (-0.382683 - .92388 - .92388 - 0.382683) \approx -1.02617.$$

5.2.15

a.

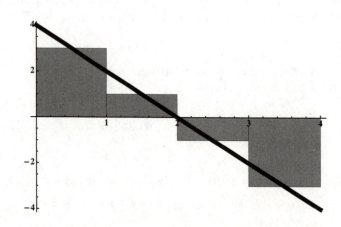

b. The left Riemann sum is $\sum_{i=0}^{3} f(x_i) \cdot 1 = 4$. The right Riemann sum is $\sum_{i=1}^{4} f(x_i) \cdot 1 = -4$. The midpoint Riemann sum is $\sum_{i=0}^{3} f\left(\frac{x_i + x_{i+1}}{2}\right) \cdot 1 = 0$.

c. The rectangles whose height is $f(x_i)$ contribute positively when $x_i < 2$ and negatively when $x_i > 2$.

5.2.17

a.

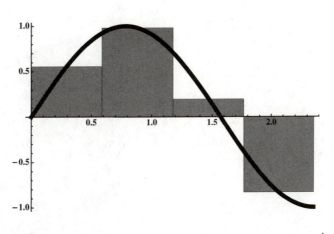

b. The left Riemann sum is $\sum_{i=0}^{3} f(x_i) \cdot \frac{3\pi}{16} \approx .735311$. The right Riemann sum is $\sum_{i=1}^{4} f(x_i) \cdot \frac{3\pi}{16} \approx .146262$. The midpoint Riemann sum is $\sum_{i=0}^{3} f\left(\frac{x_i + x_{i+1}}{2}\right) \cdot \frac{3\pi}{16} \approx .53013$.

c. The rectangles whose height is $f(x_i)$ contribute positively when $x_i < \frac{\pi}{2}$ and negatively when $x_i > \frac{\pi}{2}$.

5.2.19 This is $\int_{0}^{2} (x^2 + 1)\, dx$.

5.2.21 This is $\int_{1}^{2} x\cos(x)\, dx$.

5.2.23

The region in question is a triangle with base 4 and height 8, so the area is $\frac{1}{2} \cdot 8 \cdot 4 = 16$, and this is therefore the value of the definite integral as well.

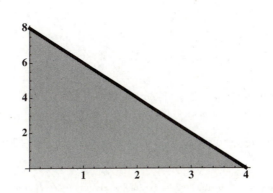

5.2.25

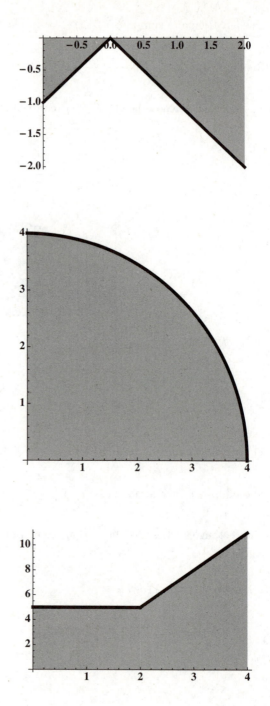

The region consists of two triangles, both below the axis. One has base 1 and height 1, the other has base 2 and height 2, so the net area is $-\frac{1}{2} \cdot 1 \cdot 1 - \frac{1}{2} \cdot 2 \cdot 2 = -2.5$.

5.2.27

The region consists of a quarter circle of radius 4, situated above the axis. So the net area is $\frac{\pi \cdot 4^2}{4} = 4\pi$.

5.2.29

The region consists of a rectangle of area 10 above the axis, and a trapezoid of area 16 above the axis, so the net area is $10 + 16 = 26$.

5.2.31 $\displaystyle\int_0^a f(x)\,dx = 16.$

5.2.33 $\displaystyle\int_a^c f(x)\,dx = 11 - 5 = 6.$

5.2.35 $\displaystyle\int_0^\pi x \sin x\,dx = A(R_1) + A(R_2) = 1 + \pi - 1 = \pi.$

5.2.37 $\displaystyle\int_0^{2\pi} x \sin x\,dx = A(R_1) + A(R_2) - A(R_3) - A(R_4) = 1 + \pi - 1 - \pi - 1 - 2\pi + 1 = -2\pi.$

5.2.39

a. $\displaystyle\int_4^0 3x(4-x)\,dx = -\int_0^4 3x(4-x)\,dx = -32.$

b. $\displaystyle\int_0^4 x(x-4)\,dx = \frac{-1}{3}\int_0^4 3x(4-x)\,dx = \frac{-1}{3}\cdot 32 = \frac{-32}{3}.$

c. $\displaystyle\int_4^0 6x(4-x)\,dx = -2\cdot\int_0^4 3x(4-x)\,dx = -2(32) = -64.$

d. $\displaystyle\int_0^8 3x(4-x)\,dx = \int_0^4 3x(4-x)\,dx + \int_4^8 3x(4-x)\,dx = 32 + \int_4^8 3x(4-x)\,dx.$

5.2.41

a. $\displaystyle\int_0^3 (5f(x))\,dx = 5\int_0^3 f(x)\,dx = 5\cdot 2 = 10.$

b. $\displaystyle\int_3^6 (-3g(x))\,dx = -3\int_3^6 g(x)\,dx = -3\cdot 1 = -3.$

c. $\displaystyle\int_3^6 (3f(x)-g(x))\,dx = 3\int_3^6 f(x)\,dx - \int_3^6 g(x)\,dx = 3(-5)-1 = -16.$

d. $\displaystyle\int_6^3 [f(x)+2g(x)]\,dx = -\left[\int_3^6 f(x)\,dx + 2\int_3^6 g(x)\,dx\right] = -[(-5)+2(1)] = 3.$

5.2.43

a. $\displaystyle\int_0^1 (4x-2x^3)\,dx = -2\int_0^1 (x^3-2x)\,dx = -2\cdot\frac{-3}{4} = \frac{3}{2}.$

b. $\displaystyle\int_1^0 (2x-x^3)\,dx = \int_0^1 x^3 - 2x\,dx = \frac{-3}{4}.$

5.2.45

$$\int_0^2 (2x+1)\,dx = \lim_{n\to\infty}\sum_{n=1}^{\infty} f(x_i)\Delta x = \lim_{n\to\infty}\sum_{i=1}^{n}\left[2\left(\frac{2i}{n}\right)+1\right]\frac{2}{n} = \lim_{n\to\infty}\left[\frac{8}{n^2}\sum_{i=1}^{n}i + \frac{2}{n}\sum_{i=1}^{n}1\right]$$

$$= \lim_{n\to\infty}\left[\frac{8}{n^2}\cdot\frac{n(n+1)}{2} + \frac{2}{n}\cdot n\right] = \lim_{n\to\infty}\left[\frac{4(n+1)}{n}+2\right] = 4+2 = 6.$$

5.2.47

$$\int_3^7 (4x+6)\,dx = \lim_{n\to\infty}\sum_{n=1}^{\infty} f(x_i)\Delta x = \lim_{n\to\infty}\sum_{i=1}^{n}\left[4\left(3+\frac{4i}{n}\right)+6\right]\frac{4}{n} = \lim_{n\to\infty}\left[\frac{64}{n^2}\sum_{i=1}^{n}i + \frac{72}{n}\sum_{i=1}^{n}1\right]$$

$$= \lim_{n\to\infty}\left[\frac{64}{n^2}\cdot\frac{n(n+1)}{2} + \frac{72}{n}\cdot n\right] = \lim_{n\to\infty}\left[\frac{32(n+1)}{n}+72\right] = 104.$$

5.2.49

$$\int_1^4 (x^2 - 1)\, dx = \lim_{n \to \infty} \sum_{n=1}^{\infty} f(x_i)\Delta x = \lim_{n \to \infty} \sum_{i=1}^{n} \left[\left(1 + \frac{3i}{n}\right)^2 - 1\right] \frac{3}{n} = \lim_{n \to \infty}\left[\frac{18}{n^2}\sum_{i=1}^{n} i + \frac{27}{n^3}\sum_{i=1}^{n} i^2 \right]$$

$$= \lim_{n \to \infty}\left[\frac{18}{n^2}\cdot\frac{n(n+1)}{2} + \frac{27}{n^3}\cdot\frac{n(n+1)(2n+1)}{6}\right] = \lim_{n \to \infty}\left[\frac{9(n+1)}{n} + \frac{18n^2 + 27n + 9}{2n^2}\right]$$

$$= 9 + 9 = 18.$$

5.2.51

a. True. See the penultimate problem in the previous section for a proof.

b. True. See the last problem in the previous section for a proof.

c. True. Since both of those function are periodic with period $\frac{2\pi}{a}$, and both have the same amount of area above the axis as below for one period, the net area of each between 0 and $\frac{2\pi}{a}$ is zero.

d. False. For example $\int_0^{2\pi} \sin x\, dx = 0 = \int_{2\pi}^0 \sin x\, dx$, but $\sin x$ is not a constant function.

e. False. Since x is not a constant, it can not be factored outside of the integral. For example $\int_0^1 x\cdot 1\, dx \neq$ $x\int_0^1 1\, dx$.

5.2.53

a.

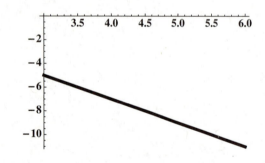

b. $\Delta x = \frac{1}{2}$, so the grid points are at 3, 3.5, 4, 4.5, 5, 5.5, and 6.

c. The left Riemann sum is $.5(-5-6-7-8-9-10) = -22.5$. The right Riemann sum is $.5(-6 - 7 - 8 - 9 - 10 - 11) = -25.5$.

d. The left Riemann sum overestimates the true value, while the right Riemann sum underestimates it.

5.2.55

a.

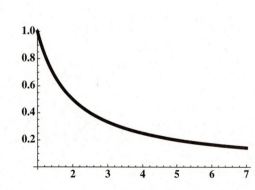

b. $\Delta x = 1$, so the grid points are at 0, 1, 2, 3, 4, 5, 6, and 7.

c. The left Riemann sum is approximately $1 + .5 + .333333 + .25 + .2 + .166666 \approx 2.45$. The right Riemann sum is approximately $.5 + .333333 + .25 + .2 + .166666 + .142857 \approx 1.59286$.

d. The left Riemann sum overestimates the true value, while the right Riemann sum underestimates it.

5.2.57

a.

$$\sum_{i=0}^{19} \left(\left(\frac{i}{20}\right)^2 + 1\right) \cdot \frac{1}{20} \approx 1.30875 \qquad \sum_{i=1}^{20} \left(\left(\frac{i}{20}\right)^2 + 1\right) \cdot \frac{1}{20} \approx 1.35875$$

$$\sum_{i=0}^{49} \left(\left(\frac{i}{50}\right)^2 + 1\right) \cdot \frac{1}{50} \approx 1.3234 \qquad \sum_{i=1}^{50} \left(\left(\frac{i}{50}\right)^2 + 1\right) \cdot \frac{1}{50} \approx 1.3434$$

$$\sum_{i=0}^{99} \left(\left(\frac{i}{100}\right)^2 + 1\right) \cdot \frac{1}{100} \approx 1.32835 \qquad \sum_{i=1}^{100} \left(\left(\frac{i}{100}\right)^2 + 1\right) \cdot \frac{1}{100} \approx 1.33835$$

b. It appears that the integral's value is about $\frac{4}{3}$.

5.2.59

a.

$$\frac{3}{20} \sum_{i=0}^{19} \frac{1}{2\left(1 + \frac{3}{20}i\right)} \approx 0.722 \qquad \frac{3}{20} \sum_{i=1}^{20} \frac{1}{2\left(1 + \frac{3}{20}i\right)} \approx 0.666$$

$$\frac{3}{50} \sum_{i=0}^{49} \frac{1}{2\left(1 + \frac{3}{50}i\right)} \approx 0.705 \qquad \frac{3}{50} \sum_{i=1}^{50} \frac{1}{2\left(1 + \frac{3}{50}i\right)} \approx 0.682$$

$$\frac{3}{100} \sum_{i=0}^{99} \frac{1}{2\left(1 + \frac{3}{100}i\right)} \approx 0.699 \qquad \frac{3}{100} \sum_{i=1}^{100} \frac{1}{2\left(1 + \frac{3}{100}i\right)} \approx 0.688$$

b. It appears that the integral's value is about .69.

5.2.61

a. $\displaystyle\sum_{i=0}^{n-1} 2\sqrt{1 + \frac{3}{2n} + \frac{3i}{n}} \cdot \frac{3}{n}$.

b.

n	20	50	100
Midpoint Sum	9.3338	9.33341	9.33335

It appears that the integral's value is about $\frac{28}{3}$.

5.2.63

a. $\displaystyle\sum_{i=0}^{n-1} \left(4\left(\frac{2}{n} + \frac{4i}{n}\right) - \left(\frac{2}{n} + \frac{4i}{n}\right)^2\right) \cdot \frac{4}{n}$.

b.

n	20	50	100
Midpoint Sum	10.68	10.6688	10.6672

It appears that the integral's value is about $\frac{32}{3}$.

5.2.65

a. $\displaystyle\int_1^4 3f(x)\,dx = 3\int_1^4 f(x)\,dx = 3\cdot\left(\int_1^6 f(x)\,dx - \int_4^6 f(x)\,dx\right) = 3\cdot(10 - 5) = 15.$

b. $\displaystyle\int_1^6 (f(x) - g(x))\,dx = \int_1^6 f(x)\,dx - \int_1^6 g(x)\,dx = 10 - 5 = 5.$

c. $\int_1^4 (f(x) - g(x))\, dx = \int_1^4 f(x)\, dx - \int_1^4 g(x)\, dx = \left(\int_1^6 f(x)\, dx - \int_4^6 f(x)\, dx \right) - 2 = (10 - 5) - 2 = 3.$

d. $\int_4^6 (g(x) - f(x))\, dx = \int_4^6 g(x)\, dx - \int_4^6 f(x)\, dx = \left(\int_1^6 g(x)\, dx - \int_1^4 g(x)\, dx \right) - 5 = (5 - 2) - 5 = -2.$

e. $\int_4^6 8g(x)\, dx = 8 \left(\int_1^6 g(x)\, dx - \int_1^4 g(x)\, dx \right) = 8(5 - 2) = 24.$

f. $\int_4^1 2f(x)\, dx = -1 \int_1^4 f(x)\, dx = -2 \cdot \left(\int_1^6 f(x)\, dx - \int_4^6 f(x)\, dx \right) = -2(10 - 5) = -10.$

5.2.67

The region above the axis is a triangle with base 2 and height $f(-2) = 6$, and the region below the axis is a triangle with base 2 and height $-f(2) = 6$, so the net area is 0, and the area is 12.

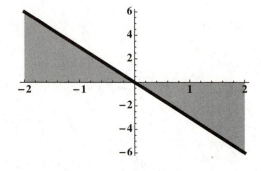

5.2.69

The region above the axis is a triangle with base 2 and height $f(0) = 1$, while the region below the axis consists of two triangles each with base 1 and height 1, so the net area is 0, and the area is 2.

5.2.71

The region in question consists of two triangles above the axis, one with base 1 and height 2, and one with base 4 and height 8, so the net area is $\frac{1}{2} \cdot 1 \cdot 2 + \frac{1}{2} \cdot 4 \cdot 8 = 17.$

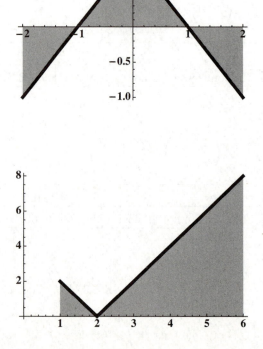

5.2.73

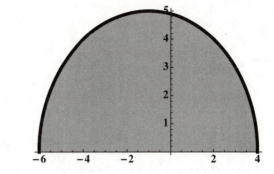

The region in question is a semicircle above the axis with radius 5, so the area is $\frac{1}{2}\pi \cdot 5^2 = \frac{25\pi}{2}$.

5.2.75 $\displaystyle\int_0^{10} f(x)\,dx = \int_0^5 2\,dx + \int_5^{10} 3\,dx = 10 + 15 = 25.$

5.2.77

$$\int_a^b cf(x)\,dx = \lim_{\Delta \to 0} \sum_{k=1}^n cf(\overline{x}_k)\Delta x_k = \lim_{\Delta \to 0} c\sum_{k=1}^n f(\overline{x}_k)\Delta x_k = c\lim_{\Delta \to 0} \sum_{k=1}^n f(\overline{x}_k)\Delta x_k = c\int_a^b f(x)\,dx.$$

5.2.79 Let n be a positive integer. Let $\Delta x = \frac{1}{n}$. Note that each grid point $\frac{i}{n}$ for $0 \le i \le n$ where i is an integer is a rational number. So $f(x_i) = 1$ for each grid point. So the right Riemann sum is $\sum_{i=1}^n f(x_i)\frac{1}{n} = \frac{1}{n}\sum_{i=1}^n 1 = \frac{1}{n}\cdot n = 1$. The left Riemann sum calculation is similar, as is the midpoint Riemann sum calculation (since the grid midpoints are also rational numbers – they are the average of two rational numbers and hence are rational as well.)

5.3   Fundamental Theorem of Calculus

5.3.1 A is also an antiderivative of f.

5.3.3 The fundamental theorem says that $\displaystyle\int_a^b f(x)\,dx = F(b) - F(a)$ where F is any antiderivative of f. So to evaluate $\displaystyle\int_a^b f(x)\,dx$, one could find an antiderivative $F(x)$, and then evaluate this at a and b and then subtract, obtaining $F(b) - F(a)$.

5.3.5

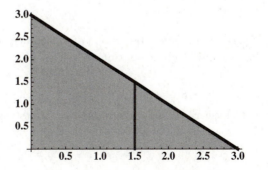

$A(x) = \displaystyle\int_0^x 3 - t\,dt$ represents the area between 0 and x and below this curve. As x increases (but remains less than 3), the trapezoidal region's area increases, so the area function increases until x is 3.

5.3.7 $\dfrac{d}{dx}\displaystyle\int_a^x f(t)\,dt = f(x)$, and $\displaystyle\int f'(x)\,dx = f(x) + C$.

5.3.9 $\dfrac{d}{dx}\displaystyle\int_a^x f(t)\,dt = f(x)$, and $\dfrac{d}{dx}\displaystyle\int_a^b f(t)\,dt = 0$. The latter is the derivative of a constant, the former follows from the Fundamental Theorem.

5.3.11

   a. $A(-2) = \int_{-2}^{-2} f(t)\,dt = 0$.   b. $F(8) = \int_4^8 f(t)\,dt = -9$.   c. $A(4) = \int_{-2}^4 f(t)\,dt = 8 + 17 = 25$.

   d. $F(4) = \int_4^4 f(t)\,dt = 0$.   e. $A(8) = \int_{-2}^8 f(t)\,dt = 25 - 9 = 16$.

5.3.13

   a. $A(x) = \int_0^x f(t)\,dt = \int_0^x 5\,dt = 5x$.

   b. $A'(x) = 5 = f(x)$.

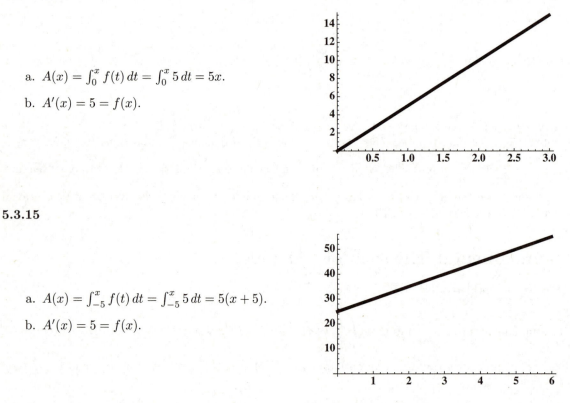

5.3.15

   a. $A(x) = \int_{-5}^x f(t)\,dt = \int_{-5}^x 5\,dt = 5(x + 5)$.

   b. $A'(x) = 5 = f(x)$.

5.3.17

   a. $A(2) = \int_0^2 t\,dt = 2$. $A(4) = \int_0^4 t\,dt = 8$. Since the region whose area is $A(x) = \int_0^x t\,dt$ is a triangle with base x and height x, its value is $\frac{1}{2} \cdot x^2$.

   b. $F(4) = \int_2^4 t\,dt = 6$. $F(6) = \int_2^6 t\,dt = 16$. Since the region whose area is $A(x) = \int_2^x t\,dt$ is a trapezoid with base $x - 2$ and $h_1 = 2$ and $h_2 = x$, its value is $(x - 2)\frac{2+x}{2} = \frac{x^2 - 4}{2} = \frac{x^2}{2} - 2$.

   c. We have $A(x) - F(x) = \frac{x^2}{2} - (\frac{x^2}{2} - 2) = 2$, a constant.

5.3.19

a.

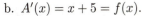

The region is a triangle with base $x + 5$ and height $x + 5$, so its area is $A(x) = \frac{1}{2}(x+5)^2$.

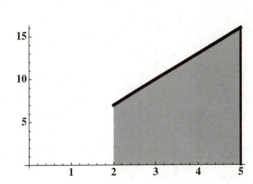

b. $A'(x) = x + 5 = f(x)$.

5.3.21

a.

The region is a trapezoid with base $x - 2$ and heights $h_1 = f(2) = 7$ and $h_2 = f(x) = 3x + 1$, so its area is $A(x) = (x-2) \cdot \frac{7+3x+1}{2} = (x-2) \cdot (\frac{3}{2}x + 4) = \frac{3}{2}x^2 + x - 8$.

b. $A'(x) = 3x + 1 = f(x)$.

5.3.23 $\int_0^1 (x^2 - 2x + 3)\,dx = \frac{x^3}{3} - x^2 + 3x \Big|_0^1 = \frac{1}{3} - 1 + 3 - (0 - 0 + 0) = \frac{7}{3}$. It does appear that the area is between 2 and 3.

5.3.25

$$\int_1^4 (1-x)(x-4)\,dx = \int_1^4 -x^2 + 5x - 4\,dx$$

$$= \frac{-x^3}{3} + \frac{5x^2}{2} - 4x \Big|_1^4 = \frac{9}{2}.$$

5.3.27

$$\int_{-2}^{3} (x^2 - x - 6)\,dx = \frac{x^3}{3} - \frac{x^2}{2} - 6x\bigg|_{-2}^{3} = \frac{-125}{6}.$$

5.3.29

$$\int_{0}^{5} (x^2 - 9)\,dx = \frac{x^3}{3} - 9x\bigg|_{0}^{5} = \frac{125}{3} - 45 - (0 - 0) = \frac{-10}{3}.$$

5.3.31 $\displaystyle\int_{-2}^{2} (x^2 - 4)\,dx = \frac{x^3}{3} - 4x\bigg|_{-2}^{2} = \frac{8}{3} - 8 - \left(\frac{-8}{3} + 8\right) = \frac{16}{3} - 16 = \frac{-32}{3}.$

5.3.33 $\displaystyle\int_{0}^{4} x(x-2)(x-4)\,dx = \int_{0}^{4} x^3 - 6x^2 + 8x\,dx = \frac{x^4}{4} - 2x^3 + 4x^2\bigg|_{0}^{4} = 64 - 128 + 64 - 0 = 0.$

5.3.35 $\displaystyle\int_{-2}^{-1} x^{-3}\,dx = \frac{x^{-2}}{-2}\bigg|_{-2}^{-1} = \frac{-1}{2x^2}\bigg|_{-2}^{-1} = \frac{-1}{2} - \frac{-1}{8} = \frac{-3}{8}.$

5.3.37

$$\int_{1}^{4} \frac{5t^6 - \sqrt{t}}{t^2}\,dt = \int_{1}^{4} (5t^4 - t^{-3/2})\,dt = (t^5 + 2t^{-1/2})\bigg|_{1}^{4} = \left(4^5 + \frac{2}{\sqrt{4}}\right) - \left(1^5 + \frac{2}{\sqrt{1}}\right) = 1022$$

5.3.39

The area (and net area) of this region is given by

$$\int_{1}^{4} \sqrt{x}\,dx = \frac{2}{3}x^{3/2}\bigg|_{1}^{4} = \frac{16}{3} - \frac{2}{3} = \frac{14}{3}.$$

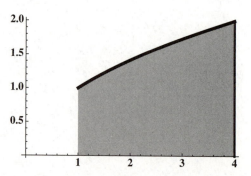

5.3.41

The net area of this region is given by

$$\int_{-2}^{2} x^4 - 16\,dx = \frac{x^5}{5} - 16x\Big|_{-2}^{2} =$$

$$\frac{32}{5} - 32 - \left(\frac{-32}{5} + 32\right) = \frac{64}{5} - 64 = \frac{-256}{5}.$$

Thus the area is $\frac{256}{5}$.

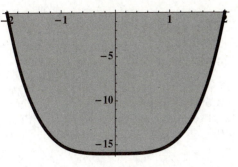

5.3.43

Since this region is below the axis, the area of it is given by

$$-\int_{2}^{4} (x^2 - 25)\,dx = -\left(\frac{x^3}{3} - 25x\Big|_{2}^{4}\right) =$$

$$-\left(\frac{64}{3} - 100 - \left(\frac{8}{3} - 50\right)\right) = 50 - \frac{56}{3} = \frac{94}{3}.$$

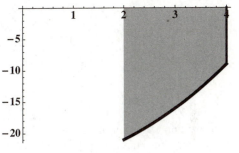

5.3.45

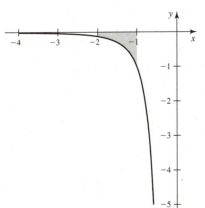

The area is given by the absolute value of the integral

$$\int_{-2}^{-1} x^{-3}\,dx = \left(-\frac{1}{2}x^{-2}\right)\Big|_{-2}^{-1} = -\frac{3}{8}$$

so the area is $\frac{3}{8}$.

5.3.47

The area is given by

$$-\int_{-\pi/4}^{0} \sin x\,dx + \int_{0}^{3\pi/4} \sin x\,dx$$

$$= \left(\cos x\Big|_{-\pi/4}^{0}\right) + \left(-\cos x\Big|_{0}^{3\pi/4}\right)$$

$$= \left(1 - \frac{\sqrt{2}}{2}\right) + \left(1 + \frac{\sqrt{2}}{2}\right) = 2.$$

5.3.49 By a direct application of the Fundamental Theorem, this is $x^2 + x + 1$.

5.3.51 By the Fundamental Theorem and the chain rule, this is $\frac{1}{x^6} \cdot 3x^2 = \frac{3}{x^4}$.

5.3.53 This is $-\dfrac{d}{dx} \displaystyle\int_1^x \sqrt{t^4 + 1} \, dt = -\sqrt{x^4 + 1}$.

5.3.55 (a) matches with (C) – its area function is increasing linearly. (b) matches with (B) – its area function increases then decreases. (c) matches with (D) – its area function is always increasing on $[0, b]$, although not linearly. (d) matches with (A) – its area function decreases at first and then eventually increases.

5.3.57

c.

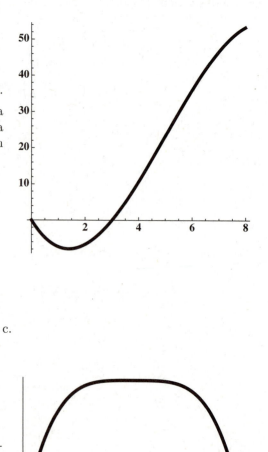

a. It appears that $A(x) = 0$ for $x = 0$ and at about $x = 3$.

b. A has a local minimum at about $x = 1.5$ where the area function changes from deceasing to increasing, and a local max at around $x = 8.5$ where the area function changes from increasing to decreasing.

5.3.59

c.

a. It appears that $A(x) = 0$ for $x = 0$ and $x = 10$.

b. A has a local maximum at $x = 5$ where the area function changes from increasing to decreasing.

5.3.61 $A(2) = -\frac{1}{4}\pi \cdot 2^2 = -\pi$. $A(5) = -\pi + \frac{1}{2} \cdot 3 \cdot 3 = \frac{9}{2} - \pi$. $A(8) = \frac{9}{2} - \pi + \frac{1}{2} \cdot 3 \cdot 3 = 9 - \pi$. $A(12) = 9 - \pi - \frac{1}{2} \cdot 4 \cdot 2 = 5 - \pi$.

5.3.63

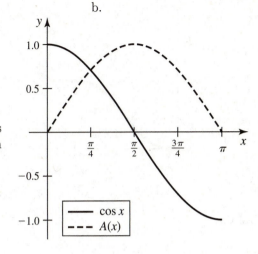

b.

a. $A(x) = \int_0^x \cos t\, dt = \sin x - \sin 0 = \sin x$

c. $A(\pi/2) = \sin \pi/2 = 1$ and $A(\pi) = \sin \pi = 0$. This reflects the net signed area under the curve $\cos x$ from 0 to $\pi/2$ and π respectively.

5.3.65

b.

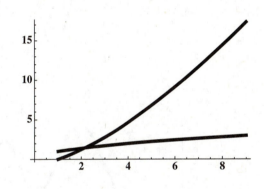

a. $A(x) = \int_1^x \sqrt{t}\, dt = \left.\frac{2}{3}t^{3/2}\right|_1^x = \frac{2}{3}x^{3/2} - \frac{2}{3}.$

c. $A(4) = \frac{2}{3}(8-1) = \frac{14}{3}$. $A(9) = \frac{2}{3}(27-1) = \frac{52}{3}$.

5.3.67

a.

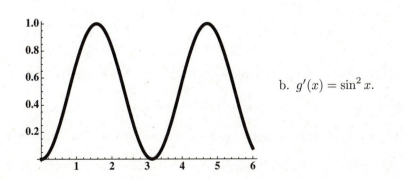

b. $g'(x) = \sin^2 x$.

c.

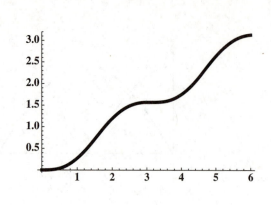

Note that g' is always positive, so g is always increasing. There are inflection points where g' changes from increasing to decreasing, and vice versa.

5.3.69

a.

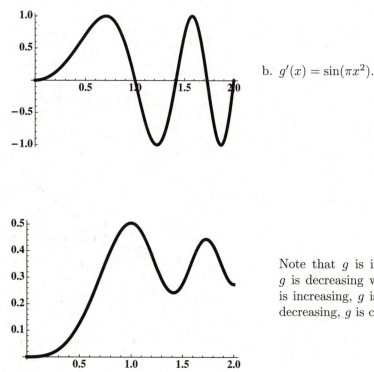

b. $g'(x) = \sin(\pi x^2)$.

c.

Note that g is increasing where $g' > 0$ and g is decreasing when $g' < 0$. Also, where g' is increasing, g is concave up and where g' is decreasing, g is concave down.

5.3.71

a. True. The net area under the curve increases as x increases, as long as f is above the axis.

b. True. The net area decreases as x increases, as long as f is below the axis.

c. False. These do not have the same derivative, so they are not antiderivatives of the same function.

d. True, since the two functions differ by a constant, and thus have the same derivative.

5.3.73

$$\int_1^4 \frac{x-2}{\sqrt{x}}\,dx = \int_1^4 \frac{x}{\sqrt{x}} - \frac{2}{\sqrt{x}}\,dx = \int_1^4 x^{1/2} - 2x^{-1/2}\,dx = \left(\frac{2}{3}x^{3/2} - 4x^{1/2}\bigg|_1^4\right)$$

$$= \frac{16}{3} - 8 - \left(\frac{2}{3} - 4\right) = \frac{14}{3} - \frac{12}{3} = \frac{2}{3}.$$

5.3.75 $\displaystyle\int_0^{\pi/3} \sec x \tan x\,dx = \left(\sec x\big|_0^{\pi/3}\right) = (2-1) = 1.$

5.3.77 $\displaystyle\int_1^8 \sqrt[3]{y}\,dy = \left(\frac{3}{4}y^{4/3}\bigg|_1^8\right) = 12 - \frac{3}{4} = \frac{45}{4}.$

5.3.79

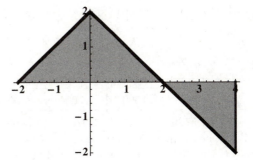

We can use geometry – there is a triangle with base 4 and height 2 and a triangle with base 2 and height 2, so the total area is $\frac{1}{2}\cdot 4\cdot 2 + \frac{1}{2}\cdot 2\cdot 2 = 6.$

5.3.81
Since the region is below the axis on $[1,\sqrt{2}]$ and above on $[\sqrt{2},4]$ we need to compute

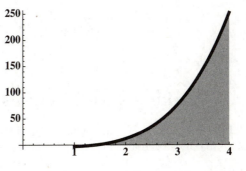

$$\int_{\sqrt{2}}^4 (x^4 - 4)\,dx - \int_1^{\sqrt{2}} (x^4 - 4)\,dx =$$

$$\left(\frac{x^5}{5} - 4x\bigg|_{\sqrt{2}}^4\right) - \left(\frac{x^5}{5} - 4x\bigg|_1^{\sqrt{2}}\right) =$$

$$\frac{1024}{5} - 16 - (4\sqrt{2}/5 - 4\sqrt{2}) - (4\sqrt{2}/5 - 4\sqrt{2}) + \frac{1}{5} - 4$$

$$= 185 + \frac{32\sqrt{2}}{5}$$

5.3.83 $\displaystyle\int_3^8 f'(t)\,dt = f(8) - f(3).$

5.3.85 $\displaystyle\frac{d}{dx}\int_0^{\cos x}(t^4 + 6)\,dt = (\cos^4 x + 6)\cdot(-\sin x).$

5.3.87

a.

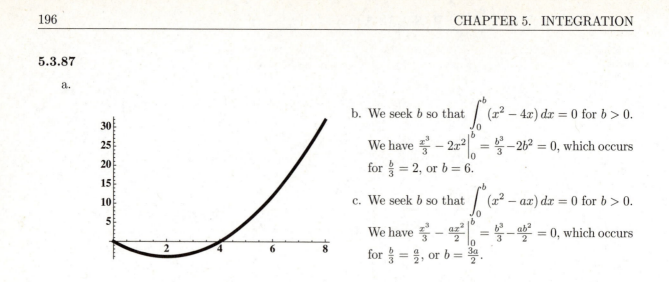

b. We seek b so that $\int_0^b (x^2 - 4x)\,dx = 0$ for $b > 0$.

We have $\frac{x^3}{3} - 2x^2 \Big|_0^b = \frac{b^3}{3} - 2b^2 = 0$, which occurs for $\frac{b}{3} = 2$, or $b = 6$.

c. We seek b so that $\int_0^b (x^2 - ax)\,dx = 0$ for $b > 0$.

We have $\frac{x^3}{3} - \frac{ax^2}{2} \Big|_0^b = \frac{b^3}{3} - \frac{ab^2}{2} = 0$, which occurs for $\frac{b}{3} = \frac{a}{2}$, or $b = \frac{3a}{2}$.

5.3.89 Since $\frac{d}{db} \int_{-1}^b x^2(3 - x)\,dx = b^2(3 - b)$ we see that this function of b has critical points at $b = 0$ and $b = 3$. Note also that the integrand is positive on $[0, 3]$, but is negative on $[3, \infty)$. So there must be a maximum for this area function at $b = 3$.

5.3.91 Differentiating both sides of the given equation yields $f(x) = -2\sin x + 3$.

5.3.93

Using a computer or calculator, we obtain:

x	500	1000	1500	2000
$S(x)$	1.5726	1.57023	1.57087	1.57098

This appears to be approaching $\frac{\pi}{2}$. Note that between 0 and π, the area is approximately half the area of a rectangle with height 1 and base π, and then from π on there is approximately as much area above the axis as below.

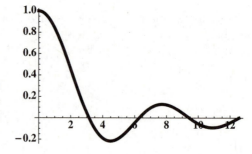

5.3.95 By the Fundamental Theorem, $S'(x) = \sin(x^2)$, so $S''(x) = 2x\cos(x^2)$, so

$$(S'(x))^2 + \left(\frac{S''(x)}{2x}\right)^2 = \sin^2(x^2) + \cos^2(x^2) = 1.$$

5.4   Working with Integrals

5.4.1 If f is odd, it is symmetric about the origin, which guarantees that between $-a$ and a, there is as much area above the axis and under f as there is below the axis and above f, so the net area must be 0.

5.4.3 $f(x) = x^{12}$ is an even function, since $f(-x) = (-x)^{12} = x^{12} = f(x)$. $g(x) = \sin(x^2)$ is also even, since $g(-x) = \sin((-x)^2) = \sin(x^2) = g(x)$.

5.4.5 The average value of a function on a closed interval $[a, b]$ will always be between the maximum and the minimum value of f on that interval. Since the function is continuous, the Mean Value Theorem assures us that the function will take on each value between the maximum and the minimum somewhere on the interval.

5.4.7 $\int_{-2}^2 (3x^8 - 2)\,dx = 2\int_0^2 3x^8 - 2\,dx = 2\left(\frac{x^9}{3} - 2x \Big|_0^2\right) = \cdot\frac{1024}{3} - 8 = \frac{1000}{3}$.

5.4.9 Note that the first two terms of the integrand form an odd function, and the last two terms form an even function. $\int_{-2}^{2} (x^9 - 3x^5 + 2x^2 - 10)\,dx = 2\int_{0}^{2} (2x^2 - 10)\,dx = 2\left(\frac{2x^3}{3} - 10x\Big|_{0}^{2}\right) = \frac{32}{3} - 40 = \frac{-88}{3}$.

5.4.11 $\int_{-10}^{10} \frac{x}{\sqrt{200 - x^2}}\,dx = 0$ since the integrand is an odd function.

5.4.13 Since the integrand is an odd function and the interval is symmetric about 0, this integral's value is 0.

5.4.15

Since the integrand is an odd function and the interval is symmetric about 0, this integral's value is 0.

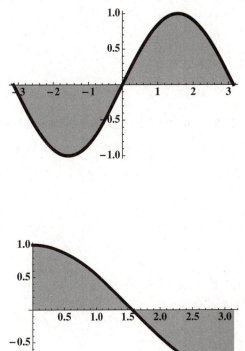

5.4.17

Because of the symmetry of the cosine function, the net area is zero between 0 and π.

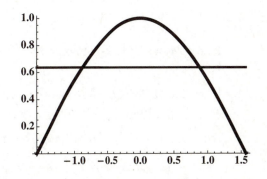

5.4.19

The average value is

$$\frac{1}{\pi}\int_{-\pi/2}^{\pi/2} \cos x\,dx = \frac{1}{\pi}\left(\sin x\big|_{-\pi/2}^{\pi/2}\right)$$

$$= \frac{1}{\pi}\cdot(1 - -1) = \frac{2}{\pi} \approx .6366.$$

5.4.21

The average value is $\dfrac{1}{1}\displaystyle\int_0^1 x^n\,dx = \left(\dfrac{x^{n+1}}{n+1}\Big|_0^1\right) = \dfrac{1}{n+1}$. The picture shown is for the case $n = 3$.

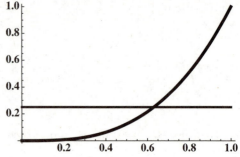

5.4.23 The average distance to the axis is given by $\dfrac{1}{20}\displaystyle\int_0^{20} 10x(20 - x)\,dx$. This is equal to $\dfrac{1}{20}\displaystyle\int_0^{20} 200x - 10x^2\,dx = \dfrac{1}{20}\left(100x^2 - \dfrac{10x^3}{3}\Big|_0^{20}\right) = \dfrac{2000}{3}$.

5.4.25 The average height is $\dfrac{1}{\pi}\displaystyle\int_0^\pi 10\sin x\,dx = \dfrac{1}{\pi}\left(-10\cos x|_0^\pi\right) = \dfrac{1}{\pi}\left(10 - -10\right) = \dfrac{20}{\pi}$.

5.4.27 Think of the semicircle as consisting of all the points $(2\cos\theta, 2\sin\theta)$ where $0 \le \theta \le \pi$. The distance to the base at any point is $2\sin\theta$. So the average distance from the base is $\dfrac{1}{\pi}\displaystyle\int_0^\pi 2\sin\theta\,d\theta = \dfrac{2}{\pi}\left(-\cos\theta\,\big|_0^\pi\right) = \dfrac{4}{\pi}$.

5.4.29 The average value is $\dfrac{1}{4}\displaystyle\int_0^4 (8 - 2x)\,dx = \dfrac{1}{4}\left(8x - x^2\big|_0^4\right) = 4$. The function has a value of 4 when $8 - 2x = 4$, which occurs when $x = 2$.

5.4.31 The average value is $\dfrac{1}{a}\displaystyle\int_0^a 1 - \dfrac{x^2}{a^2}\,dx = \dfrac{1}{a}\left(x - \dfrac{x^3}{3a^2}\Big|_0^a\right) = \dfrac{2}{3}$. The function attains this value when $\dfrac{2}{3} = 1 - \dfrac{x^2}{a^2}$, which is when $x^2 = \dfrac{a^2}{3}$, which on the given interval occurs for $x = \sqrt{3}a/3$.

5.4.33 The average value is $\dfrac{1}{2}\displaystyle\int_{-1}^1 (1 - |x|)\,dx = \dfrac{1}{2}\displaystyle\int_{-1}^0 (1 + x)\,dx + \dfrac{1}{2}\displaystyle\int_0^1 (1 - x)\,dx = \dfrac{1}{2}\left(x + \dfrac{x^2}{2}\Big|_{-1}^0\right) + \dfrac{1}{2}\left(x - \dfrac{x^2}{2}\Big|_0^1\right) = \dfrac{1}{4} + \dfrac{1}{4} = \dfrac{1}{2}$. The function attains this value twice, once on $[-1, 0]$ when $1 + x = \frac{1}{2}$ which occurs when $x = \frac{-1}{2}$, and once on $[0, 1]$ when $1 - x = \frac{1}{2}$ which occurs when $x = \frac{1}{2}$.

5.4.35

a. True. Because of the symmetry, the net area between 0 and 2 will be twice the net area between 0 and 4.

b. True. This follows because the symmetry implies that the net area from a to $a + 2$ is the opposite of the net area from $a - 2$ to a.

c. True. If $f(x) = cx + d$ on $[a, b]$ the value at the midpoint is $c \cdot \frac{a+b}{2} + d$, and the average value is
$$\dfrac{1}{b - a}\displaystyle\int_a^b cx + d\,dx = \dfrac{1}{b - a}\left(\dfrac{cx^2}{2} + dx\Big|_a^b\right) = \dfrac{1}{b - a}\left(\dfrac{cb^2}{2} + db - \left(\dfrac{ca^2}{2} + da\right)\right) = \dfrac{c}{2} \cdot (a + b) + d.$$

d. False, for example, when $a = 1$, we have that the maximum value of $x - x^2$ on $[0, 1]$ occurs at $\frac{1}{2}$ and is equal to $\frac{1}{4}$, but the average value is $\int_0^1 x - x^2 \, dx = \left(\frac{x^2}{2} - \frac{x^2}{3} \Big|_0^1 \right) = \frac{1}{2} - \frac{1}{3} = \frac{1}{6}$.

5.4.37 $\sec^2 x$ is even, so the value of this integral is $2 \int_0^{\pi/4} \sec^2 x \, dx = 2 \left(\tan x \big|_0^{\pi/4} \right) = 2 \cdot (1 - 0) = 2$.

5.4.39 The integrand is an odd function, so the value of this integral is zero.

5.4.41 The average height of the arch is given by

$$\frac{1}{630} \int_{-315}^{315} 630 - \frac{630}{315^2} x^2 \, dx = \frac{630}{630} \left(x - \frac{x^3}{3 \cdot 315^2} \Big|_{-315}^{315} \right) = (315 - 105 - (-315 + 105)) = 420 \text{ feet}.$$

5.4.43

a. $d^2 = x^2 + y^2 = x^2 + b^2(1 - (x^2/a^2))$. The average value of d^2 is $\frac{1}{2a} \int_{-a}^a b^2 + \left(1 - \frac{b^2}{a^2} \right) x^2 \, dx =$
$\frac{1}{2a} \left(b^2 x + \frac{(1 - (b^2/a^2)) x^3}{3} \Big|_{-a}^a \right) = \frac{1}{2a} \left(b^2 a + \frac{a^3}{3} - \frac{b^2 a}{3} - \left(-b^2 a - \frac{a^3}{3} + \frac{b^2 a}{3} \right) \right) = \frac{2b^2}{3} + \frac{a^2}{3}$.

b. If $a = b = R$, the above becomes $\frac{2R^2}{3} + \frac{R^2}{3} = R^2$.

c. $D^2 = (x - \sqrt{a^2 - b^2})^2 + y^2 = x^2 - 2\sqrt{a^2 - b^2}x + y^2 + a^2 - b^2 = \left(1 - \frac{b^2}{a^2} \right) x^2 - 2\sqrt{a^2 - b^2}x + a^2$. So the average value of D^2 is $\frac{1}{2a} \int_{-a}^a D^2 \, dx = \frac{1}{2a} \int_{-a}^a \left[\left(1 - \frac{b^2}{a^2} \right) x^2 + a^2 \right] dx - \frac{1}{a} \int_{-a}^a \sqrt{a^2 - b^2}x \, dx =$
$\frac{1}{a} \int_0^a \left[\left(1 - \frac{b^2}{a^2} \right) x^2 + a^2 \right] dx + 0 = \frac{1}{3}(a^2 - b^2) + a^2 = \frac{4a^2 - b^2}{3}$.

5.4.45 $f(g(-x)) = f(g(x))$, so $f(g(x))$ is an even function, and $\int_{-a}^a f(g(x)) \, dx = 2 \int_0^a f(g(x)) \, dx$.

5.4.47 $p(g(-x)) = p(g(x))$, so $p(g(x))$ is an even function, and $\int_{-a}^a p(g(x)) \, dx = 2 \int_0^a p(g(x)) \, dx$.

5.4.49

a. The average value is $\int_0^1 (ax - ax^2) \, dx = \left(\frac{ax^2}{2} - \frac{ax^3}{3} \Big|_0^1 \right) = \frac{a}{2} - \frac{a}{3} = \frac{a}{6}$.

b. The function is equal to its average value when $\frac{a}{6} = ax - ax^2$ which occurs when $6x - 6x^2 = 1$, so when $6x^2 - 6x + 1 = 0$. On the given interval, this occurs for $x = \frac{6 - \sqrt{12}}{12} = \frac{3 - \sqrt{3}}{6}$.

5.4.51

a.

The area of the triangle is $\frac{1}{2} \cdot 2a \cdot a^2 = a^3$. The area under

the parabola is $\int_{-a}^{a} a^2 - x^2 \, dx = \left(a^2 x - \frac{x^3}{3} \Big|_{-a}^{a} \right) =$

$a^3 - \frac{a^3}{3} - \left(-a^3 + \frac{a^3}{3} \right) = 2a^3 - \frac{2a^3}{3} = \frac{4a^3}{3}$, as desired.

The diagram shown is for $a = 2$.

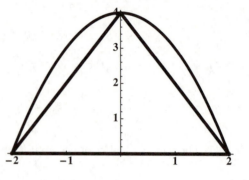

b. The area of the rectangle desribed is $2a \cdot a^2 = 2a^3$, and $\frac{2}{3}$ of this is $\frac{4a^3}{3}$, which is the area under the parabola derived above.

5.4.53 $\int_0^c x(x - c)^2 \, dx = \int_0^c x^3 - 2cx^2 + c^2 x \, dx = \left(\frac{x^4}{4} - 2c \frac{x^3}{3} + \frac{c^2 x^2}{2} \Big|_0^c \right) = \frac{c^4}{4} - \frac{2c^4}{3} + \frac{c^4}{2} = \frac{c^4}{12}$. This is one when $c = \sqrt[4]{12}$.

5.4.55

a. The left Riemann sum is given by $\dfrac{\pi}{2n} \displaystyle\sum_{k=0}^{n-1} \sin((k\pi)/(2n))$.

b.
$$\lim_{\theta \to 0} \theta \left(\frac{\cos\theta + \sin\theta - 1}{2(1 - \cos\theta)} \right) \cdot \left(\frac{1 + \cos\theta}{1 + \cos\theta} \right) = \lim_{\theta \to 0} \frac{\theta}{2} \left(\frac{(1 + \cos\theta)(\cos\theta + \sin\theta - 1)}{\sin^2\theta} \right)$$
$$= \left(\frac{1}{2} \lim_{\theta \to 0} \frac{\theta}{\sin\theta} \cdot \frac{1 + \cos\theta}{1} \right) \cdot \left(\lim_{\theta \to 0} \frac{\cos\theta - 1}{\sin\theta} + \lim_{\theta \to 0} \frac{\sin\theta}{\sin\theta} \right)$$
$$= 2 \left(\frac{1}{2} \right) \cdot \left(\lim_{\theta \to 0} \frac{\frac{\cos -1\theta}{\theta}}{\frac{\sin\theta}{\theta}} + 1 \right) = 1 \cdot (0 + 1) = 1.$$

c. Using the previous result, the left Riemann sum is given by $\dfrac{\pi}{2n} \left(\dfrac{\cos(\pi/(2n)) + \sin(\pi/(2n)) - 1}{2(1 - \cos(\pi/(2n)))} \right)$. Let $\theta = \frac{\pi}{2n}$. Then as $n \to \infty$, $\theta \to 0$, and the limit of the left Riemann sum as $n \to \infty$ is 1.

5.4.57 Suppose f is even, so that $f(-x) = f(x)$. Then $f^n(x) = f^n(-x)$, so that f^n is an even function, no matter what the parity of n is. Suppose g is an odd function, so that $g(-x) = -g(x)$. Then $g^n(-x) = (-1)^n g^n(x)$, so g^n is even when n is even, and is odd when n is odd. Summarizing, we have:

	f is even	f is odd
n is even	f^n is even	f^n is even
n is odd	f^n is even	f^n is odd

5.4.59

a. Because of the symmetry, $\int_{c-a}^{c} (f(x) - d)\, dx = \int_{c}^{c+a} (d - f(x))\, dx$. Thus,

$$\int_{c-a}^{c+a} f(x)\, dx = \int_{c-a}^{c} (f(x) - d + d)\, dx + \int_{c}^{c+a} f(x)\, dx = \int_{c-a}^{c} d\, dx + \int_{c-a}^{c} (f(x) - d)\, dx + \int_{c}^{c+a} f(x)\, dx$$

$$= ad + \int_{c}^{c+a} (d - f(x))\, dx + \int_{c}^{c+a} f(x)\, dx = ad + ad + 0 = 2ad.$$

b.

The curve is rotationally symmetric about $(\pi/4, 1/2)$. To see this, we will show that if $\sin^2(\pi/4 - x) = (1/2) - r$, then $\sin^2(\pi/4 + x) = (1/2) + r$. Using a double angle identity, we have

$$\sin^2(\pi/4 - x) = \frac{1}{2} - \frac{1}{2}\cos\left(\frac{\pi}{2} - 2x\right)$$

$$= \frac{1}{2} - \frac{1}{2}\sin(2x).$$

On the other hand,

$$\sin^2(\pi/4 + x) = \frac{1}{2} - \frac{1}{2}\cos\left(\frac{\pi}{2} + 2x\right)$$

$$= \frac{1}{2} + \frac{1}{2}\sin(2x).$$

c. Using the idea from part a), The area under f over this interval must be equal to the area of the rectangle over the interval $[\pi/4, \pi/2]$ with height $1/2 + 1/2 = 1$. Thus the area is $\frac{\pi}{4} \cdot 1 = \frac{\pi}{4}$.

5.5 Substitution Rule

5.5.1 It is based on the Chain Rule for differentiation.

5.5.3 Typically u is substituted for the inner function, so $u = g(x)$.

5.5.5 The new integral is $\int_{g(a)}^{g(b)} f(u)\, du$.

5.5.7 Using the identity $\cos^2 x = \frac{1}{2} + \frac{\cos 2x}{2}$, we have $\int \cos^2 x\, dx = \int \frac{1}{2} + \frac{\cos 2x}{2}\, dx = \frac{x}{2} + \frac{\sin 2x}{4} + C.$

5.5.9 $\int (x+1)^{12}\, dx = \frac{(x+1)^{13}}{13} + C$, since $\frac{d}{dx}\frac{(x+1)^{13}}{13} + C = (x+1)^{12}.$

5.5.11 $\int \sqrt{2x+1}\, dx = \frac{(2x+1)^{3/2}}{3} + C$, since $\frac{d}{dx}\frac{(2x+1)^{3/2}}{3} + C = \frac{3}{2} \cdot \frac{1}{3} \cdot (2x+1)^{1/2} \cdot 2 = \sqrt{2x+1}.$

5.5.13 Since $u = x^2 + 1$, $du = 2x\, dx$. Substituting yields $\int u^4\, du = \frac{u^5}{5} + C = \frac{(x^2+1)^5}{5} + C.$

5.5.15 Since $u = \sin x$, $du = \cos x\, dx$. Substituting yields $\int u^3\, du = \frac{u^4}{4} + C = \frac{\sin^4(x)}{4} + C.$

5.5.17 Let $u = x^2 - 1$. Then $du = 2x \, dx$. Substituting yields $\int u^{99} \, du = \dfrac{u^{100}}{100} + C = \dfrac{(x^2 - 1)^{100}}{100} + C$.

5.5.19 Let $u = 1 - 4x^3$. Then $du = -12x^2 \, dx$, so $\frac{-1}{6} du = 2x^2 \, dx$. Substituting yields

$$\frac{-1}{6} \int \frac{1}{\sqrt{u}} \, du = \frac{-1}{3} \cdot \sqrt{u} + C = \frac{-1}{3} \cdot \sqrt{1 - 4x^3} + C.$$

5.5.21 Let $u = x^2 + x$. Then $du = 2x + 1 \, dx$. Substituting yields $\int u^{10} \, du = \dfrac{u^{11}}{11} + C = \dfrac{(x^2 + x)^{11}}{11} + C$.

5.5.23 Let $u = x^4 + 16$. Then $du = 4x^3 \, dx$, so $\frac{1}{4} du = x^3 \, dx$. Substituting yields

$$\frac{1}{4} \int u^6 \, du = \frac{1}{4} \cdot \frac{u^7}{7} + C = \frac{(x^4 + 16)^7}{28} + C.$$

5.5.25 Let $u = 4 - 9x^2$; then $du = -18x \, dx$ so that

$$\int \frac{x}{\sqrt{4 - 9x^2}} \, dx = -\frac{1}{18} \int u^{-1/2} \, du = -\frac{\sqrt{u}}{9} + C = -\frac{\sqrt{4 - 9x^2}}{9} + C$$

5.5.27 Let $u = x^6 - 3x^2$. Then $du = (6x^5 - 6x) \, dx$, so $\frac{1}{6} du = (x^5 - x) \, dx$. Substituting yields

$$\frac{1}{6} \int u^4 \, du = \frac{1}{6} \cdot \frac{u^5}{5} + C = \frac{(x^6 - 3x^2)^5}{30} + C.$$

5.5.29 Let $u = x - 4$, so that $u + 4 = x$. Then $du = dx$. Substituting yields

$$\int \frac{u + 4}{\sqrt{u}} \, du = \int \frac{u}{\sqrt{u}} + \frac{4}{\sqrt{u}} \, du = \int u^{1/2} + 4u^{-1/2} \, du = \frac{2}{3} u^{3/2} + 8u^{1/2} + C$$

$$= \frac{2}{3} \cdot (x - 4)^{3/2} + 8\sqrt{x - 4} + C.$$

5.5.31 Let $u = x + 4$, so that $u - 4 = x$. Then $du = dx$. Substituting yields

$$\int \frac{(u - 4)}{\sqrt[3]{u}} \, du = \int u^{2/3} - 4u^{-1/3} \, du = \frac{3}{5} u^{5/3} + -6u^{2/3} + C = \frac{3}{5}(x + 4)^{5/3} - 6(x + 4)^{2/3} + C.$$

5.5.33 Let $u = 2x + 1$. Then $du = 2dx$ and $x = \frac{u-1}{2}$. Substituting yields

$$\frac{1}{2} \int \frac{u - 1}{2} \cdot \sqrt[3]{u} \, du = \frac{1}{4} \int u^{4/3} - u^{1/3} \, du = \frac{1}{4} \left(\frac{3}{7} u^{7/3} - \frac{3}{4} u^{4/3} \right) + C = \frac{3(2x + 1)^{7/3}}{28} - \frac{3(2x + 1)^{4/3}}{16} + C.$$

5.5.35 Let $u = 4 - x^2$. Then $du = -2x \, dx$. Also, when $x = 0$ we have $u = 4$ and when $x = 1$ we have $u = 3$. Substituting yields $-\int_4^3 u \, du = \int_3^4 u \, du = \left(\frac{u^2}{2} \Big|_3^4 \right) = 8 - 4.5 = 3.5$.

5.5.37 Let $u = \sin\theta$. Then $du = \cos\theta\,d\theta$. Also, when $\theta = 0$ we have $u = 0$ and when $\theta = \pi/2$ we have $u = 1$. Substituting yields $\int_0^1 u^2\,du = \left(\left.\frac{u^3}{3}\right|_0^1\right) = \frac{1}{3}$.

5.5.39 Let $u = 2y$; then $du = 2\,dy$. As y goes from $-\pi/12$ to $\pi/8$, $u = 2y$ goes from $-\pi/6$ to $\pi/4$. Thus

$$\int_{-\pi/12}^{\pi/8} \sec^2 2y\,dy = \frac{1}{2}\int_{-\pi/6}^{\pi/4} \sec^2 u\,du = \frac{1}{2}(\tan\pi/4 - \tan(-\pi/6)) = \frac{\sqrt{3}}{6} + \frac{1}{2}$$

5.5.41 Let $u = \sin x$. Then $du = \cos x\,dx$. Also, when $x = \pi/4$ we have $u = \sqrt{2}/2$ and when $x = \pi/2$ we have $u = 1$. Substituting yields $\int_{\sqrt{2}/2}^1 \frac{1}{u^2}\,du = \left(\left.\frac{-1}{u}\right|_{\sqrt{2}/2}^1\right) = \left(-1 - \frac{-2}{\sqrt{2}}\right) = \sqrt{2} - 1$.

5.5.43 Let $u = 2x - 3$; then $du = 2\,dx$ and $x = (u+3)/2$. As x goes from 2 to 6, u goes from 1 to 9. Thus

$$\int_2^6 \frac{x}{\sqrt{2x-3}}\,dx = \frac{1}{4}\int_1^9 \frac{u+3}{\sqrt{u}}\,du = \frac{1}{4}\left(\left.\frac{2}{3}u^{3/2} + 6\sqrt{u}\right)\right|_1^9 = \frac{22}{3}$$

5.5.45 $\int_{-\pi}^{\pi} \cos^2 x\,dx = 2\int_0^{\pi} \frac{1 + \cos 2x}{2}\,dx = \left(\left.x + \frac{\sin 2x}{2}\right|_0^{\pi}\right) = \pi$.

5.5.47 $\int \sin^2\left(\theta + \frac{\pi}{6}\right)\,d\theta = \frac{1}{2}\int\left(1 - \cos\left(2\theta + \frac{\pi}{3}\right)\right)\,d\theta = \frac{\theta}{2} - \frac{\sin\left(2\theta + \frac{\pi}{3}\right)}{4} + C$.

5.5.49 $\int_{-\pi/4}^{\pi/4} \sin^2 2\theta\,d\theta = 2\int_0^{\pi/4} \sin^2(2\theta)\,d\theta = 2\int_0^{\pi/4} \frac{1 - \cos 4\theta}{2}\,d\theta = \left(\left.\theta - \frac{\sin 4\theta}{4}\right|_0^{\pi/4}\right) = \frac{\pi}{4}$.

5.5.51

a. True. This follows by substituting $u = f(x)$ to obtain the integral $\int u\,du = \frac{u^2}{2} + C = \frac{f(x)^2}{2} + C$.

b. True. Again, this follows from substituting $u = f(x)$ to obtain the integral $\int u^n\,du = \frac{u^{n+1}}{n+1} + C = \frac{(f(x))^{n+1}}{n+1} + C$ where $n \neq -1$.

c. False. If this were true, then $\sin(2x)$ and $2\sin x$ would have to differ by a constant, which they do not. In fact, $\sin 2x = 2\sin x\cos x$.

d. False. The derivative of the right hand side is $(x^2 + 1)^9 \cdot 2x$ which is not the integrand on the left hand side.

e. False. If we let $u = f'(x)$, then $du = f''(x)\,dx$. Substituting yields $\int_{f'(a)}^{f'(b)} u\,du = \left(\left.\frac{u^2}{2}\right|_{f'(a)}^{f'(b)}\right) = \frac{(f'(b))^2}{2} - \frac{(f'(a))^2}{2}$.

5.5.53 Let $u = 10x$. Then $du = 10\,dw$. Substituting yields $\frac{1}{10}\int \sec^2 u\,du = \frac{1}{10}\tan(u) + C = \frac{1}{10}\tan(10x) + C$.

5.5.55 Let $u = \cot x$. Then $du = -\csc^2 x\,dx$. Substituting yields $-\int u^{-3}\,du = \frac{1}{2u^2} + C = \frac{1}{2\cot^2 x} + C$.

5.5.57 Note that $\sin x \sec^8 x = \frac{\sin x}{\cos^8 x}$. Let $u = \cos x$, so that $du = -\sin x \, dx$. Substituting yields
$-\int u^{-8} \, du = \frac{1}{7u^7} + C = \frac{1}{7\cos^7 x} + C = \frac{\sec^7 x}{7} + C.$

5.5.59 Let $u = x^2 - 1$, so that $du = 2x \, dx$. Also note that when $x = 2$ we have $u = 3$, and when $x = 3$ we have $u = 8$. Substituting yields $\frac{1}{2} \int_3^8 u^{-1/3} \, du = \frac{1}{2} \left(\frac{3u^{2/3}}{2} \Big|_3^8 \right) = \frac{3}{4} \left(4 - \sqrt[3]{9} \right).$

5.5.61 Let $u = 16 - x^4$. Then $du = -4x^3 \, dx$. Also note that when $x = 0$ we have $u = 16$, and when $x = 2$ we have $u = 0$. Substituting yields $\frac{1}{4} \int_0^{16} \sqrt{u} \, du = \frac{1}{4} \left(\frac{2u^{3/2}}{3} \Big|_0^{16} \right) = \frac{32}{3}.$

5.5.63 $A(x) = \int_0^{\sqrt{\pi}} x \sin(x^2) \, dx$. Let $u = x^2$, so that $du = 2x \, dx$. Also, when $x = 0$ we have $u = 0$ and when $x = \sqrt{\pi}$ we have $u = \pi$. Substituting yields $\frac{1}{2} \int_0^{\pi} \sin u \, du = \frac{1}{2} \left(-\cos u \big|_0^{\pi} \right) = 1.$

5.5.65 $A(x) = \int_2^6 (x-4)^4 \, dx = \left(\frac{(x-4)^5}{5} \Big|_2^6 \right) = \frac{2^5}{5} - \frac{(-2)^5}{5} = \frac{64}{5}.$

5.5.67 $A(a) = \int_0^a \left(\frac{1}{a} - \frac{x^2}{a^3} \right) dx = \left(\frac{x}{a} - \frac{x^3}{3a^3} \Big|_0^a \right) = 1 - \frac{1}{3} = \frac{2}{3}.$ This is a constant function.

5.5.69

a. $\int_0^4 \frac{200}{(t+1)^2} \, dt = \left(\frac{-200}{t+1} \Big|_0^4 \right) = -40 + 200 = 160.$

b. $\int_0^6 \frac{200}{(t+1)^3} \, dt = \left(\frac{-200}{2(t+1)^2} \Big|_0^6 \right) = \frac{-100}{49} + 100 = \frac{4800}{49}.$

c. $\Delta P = \int_0^T \frac{200}{(t+1)^r} \, dt$. This decreases as r increases, since $\frac{200}{(t+1)^r} > \frac{200}{(t+1)^{r+1}}$.

d. Suppose $\int_0^{10} \frac{200}{(t+1)^r} \, dt = 350$. Then $\left(\frac{200(t+1)^{-r+1}}{1-r} \Big|_0^{10} \right) = 350$, so $11^{1-r} - 1 = \frac{350(1-r)}{200}$, and thus $\frac{11}{11^r} = \frac{7-7r}{4} + \frac{4}{4} = \frac{11-7r}{4}$, and $11^r = \frac{44}{11-7r}$. Using trial and error to find r, we arrive at $r \approx 1.28$.

e. $\int_0^T \frac{200}{(t+1)^3} \, dt = \left(\frac{-200}{2(t+1)^2} \Big|_0^T \right) = \frac{-100}{(T+1)^2} + 100.$ As $T \to \infty$, this expression $\to 100$, so in the long run, the bacteria approaches a finite limit.

5.5.71 $\frac{1}{\pi/k - 0} \int_0^{\pi/k} \sin(kx) \, dx = \frac{k}{\pi} \cdot \left(\frac{-\cos kx}{k} \Big|_0^{\pi/k} \right) = \frac{1}{\pi} (1 - -1) = \frac{2}{\pi}.$

5.5.73 The area on the left is given by $\int_4^9 \frac{(\sqrt{x}-1)^2}{2\sqrt{x}} \, dx$. If we let $u = \sqrt{x} - 1$ so that $du = \frac{1}{2\sqrt{x}} \, dx$, we obtain the equivalent integral $\int_1^2 u^2 \, du$ which represents the area on the right.

5.5.75 Let $u = f(x)$, so that $du = f'(x)\,dx$. Substituting yields

$$\int_4^5 (5u^3 + 7u^2 + u)\,du = \left(\frac{5u^4}{4} + \frac{7u^3}{3} + \frac{u^2}{2} \Big|_4^5 \right) = \frac{7297}{12}.$$

5.5.77 Let $u = f^{(p)}(x)$ so that $du = f^{(p+1)}(x)\,dx$. Substituting yields

$$\int u^n\,du = \frac{u^{n+1}}{n+1} + C = \frac{1}{n+1}\left(f^{(p)}(x)\right)^{n+1} + C.$$

5.5.79 If we let $u = \sqrt{x+a}$, then $u^2 = x + a$ and $2u\,du = dx$. Substituting yields

$$\int_{\sqrt{a}}^{\sqrt{1+a}} (u^2 - a) \cdot u \cdot 2u\,du = \int_{\sqrt{a}}^{\sqrt{1+a}} 2u^4 - 2au^2\,du = \left(\frac{2u^5}{5} - \frac{2au^3}{3} \Big|_{\sqrt{a}}^{\sqrt{1+a}} \right)$$

$$= \frac{2(\sqrt{1+a})^5}{5} - \frac{2a(\sqrt{1+a})^3}{3} - \frac{2a^{5/2}}{5} + \frac{2a^{5/2}}{3}.$$

If we let $u = x + a$, then $u - a = x$ and $du = dx$. Substituting yields

$$\int_a^{a+1} (u - a)\sqrt{u}\,du = \left(\frac{2u^{5/2}}{5} - \frac{2au^{3/2}}{3} \Big|_a^{a+1} \right) = \frac{2(a+1)^{5/2}}{5} - \frac{2a(a+1)^{3/2}}{3} - \frac{2a^{5/2}}{5} + \frac{2a^{5/2}}{3}.$$

Note that the two results are the same.

5.5.81 If we let $u = \cos\theta$, then $du = -\sin\theta\,d\theta$. Substituting yields $\int \frac{-1}{u^2}\,du = \frac{1}{u} + C = \sec\theta + C$. If we let $u = \sec\theta$, then $du = \sec\theta\tan\theta\,d\theta$. Substituting yields $\int du = u + C = \sec\theta + C$. Note that the two results are the same.

5.5.83

a. Since $\sin 2x = 2\sin x\cos x$, we can write $(\sin x\cos x)^2 = \left(\frac{\sin 2x}{2}\right)^2 = \frac{\sin^2 2x}{4}$. Thus $I = \frac{1}{4}\int \sin^2 2x\,dx = \frac{1}{4}\left(\frac{x}{2} - \frac{\sin(4x)}{8}\right) + C = \frac{x}{8} - \frac{\sin(4x)}{32} + C$. Note that we used the result of the previous problem during this derivation.

b. $I = \frac{1}{4}\int (1 - \cos 2x)(1 + \cos 2x)\,dx = \frac{1}{4}\int (1 - \cos^2 2x)\,dx = \frac{1}{4}\int \sin^2 2x\,dx = \frac{1}{4}\left(\frac{x}{2} - \frac{\sin(4x)}{8}\right) + C = \frac{x}{8} - \frac{\sin(4x)}{32} + C$.

c. The results are clearly consistent. The work involved is similar in each method.

5.5.85

a.

   Given $\int_a^b f(cx)\,dx$, let $u = cx$. Note that $du = c \cdot dx$.

   A direct substitution yields $\frac{1}{c}\int_{ac}^{bc} f(u)\,du$.

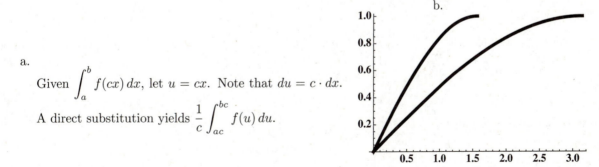

5.5.87 Let $u = \sqrt{x+1}$ so that $u^2 = x + 1$. Then $2u\,du = dx$. Substituting yields $\int 2 \cdot \dfrac{u\,du}{\sqrt{1+u}}$. Now let

$v = \sqrt{1+u}$ so that $v^2 = 1 + u$ and $2v\,dv = du$. Now a substitution yields $4 \int \dfrac{(v^2-1)v}{v}\,dv = 4\int v^2 - 1\,dv =$

$\dfrac{4v^3}{3} - 4v + C = \dfrac{4}{3}(1+u)^{3/2} - 4\sqrt{1+u} + C = \dfrac{4}{3}\left(1+\sqrt{x+1}\right)^{3/2} - 4\sqrt{1+\sqrt{x+1}} + C.$

5.5.89 Let $u = \cos\theta$, so that $du = -\sin\theta\,d\theta$. This substitution yields $\displaystyle\int_0^1 \dfrac{u}{\sqrt{u^2+16}}\,du$. Now let $v = u^2 + 16$,

so that $dv = 2u\,du$. Now a substitution yields $\dfrac{1}{2}\displaystyle\int_{16}^{17} v^{-1/2}\,dv = \left(\sqrt{v}\big|_{16}^{17}\right) = \sqrt{17} - 4.$

5.6   Chapter Five Review

5.6.1

  a. True. The antiderivative of a linear function is a quadratic function.

  b. False. $A'(x) = f(x)$, not $F(x)$.

  c. True. Note that f is an antiderivative of f', so this follows from the Fundamental Theorem.

  d. True. Since $|f(x)| \geq 0$ for all x, this integral must be positive, unless f is constantly 0.

  e. False. For example, the average value of $\sin x$ on $[0, 2\pi]$ is zero.

  f. True. This is equal to $2\int_a^b f(x)\,dx - 3\int_a^b g(x)\,dx = 2\int_a^b f(x)\,dx + 3\int_b^a g(x)\,dx.$

  g. True. The derivative of the right hand side is $f'(g(x))g'(x)$ by the Chain Rule.

5.6.3 The region can easily be divided up into rectangles and triangles. Above the axis there are 11 1×1 squares, a triangle of area $1/2$, and a triangle of area $3/2$. Below the axis there is a triangle of area $3/2$, so the net area is 11.5.

5.6.5 $\displaystyle\int_0^4 \sqrt{8x - x^2}\,dx = \int_0^4 \sqrt{16 - (x^2 - 8x + 16)}\,dx = \int_0^4 \sqrt{16 - (x-4)^2}\,dx$. This represents one quarter of the area inside the circle centered at $(4,0)$ with radius 4, so its value is $\dfrac{1}{4} \cdot 16\pi = 4\pi.$

5.6.7

  a. The right Riemann sum is $4(1) + 7(1) + 10(1) = 21.$

  b. The right Riemann sum is $\displaystyle\sum_{i=1}^n \left(3\left(1 + \dfrac{3i}{n}\right) - 2\right) \cdot \dfrac{3}{n}.$

  c. The sum evaluates as $\displaystyle\sum_{i=1}^n \dfrac{3}{n} + \sum_{i=1}^n \dfrac{27i}{n^2} = 3 + \dfrac{27}{n^2} \cdot \dfrac{n(n+1)}{2}$. As $n \to \infty$, the limit of this expression is
  $3 + 13.5 = 16.5.$

5.6.9 This sum is equal to $\displaystyle\int_0^4 (x^8 + 1)\,dx = \left(\dfrac{x^9}{9} + x\,\bigg|_0^4\right) = \dfrac{4^9}{9} + 4 = \dfrac{262180}{9}.$

5.6.11 $\displaystyle\int_{-2}^2 (3x^4 - 2x + 1)\,dx = \left(\dfrac{3x^5}{5} - x^2 + x\,\bigg|_{-2}^2\right) = \dfrac{96}{5} - 4 + 2 - \left(\dfrac{-96}{5} - 4 - 2\right) = \dfrac{192}{5} + 4 = \dfrac{212}{5}.$

5.6.13 $\displaystyle\int_0^2 (x+1)^3\,dx = \left(\frac{(x+1)^4}{4}\Big|_0^2\right) = \frac{81}{4} - \frac{1}{4} = 20.$

5.6.15 $\displaystyle\int (9x^8 - 7x^6)\,dx = x^9 - x^7 + C.$

5.6.17 $\displaystyle\int_0^1 x + \sqrt{x}\,dx = \left(\frac{x^2}{2} + \frac{2x^{3/2}}{3}\Big|_0^1\right) = \frac{1}{2} + \frac{2}{3} = \frac{7}{6}.$

5.6.19 $\displaystyle\int_0^1 \frac{6x}{(4-x^2)^{3/2}}\,dx = -3\int_0^1 \frac{-2x\,dx}{(4-x^2)^{3/2}} = 6(4-x^2)^{-1/2}\Big|_0^1 = \frac{6}{\sqrt{3}} - 3 = 2\sqrt{3} - 3$

5.6.21 Let $u = 25 - x^2$, and note that $du = -2x\,dx$. Substituting yields

$$\frac{-1}{2}\int_{25}^{16} u^{-1/2}\,du = -\left.\sqrt{u}\right|_{25}^{16} = 5 - 4 = 1.$$

5.6.23 $\displaystyle\int \sin^2(5\theta)\,d\theta = \int \frac{1 - \cos(10\theta)}{2}\,d\theta = \frac{\theta}{2} - \frac{\sin(10\theta)}{20} + C.$

5.6.25 $\displaystyle\int \frac{x^2 + 2x - 2}{(x^3 + 3x^2 - 6x)^2}\,dx = \frac{1}{3}\int \frac{(3x^2 + 6x - 6)\,dx}{(x^3 + 3x^2 - 6x)^2} = -\frac{1}{3(x^3 + 3x^2 - 6x)} + C$

5.6.27

a. $\displaystyle\int_{-4}^4 f(x)\,dx = 2\int_0^4 f(x)\,dx = 2 \cdot 10 = 20.$

b. $\displaystyle\int_{-4}^4 3g(x)\,dx = 3 \cdot 0 = 0.$

c. $\displaystyle\int_{-4}^4 4f(x) - 3g(x)\,dx = 2 \cdot 4 \cdot \int_0^4 f(x)\,dx - 3 \cdot 0 = 8 \cdot 10 - 0 = 80.$

5.6.29 $\displaystyle\int_1^4 3f(x)\,dx = 3\int_1^4 f(x)\,dx = 3(6) = 18.$

5.6.31 $\displaystyle\int_1^4 (3f(x) - 2g(x))\,dx = 3\int_1^4 f(x)\,dx - 2\int_1^4 g(x)\,dx = 3(6) - 2(4) = 18 - 8 = 10.$

5.6.33 There is not enough information to compute this integral.

5.6.35 The displacement is $\displaystyle\int_0^2 5\sin(\pi t)\,dt = \left(\frac{-5}{\pi}\cos(\pi t)\Big|_0^2\right) = 0.$ The distance traveled is

$$\int_0^2 5|\sin(\pi t)|\,dt = 5\int_0^1 \sin\pi t\,dt + 5\int_1^2 -\sin\pi t\,dt = \left(\frac{-5}{\pi}\cos(\pi t)\Big|_0^1\right) + \left(\frac{5}{\pi}\cos(\pi t)\Big|_1^2\right) = \frac{10}{\pi} + \frac{10}{\pi} = \frac{20}{\pi}.$$

5.6.37

a. The average value is 2.5. This is because for a straight line, the average value occurs at the midpoint of the interval, which is at the point $(3.5, 2.5)$.

b. The average value is 3 over the interval $[2, 4]$ and 3 over the interval $[4, 6]$, so is 3 over the interval $[2, 6]$.

5.6.39 Let $u = 2x$, so that $du = 2\,dx$. We have $\dfrac{1}{2}\displaystyle\int_2^4 f'(u)\,du = \dfrac{1}{2}\cdot(f(4) - f(2)) = \dfrac{f(4)}{2} - 2$. Since we are given that this quantity is 10, we have $f(4) = 24$.

5.6.41

By the Fundamental Theorem, $f'(x) = \frac{1}{x}$, which is always positive for $x > 1$. Thus f is always increasing. Also, $f(1) = \displaystyle\int_1^1 \frac{1}{t}\,dt = 0$. Also, $f''(x) = \frac{-1}{x^2}$ which is always negative, so f is always concave down.

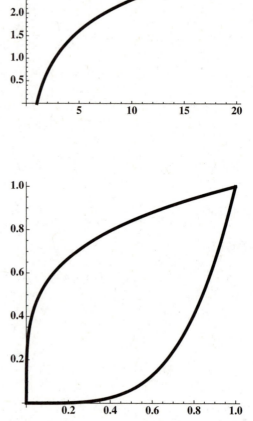

5.6.43

Because x^n and $\sqrt[n]{x}$ are inverse functions of each other, they are symmetric in the square $[0,1]\times[0,1]$ about the line $y = x$. Together, the two regions completely fill up the 1×1 square, so these two areas add to one.

5.6.45 Let $u = x^2$; then $du = 2x\,dx$ so that

$$\int \sec^8(\tan x^2)\sin(\tan x^2)x\sec^2 x^2\,dx = \frac{1}{2}\int \sec^8(\tan u)\sin(\tan u)\sec^2 u\,du$$

Now substitute $v = \tan u$. Then $dv = (1 + \tan^2 u)\,du = (1 + v^2)\,du$, so that (using the fact that $\sec^2 u = 1 + \tan^2 u$)

$$\frac{1}{2}\int \sec^8(\tan u)\sin(\tan u)\sec^2 u\,du = \frac{1}{2}\int \sec^8 v\sin v(1 + v^2)\frac{1}{1+v^2}\,dv = \frac{1}{2}\int \frac{\sin v}{\cos^8 v}\,dv$$
$$= \frac{1}{14}\sec^7 v + C = \frac{1}{14}\sec^7(\tan u) + C = \frac{1}{14}\sec^7(\tan x^2) + C$$

5.6.47 This follows by differentiating each side of the given equation. $\dfrac{d}{dx}\left(u(x) + 2\displaystyle\int_0^x u(t)\,dt\right) = u'(x) + 2u(x)$, and $\dfrac{d}{dx}10 = 0$.

5.6.49 Note that $f'(x) = (x-1)^{15}(x-2)^9$, and that the zeros of f' are at $x = 1$ and $x = 2$.

a. f' is positive and thus f is increasing on $(-\infty, 1)$ and on $(2, \infty)$, while f' is negative and f is decreasing on $(1, 2)$.

b. $f''(x) = 15(x-1)^{14}(x-2)^9 + (x-1)^{15} \cdot 9(x-2)^8 = 3(x-1)^{14}(x-2)^8(8x-13)$. f is concave up on $\left(\frac{13}{8}, \infty\right)$ and concave down on $\left(-\infty, \frac{13}{8}\right)$.

c. f has a local maximum at $x = 1$ and a local minimum at $x = 2$.

d. f has an inflection point at $x = \frac{13}{8}$.

Chapter 6

Applications of Integration

6.1 Velocity and Net Change

6.1.1 The position of an object is the coordinate of the object on the line at a given time, often denoted $s(t)$. The displacement over an interval $[a, b]$ is $s(b) - s(a)$, the difference of the object's ending position and beginning position. It can be written as $\int_a^b v(t)\,dt$ where $v(t)$ is the object's velocity at time t. The distance traveled by the object is $\int_a^b |v(t)|\,dt$, the sum of the distance traveled along the line to the right and the distance traveled along the line to the left over the given time interval.

6.1.3 The displacement is given by $\int_a^b v(t)\,dt$, since this quantity is equal to $s(b) - s(a)$.

6.1.5 The value of Q at time t will be given by $Q(t) = Q(0) + \int_0^t Q'(z)\,dz$.

6.1.7

a.

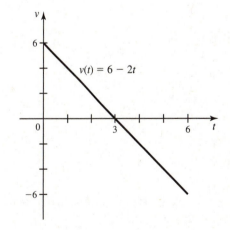

The motion is positive for $0 < t < 3$ and negative for $3 < t < 6$.

b. The displacement is $\int_0^6 (6 - 2t)\,dt = \left(6t - t^2\right)\big|_0^6 = 0$ meters.

c. The distance traveled is $\int_0^3 (6 - 2t)\,dt + \int_3^6 (2t - 6)\,dt = 9 + 9 = 18$ meters.

6.1.9

a.

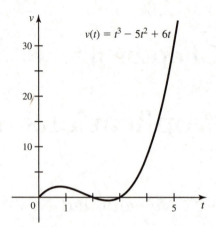

The motion is positive for $0 < t < 2$ and $3 < t < 5$, and negative for $2 < t < 3$.

b. The displacement is $\int_0^5 (t^3 - 5t^2 + 6t)\, dt = \left(\frac{t^4}{4} - \frac{5t^3}{3} + 3t^2 \right)\Big|_0^5 = \frac{275}{12}$ meters.

c. The distance traveled is

$$\int_0^2 v(t)\, dt - \int_2^3 v(t)\, dt + \int_3^5 v(t)\, dt$$
$$= \left(\frac{t^4}{4} - \frac{5t^3}{3} + 3t^2 \right)\Big|_0^2 - \left(\frac{t^4}{4} - \frac{5t^3}{3} + 3t^2 \right)\Big|_2^3 + \left(\frac{t^4}{4} - \frac{5t^3}{3} + 3t^2 \right)\Big|_3^5$$
$$= \frac{8}{3} + \frac{5}{12} + \frac{62}{3} = \frac{95}{4} \text{ meters.}$$

6.1.11

a.

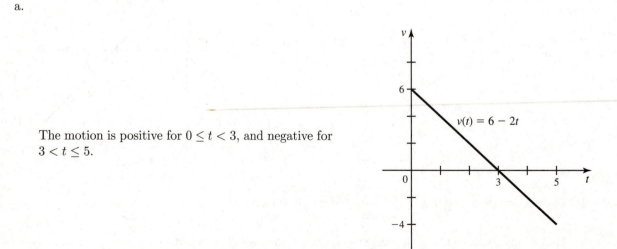

The motion is positive for $0 \le t < 3$, and negative for $3 < t \le 5$.

c.

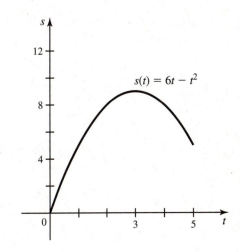

b. $s(t) = \int 6 - 2t\, dt = 6t - t^2 + C$, and since $s(0) = 0$, we must have $C = 0$. Thus, $s(t) = 6t - t^2$. Also, $s(t) = s(0) + \int_0^t (6 - 2x)\, dx = \left(6x - x^2\right)\big|_0^t = 6t - t^2$.

6.1.13

a.

The motion is positive for $0 < t < 3$, and negative for $3 < t < 4$.

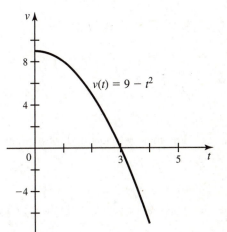

c.

b. $s(t) = \int (9 - t^2)\, dt = 9t - t^3/3 + C$, and since $s(0) = -2$, we must have $C = -2$. Thus,

$$s(t) = 9t - \frac{t^3}{3} - 2.$$

Also,

$$s(t) = s(0) + \int_0^t (9 - x^2)\, dx = -2 + \left(9x - x^3/3\right)\Big|_0^t$$

$$= 9t - \frac{t^3}{3} - 2.$$

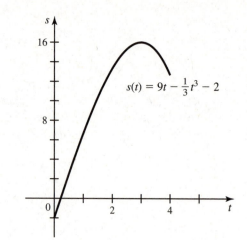

6.1.15

b.

a. $s(t) = s(0) + \int_0^t 2\pi \cos \pi x\, dx = 2 \sin \pi x\big|_0^t = 2 \sin \pi t.$

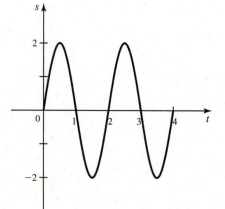

c. The mass reaches its lowest point at $t = 1.5$, $t = 3.5$ and $t = 5.5$.

d. The mass reaches its highest point at $t = .5$, $t = 2.5$, and $t = 4.5$.

6.1.17

a.

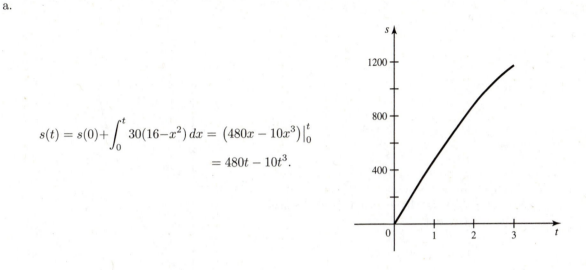

$$s(t) = s(0) + \int_0^t 30(16-x^2)\, dx = \left(480x - 10x^3\right)\Big|_0^t$$
$$= 480t - 10t^3.$$

b. Since the velocity is positive, this is given by $s(2) - s(0) = 960 - 80 = 880$ miles.

c. The velocity is 400 when $480 - 30t^2 = 400$, or $t = \sqrt{8/3}$. At this point the plane has traveled $s(\sqrt{8/3}) = 480\sqrt{8/3} - 10\sqrt{8/3}^3 \approx 740.29$ miles.

6.1.19

a.

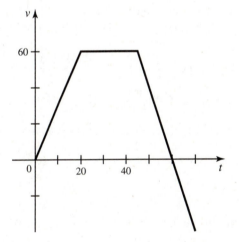

The velocity has a maximum of 60 for $20 \le t \le 45$.
The velocity is 0 at $t = 0$ and at $t = 60$.

b. $\int_0^{20} 3t\, dt + \int_{20}^{30} 60\, dt = 1200$ meters.

c. $1200 + \int_{30}^{45} 60\, dt + \int_{45}^{60}(240 - 4t)\, dt = 1200 + 900 + \left(240t - 2t^2\right)\Big|_{45}^{60} = 2550$ meters.

d. At time $t = 60$ the automobile is at position 2550. In the following 15 seconds, it moves $\int_{60}^{75}(240 - 4t)\, dt = 450$ feet in the opposite direction, so it is at position $2550 - 450 = 2100$.

6.1.21 $v(t) = \int a(t)\, dt = \int -9.8\, dt = -9.8t + C$, and since $v(0) = 20$, we have $C = 20$, so $v(t) = -9.8t + 20$. $s(t) = \int v(t)\, dt = \int(20 - 9.8t)\, dt = 20t - 4.9t^2 + D$, and since $s(0) = 0$, we must have $D = 0$. Thus $s(t) = 20t - 4.9t^2$.

6.1.23 $v(t) = \int a(t)\,dt = \int (-.01t)\,dt = -.005t^2 + C$, and since $v(0) = 10$, we have $C = 10$, so $v(t) = 10 - .005t^2$. $s(t) = \int v(t)\,dt = \int (10 - .005t^2)\,dt = 10t - \frac{1}{600}t^3 + D$, and since $s(0) = 0$, we must have $D = 0$. Thus $s(t) = 10t - \frac{1}{600}t^3$.

6.1.25

a.

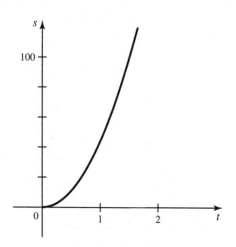

The velocity is given by $\int 88\,dt = 88t + C$, and $C = 0$ since $v(0) = 0$, so $v(t) = 88t$. The position is given by $\int 88t\,dt = 44t^2 + D$, but $D = 0$ since $s(0) = 0$, so $s(t) = 44t^2$.

b. The car travels $s(4) = 44 \cdot 16 = 704$ feet.

c. Since a quarter mile is 1320 feet, we need $44t^2 = 1320$, so $t = \sqrt{30} \approx 5.477$ seconds.

d. We need $44t^2 = 300$, so $t \approx 2.611$ seconds.

e. It reaches that speed when $88t = 178$, or $t = 89/44$ seconds. At that time the racer has traveled $s(89/44) = 44(89/44)^2 = 89^2/44 \approx 180.023$ feet.

6.1.27 $v(t) = \int a(t)\,dt = \int \dfrac{-1280}{(1 + 8t)^3}\,dt = \dfrac{80}{(1 + 8t)^2} + C$, and $C = 0$, since $v(0) = 80$. $s(t) = \int v(t)\,dt = \int \dfrac{80}{(1 + 8t)^2}\,dt = \dfrac{-10}{1 + 8t} + D$, but we can take $D = 0$ since the initial position is unspecified. Then in the first .2 seconds the train travels $s(.2) - s(0) = \dfrac{-10}{2.6} - (-10) = 10 - \dfrac{50}{13} = \dfrac{80}{13} \approx 6.154$ miles. Between time 0.2 and 0.4 the train travels $s(.4) - s(.2) = \dfrac{-10}{4.2} - \left(\dfrac{-10}{2.6}\right) = \dfrac{50}{13} - \dfrac{50}{21} = \dfrac{400}{273} \approx 1.465$ miles

6.1.29

a. In the first 35 days the number of barrels produced is $\displaystyle\int_0^{30} 800\,dt + \int_{30}^{35} (2600 - 60t)\,dt = 24000 + 3250 = 27250$.

b. In the first 50 days the number of barrels produced is $27250 + \int_{35}^{40} (2600 - 60t)\,dt + \int_{40}^{50} 200\,dt = 27250 + 1750 + 2000 = 31000$.

c. A constant 200 barrels per day times 20 days yields 4000 barrels.

6.1.31

a. $P(20) = 250 + \int_0^{20} (30 + 30\sqrt{t})\,dt = 250 + \left(30t + 20t^{3/2}\right)\Big|_0^{20} = 250 + 600 + 800\sqrt{5} = 850 + 800\sqrt{5} \approx 2639$ people.

b. $P(t) = 250 + \int_0^t (30 + 30\sqrt{x})\,dx = 250 + \left(30x + 20x^{3/2}\right)\Big|_0^t = 250 + 30t + 20t^{3/2}$ people.

6.1.33 $N'(t) = 200(t+2)^{-1/2}$ gives $N(t) = 400\sqrt{t+2} + C$; the initial condition gives $1500 = 400\sqrt{2} + C$ so that $C = 1500 - 400\sqrt{2}$ and $N(t) = 400\sqrt{t+2} + 1500 - 400\sqrt{2}$.

  a. After 14 days the population is $N(14) = 400\sqrt{16} + 1500 - 400\sqrt{2} = 3100 - 400\sqrt{2}$; after 34 days the population is $N(34) = 400\sqrt{36} + 1500 - 400\sqrt{2} = 3900 - 400\sqrt{2}$.

  b. See above.

6.1.35

  a. The additional cost is $\int_{100}^{150} (2000 - .5x)\,dx = \left(2000x - \frac{x^2}{4}\right)\Big|_{100}^{150} = 96875$ dollars.

  b. The additional cost is $\int_{500}^{550} (2000 - .5x)\,dx = \left(2000x - \frac{x^2}{4}\right)\Big|_{500}^{550} = 86875$ dollars.

6.1.37

  a. The additional cost is $\int_{100}^{150} (300 + 10x - .01x^2)\,dx = \left(300x + 5x^2 - \frac{x^3}{300}\right)\Big|_{100}^{150} = 69583.33$ dollars.

  b. The additional cost is $\int_{500}^{550} (300 + 10x - .01x^2)\,dx = \left(300x + 5x^2 - \frac{x^3}{300}\right)\Big|_{500}^{550} = 139583.33$ dollars.

6.1.39

  a. False. This would only be the case if the motion was all in the same direction. If the object changes direction at all, then the distance traveled is greater than the displacement.

  b. True. This is because $v(t) = |v(t)|$ in this case.

  c. True. This is because $V'(t) > 0$ for $0 < t < 10$, but $V'(t) < 0$ for $t > 10$.

  d. True. The cost of increasing production from A to B is given by $\int_A^B C'(t)\,dt$, which is geometrically the area under the curve $y = C'(x)$ from A to B. If C' is positive and decreasing, there is more area under the curve from A to $2A$ than from $2A$ to $3A$.

6.1.41

  a. The displacement is the net area, which is $\frac{1}{2} \cdot 2 \cdot 2 - \frac{1}{2} \cdot \frac{4}{3} \cdot 1 + \frac{1}{2} \cdot \frac{5}{3} \cdot 2 = 3$.

  b. The distance traveled is $\frac{1}{2} \cdot 2 \cdot 2 + \frac{1}{2} \cdot \frac{4}{3} \cdot 1 + \frac{1}{2} \cdot \frac{5}{3} \cdot 2 = \frac{13}{3}$

  c. $s(5) = s(0) + \int_0^5 v(t)\,dt = 0 + 3 = 3$.

  d. $s(t) = \begin{cases} \int_0^t (-x+2)\,dx & \text{if } 0 \le t \le 3, \\ \frac{3}{2} + \int_3^t (3x - 10)\,dx & \text{if } 3 < t \le 4, \\ 2 + \int_4^t (-2x + 10)\,dx & \text{if } 4 < t \le 5 \end{cases} = \begin{cases} \frac{-t^2}{2} + 2t & \text{if } 0 \le t \le 3, \\ \frac{3t^2}{2} - 10t + 18 & \text{if } 3 < t \le 4, \\ -t^2 + 10t - 22 & \text{if } 4 < t \le 5. \end{cases}$

6.1.43 The distance traveled is $\int_0^4 (1 - (t^2/16))\,dt = \left(t - (t^3/48)\right)\Big|_0^4 = \frac{8}{3}$. So the same distance could have been traveled over the given time period at a constant velocity of $\frac{8/3}{4} = \frac{2}{3}$.

6.1.45 The distance traveled is $\int_0^5 t\sqrt{25 - t^2}\,dt = \frac{1}{2}\int_0^{25} \sqrt{u}\,du = \left(\frac{1}{3}u^{3/2}\right)\Big|_0^{25} = \frac{125}{3}$. So the same distance could have been traveled over the given time period at a constant velocity of $\frac{125/3}{5} = \frac{25}{3}$.

6.1.47

a.

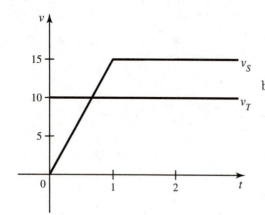

b. After 1 hour, Theo has ridden $1 \cdot 10 = 10$ miles, and Sasha has ridden $\frac{1}{2} \cdot 15 = 7.5$ miles, so Theo has ridden farther.

c. After 2 hours, Theo has ridden $2 \cdot 10 = 20$ miles, and Sasha has ridden $7.5 + (15)(1) = 22.5$ miles, so Sasha has ridden farther.

d. The times when they arrive at the various mile markers are in the following table:

	10	15	20
Theo	1	3/2	2
Sasha	7/6	3/2	11/6

Note that Theo hits the 10 mile marker first, then they are tied as they hit the 15 mile marker, and Sasha hits the 20 mile marker first. The area under v_S is the same as the area under v_T for $t = 1.5$, for $t < 1.5$ the area under v_T is greater, and for $t > 1.5$, the area under v_S is greater.

e. Theo will then hit the 20 mile mark in $18.8/10 = 1.88$ hours. Sasha hits the 20 mile mark at $t = 11/6 \approx 1.833$ hours, so Sasha will win.

f. A head start of .2 hours is equivalent for Theo of $10 \cdot .2 = 2$ miles. It will take him $18/10 = 1.8$ hours to ride the other 18 miles, while it still takes Sasha about 1.83 hours to cover 20 miles, so Theo will win.

6.1.49

a.

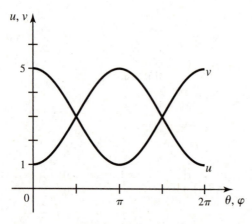

Abe starts out running into a headwind, Bess starts out running with a tailwind.

b. Abe's average speed is

$$\frac{1}{2\pi} \int_0^{2\pi} 3 - 2\cos\phi \, d\phi$$

$$= \frac{1}{2\pi} \left(3\theta - 2\sin\phi\right)\big|_0^{2\pi} = 3 \text{ mph},$$

and Bess's average speed is

$$\frac{1}{2\pi} \int_0^{2\pi} 3 + 2\cos\theta \, d\theta$$

$$= \frac{1}{2\pi} \left(3\theta + 2\sin\theta\right)\big|_0^{2\pi} = 3 \text{ mph},$$

so they have the same average speed.

c. If the radius is 50 meters, then the circumference of the track is 100π meters which is approximately .1952 miles. Since they both average 3 miles per hour, this should take them about $.1952/3 \approx .065$ hours, or about 3.9 minutes. They should tie the race.

6.1.51

a. $\int_0^2 20(1 + \cos(\pi t/12))\,dt = \left(20t + \frac{240}{\pi}\sin(\pi t/12)\right)\big|_0^2 = 40 + \frac{240}{\pi} \cdot \frac{1}{2} = 40 + \frac{120}{\pi} \approx 78.2\,m^3.$

b.

$$\int_0^t 20(1 + \cos(\pi x/12))\,dx$$
$$= \left(20x + \frac{240}{\pi}\sin(\pi x/12)\right)\Big|_0^t$$
$$= 20t + \frac{240}{\pi} \cdot \sin(\pi t/12)\,m^3.$$

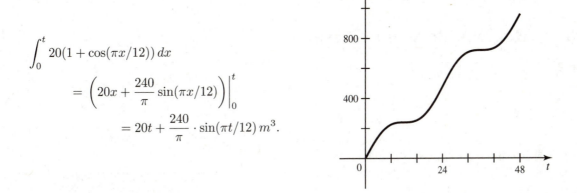

c. The reservoir is full when $20T + \frac{240}{\pi}\sin(\pi T/12) = 2500$, which occurs for $T \approx 122.6$ hours.

6.1.53

a.   b.

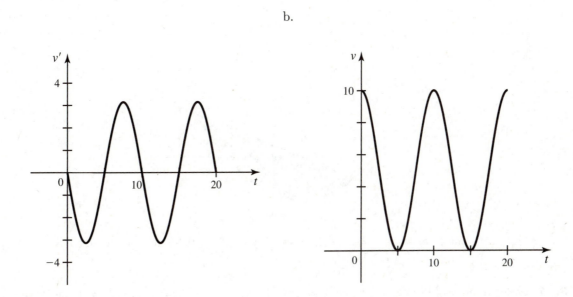

$V(t) = \int(-\pi L\sin(\pi t/5))\,dt = 5L\,(\cos(\pi t/5)) + C$. And since $V(0) = 10L$, we have $10L = 5L + C$, so $C = 5L$. Thus, $V(t) = 5L(1 + \cos(\pi t/5))$.

c. One breath takes $\frac{2\pi}{\pi/5} = 10$ seconds, so the breathing rate is 6 breaths per minute.

6.1.55

a. $E = \int_0^{24}(300 - 200\sin(\pi t/12))\,dt = \left(300t + \frac{2400}{\pi}\cos(\pi t/12)\right)\big|_0^{24} = 7200$ MWh. This is equivalent to $7.2 \times 10^6 \cdot 3.6 \times 10^6 = 2.592 \times 10^{13}$ Joules.

b. For one day, $\dfrac{7.2 \times 10^6\,\text{KWh}}{450\,\text{Kwh/kg}} = 16,000$ kg coal needed. For one year, $16000\,\text{kg} \times 365 = 5,840,000$ kg coal needed.

c. For one day, $\dfrac{7.2 \times 10^6\,\text{KWh}}{1.6 \times 10^4\,\text{Kwh/g}} = 450$ g U-235 need. For one year, $450 \times 365 = 164,250$ g needed.

d. $\dfrac{7.2 \times 10^6\,\text{KWh/day}}{(200\,\text{KW/turbine}) \cdot (24\,\text{hours/day})} \approx 1500$ turbines.

6.2   Regions Between Curves

6.2.1

If f and g intersect at $x = a$ and $x = b$ with $a < b$ and if $f(x) \geq g(x)$ on $[a, b]$, then the area between these curves is given by $\int_a^b (f(x) - g(x))\,dx$.

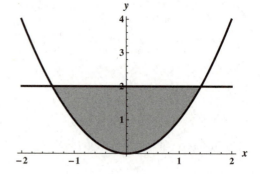

6.2.3

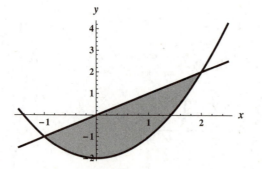

6.2.5 The curves intersect when $x = x^2 - 2$, or $(x + 1)(x - 2) = 0$, so at $x = -1$ and $x = 2$. The area is $\int_{-1}^{2}(x - (x^2 - 2)) = \left(2x + x^2/2 - x^3/3\right)\big|_{-1}^{2} = 4.5$.

6.2.7 The two curves intersect when $3 - x = x^2 + 1$ or $x^2 + x - 2 = 0$, so at $x = 1$ and $x = -2$. The intersection point given in the picture is clearly the one at $x = 1$. Thus the area is

$$\int_0^1 (3 - x) - (x^2 + 1)\,dx = \left(-\frac{1}{3}x^3 - \frac{1}{2}x^2 + 2x\right)\bigg|_0^1 = \frac{7}{6}$$

6.2.9

$$\int_{-1}^{4} (3x + 3 - (2x + 2))\,dx = \int_{-1}^{4} (x + 1)\,dx$$
$$= \left(x^2/2 + x\right)\Big|_{-1}^{4} = 8 + 4 - ((1/2) - 1) = 12.5.$$

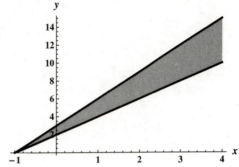

6.2.11

These curves intersect where $2x^2 = x^2 + 4$, so at $x = \pm 2$. Additionally, $x^2 + 4 \geq 2x^2$ on $[-2, 2]$, so the area between the curves is

$$\int_{-2}^{2} (x^2 + 4) - 2x^2\,dx = \left(-\frac{1}{3}x^3 + 4x\right)\Big|_{-2}^{2} = \frac{32}{3}$$

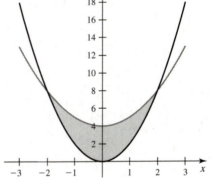

6.2.13

$x^4 - 9 = 7$ when $x = \pm 2$, so the area is

$$\int_{-2}^{2} (16 - x^4)\,dx = \left(16x - \frac{1}{5}x^5\right)\Big|_{-2}^{2} = \frac{256}{5}$$

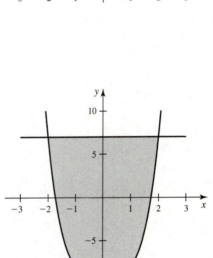

6.2.15 The curves intersect at $\pi/4$, so the area is given by $\int_{0}^{\pi/4} \sin x\,dx + \int_{\pi/4}^{\pi/2} \cos x\,dx = (-\cos x)\Big|_{0}^{\pi/4} + (\sin x)\Big|_{\pi/4}^{\pi/2} = (1 - \sqrt{2}/2) + (1 - \sqrt{2}/2) = 2 - \sqrt{2}.$

6.2.17

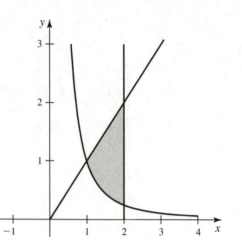

This is the region bounded by $x = 1$ and $x = 2$ above $y = 1/x^2$ and below $y = x$. The area is

$$\int_1^2 \left(x - \frac{1}{x^2} \right) dx = \left(\frac{1}{2}x^2 + \frac{1}{x} \right) \Big|_1^2 = 1$$

6.2.19

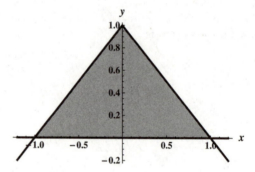

Note that this represents a triangle with base 2 and height 1, so the area is 1.

6.2.21

$$\int_2^6 (x/2 - |x - 3|)\, dx$$

$$= \int_2^3 (x/2 - (3 - x))\, dx + \int_3^6 (x/2 - (x - 3))\, dx$$

$$= \int_2^3 (3x/2 - 3)\, dx + \int_3^6 (3 - x/2)\, dx$$

$$= \left(3x^2/4 - 3x \right) \Big|_2^3 + \left(3x - x^2/4 \right) \Big|_3^6$$

$$= 27/4 - 9 - (3 - 6) + (18 - 9 - (9 - 9/4)) = 3/4 + 9/4 = 3.$$

6.2.23 Since the lines intersect at $y = 8$, the area is given by $\int_0^8 (-y/2 + 4 - (y - 8))\, dy = \int_0^8 (-3y/2 + 12)\, dy = \left(-3y^2/4 + 12y \right) \Big|_0^8 = -48 + 96 = 48.$

6.2.25

The area is

$$\int_0^3 (y-2)-(-\sqrt{4-y})\,dy + \int_3^4 (\sqrt{4-y}-(-\sqrt{4-y}))\,dy$$

$$= \left(\frac{y^2}{2} - 2y + \frac{-2}{3}(4-y)^{3/2}\right)\Big|_0^3 + \frac{-4}{3}\left((4-y)^{3/2}\right)\Big|_3^4$$

$$= \frac{9}{2} - 6 - \frac{2}{3} - (0 - 0 - \frac{-16}{3}) + \frac{-4}{3}(0-1) = \frac{9}{2}.$$

6.2.27

a. The area is given by $\int_{-\sqrt{2}}^{-1}(-(x^2-2))\,dx + \int_{-1}^0 (-x)\,dx$.

b. The area can also be written as $\int_{-1}^0 (y - (-\sqrt{y+2}))\,dy$.

6.2.29

a. The area is given by $\int_{-3}^{-2}(\sqrt{x+3} - (-\sqrt{x+3}))\,dx + \int_{-2}^6 (\sqrt{x+3} - x/2)\,dx$.

b. The area can also be written $\int_{-1}^3 (2y - (y^2 - 3))\,dy$.

6.2.31

a. The area is given by

$$\int_0^{7.5} \sqrt{x}\,dx + \int_{7.5}^9 (\sqrt{x} - (2x-15))\,dx = \left(2x^{3/2}/3\right)\Big|_0^{7.5} + \left(2x^{3/2}/3 - x^2 + 15x\right)\Big|_{7.5}^9$$

$$= \frac{2(15/2)^{3/2}}{3} + (18 - 81 + 135) - (\frac{2(15/2)^{3/2}}{3} - (7.5)^2 - 15 \cdot 7.5) = 15.75.$$

b. The area can also be written as

$$\int_0^3 (y/2 + 7.5 - y^2)\,dy = \left(y^2/4 + 7.5y - y^3/3\right)\Big|_0^3 = (9/4 + 45/2 - 9) - 0 = 15.75.$$

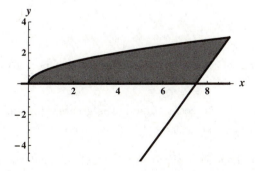

6.2.33

The area is given by

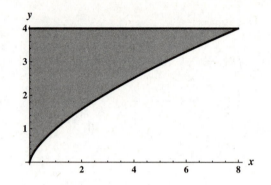

$$\int_0^8 (4 - x^{2/3}) \, dx = \left(4x - 3x^{5/3}/5 \right) \Big|_0^8$$
$$= 32 - 96/5 = 64/5.$$

6.2.35

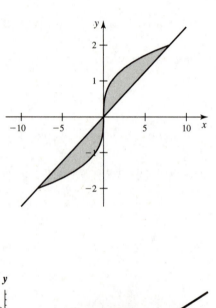

The curves intersect where $x = 4y = y^3$, so at $y = -2, 0, 2$. The area is twice the area from $y = 0$ to $y = 2$, so is

$$2 \int_0^2 (4y - y^3) \, dy = \left(4y^2 - \frac{1}{2}y^4 \right) \Big|_0^2 = 8$$

6.2.37

The area is given by

$$\int_0^{\sqrt{3}/2} (2x\sqrt{1 - x^2} - x) \, dx$$

$$= \left(\frac{-2}{3}(1 - x^2)^{3/2} - \frac{x^2}{2} \right) \Big|_0^{\sqrt{3}/2}$$

$$= \frac{-2}{24} - \frac{3}{8} + \frac{2}{3} = \frac{5}{24}.$$

6.2.39

a. False. This can be done either with respect to x or with respect to y. For the latter, the relevant integral is $\int_0^1 (y - y^2) \, dy$.

b. False. On the interval $(0, \pi/4)$ the cosine function is greater, but on $(\pi/4, \pi/2)$ the sine function is greater. The area is $\int_0^{\pi/4} (\cos x - \sin x) \, dx + \int_{\pi/4}^{\pi/2} (\sin x - \cos x) \, dx$.

c. True. They both represent the area of the region in the first quadrant under $y = x$ and above $y = x^2$.

6.2.41

The area is given by

$$\int_4^5 (7x - 19 - (x-1)^2)\, dx = \left. \left(7x^2/2 - 19x - (x-1)^3/3\right)\right|_4^5$$
$$= (175/2 - 95 - 64/3) - (56 - 76 - 9)$$
$$= \frac{-173}{6} + 29 = \frac{1}{6}.$$

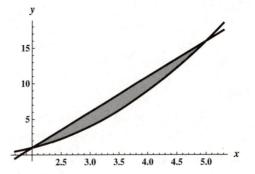

6.2.43

The area is given by

$$\int_2^5 (5x - 9 - (x-1)^2)\, dx = \left. \left(5x^2/2 - 9x - (x-1)^3/3\right)\right|_2^5$$
$$= 125/2 - 45 - 64/3 - (10 - 18 - (1/3)) = 4.5.$$

6.2.45 This is given by $\int_0^4 (3y - (y^2 - y))\, dy = \int_0^4 (4y - y^2)\, dy = \left. \left(2y^2 - y^3/3\right)\right|_0^4 = 32 - 64/3 = 32/3.$

6.2.47 The area is given by $\int_0^3 (y/2 + 7.5 - y^2)\, dy = \left. \left(y^2/4 + 7.5y - y^3/3\right)\right|_0^3 = (2.25 + 22.5 - 9) = 15.75.$

6.2.49 On $[0, \pi/2]$, $\sin x \geq 1 - \cos x$, so the area is

$$\int_0^{\pi/2} (\sin x + \cos x - 1)\, dx = \left. (\sin x - \cos x - x)\right|_0^{\pi/2} = 2 - \frac{\pi}{2}$$

On $[\pi/2, 2\pi]$, $1 - \cos x \geq \sin x$, so the area is

$$\int_{\pi/2}^{2\pi} (1 - \cos x - \sin x)\, dx = \left. (x - \sin x + \cos x)\right|_{\pi/2}^{2\pi} = \frac{3\pi}{2} + 2$$

6.2.51

a. The area of R_1 is $\int_0^1 (x - x^p)\, dx = \left. \left(x^2/2 - \frac{x^{p+1}}{p+1}\right)\right|_0^1 = \frac{1}{2} - \frac{1}{p+1} = \frac{p-1}{2p+2}$. The area of R_2 is $\int_0^1 (x^{1/q} -$

$x)\, dx = \left. \left(\frac{q}{q+1} x^{(q+1)/q} - x^2/2\right)\right|_0^1 = \frac{q}{q+1} - \frac{1}{2} = \frac{q-1}{2q+2}$. Clearly, if $q = p$ then $R_1 = R_2$.

b. Using the results above, if $p > q$, then

$$R_1 - R_2 = \frac{(p-1)(2q+2) - (q-1)(2p+2)}{(2p+2)(2q+2)} = \frac{4p - 4q}{(2p+2)(2q+2)} = \frac{p-q}{(p+1)(q+1)} > 0,$$

so $R_1 > R_2$.

c. If $p < q$, then $R_1 - R_2$ computed above is less than 0, so $R_1 < R_2$.

6.2.53

$$A = \int_0^{\sqrt{2}-1} (2x - x/2)\,dx + \int_{\sqrt{2}-1}^{(\sqrt{17}-1)/4} (1 - x^2 - x/2)\,dx = \left(3x^2/4\right)\big|_0^{\sqrt{2}-1} + \left(x - x^3/3 - x^2/4\right)\big|_{\sqrt{2}-1}^{(\sqrt{17}-1)/4}$$

$$= \frac{3}{4}\left(3 - 2\sqrt{2}\right) - \left(\sqrt{2}-1\right)\left(1 + \frac{1}{4}\left(1 - \sqrt{2}\right) + \frac{1}{3}\left(2\sqrt{2} - 3\right)\right)$$

$$+ \frac{1}{4}\left(\sqrt{17} - 1\right)\left(1 + \frac{1}{16}\left(1 - \sqrt{17}\right) - \frac{1}{48}\left(\sqrt{17} - 1\right)^2\right)$$

$$= \frac{1}{96}\left(135 - 128\sqrt{2} + 17\sqrt{17}\right) \approx .2508.$$

6.2.55 $A = \int_{-4}^{5}(8 - y - (y-2)^2/3)\,dy = \left(8y - y^2/2 - (y-2)^3/9\right)\big|_{-4}^{5} = 40 - 25/2 - 3 - (-32 - 8 + 24) = 40.5.$

6.2.57 $A_n = \int_0^1 (x^{1/n} - x)\,dx = \left(\dfrac{nx^{(n+1)/n}}{n+1} - x^2/2\right)\bigg|_0^1 = \dfrac{n}{n+1} - \dfrac{1}{2} = \dfrac{n-1}{2n+2}.$

6.2.59 Using the result of the previous problem, $\lim_{n\to\infty} A_n = \lim_{n\to\infty} \frac{n-1}{n+1} = 1$. As $n \to \infty$, the region in question approaches the 1×1 square over the interval $[0,1]$, which has area 1.

6.2.61

a. The point (n,n) on the curve $y = x$ would represent the notion that the lowest $100n\%$ of the society owns $100n\%$ of the wealth, which would represent a form of equality.

b. Since the lowest 0% of society can't own anything, we must have $L(0) = 0$, and since the lowest 100% of society must own everything, we have $L(1) = 1$. Since L is increasing, $L'(x) > 0$, and since by definition, the lowest $100n\%$ of the society can't own more than $100n\%$ of the goods (otherwise the definition of "lowest" would be violated), our curve must lie below the line $y = x$, and must thus be concave up, so $L''(x) > 0$.

c.

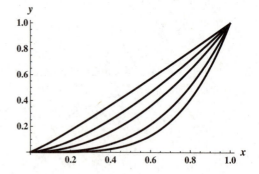

$y = x^{1.1}$ is closest to $y = x$, and $y = x^4$ is furthest from $y = x$.

d. Note that $B = \int_0^1 L(x)\,dx$, and $A + B = 1/2$, so $A = \frac{1}{2} - \int_0^1 L(x)\,dx$. Then $G = \frac{A}{A+B} = \frac{A}{1/2} = 2A = 1 - 2\int_0^1 L(x)\,dx$.

e. For $L(x) = x^p$, we have $G = 1 - 2\int_0^1 x^p\,dx = 1 - 2\left(\frac{x^{p+1}}{p+1}\right)\big|_0^1 = 1 - \frac{2}{p+1} = \frac{p-1}{p+1}$. So we have

p	1.1	1.5	2	3	4
G	1/21	1/5	1/3	1/2	3/5

f. For $p = 1$ we have $G = 0$. Since $\lim_{p\to\infty} \frac{p-1}{p+1} = 1$, the largest possible value of G approaches 1.

g. For $L(x) = 5x^2/6 + x/6$, note that $L(0) = 0$, $L(1) = 1$, $L'(x) = 5x/3 + 1/6 > 0$ on $[0,1]$, and $L''(x) = 5/3 > 0$ as well. The Gini index is $G = 1 - 2\int_0^1 (5x^2/6 + x/6)\, dx = 1 - 2\left(5x^3/18 + x^2/12\right)\big|_0^1 = 1 - 2\left(5/18 + 1/12\right) = 1 - 5/9 - 1/6 = 5/18$.

6.2.63

Some experimental data:

a	-2	-1	0	1
Approximate Area	7.812	6.928	7.812	10.667

The area seems to be minimized for $a = -1$.

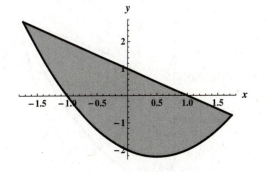

6.2.65 $A = 2\int_0^1 y^2\sqrt{1 - y^3}\, dy = \frac{2}{3}\int_0^1 \sqrt{u}\, du$ (where $u = 1 - y^3$). So $A = \frac{2}{3}\left(\frac{2}{3}u^{3/2}\right)\big|_0^1 = \frac{4}{9}$.

6.2.67

a. $F(a) = \int_0^b x(x - a)(x - b)\, dx = \int_0^b (abx - ax^2 - bx^2 + x^3)\, dx = \left(abx^2/2 - ax^3/3 - bx^3/3 + x^4/4\right)\big|_0^b = ab^3/2 - ab^3/3 - b^4/3 + b^4/4 = ab^3/6 - b^4/12$. $F(a) = 0$ when $a = b/2$.

b.

$$A(a) = \int_0^a x(x - a)(x - b)\, dx - \int_a^b x(x - a)(x - b)\, dx$$

$$= \left(abx^2/2 - ax^3/3 - bx^3/3 + x^4/4\right)\big|_0^a - \left(abx^2/2 - ax^3/3 - bx^3/3 + x^4/4\right)\big|_a^b$$

$$= -a^4/6 + a^3b/3 - ab^3/6 + b^4/12$$

$$A'(a) = -2a^3/3 + a^2b - b^3/6 = -\frac{1}{6}(2a - b)\left(2a^2 - 2ab - b^2\right).$$

Also, $A''(a) = -2a^2 + 2ab$. Note that $A'(b/2) = 0$, and $A''(b/2) = -b^2/2 + b^2 > 0$, so there is a minimum at $a = b/2$. The other critical number is $a = \dfrac{2b + \sqrt{4b^2 - (4)(2)(-b^2)}}{4} = \dfrac{b + \sqrt{3b^2}}{2} = \left(\dfrac{1 + \sqrt{3}}{2}\right)b$, where there is a local maximum. The graph shown is $A(a)$ for $b = 4$.

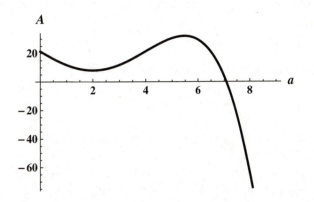

6.2.69

a.

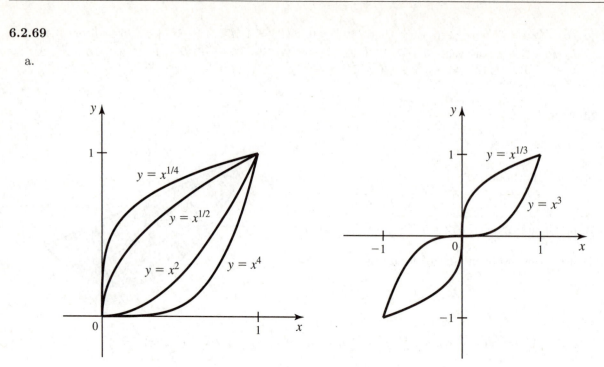

b. $A_n(x)$ is the net area of the region between the graphs of f and g from 0 to x.

c. Note that

$$\int_0^1 (f(s) - g(s))\, ds = \left(\frac{s^{n+1}}{n+1} - \frac{s^{(1/n)+1}}{(1/n)+1} \right)\Bigg|_0^1 = \frac{1}{n+1} - \frac{n}{n+1} = \frac{1-n}{n+1} < 0.$$

So we seek the smallest c so that $\int_1^c (f(s) - g(s))\, ds = \frac{n-1}{n+1}$. This occurs when

$$\left(\frac{s^{n+1}}{n+1} - \frac{s^{(1/n)+1}}{(1/n)+1} \right)\Bigg|_1^c = \frac{n-1}{n+1},$$

or $\dfrac{c^{n+1}}{n+1} = \dfrac{c^{(n+1)/n}}{(n+1)/n}$, or $c^{n+1} = nc^{(n+1)/n}$, or $c^{n-(1/n)} = n$, so $c = n^{n/(n^2-1)}$. As n increases, this root increases as well.

6.3   Volume by Slicing

6.3.1 $A(x)$ is the area of the cross section through the solid at the point x.

6.3.3 $V = \int_0^2 \pi(4x^2 - x^4)\, dx.$

6.3.5 The cross sections are disks and $A(x)$ is the area of a disk.

6.3.7 $A(x) = \frac{1}{2} \cdot 4 \cdot 3 = 6.$ $V = \int_0^5 6\, dx = 30.$

6.3.9 $A(x) = (2\sqrt{25 - x^2})^2 = 100 - 4x^2.$ $V = \int_0^5 A(x)\, dx = \int_0^5 (100 - 4x^2)\, dx = \left(100x - 4x^3/3 \right)\Big|_0^5 = 500 - \frac{500}{3} = \frac{1000}{3}.$

6.3.11 $A(x) = \frac{\pi}{2}\left(\frac{2-x}{2} \right)^2 = \frac{\pi}{8}(4 - 4x + x^2).$ $V = \int_0^2 A(x)\, dx = \int_0^2 \frac{\pi}{8}(4 - 4x + x^2)\, dx = \frac{\pi}{8}\left(4x - 2x^2 + x^3/3 \right)\Big|_0^2 = \frac{\pi}{3}.$

6.3.13 The relationship between the height h of tetrahedron and the edge length l is $h = l\sqrt{2/3}$, which can be deduced using triangle geometry and the Pythagorean theorem. Let z be the distance from the top vertex of the tetrahedron down toward the base perpendicularly. The cross sections perpendicular to this axis are all equilateral triangles with height z, so their side length is $\sqrt{3/2}z$, and their area is each $\sqrt{3/2}z/2 \cdot \frac{\sqrt{3}}{2}(\sqrt{3/2}z) = \frac{3\sqrt{3}z^2}{8}$. The Volume is thus $\int_0^{4\sqrt{2/3}} \frac{3\sqrt{3}}{8} z^2 \, dz = \frac{3\sqrt{3}}{8} (z^3/3)\big|_0^{4\sqrt{2/3}} = \frac{16\sqrt{2}}{3}$.

6.3.15 $V = \pi \int_0^3 4x^2 \, dx = 4\pi \left(x^3/3 \right)\big|_0^3 = 36\pi$.

6.3.17 The volume of the solid, using the disk method, is

$$\int_0^2 \pi f(x)^2 \, dx = \int_0^2 \pi(4 - x^2)^2 \, dx = \pi \left(16x - \frac{8}{3}x^3 + \frac{1}{5}x^5 \right) \bigg|_0^2 = \frac{256}{15}\pi$$

6.3.19 $V = \pi \int_0^\pi \sin^2 x \, dx = \frac{\pi}{2} \int_0^\pi (1 - \cos(2x)) \, dx = \frac{\pi}{2} (x - \sin(2x)/2)\big|_0^\pi = \frac{\pi^2}{2}$.

6.3.21 $\displaystyle\int_0^{1/2} \pi f(x)^2 \, dx = \int_0^{1/2} \pi \frac{1}{(1-x)^{1/2}} \, dx = -2\pi\sqrt{1-x}\,\bigg|_0^{1/2} = \pi(2 - \sqrt{2})$

6.3.23 $V = \pi \int_0^4 ((2\sqrt{x})^2 - x^2) \, dx = \pi \left(2x^2 - x^3/3 \right)\big|_0^4 = \pi(32 - 64/3) = \frac{32\pi}{3}$.

6.3.25

$$\int_{\pi/6}^{5\pi/6} \pi(f(x)^2 - g(x)^2) \, dx = \int_{\pi/6}^{5\pi/6} \pi(2\sin x - 1) \, dx = \pi(-2\cos x - x)\bigg|_{\pi/6}^{5\pi/6} = \frac{2\pi}{3}(3\sqrt{3} - \pi)$$

6.3.27

$$V = \pi \int_0^2 ((4x)^2 - (4x^2 - x^3)^2) \, dx = \pi \int_0^2 (-x^6 + 8x^5 - 16x^4 + 16x^2) \, dx$$
$$= \pi \left(-x^7/7 + 4x^6/3 - 16x^5/5 + 16x^3/3 \right)\bigg|_0^2 = \frac{256\pi}{35}.$$

6.3.29

$$V = \pi \int_0^{\pi/2} (\sin x - \sin^2 x) \, dx = \pi \int_0^{\pi/2} \left(\sin x - \frac{1}{2}(1 - \cos(2x)) \right) dx = \pi \left(\sin(2x)/4 - \cos x - x/2 \right)\big|_0^{\pi/2}$$
$$= \pi(1 - \pi/4) = \frac{4\pi - \pi^2}{4}.$$

6.3.31 $V = \pi \int_0^6 (y^2 - y^2/4) \, dy = \frac{3\pi}{4} \left(y^3/3 \right)\big|_0^6 = 54\pi$.

6.3.33 $V = \pi \int_0^8 (4 - y^{2/3}) \, dy = \pi \left(4y - (3/5)y^{5/3} \right)\big|_0^8 = \frac{64\pi}{5}$.

6.3.35 $V = 2\pi \int_0^4 (\sqrt{16 - y^2})^2 \, dy = 2\pi \left(16y - y^3/3 \right)\big|_0^4 = \frac{256\pi}{3}$.

6.3.37 About the x-axis: $V_x = \pi \int_0^5 4x^2 \, dx = \pi \left(4x^3/3 \right)\big|_0^5 = \frac{500\pi}{3}$. About the y-axis: $V_y = \pi \int_0^{10} (25 - y^2/4) \, dy = \pi \left(25y - y^3/12 \right)\big|_0^{10} = \frac{500\pi}{3}$. The volumes are the same.

6.3.39 About the x-axis: $V_x = \pi \int_0^1 (1 - x^3)^2 \, dx = \pi \int_0^1 (1 - 2x^3 + x^6) \, dx = \pi \left(x - x^4/2 + x^7/7 \right)\big|_0^1 = \frac{9\pi}{14}$.
About the y-axis: $V_y = \pi \int_0^1 (\sqrt[3]{1-y})^2 \, dy = -\pi \left(\frac{3}{5}(1-y)^{5/3} \right)\big|_0^1 = -\pi(0 - 3/5) = \frac{3\pi}{5}$. The volume V_x is
bigger.

6.3.41

a. False. The cross sections are not disks or washers.

b. True. It is given by $V = \pi \int_0^R (R^2 - x^2) \, dx$.

c. True. This is because if we shift the sine function horizontally by $\pi/2$ units, we obtain the cosine
function.

6.3.43

Using the washer method, the volume is

$$\int_1^6 \pi(1 - x^{-3}) \, dx = \left(\pi x + \frac{\pi}{2x^2} \right)\bigg|_1^6 = \frac{325}{72}\pi$$

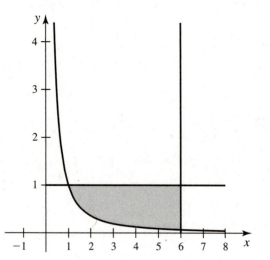

6.3.45

This region is symmetric around $y = 2$, so we find
the volume from $y = 0$ to $y = 2$ and double it.
Solving for x in $y = x^{1/3}$ gives $x = y^3$, so we have

$$2 \int_0^2 \pi y^6 \, dy = \frac{2\pi}{7}y^7 \bigg|_0^2 = \frac{256}{7}\pi$$

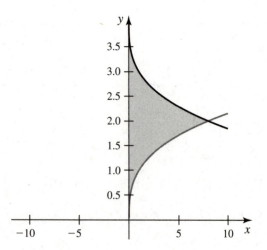

6.3.47 The volume V_S is $\pi \int_0^{\sqrt{a}} (y^2 - a)^2 \, dy = \pi \left(y^5/5 - 2ay^3/3 + a^2 y \right)\big|_0^{\sqrt{a}} = \frac{8\pi}{15}a^{5/2}$. The volume V_T is
$\frac{1}{3}\pi a^2 \cdot a^{1/2} = \frac{1}{3}\pi a^{5/2}$. The ratio of $V_S/V_T = \frac{8/15}{1/3} = \frac{8}{5}$.

6.3.49

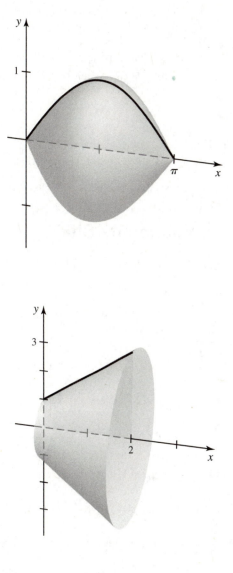

a. This comes from revolving the region in the first quadrant bounded by $\sin x$ and the line $y = 0$ between $x = 0$ and $x = \pi$ around the x axis.

b. This comes from revolving the region in the first quadrant bounded by $y = x+1$ and the line $y = 0$ between $x = 0$ and $x = 2$ around the x axis.

6.3.51

a. Think of the cone as being obtained by revolving the region under $y = x$ in the first quadrant from $x = 0$ to $x = R$ around the x-axis. The volume is $\pi \int_0^R x^2 \, dx = \pi R^3/3 = \frac{1}{3}V_C$.

b. Think of the hemisphere as being obtained by revolving the region under $y = \sqrt{R^2 - x^2}$ between $x = 0$ and $x = R$ around the x-axis. The volume is $\pi \int_0^R (R^2 - x^2) \, dx = \pi \left(R^2 x - x^3/3 \right)\big|_0^R = \frac{2\pi R^3}{3} = \frac{2}{3}V_C$.

6.3.53 $V = \pi \int_{-2}^{2}((3 + \sqrt{4 - y^2})^2 - (3 - \sqrt{4 - y^2})^2) \, dy = 24\pi \int_0^2 \sqrt{4 - y^2} \, dy$. Note that the last integral represents $1/4$ the area of a circle of radius 2, so has value π. Thus $V = 24\pi(\pi) = 24\pi^2$.

6.3.55 $V = \pi \int_0^2 (36 - (2 + x^2)^2) \, dx = \pi \int_0^2 (32 - 4x^2 - x^4) \, dx = \pi \left(32x - 4x^3/3 - x^5/5 \right)\big|_0^2 = \pi(64 - 32/3 - 32/5) = \frac{704\pi}{15}$.

6.3.57 $V = \pi \int_0^2 ((6 - x^2)^2 - 4) \, dx = \pi \int_0^2 (x^4 - 12x^2 + 32) \, dx = \pi \left(x^5/5 - 4x^3 + 32x \right)\big|_0^2 = \frac{192\pi}{5}$.

6.3.59

a. The cross sections of such an object are washers, with inner radius given b $g(x) - y_0$ and outer radius given by $f(x) - y_0$, so the volume is given by $\pi \int_a^b ((f(x) - y_0)^2 - (g(x) - y_0)^2)\,dx$.

b. The cross section of this object are also washers, and this time the inner radius is $y_0 - f(x)$ and the outer radius is $y_0 - g(x)$, so the volume is given by $\pi \int_a^b ((y_0 - g(x))^2 - (y_0 - f(x))^2)\,dx$.

6.3.61

a. By the general slicing method, $V(x) = \int_a^b A(x)\,dx$. Since the two figures have the same cross sections $A(x)$, they must therefore have the same volumes.

b. We are seeking the value of r so that $10\pi r^2 = 40$, so $r^2 = \frac{4}{\pi}$, and $r = \frac{2}{\sqrt{\pi}}$ meters.

6.4   Volume by Shells

6.4.1 $V = 2\pi \int_a^b x(f(x) - g(x))\,dx$.

6.4.3 ...revolved about the x axis. ...using the disk/washer method and integrating with respect to $\underline{x}$ or using the shell method and integrating with respect to $\underline{y}$.

6.4.5 $\displaystyle\int_0^2 2\pi x\left(\frac{x}{2} + 1\right)dx = 2\pi\left(\frac{x^3}{6} + \frac{x^2}{2}\right)\Big|_0^2 = \frac{20}{3}\pi$

6.4.7 $V = 2\pi \int_0^1 x(3 - 3x)\,dx = 2\pi\left(3x^2/2 - x^3\right)\big|_0^1 = \pi$.

6.4.9 $V = 2\pi \int_0^{\sqrt{\pi/2}} x\cos(x^2)\,dx = \pi\left(\sin(x^2)\right)\big|_0^{\sqrt{\pi/2}} = \pi$.

6.4.11 $V = 2\pi \int_0^2 y(4 - y^2)\,dy = 2\pi\left(2y^2 - y^4/4\right)\big|_0^2 = 8\pi$.

6.4.13 $V = 2\pi \int_2^4 y(4 - y)\,dy = 2\pi\left(2y^2 - y^3/3\right)\big|_2^4 = \frac{32\pi}{3}$.

6.4.15 Solve for x to get $x = \sqrt[3]{4}/y^{2/3}$; then the volume is

$$\int_2^{16} 2\pi y \frac{\sqrt[3]{4}}{y^{2/3}}\,dy = 2\sqrt[3]{4}\pi \int_2^{16} y^{1/3}\,dy = \frac{3\sqrt[3]{4}}{2}\pi y^{4/3}\Big|_2^{16} = \frac{3\sqrt[3]{4}}{2} \cdot 15 \cdot 2^{4/3}\pi = 90\pi$$

6.4.17 $V = 2\pi \int_0^1 (x + 2)x^2\,dx = 2\pi\left(x^4/4 + 2x^3/3\right)\big|_0^1 = \frac{11\pi}{6}$.

6.4.19 $V = 2\pi \int_0^1 (y+2)(1-\sqrt{y})\,dy = 2\pi \int_0^1 (y+2-y^{3/2}-2y^{1/2})\,dy = 2\pi\left(y^2/2 + 2y - 2y^{5/2}/5 - 4y^{3/2}/3\right)\big|_0^1 = 2\pi(1/2 + 2 - 2/5 - 4/3) = \frac{23\pi}{15}$.

6.4.21 Consider the region in the first quadrant bounded by the coordinate axes and the line $y = 8 - (8/3)x$. We can generate the desired cone by revolving this region around the y-axis. We then have $V = 2\pi \int_0^3 x(8 - (8/3)x)\,dx = \frac{16\pi}{3} \int_0^3 (3x - x^2)\,dx = \frac{16\pi}{3}\left(3x^2/2 - x^3/3\right)\big|_0^3 = 24\pi$.

6.4.23 Consider the triangle in the first quadrant bounded by $y = 0$, $x = 3$, and $y = 9 - (3/2)x$. The solid in question is formed by revolving this region around the y-axis. The volume is $V = 2\pi \int_3^6 x(9 - (3/2)x)\,dx = 2\pi\left(9x^2/2 - x^3/2\right)\big|_3^6 = 54\pi$.

6.4.25 Consider the part of the ellipse which lies in the first quadrant. We can obtain half the ellipsoid by revolving this region around the y-axis. So the volume of the whole ellipsoid is $V = 4\pi \int_0^2 x\sqrt{2 - (x^2/2)}\, dx$. Let $u = 2 - x^2/2$. Then $V = 4\pi \int_0^2 u^{1/2}\, du = 8\pi/3 \left(u^{3/2}\right)\big|_0^2 = \frac{16\pi\sqrt{2}}{3}$.

6.4.27 Washers: $V = \pi \int_0^1 (x^{2/3} - x^2)\, dx = \pi \left(3x^{5/3}/5 - x^3/3\right)\big|_0^1 = \frac{4\pi}{15}$. Shells: $V = 2\pi \int_0^1 y(y - y^3)\, dy = 2\pi \left(y^3/3 - y^5\right)\big|_0^1 = 2\pi(1/3 - 1/5) = \frac{4\pi}{15}$. The two methods are equally easy to apply.

6.4.29 Washers:

$$V = \pi \int_0^2 ((1 - x/3)^2 - (1/(x+1))^2)\, dx = \pi \left(x - x^2/3 + x^3/27 + 1/(x+1)\right)\big|_0^2 = \frac{8\pi}{27}.$$

Shells:

$$V = 2\pi \int_{1/3}^1 y(3 - 3y - (1/y - 1))\, dy = 2\pi \int_{1/3}^1 (4y - 3y^2 - 1)\, dy = 2\pi \left(2y^2 - y^3 - y\right)\big|_{1/3}^1$$

$$= 2\pi(0 - (2/9 - 1/27 - 1/3)) = \frac{8\pi}{27}.$$

The shell method seems a little easier to apply.

6.4.31 The curves intersect when $8 - 2x = 16 - x^2$, or $x^2 - 2x - 8 = 0$, so at $x = -2, 4$. Using the washer method, the area is

$$\int_{-2}^4 \pi((16 - x^2)^2 - (8 - 2x)^2)\, dx = \pi \int_{-2}^4 x^4 - 36x^2 + 32x + 192\, dx = \frac{3456}{5}\pi$$

To use the shell method, solve for x to get $x = \pm\sqrt{16 - y}$ and $x = (8 - y)/2$. We need two integrals, one for y from 0 to 12 and one for y from 12 to 16:

$$\int_0^{12} 2\pi y \left(\sqrt{16 - y} - \frac{8 - y}{2}\right) dy + \int_{12}^{16} 2\pi y \cdot 2\sqrt{16 - y}\, dy$$

$$= 2\pi \left[\int_0^{12} \left(y\sqrt{16 - y} - 4y + \frac{y^2}{2}\right) dy + \int_{12}^{16} y \cdot 2\sqrt{16 - y}\, dy\right]$$

$$= 2\pi \left[\left(-\frac{2}{15}(32 + 3y)(16 - y)^{3/2} - 2y^2 + \frac{y^3}{6}\right)\bigg|_0^{12} + \left(-\frac{4}{15}(32 + 3y)(16 - y)^{3/2}\right)\bigg|_{12}^{16}\right]$$

$$= 2\pi \left[\left(-\frac{2}{15}(68)(8) - 288 + 288 + \frac{2}{15}(32)(64)\right) - \frac{4}{15}(0 - 68(8))\right]$$

$$= 2\pi \left(\frac{3008}{15} + \frac{2176}{15}\right) = 2\pi \cdot \frac{1728}{5} = \frac{3456}{5}\pi$$

Clearly for this example the washer method is easier to apply.

6.4.33

   a. True. Otherwise, we wouldn't have shells!

   b. False. Either method can be used when revolving around either axis.

   c. True.

6.4.35

We use the shell method to avoid two separate integrals:

$$\int_2^4 2\pi x \cdot x^{-3}\, dx = 2\pi \int_2^4 x^{-2}\, dx = \frac{\pi}{2}$$

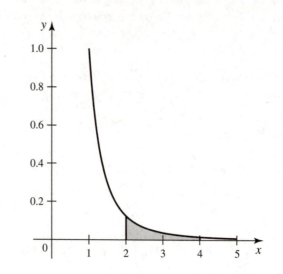

6.4.37 $V = 2\pi \int_0^1 ((2-x^2)^2 - (x^2)^2)\, dx = 2\pi \int_0^1 (4-4x^2)\, dx = 2\pi \left(4x - 4x^3/3\right)\big|_0^1 = \frac{16\pi}{3}.$

6.4.39 $V = 2\pi \int_2^6 x(2x+2-x)\, dx = 2\pi \int_2^6 (x^2+2x)\, dx = 2\pi \left(x^3/3 + x^2\right)\big|_2^6 = \frac{608\pi}{3}.$

6.4.41 $V = \int_0^1 A(y)\, dy = \int_0^1 \frac{\pi}{2}(\sqrt{y})^2\, dy = \frac{\pi}{2}\left(y^2/2\right)\big|_0^1 = \frac{\pi}{4}.$

6.4.43 By symmetry, $V = 4\pi \int_0^1 ((1-x)/2)^2\, dx = \pi \int_0^1 (1-2x+x^2)\, dx = \pi \left(x - x^2 + x^3/3\right)\big|_0^1 = \frac{\pi}{3}.$

6.4.45

a. $V_1 = \pi \int_0^1 (ax^2+1)^2\, dx = \pi \int_0^1 (a^2x^4 + 2ax^2 + 1)\, dx = \pi(a^2/5 + 2a/3 + 1).$ $V_2 = 2\pi \int_0^1 x(ax^2+1)\, dx = 2\pi(a/4 + 1/2) = \pi a/2 + \pi.$

b. These are equal when $\pi a/2 + \pi = \pi a^2/5 + 2a\pi/3 + \pi$, or when $\pi a^2/5 + a\pi/6 = 0$, which occurs when $a = 0$ and when $a = -5/6$.

6.4.47

a. $V = \pi \int_0^h (rx/h)^2\, dx = \frac{\pi r^2}{h^2}\left(x^3/3\right)\big|_0^h = \frac{\pi r^2 h}{3}.$

b. $V = 2\pi \int_0^r y(h - (h/r)y)\, dy = 2\pi h\left(y^2/2 - y^3/(3r)\right)\big|_0^r = \frac{\pi r^2 h}{3}.$

6.4.49

$$V = 2\pi \int_0^{\sqrt{16h-h^2}} x(h - 8 + \sqrt{64 - x^2})\, dx = 2\pi \left(hx^2/2 - 4x^2 - \frac{1}{3}\left(64 - x^2\right)^{3/2}\right)\Bigg|_0^{\sqrt{16h-h^2}}$$

$$= \pi(16h^2 - h^3 - 8(16h - h^2) - \frac{2}{3}(8-h)^3 - (0 - 0 - 1024/3)) = -\frac{1}{3}(h - 24)\pi h^2.$$

6.4.51 $V = 2\pi \int_0^2 ((3 + \sqrt{4-x^2})^2 - (3 - \sqrt{4-x^2})^2)\, dx = 24\pi \int_0^2 \sqrt{4-x^2}\, dx = 24\pi(\pi) = 24\pi^2.$ The last integral can be computed with a computer, or one can note that it represents $1/4$ of the area of the circle of radius 2 centered at the origin, so its value is $(1/4)\pi \cdot 4 = \pi.$

6.4.53

a. $V = 2\pi \int_0^a y^2\, dx = 2\pi \int_0^a b^2(1 - (x^2/a^2)\, dx = 2\pi b^2 \left(x - x^3/(3a^2) \right)\big|_0^a = \frac{4\pi ab^2}{3}$.

b.

$$V = 2 \cdot 2\pi \int_0^a (xb\sqrt{1 - (x^2/a^2)})\, dx = \frac{4\pi b}{a} \int_0^a x\sqrt{a^2 - x^2}\, dx = \frac{4\pi b}{a} \left(\frac{-1}{3}(a^2 - x^2)^{3/2} \right)\bigg|_0^a$$

$$= \frac{4\pi b}{a} \left(\frac{-1}{3}(0 - a^3) \right) = \frac{4\pi ba^2}{3}.$$

c. The ellipsoids generated are different when $a \neq b$, so there isn't any reason to expect them to have the same volumes.

6.4.55

a. Consider the region in the first quadrant bounded by the coordinate axes and $y = 8 - 2x$. The integral on the left represents the volume of the solid obtained when this region is revolved around the x-axis, using the disk/washer method, and the integral on the right represents the same volume calculated using the shell method. Hence, they are equal.

b. Consider the region in the first quadrant bounded by the line $x = 0$, the line $y = 5$, and the curve $y + x^2 + 1$. The integral on the left represents $\frac{1}{\pi}$ times the volume of the solid obtained when this region is revolved around the x-axis, using the disk/washer method, and the integral on the right represents $\frac{1}{\pi}$ times the same volume calculated using the shell method. Hence, they are equal.

6.5   Length of Curves

6.5.1 Given $f(x)$ and a and b, compute $f'(x)$ and then compute $\int_a^b \sqrt{1 + f'(x)^2}\, dx$.

6.5.3 $L = \int_1^5 \sqrt{1 + 2^2}\, dx = 4 \cdot \sqrt{5}$.

6.5.5 $y' = \sqrt{x}/2$, so $1 + y'^2 = 1 + x/4$. Thus,

$$L = \int_0^{60} \sqrt{1 + (x/4)}\, dx = \int_0^{60} \frac{1}{2}\sqrt{x + 4}\, dx = \left(\frac{1}{3}(x + 4)^{3/2} \right)\bigg|_0^{60} = \frac{504}{3} = 168$$

6.5.7 $y' = x(x^2 + 2)^{1/2}$, so $1 + y'^2 = 1 + x^2(x^2 + 2) = x^4 + 2x^2 + 1 = (x^2 + 1)^2$. Thus, $L = \int_0^1 (x^2 + 1)\, dx = \left(x^3/3 + x \right)\big|_0^1 = \frac{4}{3}$.

6.5.9 $y' = x^3 - \frac{1}{4x^3}$, so $1 + y'^2 = 1 + x^6 - \frac{1}{2} + \frac{1}{16x^6} = \left(x^3 + \frac{1}{4x^3} \right)^2$. Thus,

$$L = \int_1^2 \left(x^3 + \frac{1}{4x^3} \right) dx = \left(x^4/4 + \frac{-1}{8}x^{-2} \right)\bigg|_1^2 = 4 - (1/32) - (1/4 - (1/8)) = \frac{123}{32}.$$

6.5.11

a. $y' = 2x$, so $1 + y'^2 = 1 + 4x^2$, so $L = \int_{-1}^1 \sqrt{1 + 4x^2}\, dx$.

b. $L = \int_{-1}^1 \sqrt{1 + 4x^2}\, dx \approx 2.96$.

6.5.13

a. $f'(x) = \sec^2 x$, so $L = \int_0^{\pi/4} \sqrt{1 + \sec^4 x}\, dx$

b. Using technology, the value of this integral is approximately 1.277978.

6.5.15

a. $y' = \dfrac{1}{2\sqrt{x-2}}$, so $1 + y'^2 = 1 + \dfrac{1}{4(x-2)} = \dfrac{4x-7}{4x-8}$, so $L = \int_3^4 \sqrt{\dfrac{4x-7}{4x-8}}\, dx$.

b. $L = \int_3^4 \sqrt{\dfrac{4x-7}{4x-8}}\, dx \approx 1.08$.

6.5.17

a. $y' = -2\sin(2x)$, so $1 + y'^2 = 1 + 4\sin^2(2x)$, so $L = \int_0^\pi \sqrt{1 + 4\sin^2(2x)}\, dx$.

b. $L = \int_0^\pi \sqrt{1 + 4\sin^2(2x)}\, dx \approx 5.27$.

6.5.19

a. $y' = -1/x^2$, so $1 + y'^2 = 1 + \frac{1}{x^4} = \frac{x^4+1}{x^4}$. Thus, $L = \int_1^{10} \frac{\sqrt{x^4+1}}{x^2}\, dx$.

b. $L = \int_1^{10} \frac{\sqrt{x^4+1}}{x^2}\, dx \approx 9.15$.

6.5.21 $\frac{dx}{dy} = 2$, so $1 + \left(\frac{dx}{dy}\right)^2 = 1 + 4 = 5$, so $L = \int_{-3}^4 \sqrt{5}\, dy = 7\sqrt{5}$.

6.5.23 $\frac{dx}{dy} = y^3 - 1/(4y^3)$, so $1 + \left(\frac{dx}{dy}\right)^2 = 1 + y^6 - 1/2 + 1/(16y^6) = \left(y^3 + \frac{1}{4y^3}\right)^2$. So

$$L = \int_1^2 \left(y^3 + \frac{1}{4} \cdot y^{-3}\right) dy = \left(y^4/4 - \frac{1}{8y^2}\right)\Big|_1^2 = 4 - (1/32) - (1/4 - 1/8) = \frac{123}{32}.$$

6.5.25

a. False. For example, if $f(x) = x^2$, the first integrand is $\sqrt{1 + 4x^2}$ and the second is $1 + 2x$, which clearly yield different values for (for example) $a = 0$ and $b = 1$.

b. True. They are both equal to $\int_a^b \sqrt{1 + f'(x)^2}\, dx$.

c. False. Since $\sqrt{1 + f'(x)^2} > 0$, arc length can't be negative.

6.5.27

a. We are seeking functions $f(x)$ so that $f'(x) = \pm 4x^2$, so any function of the form $f(x) = \pm 4x^3/3 + C$ will work.

b. We are seeking functions $f(x)$ so that $f'(x) = \pm 6\cos(2x)$, so any function of the form $f(x) = \pm 3\sin(2x) + C$ will work.

6.5.29 The length of the parabola is $\int_{-1}^1 \sqrt{1 + 4x^2}\, dx \approx 2.9597$. The length of the given cosine function is $\int_{-1}^1 \sqrt{1 + \frac{\pi^2}{4}\sin^2(\pi x/2)}\, dx \approx 2.924$, so the parabola is longer.

6.5.31 $f'(x) = 0.00074x$, so $L = \int_{-640}^{640} \sqrt{1 + (0.00074x)^2}\, dx \approx 1326.4$ meters.

6.5.33

a. Let $u = 2x$. Then the given integral is equal to $\frac{1}{2} \int_a^b \sqrt{1 + f'(u)^2}\, du = \frac{L}{2}$.

b. Let $u = cx$. Then the given integral is equal to $\frac{1}{c} \int_a^b \sqrt{1 + f'(u)^2}\, du = \frac{L}{c}$.

6.5.35

a.

$$f'(x) = nax^{n-1} - \frac{1}{4anx^{n-1}}$$

so that

$$1 + f'(x)^2 = 1 + n^2 a^2 x^{2n-2} + \frac{1}{16 a^2 n^2 x^{2n-2}} - \frac{1}{2} = (nax^{n-1})^2 + \frac{1}{2} + \frac{1}{(4nax^{n-1})^2}$$

$$= \left(nax^{n-1} + \frac{1}{4nax^{n-1}} \right)^2$$

and then the arc length is

$$\int_c^d \sqrt{1 + f'(x)^2}\, dx = \int_c^d \left(nax^{n-1} + \frac{1}{4na} x^{-(n-1)} \right) dx$$

which is a rational function that we know how to integrate.

b.

$$\int_1^2 anx^{n-1} + \frac{1}{4an} x^{1-n}\, dx = ax^n + \frac{1}{4an(2-n)} x^{2-n} \Big|_1^2 = a(2^n - 1) + \frac{2^{2-n} - 1}{4an(2-n)}$$

6.6 Physical Applications

6.6.1 $m = \rho_1 \cdot l_1 + \rho_2 \cdot l_2 = 1\,\text{g/cm} \cdot 50\,\text{cm} + 2\,\text{g/cm} \cdot 50\,\text{cm} = 150\,\text{g}$.

6.6.3 Work is the force times the distance moved.

6.6.5 Different volumes of water are moved different distances.

6.6.7 $F = \rho g h = 1000 \cdot 9.8 \cdot 4 = 39,200\,\text{N/m}^2$.

6.6.9 $m = \int_0^\pi (1 + \sin x)\, dx = (x - \cos x)\big|_0^\pi = (\pi - (-1)) - (0 - 1) = \pi + 2$.

6.6.11 $= \int_0^2 (2 - x/2)\, dx = \left(2x - x^2/4 \right)\big|_0^2 = 3$.

6.6.13 $m = \int_0^1 x\sqrt{2 - x^2}\, dx = \left(\frac{-1}{3}(2 - x^2)^{3/2} \right)\big|_0^1 = (-1/3) + (2\sqrt{2}/3) = \frac{2\sqrt{2}-1}{3}$.

6.6.15 $m = \int_0^2 1\, dx + \int_2^4 (1 + x)\, dx = 2 + \left(x + x^2/2 \right)\big|_2^4 = 2 + (4 + 8) - (2 + 2) = 10$.

6.6.17 $W = \int_0^5 5\, dx = 25$ J.

6.6.19

a. $f(x) = kx$, and $f(.5) = 50$, so $k(.5) = 50$, so $k = 100$. Therefore $W = \int_0^{1.5} 100x\, dx = \left(50x^2 \right)\big|_0^{1.5} = 112.5$ J.

b. $\int_0^{-.5} 100x\, dx = \left(50x^2 \right)\big|_0^{-.5} = 12.5$ J.

6.6.21 $100 = \int_0^{.5} kx \, dx = \frac{1}{8} k$, so $k = 800$. $W = \int_{.5}^{1.25} kx \, dx = \left(400x^2 \right) \big|_{.5}^{1.25} = 525$ J.

6.6.23

$$W = \int_0^{2.5} \rho g A(y)(2.5 - y) \, dy = 1000 \cdot 9.8 \cdot 25 \cdot 15 \int_0^{2.5} (2.5 - y) \, dy = 3675000 \left(2.5y - y^2/2 \right) \big|_0^{2.5}$$

$$= 3675000 \cdot (2.5)^2/2 = 11,484,375 \text{ J.}$$

6.6.25

a. Let the vertex of the cone be at $(0,0)$, with the y-axis vertically oriented. Note that the area of a horizontal slice at height y is $\pi y^2/16$, and it must move $6 - y$ meters to get to the top. $W = \int_0^6 \rho g \pi \frac{y^2}{16} (6 - y) \, dy = (\pi \rho g/16) \int_0^6 (6y^2 - y^3) \, dy = (\pi \rho g/16) \left(2y^3 - y^4/4 \right) \big|_0^6 = (\pi \rho g/16)(108) = 66,150\pi$.

b. Not true. $\int_3^6 \rho g \pi \frac{y^2}{16} (6 - y) \, dy = \frac{\rho g \pi 297}{64} \neq \int_0^3 \rho g \pi \frac{y^2}{16} (6 - y) \, dy = \frac{\rho g \pi 135}{64}$.

6.6.27

a. Orient the axes so that the south pole of the tank is at $(0,0)$ and the north pole is at $(0,16)$. The cross section of the tank which contains the xy plane intersects the tank in the circle centered at $(0,8)$ with radius 8, so the curve is $x^2 + (y - 8)^2 = 8^2$, A slice at height y has area $\pi x^2 = \pi(16y - y^2)$. $W = \int_0^{16} \rho g \pi (16y - y^2) \, dy = \pi \rho g \left(8y^2 - y^3/3 \right) \big|_0^{16} = \pi \rho g \frac{16^3}{6} = 200,704,000 \cdot \pi/3$ J.

b. Total weight of the water lifted up for 18 meters is $W = \frac{4\pi}{3} R^3 \rho g h = \frac{4\pi}{3} 8^3 \cdot 1000 \cdot 9.8 \cdot 18 = 120,422,400\pi$ J.

6.6.29

a. Orient the axes so that the lower corners of the trough are at $(-0.25, 0)$ and at $(.25, 0)$. Then the upper corners are at $(-.5, 1)$ and at $(.5, 1)$. Note that the line between $(.25, 0)$ and $(.5, 1)$ is given by $y = 4x - 1$. The area of a slice at height y is $2x \cdot 10 = 20 \cdot \frac{1}{4}(y + 1) = 5(y + 1)$. Thus, $W = \rho g \int_0^1 5(y + 1)(1 - y) \, dy = 5\rho g \int_0^1 (1 - y^2) \, dy = 5\rho g \left(y - y^3/3 \right) \big|_0^1 = \frac{10\rho g}{3} = 32,666.\overline{6}$ J.

b. Yes. If the length is doubled, the area of each slice is doubled, so the work integral is doubled as well.

6.6.31 Orient the axes so that the lower corners of the trapezoid are at $(5, 0)$ and $(-5, 0)$, and the upper corners are at $(10, 15)$ and $(-10, 15)$. Note that the line between the corners for $x > 0$ is given by $y = 3(x-5)$, so at level y, we have a width of $2x = \frac{2y}{3} + 10$. $F = \rho g \int_0^{15} (15 - y) \left(\frac{2y+30}{3} \right) dy = \frac{2\rho g}{3} \int_0^{15} (225 - y^2) \, dy = \frac{2\rho g}{3} \left(225y - y^3/3 \right) \big|_0^{15} = \frac{2\rho g}{3} (2250) = 1500\rho g = 14,700,000$ N.

6.6.33 Orient the axes so that the bottom vertex is at $(0, 0)$. The other vertices are at $(\pm 10, 30)$, and the line between $(0, 0)$ and $(10, 30)$ is given by $y = 3x$. Thus, a slice at height y has width $2x = 2y/3$. $F = \int_0^{30} \rho g (30 - y) \frac{2y}{3} \, dy = \frac{2\rho g}{3} \left(15y^2 - y^3/3 \right) \big|_0^{30} = \frac{2\rho g}{3} (4500) = 3000\rho g = 29,400,000$ N.

6.6.35 $F = \int_0^{50} (150 + 2y)(80) \, dy = 80 \left(150y + y^2 \right) \big|_0^{50} = 800,000$ N.

6.6.37 $F = \int_1^{1.5} \rho g (4 - y)(.5) \, dy = \rho g \left(2y - y^2/4 \right) \big|_1^{1.5} = \frac{11\rho g}{16} = 6737.5$ N.

6.6.39

a. True. $m = \int_a^b \rho(x) \, dx = \frac{1}{b-a} \int_a^b \rho(x) \, dx \cdot (b - a) = \overline{\rho} \cdot L$.

b. True. $\int_0^L kx \, dx = \frac{kL^2}{2} = \int_0^{-L} kx \, dx$.

c. True. This follows since work is force times distance.

d. False. Although they have the same geometry, they are placed at different depths of the water, so the force is different.

6.6.41

a. Compared to the linear spring $F(x) = 16x$, the restoring force is less for large displacements.

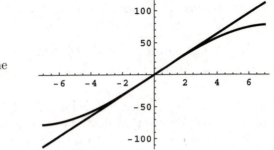

b. $W = \int_0^{1.5}(16x - 0.1x^3)\,dx = \left(8x^2 - .025x^4\right)\big|_0^{1.5} = 17.83$ J.

c. $W = \int_0^{-2}(16x - 0.1x^3)\,dx = \left(8x^2 - .025x^4\right)\big|_0^{-2} = 31.6$ J.

6.6.43 Orient the axes so that $(0,0)$ is in the middle of the bottom of the cup. Note that the line between $(.02, 0)$ and $(.025, .15)$ is given by $y = 30x - 3/5$, so $x = \frac{y}{30} + \frac{1}{50}$. The area of a cross section at height y is given by $\pi x^2 = \pi\left(\frac{5y+3}{150}\right)^2$. Note that the distance the slice must travel is $.2 - y$, since it must go $.05$ above the top of the glass. Thus, $W = \pi\rho g \int_0^{1.5}\left(\frac{5y+3}{150}\right)^2(.2 - y)\,dy = \frac{\pi\rho g}{5\cdot 150^2}\int_0^{1.5}(5y+3)^2(1 - 5y)\,dy = \frac{\pi\rho g}{5\cdot 150^2}\int_0^{1.5}(-125y^3 - 125y^2 - 15y + 9)\,dy = \frac{\pi\rho g}{5\cdot 150^2}\left(-125y^4/4 - 125y^3/3 - 15y^2/2 + 9y\right)\big|_0^{1.5} = \frac{\pi\rho g}{5\cdot 150^2}(1.0248) \approx 0.280456$ J.

6.6.45

a.

$$W = \int_0^{2500000}\frac{GMm}{(x+R)^2}\,dx = GMm\left(\frac{-1}{x+R}\right)\Big|_0^{2500000} = GMm\left(\frac{1}{R} - \frac{1}{R+2500000}\right)$$
$$= \frac{GMm2500000}{R(R+2500000)} \approx 8.87435 \times 10^9 \text{ J.}$$

b. $W(x) = \int_0^x\frac{GMm}{(t+R)^2}\,dt = GMm\left(\frac{-1}{t+R}\right)\Big|_0^x = GMm\left(\frac{1}{R} - \frac{1}{R+x}\right) = \frac{GMmx}{R(R+x)} = \frac{500GMx}{R(R+x)}.$

c. $\lim\limits_{x\to\infty}\frac{GMmx}{R(R+x)} = \lim\limits_{x\to\infty}\frac{GMm}{R\left(\left(\frac{R}{x}\right)+1\right)} = \frac{GMm}{R}.$

d. Suppose $\frac{GMmx}{R(R+x)} = \frac{1}{2}mv^2$, then $v^2 = \frac{2GMx}{R(R+x)}$, and as $x \to \infty$ we have $v^2 = \frac{2GM}{R}$, so $v = \sqrt{\frac{2GM}{R}}.$

6.6.47

a. $W_1 = \int_0^{30}\rho g(30 - y)\,dy = 5g\left(30y - y^2/2\right)\big|_0^{30} = 2250g$ J.

b. $W = W_1 + W_2$, where W_2 is the work to just lift the block. $W_2 = 50g \cdot 30 = 1500g$, so $W = 2250g + 1500g = 3750g$ J.

6.6.49

a. The acceleration due to gravity is $F = mg$, and is in the vertical direction. The tangent direction to the curve is perpendicular to the normal, which makes an angle of θ with the the vertical. If we form a right triangle with hypotenuse of length mg, and angle θ, then the two legs must have lengths $mg\sin\theta$ (parallel to the curve) and $mg\cos\theta$ (normal to the curve.)

b. Note that the angle t is $t = S/L$, where S is arc length. Then $S = Lt$ and $dS = L\,dt$. So $W = \int_0^\theta F\,ds = \int_0^\theta mg\sin\theta \cdot L\,d\theta = mgL\left(-\cos\theta\right))|_0^\theta = mg(L - L\cos\theta) = mgh.$

6.6.51   The plate on the left has more than half of its area below the horizontal line which is $3/2$ below the surface, while the plate on the right has exactly half its area below that line, so the plate on the left should have more force on it than the plate on the right. Note that the line in the first quadrant from $(0,0)$ to $(\sqrt{2}/2, \sqrt{2}/2)$ is given by $y = x$, while the line from $(\sqrt{2}/2, \sqrt{2}/2)$ to $(0, \sqrt{2})$ is given by $y = -x + \sqrt{2}$. So the width of a slice at height y in the lower part of the region is $2y$, and in the upper part of the region is $2(\sqrt{2} - y)$. The force on the plate on the left is given by

$$F = \rho g \int_0^{\sqrt{2}/2} (\sqrt{2} + 1 - y)(2y)\,dy + \rho g \int_{\sqrt{2}/2}^{\sqrt{2}} (\sqrt{2} + 1 - y)2(\sqrt{2} - y)\,dy$$

$$= \rho g \int_0^{\sqrt{2}/2} (-2y^2 + 2\sqrt{2}y + 2y)\,dy + \rho g \int_{\sqrt{2}/2}^{\sqrt{2}} (2y^2 - 4\sqrt{2}y - 2y + 2\sqrt{2} + 4)\,dy$$

$$= \rho g \left(-\frac{2y^3}{3} + \sqrt{2}y^2 + y^2\right)\Bigg|_0^{\sqrt{2}/2} + \rho g \left(2\left(\frac{y^3}{3} - \frac{1}{2}\left(1 + 2\sqrt{2}\right)y^2 + \sqrt{2}y + 2y\right)\right)\Bigg|_{\sqrt{2}/2}^{\sqrt{2}}$$

$$= \rho g \left(\frac{1}{2} + \frac{\sqrt{2}}{3} + \frac{1}{2} + \frac{1}{3\sqrt{2}}\right)$$

$$= \rho g \left(1 + \frac{\sqrt{2}}{2}\right) \text{ N.}$$

The force on the plate on the right is given by

$$F = \int_0^1 \rho g(2 - y) \cdot 1\,dy = \rho g \left(2y - y^2/2\right)\big|_0^1 = \frac{3\rho g}{2} \text{ N.}$$

So in fact the plate on the left does have more force on it than the one on the right, since $1 + \dfrac{\sqrt{2}}{2} > \dfrac{3}{2}$.

6.6.53

a. $F = \int_1^3 \rho g(4 - y) \cdot 2\,dy = \rho g \left(8y - y^2\right)\big|_1^3 = 8\rho g = 78,400$ N. This is less than 90,000 N., so the window can withstand the force.

b. $F(h) = \int_1^3 \rho g(h - y) \cdot 2\,dy = \rho g \left(2hy - y^2\right)\big|_1^3 = \rho g(6h - 9 - (2h - 1)) = 4\rho g(h - 2)$. This is less than or equal to 90,000 Newtons when $h \leq 4.296$ meters.

6.7   Chapter Six Review

6.7.1

a. True. A vertical slice would lead to shells, while a horizontal slice would lead to either disks or washers.

b. True. In order to find position, you would also need to know its position at some time.

c. True. If dV/dt is constant, then V is a linear function of time.

6.7.3 The position $s(t)$ and the displacement are the same, since the projective started on the ground at position 0. $s(t) = \int_0^t v(x)\,dx = \int_0^t (20 - 10x)\,dx = \left(20x - 5x^2\right)\big|_0^t = 20t - 5t^2$. Note that the projectile is moving up for $0 < t < 2$ and down for $2 < t < 4$. Thus the distance traveled is equal to the position for $0 < t < 2$, but for $2 < t < 4$ the distance traveled is $20 + (20 - (20t - 5t^2)) = 40 - 20t + 5t^2$.

6.7.5

a. $v(t) = \int a(t)\,dt = \int 2\sin(\pi t/4)\,dt = \frac{-8}{\pi}\cos(\pi t/4) + C$, and since $v(0) = \frac{-8}{\pi}$, we have $C = 0$. Thus, $v(t) = \frac{-8}{\pi}\cos(\pi t/4)$. $s(t) = \int v(t)\,dt = \int \frac{-8}{\pi}\cos(\pi t/4)\,dt = \frac{-32}{\pi^2}\sin(\pi t/4) + D$, and since $s(0) = 0$ we have $D = 0$. Thus, $s(t) = \frac{-32}{\pi^2}\sin(\pi t/4)$.

b. s is periodic with period 8, so we only consider $0 \le t \le 8$. There are critical numbers for s at $t = 2$ and $t = 6$. There is a maximum for s of $\frac{32}{\pi^2}$ at $t = 6$ and a minimum of $\frac{-32}{\pi^2}$ at $t = 2$.

c. The average velocity is $\frac{1}{8}\int_0^8 \frac{-8}{\pi}\cos(\pi t/4)\,dt = \frac{-4}{\pi^2}\left(\sin(\pi t/4)\right)\big|_0^8 = 0$. The average position is $\frac{1}{8}\int_0^8 \frac{-32}{\pi^2}\sin(\pi t/4)\,dt = \frac{16}{\pi^3}\left(\cos(\pi t/4)\right)\big|_0^8 = 0$.

6.7.7

a. For $0 \le t \le 8$ we have $R'(t) = 4t^{1/3}$, so $R(t) = 3t^{4/3} + C$, but $C = 0$, so $R(t) = 3t^{4/3}$.

b. Note that $R(8) = 48$, so for $t > 8$, $R(t) = 48 + \int_8^t 2\,dx = 48 + 2(t - 8)$.

c. The fuel runs out when $150 = 48 + 2(t - 8)$, which occurs for $t = 59$.

6.7.9

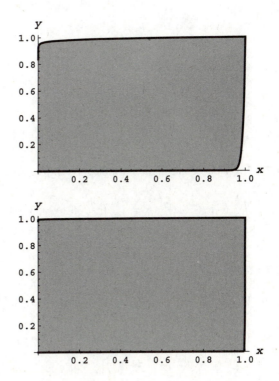

$$\int_0^1 (x^{1/p} - x^p)\,dx = \left(\frac{p}{p+1}x^{(p+1)/p} - \frac{1}{p+1}x^{p+1}\right)\bigg|_0^1$$
$$= \frac{p}{p+1} - \frac{1}{p+1} = \frac{p-1}{p+1}.$$

For $p = 100$ the area is $99/101$, and for $p = 1000$ the area is $999/1001$.

6.7.11

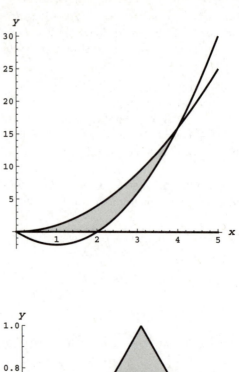

$$\int_0^2 x^2\, dx + \int_2^4 (x^2 - 2x^2 + 4x)\, dx$$
$$= \left(x^3/3\right)\big|_0^2 + \left(-x^3/3 + 2x^2\right)\big|_2^4$$
$$= (8/3) + (-64/3 + 32 - (-8/3 + 8)) = 8.$$

6.7.13

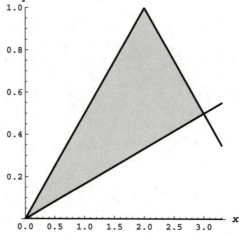

$$\int_0^2 (x/2 - x/6)\, dx + \int_2^3 ((2 - x/2) - x/6)\, dx$$
$$= \left(x^2/6\right)\big|_0^2 + \left(2x - x^2/3\right)\big|_2^3$$
$$= \frac{2}{3} + (6 - 3) - (4 - 4/3) = 1.$$

6.7.15 $A(a) = \int_0^{\sqrt[3]{a}} \left(\sqrt{x/a} - x^2/a\right) dx = \left(\frac{2}{3}\frac{x^{3/2}}{\sqrt{a}} - \frac{x^3}{3a}\right)\Big|_0^{\sqrt[3]{a}} = \frac{1}{3}.$

6.7.17 By the general slicing method, the volume is

$$\int_0^{12} (-z/6 + 2)^2\, dz = \int_0^{12} (z^2/36 - 2z/3 + 4)\, dz = \left(z^3/108 - z^2/3 + 4z\right)\big|_0^{12} = 16 - 48 + 48 = 16.$$

6.7.19

a. $V = 2\pi \int_0^1 x((1 + \sqrt{x}) - (1 - \sqrt{x}))\, dx = 4\pi \int_0^1 x^{3/2}\, dx = 4\pi \left(\frac{2x^{5/2}}{5}\right)\Big|_0^1 = \frac{8\pi}{5}.$

b. $V = \pi \int_0^2 (1^2 - ((1-y)^2)^2)\, dy = \pi \int_0^2 (1 - (y^4 - 4y^3 + 6y^2 - 4y + 1))\, dy = \pi \int_0^2 (-y^4 + 4y^3 - 6y^2 + 4y)\, dy = \pi \left(-\frac{y^5}{5} + y^4 - 2y^3 + 2y^2\right)\Big|_0^2 = \frac{8\pi}{5}$

6.7.21

$$V = \pi \int_0^h y^2 \, dx = \pi r^2 \int_0^h (1 - x/h)^2 \, dx = \pi r^2 \int_0^h (1 - 2x/h + x^2/(h^2)) \, dx$$

$$= \pi r^2 \left(x - x^2/h + x^3/(3h^2) \right)\big|_0^h = \frac{\pi r^2 h}{3}.$$

6.7.23

a. $V_x = \pi \int_1^2 \frac{1}{x^2} \, dx = \pi \left(-1/x \right)\big|_1^2 = \frac{\pi}{2}$. $V_y = 2\pi \int_1^2 x \cdot x^{-1} \, dx = 2\pi$, so $V_y > V_x$.

b. $V_x = \pi \int_1^4 \frac{1}{x^6} \, dx = \pi \left(-x^{-5}/5 \right)\big|_1^4 = \frac{\pi}{5}(1 - \frac{1}{1024})$. $V_y = 2\pi \int_1^4 x \cdot x^{-3} \, dx = 2\pi \left(-1/x \right)\big|_1^4 = \frac{3\pi}{2}$, so $V_y > V_x$.

c. $V_x = \pi \int_1^a \frac{1}{x^{2p}} \, dx = \frac{\pi}{1-2p}(-1 + a^{1-2p})$ since $p \neq 1/2$.

d. $V_y = 2\pi \int_1^a \frac{1}{x^{p-1}} \, dx = \frac{2\pi(a^{2-p}-1)}{2-p}$ since $p \neq 2$.

e. No, it appears that $V_y > V_x$ for all values of a and p.

6.7.25 $f'(x) = \frac{1}{2}(x^{-1/2} - x^{1/2})$ and $1 + f'(x)^2 = 1 + \frac{1}{4}\left(x + \frac{1}{x} - 2 \right) = \frac{1}{4}(x^{-1/2} + x^{1/2})^2$, so the arc length is

$$\int_1^3 \sqrt{\frac{1}{4}(x^{-1/2} + x^{1/2})^2} \, dx = \frac{1}{2}\int_1^3 x^{-1/2} + x^{1/2} \, dx = \frac{1}{3}\sqrt{x}(x+3)\bigg|_1^3 = 2\sqrt{3} - \frac{4}{3}$$

6.7.27 $m = \int_0^9 (3 + 2\sqrt{x}) \, dx = \left(3x + 4x^{3/2}/3 \right)\big|_0^9 = 63$ gm.

6.7.29 Since $50 = \int_0^{.2} kx \, dx = \left(kx^2/2 \right)\big|_0^{.2} = \frac{k}{50}$, we must have $k = 2500$. $W = \int_{.2}^{.7} 2500x \, dx = \left(1250x^2 \right)\big|_{.2}^{.7} = 562.5$ J.

6.7.31 The semicircle is given by $x = -\sqrt{400 - y^2}$, so the area of a slice at position $-y$ is $\pi x^2 = \pi(400 - y^2)$. $P = \int_{-20}^0 \pi \rho g (400 - y^2)(-y) \, dy = \pi \rho g \left(y^4/4 - 200y^2 \right)\big|_{-20}^0 \approx 1.2315 \times 10^9$ N.

Chapter 7

Logarithmic and Exponential Functions

7.1 Inverse Functions

7.1.1 $f(x) = 2x + 1$ is one-to-one on all of $\mathbb{R}$. If $f(a) = f(b)$, then $2a + 1 = 2b + 1$, so it must follow that $a = b$.

7.1.3

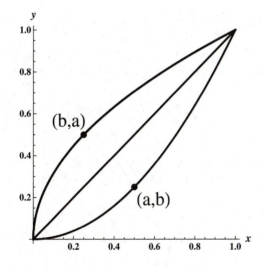

Recall that the graph of $f^{-1}(x)$ is obtained from the graph of $f(x)$ by reflecting across the line $y = x$. Thus, if (a, b) is on the graph of $y = f(x)$, then (b, a) must be on the graph of $y = f^{-1}(x)$.

7.1.5 To find the inverse of $y = 3x - 4$, we write $x = 3y - 4$ and solve for y. We have $x + 4 = 3y$, so $y = \frac{x+4}{3}$. Thus $f^{-1}(x) = \frac{x+4}{3}$.

7.1.7 By Theorem 7.3, and since f is one-to-one, we have

$$(f^{-1})'(8) = \frac{1}{f'(f^{-1}(8))} = \frac{1}{f'(2)} = \frac{1}{4}$$

7.1.9 f is one-to-one on $(-\infty, -1]$, on $[-1, 1]$, and on $[1, \infty)$.

7.1.11 f is one-to-one on $\mathbb{R}$, so it has an inverse on $\mathbb{R}$.

7.1.13 f is one-to-one on its domain, which is $(-\infty, 5) \cup (5, \infty)$, so it has an inverse on that set.

7.1.15

a. Switching the x and y, we have $x = 6 - 4y$. Solving for y in terms of x we have $4y = 6 - x$, so $y = f^{-1}(x) = \frac{6-x}{4}$.

b. $f(f^{-1}(x)) = f\left(\frac{6-x}{4}\right) = 6 - 4 \cdot \left(\frac{6-x}{4}\right) = 6 - (6 - x) = x$.
$f^{-1}(f(x)) = f^{-1}(6 - 4x) = \frac{6-(6-4x)}{4} = \frac{4x}{4} = x$.

7.1.17

a. Switching the x and y, we have $x = 3y + 5$. Solving for y in terms of x we have $y = \frac{x-5}{3}$, so $y = f^{-1}(x) = \frac{x-5}{3}$.

b. $f(f^{-1}(x)) = f\left(\frac{x-5}{3}\right) = 3\left(\frac{x-5}{3}\right) + 5 = (x - 5) + 5 = x$.
$f^{-1}(f(x)) = f^{-1}(3x + 5) = \frac{(3x+5)-5}{3} = \frac{3x}{3} = x$.

7.1.19

a. Switching the x and y, we have $x = \sqrt{y + 2}$. Solving for y in terms of x we have $y = x^2 - 2$. Note that since range of f is $[0, \infty)$, that is the domain of f^{-1}.

b. $f(f^{-1}(x)) = f(x^2 - 2) = \sqrt{x^2 - 2 + 2} = |x| = x$, since x is in the domain of f^{-1} and is thus positive.
$f^{-1}(f(x)) = f^{-1}(\sqrt{x + 2}) = \sqrt{x + 2}^2 - 2 = x + 2 - 2 = x$.

7.1.21 First note that since the expression is symmetric, switching x and y doesn't change the expression. Solving for y gives $|y| = \sqrt{1 - x^2}$. To get the four one-to-one functions, we restrict the domain and choose either the upper part or lower part of the circle as follows:

a. $f_1(x) = \sqrt{1 - x^2}$, $0 \leq x \leq 1$ $f_2(x) = \sqrt{1 - x^2}$, $-1 \leq x \leq 0$ $f_3(x) = -\sqrt{1 - x^2}$, $-1 \leq x \leq 0$ $f_4(x) = -\sqrt{1 - x^2}$, $0 \leq x \leq 1$

b. Reflecting these functions across the line $y = x$ yields the following: $f_1^{-1}(x) = \sqrt{1 - x^2}$, $0 \leq x \leq 1$ $f_2^{-1}(x) = -\sqrt{1 - x^2}$, $0 \leq x \leq 1$ $f_3^{-1}(x) = -\sqrt{1 - x^2}$, $-1 \leq x \leq 0$ $f_4^{-1}(x) = \sqrt{1 - x^2}$, $-1 \leq x \leq 0$

7.1.23

Switching x and y gives $x = 8 - 4y$. Solving this for y yields $y = \frac{8-x}{4}$.

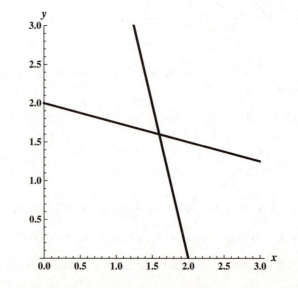

7.1.25

Switching x and y gives $x = \sqrt{y}$. Solving this for y yields $y = x^2$, but note that the range of f is $[0, \infty)$ so that is the domain of f^{-1}.

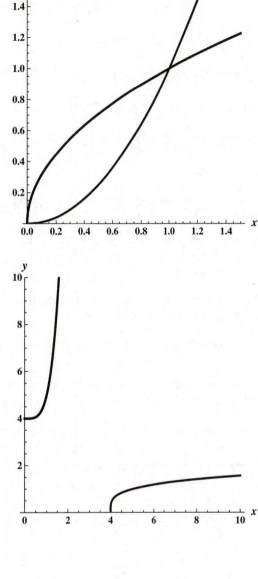

7.1.27

Switching x and y gives $x = y^4 + 4$. Solving this for y yields $y = \sqrt[4]{x - 4}$.

7.1.29

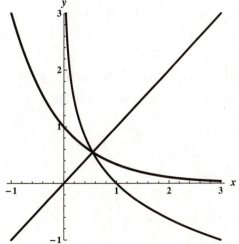

7.1.31 $f(4) = 16$ and $(f^{-1})'(16) = \frac{1}{f'(4)} = \frac{1}{3}$.

7.1.33 $f\left(\frac{\pi}{4}\right) = 1$ and $(f^{-1})'(1) = \frac{1}{f'\left(\frac{\pi}{4}\right)} = \frac{1}{\sec^2\left(\frac{\pi}{4}\right)} = \frac{1}{2}$.

7.1.35 $f(4) = 2$ and $(f^{-1})'(2) = \frac{1}{f'(4)} = \frac{1}{(1/2\sqrt{4})} = 4$.

7.1.37 $f(4) = 36$ and $(f^{-1})'(36) = \frac{1}{f'(4)} = \frac{1}{2(4+2)} = \frac{1}{12}$.

7.1.39 Note that $f(1) = 3$. So $(f^{-1})'(3) = \frac{1}{f'(1)} = \frac{1}{4}$.

7.1.41 $(f^{-1})'(4) = \frac{1}{f'(7)} = \frac{4}{5}$, so $f'(7) = \frac{5}{4}$.

7.1.43

a. False. For example $f(0) = 1$, but the alleged inverse function evaluated at 1 is not 0 (rather, it has value 1/2.)

b. True. f is its own inverse because $f(f(x)) = f(1/x) = \frac{1}{1/x} = x$.

c. False. In fact, $f(x) = x^3 + x$ is one-to-one, so has one inverse on all of $(-\infty, \infty)$.

d. True. A tenth-degree polynomial could have nine local maxima or minima, giving ten segments on which the function is one-to-one.

e. True. By part (b), $f^{-1}(x) = 1/x$, so that $(f^{-1}(x))' = -1/x^2$.

7.1.45 Note that f is one-to-one, so there is only one inverse. Switching x and y gives $x = (y+1)^3$. Then $\sqrt[3]{x} = y + 1$, so $y = f^{-1}(x) = \sqrt[3]{x} - 1$. The domain of f^{-1} is $\mathbb{R}$.

7.1.47 Note that to get a one-to-one function, we should restrict the domain to either $[0, \infty)$ or $(-\infty, 0]$. Switching x and y yields $x = \frac{2}{y^2+2}$, so $y^2 + 2 = (2/x)$. So $y = \pm\sqrt{(2/x) - 2}$. So the inverse of f when the domain of f is restricted to $[0, \infty)$ is $f^{-1}(x) = \sqrt{(2/x) - 2}$, while if the domain of f is restricted to $(-\infty, 0]$ the inverse is $f^{-1}(x) = -\sqrt{(2/x) - 2}$. In either case, the domain of f^{-1} is $(0, 1]$.

7.1.49 Let $f(y) = 3y - 4$. Then $f'(y) = 3$ for all y in the domain of f. Let $y = f^{-1}(x)$. $(f^{-1})'(x) = \dfrac{1}{f'(y)} = \dfrac{1}{3}$.

7.1.51 Let $x = f(y) = y^2 - 4$ for $y > 0$. Note that this means that $y = \sqrt{x+4}$. Then $f'(y) = 2y$. So $(f^{-1})'(x) = \dfrac{1}{f'(y)} = \dfrac{1}{2y} = \dfrac{1}{2\sqrt{x+4}}$.

7.1.53 For $y \geq -2$, let $x = \sqrt{y+2}$. Note that it then follows that $x \geq 0$. Then $1 = \frac{y'}{2\sqrt{y+2}}$, and $x^2 = y + 2$, so $y = x^2 - 2$. Thus we have $(f^{-1})'(x) = y' = 2\sqrt{x^2 - 2 + 2} = 2|x| = 2x$, since $x \geq 0$.

7.1.55 For $y > 0$, let $x = y^{-1/2}$. Then $1 = \frac{-1}{2}y^{-3/2}y'$, so $y' = -2y^{3/2} = -2(x^{-2})^{3/2} = -2x^{-3}$ where $x > 0$.

7.1.57 Solve for r to get $r^3 = \frac{3}{4\pi}V$, so that

$$r = \sqrt[3]{\frac{3}{4\pi}V}$$

7.1.59 Solve for r (remembering that we are assuming $r \geq 0$) to get $r^2 = \frac{V}{10\pi}$, so that

$$r = \sqrt{\frac{V}{10\pi}}$$

7.1.61

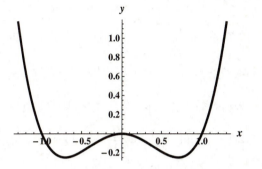

a. f is one-to-one on $(-\infty, -\sqrt{2}/2]$, on $[-\sqrt{2}/2, 0]$, on $[0, \sqrt{2}/2]$, and on $[\sqrt{2}/2, \infty)$.

b. If $u = x^2$, then our function becomes $y = u^2 - u$. Completing the square gives $y + (1/4) = u^2 - u + (1/4) = (u - (1/2))^2$. Thus $|u - (1/2)| = \sqrt{y + (1/4)}$, so $u = (1/2) \pm \sqrt{y + (1/4)}$, with the "+" applying for $u = x^2 > (1/2)$ and the "−" applying when $u = x^2 < (1/2)$. Now letting $u = x^2$, we have $x^2 = (1/2) \pm \sqrt{y + (1/4)}$, so $x = \pm\sqrt{(1/2) \pm \sqrt{y + (1/4)}}$.

c. Switching x and y gives the following inverses:

Domain of f	$(-\infty, -\sqrt{2}/2]$	$[-\sqrt{2}/2, 0]$	$[0, \sqrt{2}/2]$	$[\sqrt{2}/2, \infty)$
Range of f	$[-1/4, \infty)$	$[-1/4, 0]$	$[-1/4, 0]$	$[-1/4, \infty)$
Inverse of f	$-\sqrt{(1/2) + \sqrt{x + (1/4)}}$	$-\sqrt{(1/2) - \sqrt{x + (1/4)}}$	$\sqrt{(1/2) - \sqrt{x + (1/4)}}$	$\sqrt{(1/2) + \sqrt{x + (1/4)}}$

7.1.63 Let $y = x^3 + 2x$. This function is one-to-one, so it has an inverse. Making the suggested substitution yields $y = (z - 2/(3z))^3 + 2(z - 2/(3z))$. Expanding gives $y = z^3 - 2z + 4/(3z) - 8/(27z^3) + 2z - 4/(3z) = z^3 - 8/(27z^3)$. Thus we have $y = z^3 - 8/(27z^3)$, so $27z^3 y = 27(z^3)^2 - 8$, or $27(z^3)^2 - 27y(z^3) - 8 = 0$.

Applying the quadratic formula gives $z^3 = \frac{y}{2} \pm \frac{\sqrt{3}\sqrt{32 + 27y^2}}{18}$. We will take the "+" part and finish solving to obtain:

$$z = \sqrt[3]{\frac{y}{2} + \frac{\sqrt{3}\sqrt{32 + 27y^2}}{18}}$$

Now

$$x = z - (2/(3z)) = \frac{3z^2 - 2}{3z} = \frac{3\left(\sqrt[3]{\frac{y}{2} + \frac{\sqrt{3}\sqrt{32+27y^2}}{18}}\right)^2 - 2}{3\sqrt[3]{\frac{y}{2} + \frac{\sqrt{3}\sqrt{32+27y^2}}{18}}}.$$

So the inverse function $f^{-1}(x)$ is now obtained by switching y and x.

7.1.65   a. Since $y = ax + b$, we have $f'(x_0) = a$, so that $b = y_0 - f'(x_0)x_0$.

b. $(f^{-1})'(y_0) = c$, so

$$c = \frac{1}{f'(x_0)}, \qquad d = x_0 - \frac{y_0}{f'(x_0)}$$

c. We must show that $L(M(x)) = x$. Note that from parts (a) and (b), $ac = 1$; then

$$L(M(x)) = a(cx + d) + b = acx + ad + b = x + a(x_0 - cy_0) + (y_0 - ax_0) = x - acy_0 + y_0 = x$$

7.2   The Natural Logarithm and Exponential Functions

7.2.1 The domain is $(0, \infty)$ and the range is $(-\infty, \infty)$.

7.2.3 $y = \ln x$ if and only if $x = e^y$. Differentiating implicitly yields $1 = e^y \cdot y'$, so $y' = \frac{1}{e^y} = \frac{1}{e^{\ln x}} = \frac{1}{x}$ for $x > 0$.

7.2.5 $\frac{d}{dx} \ln(kx) = \frac{1}{kx} \cdot k = \frac{1}{x}$. This is valid for $x > 0$ if $k > 0$ and $x < 0$ if $k < 0$. Also, we can write $\ln(kx) = \ln(k) + \ln(x)$, so its derivative is $0 + \frac{1}{x} = \frac{1}{x}$.

7.2.7 $\frac{d}{dx} \ln(x^2) = \frac{1}{x^2} \cdot (2x) = \frac{2}{x}$. This is valid on $(-\infty, 0) \cup (0, \infty)$.

7.2.9 $\frac{d}{dx} \left[\ln \left(\frac{x+1}{x-1} \right) \right] = \frac{x-1}{x+1} \left(\frac{(x-1)-(x+1)}{(x-1)^2} \right) = \frac{-2}{(x+1)(x-1)}$. This is valid on $(-\infty, -1) \cup (1, \infty)$.

7.2.11 $\frac{d}{dx} (x^2 + 1) \ln x = 2x \ln x + \frac{x^2+1}{x}$. This is valid on $(0, \infty)$.

7.2.13 $\frac{d}{dx} (\ln(\ln(x))) = \frac{1}{\ln x} \cdot \frac{1}{x}$. This is valid on $(1, \infty)$, which is the domain of $\ln(\ln x)$.

7.2.15 $\displaystyle\int \frac{3}{x - 10} \, dx = 3 \ln |x - 10| + C$

7.2.17 $\displaystyle\int \left(\frac{2}{x - 4} - \frac{3}{2x + 1} \right) dx = 2 \ln |x - 4| - \frac{3}{2} \ln |2x + 1| + C$

7.2.19 $\int_0^3 \frac{2x-1}{x+1} \, dx = \int_0^3 \left(2 - \frac{3}{x+1} \right) dx = (2x - 3 \ln(x+1))|_0^3 = 6(1 - \ln 2)$.

7.2.21 Let $u = \ln(\ln x)$, so that $du = \frac{1}{x \ln x} \, dx$. Then,

$$\int_3^4 \frac{dx}{2x \ln x \ln^3(\ln x)} = \int_{\ln(\ln 3)}^{\ln(\ln 4)} \frac{1}{2u^3} \, du = \left(\frac{-1}{4u^2} \right) \Bigg|_{\ln(\ln 3)}^{\ln(\ln 4)} = \frac{1}{4} \left(\frac{1}{(\ln(\ln 3))^2} - \frac{1}{(\ln(\ln 4))^2} \right).$$

7.2.23 $f'(x) = -9e^{-x} - 10e^{2x} - 6e^x$.

7.2.25 $f'(x) = \dfrac{(e^{-x} + 2)(2e^{2x}) - e^{2x}(-e^{-x})}{(e^{-x} + 2)^2} = \dfrac{2e^x + 4e^{2x} + e^x}{(e^{-x} + 2)^2} = \dfrac{3e^x + 4e^{2x}}{(e^{-x} + 2)^2}$

7.2.27 $f'(x) = (\sin 2x)' \cdot e^{\sin 2x} = 2 \cos(2x) e^{\sin 2x}$, so that $f'(\pi/4) = 2 \cos(\pi/2) e^{\sin \pi/2} = 0$.

7.2.29 $y' = \dfrac{e^x}{4} - 1$, so that at $a = 0$ the slope is $\dfrac{1}{4} - 1 = -\dfrac{3}{4}$. At $a = 0$, $f(a) = \frac{1}{4}$, so that the tangent line is

$$y - \frac{1}{4} = -\frac{3}{4}(x - 0), \quad \text{or } y = \frac{1}{4} - \frac{3}{4}x.$$

7.2.31 $\int (e^{2x} + 1) \, dx = \frac{1}{2} e^{2x} + x + C$.

7.2.33

$$\int_0^{\ln 3} e^x (e^{3x} + e^{2x} + e^x) \, dx = \int_0^{\ln 3} (e^{4x} + e^{3x} + e^{2x}) \, dx = \left(\frac{1}{4} e^{4x} + \frac{1}{3} e^{3x} + \frac{1}{2} e^{2x} \right) \Bigg|_0^{\ln 3}$$

$$= \frac{81}{4} + 9 + \frac{9}{2} - \left(\frac{1}{4} + \frac{1}{3} + \frac{1}{2} \right) = \frac{98}{3}.$$

7.2.35 Let $u = e^x - e^{-x}$ so that $du = e^x + e^{-x}\, dx$. Then $\int \frac{e^x + e^{-x}}{e^x - e^{-x}}\, dx = \int \frac{1}{u}\, du = \ln|u| + C = \ln|e^x - e^{-x}| + C$.

7.2.37 Let $u = \sqrt{x}$ so that $du = \frac{1}{2\sqrt{x}}\, dx$. Then $\int \frac{e^{\sqrt{x}}}{\sqrt{x}}\, dx = 2\int e^u\, du = 2e^u + C = 2e^{\sqrt{x}} + C$.

7.2.39 Let $y = \frac{(x+1)^{10}}{(2x-4)^8}$, so $\ln y = \ln\left(\frac{(x+1)^{10}}{(2x-4)^8}\right) = 10\ln(x+1) - 8(\ln(2x-4))$. Then

$$\frac{1}{y}y' = \frac{10}{x+1} - \frac{8}{2x-4}\cdot 2,$$

$$y' = \frac{(x+1)^{10}}{(2x-4)^8}\cdot\left(\frac{10}{x+1} - \frac{8}{x-2}\right).$$

7.2.41 Let $y = x^{\ln x}$. Then $\ln y = (\ln x)^2$. Thus $\frac{1}{y}y' = 2\ln x \cdot \frac{1}{x}$, so $y' = x^{\ln x}\left(\frac{2\ln x}{x}\right)$.

7.2.43 Let $y = \frac{(x+1)^{3/2}(x-4)^{5/2}}{(5x+3)^{2/3}}$. Then

$$\ln y = \ln\left(\frac{(x+1)^{3/2}(x-4)^{5/2}}{(5x+3)^{2/3}}\right) = \frac{3}{2}\ln(x+1) + \frac{5}{2}\ln(x-4) - \frac{2}{3}\ln(5x+3)$$

so that

$$\frac{1}{y}y' = \frac{3}{2(x+1)} + \frac{5}{2(x-4)} - \frac{10}{3(5x+3)},$$

$$y' = \frac{(x+1)^{3/2}(x-4)^{5/2}}{(5x+3)^{2/3}}\cdot\left(\frac{3}{2(x+1)} + \frac{5}{2(x-4)} - \frac{10}{3(5x+3)}\right).$$

7.2.45 Let $y = (\sin x)^{\tan x}$, and assume $0 < x < \pi$, $x \neq \frac{\pi}{2}$. Then $\ln y = (\tan x)\ln(\sin x)$. Then

$$\frac{1}{y}y' = (\sec^2 x)\ln(\sin x) + \frac{\tan x\cos x}{\sin x},$$

$$y' = (\sin x)^{\tan x}\left((\sec^2 x)\ln(\sin x) + 1\right).$$

7.2.47

a. True. This follows because $e^{\ln xy} = xy = e^{\ln x}e^{\ln y} = e^{\ln x + \ln y}$. Because the exponential function is one-to-one, it follows that $\ln xy = \ln x + \ln y$.

b. False. Zero is not in the domain of the natural logarithm function.

c. False. For example, $\ln(1+1) = \ln(2) \neq \ln(1) + \ln(1) = 0$.

d. True, by induction. $f'(x) = f^{(1)}(x) = ke^{kx}$, so it is true for $n = 1$. If it is true for n, then

$$f^{(n+1)}(x) = f^{(n)'}(x) = (k^n e^{kx})' = k^{n+1}e^{kx}$$

e. True. $\int_1^e \frac{1}{x}\, dx = \ln x\Big|_1^e = \ln e - \ln 1 = 1$.

7.2.49

h	$(1+2h)^{1/h}$	h	$(1+2h)^{1/h}$
10^{-1}	6.1917	-10^{-1}	9.3132
10^{-2}	7.2446	-10^{-2}	7.5404
10^{-3}	7.3743	-10^{-3}	7.4039
10^{-4}	7.3876	-10^{-4}	7.3905
10^{-5}	7.3889	-10^{-5}	7.3892
10^{-6}	7.3890	-10^{-6}	7.3891

Let $y = (1+2h)^{1/h}$. Then

$$\ln y = \frac{\ln(1+2h)}{h}.$$

$$\lim_{h \to 0} \ln y = \lim_{h \to 0} \frac{\ln(1+2h)}{h} = \lim_{h \to 0} \frac{2}{1+2h} = 2,$$

so the limit of y as $h \to 0$ is e^2.

7.2.51

x	$\frac{2^x-1}{x}$	x	$\frac{2^x-1}{x}$
10^{-1}	.71773	-10^{-1}	.66967
10^{-2}	.69556	-10^{-2}	.69075
10^{-3}	.69339	-10^{-3}	.69291
10^{-4}	.69317	-10^{-4}	.69312
10^{-5}	.69315	-10^{-5}	.69314
10^{-6}	.69315	-10^{-6}	.69315

$$\lim_{x \to 0} \frac{2^x-1}{x} = \lim_{x \to 0} \frac{2^x \ln(2)}{1} = \ln(2).$$

7.2.53

a. Use the change of variables $u = \cos x$; then $du = -\sin x \, dx$, and

$$\int \tan x \, dx = \int \frac{\sin x}{\cos x} \, dx = -\int \frac{1}{u} \, du = -\ln|u| + C = -\ln|\cos x| + C$$

Since $|\sec x| = \frac{1}{|\cos x|}$, we have $\ln|\sec x| = -\ln|\cos x|$, so that the integral is also equal to $\ln|\sec x| + C$.

b. Use the change of variables $u = \sin x$; then $du = \cos x \, dx$, and

$$\int \cot x \, dx = \int \frac{\cos x}{\sin x} \, dx = \int \frac{1}{u} \, du = \ln|u| + C = \ln|\sin x| + C$$

7.2.55 The average value is given by $\frac{1}{p-1} \int_1^p \frac{1}{x} \, dx = \frac{1}{p-1} \left. (\ln x) \right|_1^p = \frac{\ln p}{p-1}$. Note that

$$\lim_{p \to \infty} \frac{\ln p}{p-1} = \lim_{p \to \infty} \frac{1/p}{1} = 0.$$

7.2.57 $\frac{d}{dx} e^{-10x^2} = e^{-10x^2} \cdot -20x.$

7.2.59 $\frac{d}{dx} \ln(\sqrt{10x}) = \frac{d}{dx} \ln((10x)^{1/2}) = \frac{1}{2} \frac{d}{dx} \ln(10x) = \frac{1}{2x}$

7.2.61 If $f(x) = \frac{(x^2+1)(x-3)}{(x+2)^3}$, then $\ln f(x) = \ln(x^2+1) + \ln(x-3) - 3\ln(x+2)$, so that using logarithmic differentiation,

$$f'(x) = f(x) \frac{d}{dx} (\ln f(x)) = f(x) \left(\frac{2x}{x^2+1} + \frac{1}{x-3} - \frac{3}{x+2} \right) = \frac{9x^2 - 14x + 11}{(x+2)^4}$$

7.2.63 Let $u = x^3$, so that $du = 3x^2\,dx$. Then $\int x^2 e^{x^3}\,dx = \int \frac{1}{3} e^u\,du = \frac{1}{3} e^u + C = \frac{1}{3} e^{x^3} + C$.

7.2.65 $e^{\ln x} = x$, so

$$\int_1^{2e} \frac{e^{\ln x}}{x}\,dx = \int_1^{2e} 1\,dx = 2e - 1$$

7.2.67 Let $u = \ln x$ so that $du = \frac{1}{x}\,dx$. Then we have $\int_0^2 u^5\,du = \left(\frac{u^6}{6} \right)\Big|_0^2 = \frac{32}{3}$.

7.2.69 Note that the area of the triangle described is $\frac{1}{2}xy$, so if $xy \geq \frac{1}{2}$, then $\frac{1}{2}xy \geq \frac{1}{4}$, which is not what we are seeking. However, if $xy < 1/2$, then the triangle formed will give us what we want. Note that for any choice of x with $0 < x < 1/2$ and any choice of possible y, we have $xy < 1/2$ since $0 < y < 1$. Thus since there is a probability of $\frac{1}{2}$ of choosing $0 < x < 1/2$, the probability we seek is at least $1/2$. In addition, for $1/2 < x < 1$, if $y < \frac{1}{2x}$, then $xy < \frac{1}{2}$. So we should add $\int_{1/2}^1 \frac{1}{2x}\,dx = \left(\frac{1}{2} \ln x \right)\Big|_{1/2}^1 = (\ln 2)/2$. Thus the probability we seek is $\frac{1}{2}(1 + \ln 2)$.

7.2.71

a.

We used a graphing rectangle of $[0, 25] \times [0, 8000]$.

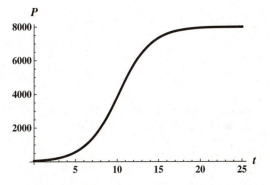

b. To find when $P(t)$ hits 5000, we solve $5000 = \dfrac{400000}{50 + 7950e^{-0.5t}}$, or $50 + 7950e^{-0.5t} = 80$. This leads to $7950e^{-0.5t} = 30$, or $-0.5t = \ln\left(\dfrac{30}{7950}\right)$. So we have $t = 2\ln(265) \approx 11.16$ years. The carrying capacity is $\lim\limits_{t\to\infty} P(t) = \dfrac{400{,}000}{50} = 8000$. Ninety percent of 8000 is 7200, so we seek the time when $P(t) = 7200$. We have $7200 = \dfrac{400000}{50 + 7950e^{-0.5t}}$, or $50 + 7950e^{-0.5t} = \dfrac{500}{9}$. This leads to $7950e^{-0.5t} = \dfrac{50}{9}$, or $-0.5t = \ln\left(\dfrac{50}{71550}\right)$. So we have $t = 2\ln\left(\dfrac{71550}{50}\right) \approx 14.53$ years.

c. $\frac{dP}{dt} = \frac{-400000}{(50 + 7950e^{-0.5t})^2} \cdot (7950)(-0.5)e^{-0.5t}$. At $t = 0$ we have $P'(0) = \frac{400,000 \cdot 7950 \cdot .5}{8000^2} = \frac{1,590,000,000}{8000^2} \approx 25$ fish per year. At $t = 5$ we have $P'(5) = \frac{1,590,000,000e^{-5/2}}{(50 + 7950e^{-5/2})^2} \approx 264$ fish per year.

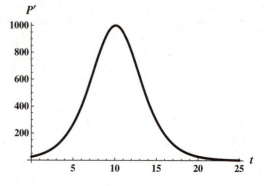

d. The maximum is at about $t = 10$ years.

7.2.73

a. $\ln(P(t)) = \ln(3 \cdot 10^{10}) - \ln(2 + 3e^{-0.025t})$. $\frac{d}{dt} \ln(P(t)) = \frac{P'(t)}{P(t)} = r(t) = \frac{0.075 \cdot e^{-0.025t}}{2 + 3e^{-0.025t}}$. $r(0) = \frac{0.075}{5} = 0.015$, so the population is growing at 1.5% per year in 1999.

b. $r(11) = \frac{0.075e^{-0.275}}{2 + 3e^{-0.275}} \approx 0.0133$. $r(21) = \frac{0.075e^{-0.5}}{2 + 3e^{-0.5}} \approx 0.0118$. The relative growth rate decreases over time.

c. $\lim\limits_{t \to \infty} r(t) = \lim\limits_{t \to \infty} \frac{0.075}{3 + 2e^{0.025t}} = 0$, since the denominator increases without bound. The relative growth rate becomes smaller and smaller as the population nears the carrying capacity.

7.2.75

a. We seek k so that $\lim_{b \to \infty} \int_0^b 10^7 e^{-kt}\, dt = 2 \times 10^9$. Thus we require $\lim_{b \to \infty} \left(\frac{e^{-kt}}{-k} \right)\Big|_0^b = 200$, or $\lim_{b \to \infty} \left(\frac{1}{-ke^{bt}} - \frac{1}{-k} \right) = 200$, so $k = \frac{1}{200} = .005$.

b. We want T so that $(2 \times 10^7) \int_0^T e^{-.005t}\, dt = 2 \times 10^9$, so $\left(-200e^{-.005t} \right)\Big|_0^T = 100$, so $1 - e^{-T/200} = 1/2$, so $T = 200 \ln(2) \approx 138.63$ years.

7.2.77

a. $\ln\left(\frac{e^x}{e^y} \right) = \ln(e^x) - \ln(e^y) = x - y$. Thus, $e^{x-y} = \frac{e^x}{e^y}$.

b. $\ln((e^x)^y) = y \ln(e^x) = yx$, so $e^{xy} = (e^x)^y$.

7.2.79 Since $1/x$ is decreasing, we know that the left Riemann sum is an overestimate. Using $n = 2$ subintervals, we have $\int_1^2 1/x\, dx = \ln 2 < \frac{1}{1} \cdot \frac{1}{2} + \frac{1}{3/2} \cdot \frac{1}{2} = \frac{5}{6} < 1$. Since $1/x$ is decreasing, we know that the right Riemann sum is an underestimate. We have

$$\int_1^3 1/x\, dx = \ln 3 > \frac{2}{7} \left(\frac{1}{8/7} + \frac{1}{10/7} + \frac{1}{12/7} + \frac{1}{14/7} + \frac{1}{16/7} + \frac{1}{18/7} + \frac{1}{20/7} \right) = \frac{2761}{2520} > 1.$$

7.2.81 Since $1/x$ is decreasing, the left Riemann sum is an overestimate to the integral. Over the interval $[1, n+1]$ with n subdivisions, we must have $\int_1^{n+1} \frac{1}{x}\, dx = \ln(n+1) < \frac{n+1-1}{n}(1 + \frac{1}{2} + \frac{1}{3} + \cdots + \frac{1}{n}) = \sum_{i=1}^n \frac{1}{i}$. Since $\ln(n+1) \to \infty$ as $n \to \infty$, it must follow that the harmonic partial sum $\left(1 + \frac{1}{2} + \frac{1}{3} + \cdots + \frac{1}{n}\right) \to \infty$ as well.

7.3 Logarithmic and Exponential Functions with Other Bases

7.3.1 $\frac{d}{dx} b^x = b^x \ln b$, for $b > 0$ and all x. In the case $b = e$, we rule states that $\frac{d}{dx} e^x = e^x \ln e = e^x$ since $\ln e = 1$.

7.3.3 $b^x = (e^{\ln b})^x = e^{(\ln b) \cdot x} = e^{x \ln b}$.

7.3.5 $\int 4^x \, dx = \int e^{x \ln 4}\, dx = \frac{1}{\ln 4} e^{x \ln 4} + C = \frac{4^x}{\ln 4} + C$.

7.3.7 $3^x = e^{x \ln 3}$. $x^\pi = e^{\pi \ln x}$. $x^{\sin x} = e^{\sin x \cdot \ln x}$.

7.3.9 If $\log_{10} x = 3$, then $10^3 = x$, so $x = 1000$.

7.3.11 If $\log_8 x = 1/3$, then $x = 8^{1/3} = 2$.

7.3.13 Since $10^{x^2 - 4} = 1 = 10^0$, we must have $x^2 - 4 = 0$. So $x = \pm 2$.

7.3.15 Since $2^{|x|} = 16 = 2^4$, we must have $|x| = 4$, so $x = \pm 4$.

7.3.17 Since $7^x = 21$, we have that $\ln 7^x = \ln 21$, so $x \ln 7 = \ln 21$, and $x = \frac{\ln 21}{\ln 7}$.

7.3.19 Since $3^{3x-4} = 15$, we have that $\ln 3^{3x-4} = \ln 15$, so $(3x - 4)\ln 3 = \ln 15$. Thus, $3x - 4 = \frac{\ln 15}{\ln 3}$, so $x = \frac{(\ln 15)/(\ln 3) + 4}{3} = \frac{\ln 15 + 4 \ln 3}{3 \ln 3}$.

7.3.21 $y' = 5 \cdot \frac{d}{dx} 4^x = 5 \cdot \ln 4 \cdot 4^x$.

7.3.23 $y' = 3x^2 3^x + x^3 3^x \ln 3$.

7.3.25 $\frac{dA}{dt} = 250(1.045)^{4t} \cdot \ln(1.045) \cdot 4 = 1000 \cdot \ln(1.045) \cdot (1.045^{4t})$.

7.3.27

   a. $T = 10 \cdot 2^{-0.274 \cdot 16}$ minutes ≈ 28.7 seconds.

   b. $\frac{\Delta T}{\Delta a} = \frac{10 \cdot 2^{-0.274 \cdot 8} - 10 \cdot 2^{-0.274 \cdot 2}}{8 - 2} \approx -0.78$ minutes per 1000 feet, which is about -46.512 seconds per 1000 feet.

   c. $\frac{dT}{da} = -2.74 \cdot 2^{-0.274 \cdot a} \cdot \ln 2$. At $a = 8$ we have $\frac{dT}{da} = -2.74 \cdot 2^{-0.274 \cdot 8} \cdot \ln 2 \approx -0.42$ minutes per 1000 feet. Every 1000 feet the airplane climbs, leaves about .42 minutes less time of consciousness, which corresponds to about 24.94 seconds.

7.3.29

   a. At $Q = 10\mu\text{Ci}$ we have $10 = 350 \cdot \left(\frac{1}{2}\right)^{t/13.1}$, so $\ln(1/35) = \frac{t}{13.1} \ln(1/2)$, so $t = 13.1 \cdot \frac{\ln 35}{\ln 2} \approx 67.19$ hours.

   b. $\frac{dQ}{dt} = \frac{350}{13.1} \cdot \ln\left(\frac{1}{2}\right) \cdot \left(\frac{1}{2}\right)^{t/13.1}$. We have $Q'(12) \approx -9.81453$, $Q'(24) \approx -5.20136$, and $Q'(48) \approx -1.46087$. The rate at which the iodine decreases is decreasing in absolute value as time increases.

7.3.31 $\int_{-1}^{1} 10^x \, dx = \left(\dfrac{10^x}{\ln 10}\right)\bigg|_{-1}^{1} = \dfrac{10 - 10^{-1}}{\ln 10} = \dfrac{99}{10 \ln 10}$.

7.3.33 $\int_{1}^{2}(1 + \ln x)x^x \, dx = (x^x)|_{1}^{2} = 4 - 1 = 3$.

7.3.35 $g'(y) = e^y y^e + e^{y+1}y^{e-1}$.

7.3.37 $s'(t) = -\sin(2^t) \cdot 2^t \ln 2$.

7.3.39 $f'(x) = 2x^{3/2} + \frac{3}{2}(2x - 3)x^{1/2} = 5x^{3/2} - \frac{9}{2}x^{1/2}$.

7.3.41 $f(x) = e^{4x \ln(2x)}$ so $f'(x) = e^{4x \ln(2x)}\left(4x \cdot (1/x) + \ln(2x) \cdot 4\right) = (2x)^{4x}(4 + 4\ln(2x))$.

7.3.43 $y' = 2^{x^2} \cdot \ln 2 \cdot (2x) = 2^{x^2+1}x \ln 2$.

7.3.45 $\ln(H(x)) = 2x\ln(x+1)$, so $\frac{1}{H(x)}H'(x) = \frac{2x}{x+1} + 2\ln(x+1)$, and thus $H'(x) = (x+1)^{2x}\left(\frac{2x}{x+1} + 2\ln(x+1)\right)$.

7.3.47 Let $y = x^{\sin x}$. Then $\ln y = \sin x \ln x$, so $\frac{1}{y}y' = \cos x \ln x + \frac{\sin x}{x}$. At the point $(1, 1)$ we have $y' = \sin 1$, so the tangent line is given by $y - 1 = (\sin 1)(x - 1)$.

7.3.49 Let $y = (x^2)^x = x^{2x}$. Then $\ln y = x \ln x^2$ and $\frac{1}{y}y' = \ln x^2 + 2$, so $y' = x^{2x}(\ln x^2 + 2)$. This quantity is zero when $\ln x^2 = -2$, or $x^2 = e^{-2}$. Thus there are horizontal tangents at $|x| = e^{-1}$, so for $x = \pm\frac{1}{e}$. The two tangent lines are given by $y = \frac{1}{e^{2/e}}$ (at $(\frac{1}{e}, \frac{1}{e^{2/e}})$) and $y = e^{2/e}$ (at $(\frac{-1}{e}, e^{2/e})$.)

7.3.51 $y' = 4 \cdot \dfrac{2x}{(x^2 - 1) \cdot \ln 3} = \dfrac{8x}{(x^2 - 1) \cdot \ln 3}$.

7.3.53 $y' = -\sin x(\ln(\cos^2 x)) + \cos x \cdot \left(\dfrac{2\cos x(-\sin x)}{\cos^2 x}\right) = (-\sin x)(\ln(\cos^2 x) + 2)$.

7.3.55 $y' = \dfrac{d}{dx}(\log_4 x)^{-1} = -(\log_4 x)^{-2} \cdot \dfrac{1}{x\ln 4} = \dfrac{-1}{x(\ln 4)(\log_4 x)^2}$.

7.3.57

a. False. $\log_2 9$ is a constant, so its derivative is 0.

b. True, at least for $x > 1$. However, if $x < -1$, then the right-hand side is defined while the left-hand side isn't.

c. False. The correct way to write that function would be $e^{(x+1)\ln 2}$.

d. False. $\frac{d}{dx}(\sqrt{2})^x = (\sqrt{2})^x \ln(\sqrt{2})$.

e. True. This follows from the generalized power rule.

7.3.59

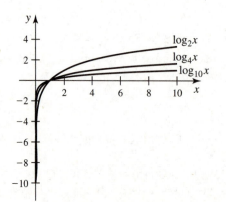

7.3.61

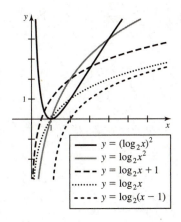

7.3.63 $\frac{d^2}{dx^2}(\log x) = \frac{d}{dx}\left(\frac{1}{x \ln 10}\right) = \frac{-1}{x^2 \ln 10}$.

7.3.65 $\frac{d^3}{dx^3}(x^2 \ln x) = \frac{d^2}{dx^2}(2x \ln x + x) = \frac{d}{dx}(2 \ln x + 2 + 1) = \frac{2}{x}$.

7.3.67

(i) $y' = \frac{d}{dx}\left(e^{x \ln 3}\right) = \left(e^{x \ln 3}\right) \cdot \ln 3 = 3^x \ln 3$.

(ii) Let $y = 3^x$. Then $\ln y = x \ln 3$. So $\frac{1}{y}y' = \ln 3$, and $y' = 3^x \ln 3$.

7.3.69 $f'(x) = \frac{d}{dx}\left(\frac{1}{2}\log 10x\right) = \frac{d}{dx}\frac{1}{2}[\log 10 + \log x] = \frac{1}{2x \ln 10}$.

7.3.71 $f'(x) = \frac{d}{dx}(\ln(2x - 1) + 3\ln(x + 2) - 2\ln(1 - 4x)) = \frac{2}{2x-1} + \frac{3}{x+2} + \frac{8}{1-4x}$.

7.3.73

$y' = \frac{d}{dx}e^{\sin x \ln 2} = (\cos x)(\ln 2)2^{\sin x}$. At $x = \pi/2$ we have $y' = 0$, so the tangent line is given by $y = 2$.

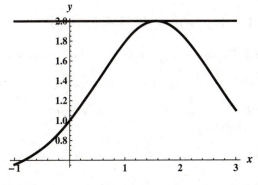

7.3.75 Let $y = x^{10x}$. Then $\ln y = 10x \ln x$, so $\frac{1}{y}y' = 10 \ln x + 10$, and $y' = x^{10x}(10)(\ln x + 1)$.

7.3.77 Let $y = x^{\cos x}$. Then $\ln y = \cos x \ln x$, and $\frac{1}{y}y' = (-\sin x)(\ln x) + \frac{\cos x}{x}$. Thus,

$$y' = x^{\cos x}\left(\frac{\cos x}{x} - \sin x \ln x\right).$$

7.3.79 Let $y = \left(1 + \frac{1}{x}\right)^x$. Then $\ln y = x \ln \left(1 + \frac{1}{x}\right)$, so

$$\frac{1}{y}y' = \ln \left(1 + \frac{1}{x}\right) + x\left(\frac{-1/x^2}{1 + 1/x}\right) = \ln \left(1 + \frac{1}{x}\right) - \frac{1}{x+1}.$$

Thus,

$$y' = \left(1 + \frac{1}{x}\right)^x \left(\ln \left(1 + \frac{1}{x}\right) - \frac{1}{x+1}\right).$$

7.3.81 Let $y = x^{x^{10}}$. Then $\ln y = x^{10} \ln x$, so $\frac{1}{y}y' = 10x^9 \ln x + \frac{x^{10}}{x} = x^9(10 \ln x + 1)$. Thus $y' = x^{x^{10}} \cdot x^9(10 \ln x + 1)$.

7.3.83 $\int 3^{-2x}\, dx = \int e^{-2x \ln 3}\, dx = e^{-2x \ln 3} \cdot \frac{1}{-2 \ln 3} + C = \frac{-3^{-2x}}{2 \ln 3} + C$.

7.3.85 Let $u = x^3$, so that $du = 3x^2\, dx$. Then we have $\int \frac{1}{3}10^u\, du = \frac{10^u}{3 \ln 10} + C = \frac{10^{x^3}}{3 \ln 10} + C$.

7.3.87 Let $u = \ln x$ so that $du = \frac{1}{x}\, dx$. Then we have $\int_0^{\ln(2)+1} 3^u\, du = \left(\frac{3^u}{\ln 3}\right)\Big|_0^{\ln(2)+1} = \frac{3 \cdot 3^{\ln 2} - 1}{\ln 3}$.

7.3.89 Let $f(x) = \ln x$ and $a = e$. Then $f'(e) = \lim\limits_{x \to e} \frac{\ln x - 1}{x - e} = \frac{1}{e}$.

7.3.91 Let $f(x) = x^x$ and $a = 3$. Then $f'(3) = \lim\limits_{h \to 0} \frac{(3+h)^{3+h} - 27}{h} = 3^3 \cdot (\ln 3 + 1) = 27(1 + \ln 3)$.

7.3.93 Let $y = u(x)^{v(x)}$. Then $\ln y = v(x) \ln(u(x))$, so $\frac{1}{y}y' = v'(x) \ln(u(x)) + v(x) \cdot \frac{u'(x)}{u(x)}$. Thus we have $y' = u(x)^{v(x)} \cdot \left(v'(x) \ln(u(x)) + v(x) \cdot \frac{u'(x)}{u(x)}\right) = u(x)^{v(x)} \cdot \left(\frac{v(x)}{u(x)}\frac{du}{dx} + \ln(u(x))\frac{dv}{dx}\right)$.

7.3.95 Using the change of base formulas $\log_b c = \frac{\ln c}{\ln b}$ and $\log_c b = \frac{\ln b}{\ln c}$ we have

$$(\log_b c) \cdot (\log_c b) = \frac{\ln c}{\ln b} \cdot \frac{\ln b}{\ln c} = 1.$$

7.4   Exponential Models

7.4.1 Exponential growth occurs for a constant relative growth rate.

7.4.3 It is the time it takes the population to double in size.

7.4.5 For exponential growth modeled by $y = y_0 e^{kt}$, the doubling time T_2 is $T_2 = \frac{\ln 2}{k}$, where $k > 0$ is the growth constant.

7.4.7 Compound interest and world population growth.

7.4.9 For $f(t)$, $\frac{df}{dt} = 10.5$, so the absolute growth rate is constant. For $g(t)$, $\frac{dg}{dt} = 100e^{t/10} \cdot \frac{1}{10} = 10e^{t/10}$, so the growth rate is not constant but the relative growth rate $\frac{1}{g(t)}g'(t) = \frac{1}{10}$ is constant.

7.4.11 The growth is modeled by $p(t) = 90,000e^{kt}$, with $t = 0$ corresponding to 2010, and time measured in years. Since $(1.024)(90,000) = 90,000e^k$, we have $k = \ln(1.024)$. The doubling time is $T_2 = \frac{\ln 2}{\ln(1.024)} \approx 29.2263$, so it should reach 180,000 around the year 2039.

7.4.13 The price of a cart of groceries is modeled by $p(t) = 100e^{kt}$ where $t = 0$ corresponds to 2005, and t is measured in years and $k = \ln(1.03)$. The price of a the groceries in 2015 when $t = 10$ should be about $p(10) = 100e^{10 \ln(1.03)} \approx \$134.39$.

7.4.15

    a. Note that $k = \ln(1.007)$. The doubling time is $T_2 = \frac{\ln 2}{\ln(1.007)} \approx 99.367$ years. In 2100 when $t = 100$ we would project $281e^{100\ln(1.007)} \approx 564.486$ million.

    b. If $k = \ln(1.005)$, then $T_2 = \frac{\ln 2}{\ln(1.005)} \approx 138.976$ years, and the population in 2100 would be about $281e^{100\ln(1.005)} \approx 462.714$ million. If $k = \ln(1.009)$, then $T_2 = \frac{\ln 2}{\ln(1.009)} \approx 77.3624$ years, and the population in 2100 would be about $281e^{100\ln(1.009)} \approx 688.372$ million.

    c. The projections are very sensitive to the growth rate.

7.4.17 The homicide rate is modeled by $y(t) = 800e^{-kt}$, where $t = 0$ corresponds to 2010, and time is measured in years, and $-k = \ln(.97)$. The rate will reach 600 when $6/8 = e^{\ln(.97)t}$, or $t = \frac{\ln(6/8)}{\ln(.97)} \approx 9.44483$ years. So it should achieve this rate in 2019.

7.4.19 Let $y = 1000e^{-kx}$ model the pressure at x feet above sea level. We know that $\frac{1}{3} = e^{-30000k}$, so $k = \frac{\ln(1/3)}{-30000}$. We want to know for what x does $\frac{1}{2} = e^{-kx}$, so we are seeking $x = \frac{\ln 2}{k} = \frac{30000\ln(2)}{\ln 3} \approx 18,928$ feet above sea level. The pressure is 1/100th of the sea-level pressure when $x = \frac{30000\ln(100)}{\ln 3} \approx 125,754$ feet.

7.4.21

    a. The amount of Valium in the bloodstream is modeled by $a(t) = 20e^{-kt}$. If the half-life is 36 hours, then $36 = \frac{\ln 2}{k}$, so $k = \frac{\ln 2}{36}$. So $a(12) = 20e^{-12\ln(2)/36} = 20e^{-\ln(2)/3} \approx 15.87$ mg.

    b. The concentration of Valium will reach 2 mg when $.1 = e^{-\ln(2)t/36}$, which is when $t = \frac{-36\ln(.1)}{\ln 2} \approx 119.59$ hours.

7.4.23 The amount of U-238 in the rock is modeled by $a(t) = a_0e^{-kt}$ with $k = \frac{\ln 2}{4.5}$, where time is measured in billions of years. We seek t so that $a_0 = a_0e^{-kt}$, so $t = \frac{\ln(.85)}{-k} = \frac{\ln(.85)(-4.5)}{\ln 2} \approx 1.055$. So the cloth was painted about 1.055 billion years ago.

7.4.25

    a. False. If that were the correct formula, then $y(1) = y_0e^{.06} = (1.06184)y_0 \neq (1.06)y_0$.

    b. False. If it increases by ten precent per year, then after 3 years it increases by a factor of $(1.1)^3 = 1.331$ which corresponds to 33.1 percent.

    c. True. The relative decay rate is constant, so the decay is exponential.

    d. True. This follows because the doubling time is related to k by the equation $T_2 = \frac{\ln 2}{k}$.

    e. True. This time would be the constant $\frac{\ln 10}{k}$.

7.4.27 As in the previous problem, the doubling time is $\frac{\ln 2}{k}$ which is constant as a function of t.

7.4.29

    a. After 5 hours, Abe has run $\int_0^5 \frac{4}{t+1}\,dt = (4\ln(t+1))\big|_0^5 = 4\ln 6 \approx 7.17$ miles. After 5 hours, Bob has run $\int_0^5 4e^{-t/2}\,dt = \left(8e^{-t/2}\right)\big|_5^0 = 8(1 - e^{-5/2}) \approx 7.34$. So after 5 hours, Bob is ahead. After 10 hours, Abe has run $\int_0^{10} \frac{4}{t+1}\,dt = (4\ln(t+1))\big|_0^{10} = 4\ln 11 \approx 9.59$ miles. After 10 hours, Bob has run $\int_0^{10} 4e^{-t/2}\,dt = \left(8e^{-t/2}\right)\big|_{10}^0 = 8(1 - e^{-5}) \approx 7.95$. So after 10 hours, Abe is ahead.

b.

Bob's distance function is given by $8(1 - e^{t/2})$ and is bounded above by 8. Abe's distance function is given by $4\ln(t + 1)$ and is unbounded.

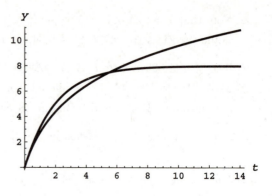

7.4.31 $(1 + .008)^{12} - 1 \approx 10.034\%$, which is more than $12 \cdot .008 \approx 9.6\%$.

7.4.33 As in the previous problem, $s(t) = \frac{v_0}{k}(1 - e^{-kt})$. Consider the equations $s(t_2) - s(t_1) = 1200 = \frac{v_0}{k}(e^{-kt_1} - e^{-kt_2})$ and $v(t_2) - v(t_1) = -100 = v_0(e^{-kt_2} - e^{-kt_1})$. Dividing these two equations yields $k = \frac{1}{12}$. Now $v(t_1) = 1000 = v_0e^{-t_1/12}$ and $v(t_2) = 900 = v_0e^{-t_2/12}$, so $\frac{1000}{900} = e^{(-1/12)(t_1 - t_2)}$, so $t_1 - t_2 = -12\ln(10/9) \approx -1.2643$ seconds. The deceleration takes about 1.2643 seconds to occur.

7.4.35 The initial volume is $V_0 = \frac{4\pi}{3}\left(\frac{5}{10000}\right)^3 = \frac{\pi}{6 \times 10^9}$ cubic cm. $k = \frac{\ln 2}{35}$. Suppose $0.5 = V_0e^{kt} = \frac{\pi}{6 \times 10^9}e^{(t\ln 2)/35}$. Then $\frac{t\ln 2}{35} = \ln\left(\frac{3 \times 10^9}{\pi}\right)$, and $t = \frac{35}{\ln 2}\ln\left(\frac{3 \times 10^9}{\pi}\right) \approx 1044$ days.

7.4.37 Revenue is given by $R(x) = 40xe^{-x/50}$. So $R'(x) = 40xe^{-x/50} \cdot \left(\frac{-1}{50}\right) + 40e^{-x/50} = 40e^{-x/50}(1 - (x/50))$. The critical number is $x = 50$, and this number yields a maximum, since $R'(x) > 0$ on $(0, 50)$, while $R'(x) < 0$ for $x > 50$. Thus, she should charge \$50.00.

7.4.39 If $y_0e^{kt} = y_0(1 + r)^t$, then $(e^k)^t = (1 + r)^t$, so $e^k = 1 + r$, and $k = \ln(1 + r)$. If $y_0e^{kt} = y_02^{t/T_2}$, then $(e^k)^t = (2^{1/T_2})^t$, so $e^k = 2^{1/T_2}$, and $k = \frac{\ln 2}{T_2}$.

7.5   Inverse Trigonometric Functions

7.5.1 The sine function is not one-to-one over its whole domain, so in order to define an inverse, it must be restricted to an interval on which it is one-to-one.

7.5.3 $\tan(\tan^{-1}(x)) = x$ for all real numbers x. (Note that the domain of the inverse tangent is $\mathbb{R}$). However, it is not always true that $\tan^{-1}(\tan x) = x$. For example, $\tan 27\pi = 0$, and $\tan^{-1}(0) = 0$. Thus $\tan^{-1}(\tan(27\pi)) \neq 27\pi$.

7.5.5 The numbers $\pm\pi/2$ are not in the range of $\tan^{-1} x$. The range is $(-\pi/2, \pi/2)$. However, it is true that as x increases without bound, the values of $\tan^{-1} x$ get close to $\pi/2$, and as x decreases without bound, the values of $\tan^{-1} x$ get close to $-\pi/2$.

7.5.7

$$\frac{d}{dx}\sin^{-1}(x) = \frac{1}{\sqrt{1 - x^2}}, \quad -1 < x < 1$$

$$\frac{d}{dx}\tan^{-1}(x) = \frac{1}{1 + x^2}, \quad -\infty < x < \infty$$

$$\frac{d}{dx}\sec^{-1}(x) = \frac{1}{|x|\sqrt{x^2 - 1}}, \quad |x| > 1.$$

7.5.9 $y' = \dfrac{1}{1 + x^2}$. At $x = -2$ we have $y'(-2) = \frac{1}{1+4} = \frac{1}{5}$.

7.5.11 $\sin^{-1}(\sqrt{3}/2) = \pi/3$, since $\sin(\pi/3) = \sqrt{3}/2$.

7.5.13 $\cos^{-1}(-1/2) = 2\pi/3$, since $\cos(2\pi/3) = -1/2$.

7.5.15 $\cos(\cos^{-1}(-1)) = \cos(\pi) = -1$.

7.5.17

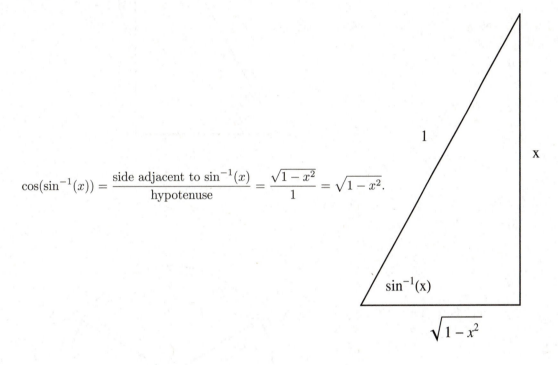

$$\cos(\sin^{-1}(x)) = \frac{\text{side adjacent to } \sin^{-1}(x)}{\text{hypotenuse}} = \frac{\sqrt{1 - x^2}}{1} = \sqrt{1 - x^2}.$$

7.5.19

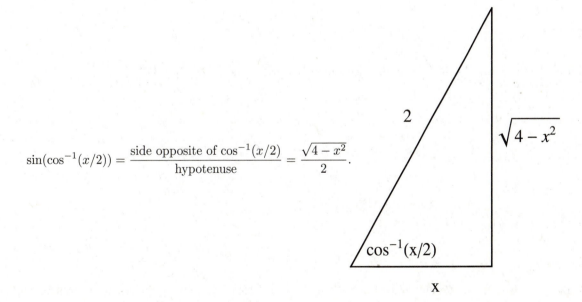

$$\sin(\cos^{-1}(x/2)) = \frac{\text{side opposite of } \cos^{-1}(x/2)}{\text{hypotenuse}} = \frac{\sqrt{4 - x^2}}{2}.$$

7.5.21

Using the identity given, we have $\sin(2\cos^{-1}(x)) = 2\sin(\cos^{-1}(x))\cos(\cos^{-1}(x)) = 2x\sin(\cos^{-1}(x)) = 2x\sqrt{1-x^2}$.

7.5.23

Let $\theta = \cos^{-1}(x)$, and note from the diagram that it then follows that $\cos^{-1}(-x) = \pi - \theta$. So $\cos^{-1}(x) + \cos^{-1}(-x) = \theta + \pi - \theta = \pi$.

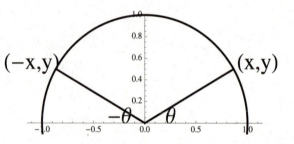

7.5.25 $\tan^{-1}(\sqrt{3}) = \tan^{-1}\left(\frac{\sqrt{3}/2}{1/2}\right) = \pi/3$, since $\sin(\pi/3) = \sqrt{3}/2$ and $\cos(\pi/3) = 1/2$.

7.5.27 $\sec^{-1}(2) = \sec^{-1}\left(\frac{1}{1/2}\right) = \pi/3$, since $\sec(\pi/3) = \frac{1}{\cos(\pi/3)} = \frac{1}{1/2} = 2$.

7.5.29 $\tan^{-1}(\tan(\pi/4)) = \tan^{-1}(1) = \pi/4$.

7.5.31 Let $\csc^{-1}(\sec 2) = z$. Then $\csc z = \sec 2$, so $\sin z = \cos 2$. Now by applying the result of problem 44, we see that $z = \sin^{-1}(\cos 2) = \pi/2 - 2 = \frac{\pi - 4}{2}$.

7.5.33

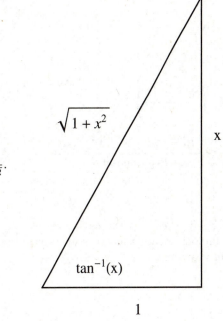

$$\cos(\tan^{-1}(x)) = \frac{\text{side adjacent to } \tan^{-1}(x)}{\text{hypotenuse}} = \frac{1}{\sqrt{1+x^2}}.$$

7.5.35

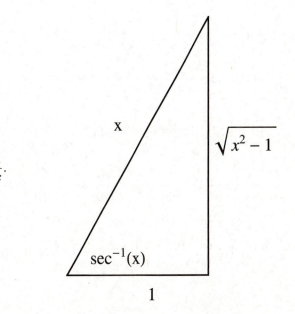

$$\cos(\sec^{-1}(x)) = \frac{\text{side adjacent to } \sec^{-1} x}{\text{hypotenuse}} = \frac{1}{x}.$$

7.5.37

Assume $x > 0$. Then

$$\sin\left(\sec^{-1}\left(\frac{\sqrt{x^2+16}}{4}\right)\right)$$

$$= \frac{\text{side opposite of } \sec^{-1}\left(\frac{\sqrt{x^2+16}}{4}\right)}{\text{hypotenuse}}$$

$$= \frac{|x|}{\sqrt{x^2+16}}.$$

Note: If $x < 0$, then the expression results in a positive number, hence the necessary absolute value sign in the result.

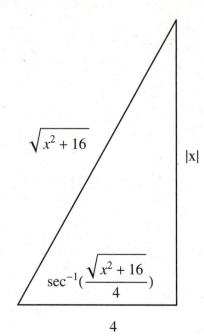

7.5.39 Since $\sin\theta = \frac{x}{6}$, $\theta = \sin^{-1}(x/6)$. Also, $\theta = \tan^{-1}\left(\frac{x}{\sqrt{36-x^2}}\right) = \sec^{-1}\left(\frac{6}{\sqrt{36-x^2}}\right)$.

7.5.41 $\dfrac{d}{dx}\sin^{-1}(2x) = \dfrac{2}{\sqrt{1-4x^2}}.$

7.5.43 $\dfrac{d}{dw}\cos(\sin^{-1}(2w)) = (-\sin(\sin^{-1}(2w)))\cdot\dfrac{d}{dw}\left(\sin^{-1}(2w)\right) = -2w\cdot\dfrac{2}{\sqrt{1-4w^2}} = \dfrac{-4w}{\sqrt{1-4w^2}}.$

7.5.45 $\dfrac{d}{dx}\sin^{-1}(e^{-2x}) = \dfrac{1}{\sqrt{1-e^{-4x}}}\cdot\dfrac{d}{dx}e^{-2x} = \dfrac{-2e^{-2x}}{\sqrt{1-e^{-4x}}}.$

7.5.47 $\dfrac{d}{dy}\tan^{-1}(2y^2-4) = \dfrac{1}{1+(2y^2-4)^2}\cdot\dfrac{d}{dy}(2y^2-4) = \dfrac{4y}{1+(2y^2-4)^2}.$

7.5.49 $\dfrac{d}{dz}\cot^{-1}(\sqrt{z}) = \dfrac{-1}{1+\sqrt{z}^2}\cdot\dfrac{d}{dz}\sqrt{z} = \dfrac{-1}{1+z}\cdot\dfrac{1}{2\sqrt{z}} = \dfrac{-1}{2\sqrt{z}(1+z)}.$

7.5.51 $\dfrac{d}{dx}\cos^{-1}\left(\dfrac{1}{x}\right) = \dfrac{-1}{\sqrt{1-\frac{1}{x^2}}}\cdot\left(\dfrac{-1}{x^2}\right) = \dfrac{1}{x^2\sqrt{\frac{x^2-1}{x^2}}} = \dfrac{|x|}{x^2\sqrt{x^2-1}} = \dfrac{1}{|x|\sqrt{x^2-1}},$ for $|x| > 1$.

7.5.53 $\dfrac{d}{du}\csc^{-1}(2u+1) = \dfrac{-1}{|2u+1|(\sqrt{(2u+1)^2-1}}\cdot 2 = \dfrac{-2}{|2u+1|(\sqrt{(2u+1)^2-1}}.$

7.5.55 $\dfrac{d}{dy}\cot^{-1}\left(\dfrac{1}{1+y^2}\right) = \left(\dfrac{-1}{1+\left(\frac{1}{1+y^2}\right)^2}\right)\cdot\left(\dfrac{-2y}{(1+y^2)^2}\right) = \dfrac{2y}{(1+y^2)^2+1}.$

7.5.57 $\dfrac{d}{dx}\sec^{-1}(\ln x) = \dfrac{1}{|\ln x|\sqrt{(\ln x)^2-1}}\cdot\dfrac{1}{x} = \dfrac{1}{x|\ln x|\sqrt{(\ln x)^2-1}}.$

7.5.59 $\frac{d}{dx} \csc^{-1}(\tan(e^x)) = \frac{-1}{|\tan e^x|\sqrt{(\tan(e^x))^2 - 1}} \cdot \sec^2(e^x) \cdot e^x.$

7.5.61 $\frac{d}{ds} \cot^{-1}(e^s) = \frac{-1}{1 + e^{2s}} \cdot e^s = \frac{-e^s}{1 + e^{2s}}.$

7.5.63

a. $\frac{x}{150} = \cot \theta$, so $\theta = \cot^{-1}\left(\frac{x}{150}\right)$. Then $\frac{d\theta}{dx} = \frac{-1}{1+\left(\frac{x}{150}\right)^2} \cdot \frac{1}{150} = \frac{-150}{(150)^2 + x^2}$. When $x = 500$, we have $\frac{d\theta}{dx} = \frac{-150}{150^2 + 500^2} \approx -0.00055$ radians per meter.

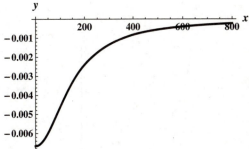

b. The most rapid change is at $x = 0$ where $\frac{d\theta}{dx} = \frac{-1}{150} \approx -0.0067$ radians per meter.

7.5.65 Using Table 4.6 (formula 9, $a = 5$), $\int \frac{6}{\sqrt{25 - x^2}} \, dx = 6\sin^{-1}\left(\frac{x}{5}\right) + C$. Check: $\frac{d}{dx}\left(6\sin^{-1}\left(\frac{x}{5}\right) + C\right) = \frac{6}{\sqrt{1-(x^2/25)}} \cdot \frac{1}{5} = \frac{6}{\sqrt{25-x^2}}.$

7.5.67 Using Table 4.6 (formula 11, $a = 10$), $\int \frac{1}{x\sqrt{x^2 - 100}} \, dx = \frac{1}{10}\sec^{-1}\left|\frac{x}{10}\right| + C$. Check: $\frac{d}{dx}\left(\frac{1}{10}\sec^{-1}\left|\frac{x}{10}\right| + C\right) = \frac{1}{10} \cdot \frac{1}{(x/10)\cdot\sqrt{(x^2/100)-1}} \cdot \frac{1}{10} = \frac{1}{x\sqrt{x^2-100}}.$

7.5.69
$$\int_0^3 \frac{dx}{\sqrt{36 - x^2}} = \sin^{-1}\left(\frac{x}{6}\right) \Big|_0^3 = \frac{\pi}{6}$$

7.5.71
$$\int_0^{3/2} \frac{dx}{\sqrt{36 - 4x^2}} = \frac{1}{2} \int_0^{3/2} \frac{dx}{\sqrt{9 - x^2}} = \frac{1}{2}\sin^{-1}\left(\frac{x}{3}\right)\Big|_0^{3/2} = \frac{\pi}{12}$$

7.5.73

a. False. In general, the two sides bear no relationship to one another. For example, if $x = \sqrt{3}$, $\tan^{-1} x = \pi/3$ while $\sin^{-1} x$ and $\cos^{-1} x$ are not even defined.

b. True. The range of $\cos^{-1}$ is $[0, \pi]$, and $15\pi/16$ is in this range.

c. False. For example if $x = \pi/2$, the right-hand side evaluates to 1 while the left-hand side is not defined.

d. True, since $\frac{d}{dx}\sin^{-1}(x) = -\frac{d}{dx}\cos^{-1}(x)$.

e. False. $\frac{d}{dx}\tan^{-1}(x) = \frac{1}{1+x^2}$ for all x, and this doesn't equal $\sec^2 x$ anywhere except at the origin (one curve is always less than or equal to one, and the other is always greater than or equal to one.)

f. True. $\frac{d}{dx}\sin^{-1}(x) = \frac{1}{\sqrt{1-x^2}}$, and this is minimal when its denominator is as big as possible, which occurs when $x = 0$. So the smallest possible slope of a tangent line for this function on $(-1, 1)$ is $\frac{1}{\sqrt{1-0^2}} = 1$.

g. True. $\frac{d}{dx}\sin x = \cos x$ and $\cos x = 1$ for $x = 0$ and $-1 \le \cos x \le 1$ for all x. Thus 1 is the largest possible slope for a tangent line to the sine function.

7.5.75 If $\cos\theta = 5/13$, then the Pythagorean identity gives $|\sin\theta| = 12/13$. But if $0 < \theta < \pi/2$, then the sine of θ is positive, so $\sin\theta = 12/13$. Thus $\tan\theta = 12/5$, $\cot\theta = 5/12$, $\sec\theta = 13/5$, and $\csc\theta = 13/12$.

7.5.77 If $\csc\theta = 13/12$, then $\sin\theta = 12/13$, and the Pythagorean identity gives $|\cos\theta| = 5/13$. But if $0 < \theta < \pi/2$, then the cosine of θ is positive, so $\cos\theta = 5/13$. Thus $\tan\theta = 12/5$, $\cot\theta = 5/12$, and $\sec\theta = 13/5$.

7.5.79

a.

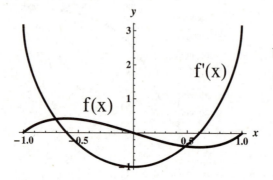

b. $f'(x) = 2x\sin^{-1}(x) + \frac{x^2-1}{\sqrt{1-x^2}}$.

c. Note that f' is zero and f has a horizontal tangent line at about $x = -0.61$ and at about $x = 0.61$.

7.5.81

a.

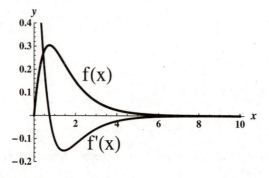

b. $f'(x) = -e^{-x}\tan^{-1}(x) + e^{-x} \cdot \frac{1}{1+x^2}$.

c. Note that f' is zero and f has a horizontal tangent line at about $x = .75$.

7.5.83 $y^2 - 4y + 5 = (y-2)^2 + 1$, so

$$\int \frac{dy}{y^2 - 4y + 5} = \int \frac{dy}{(y-2)^2 + 1} = \tan^{-1}(y-2) + C$$

7.5.85 Let $u = e^x$; then $du = e^x\,dx$ so that

$$\int \frac{e^x}{e^{2x} + 4}\,dx = \int \frac{du}{u^2 + 4} = \frac{1}{2}\tan^{-1}\left(\frac{u}{2}\right) = \frac{1}{2}\tan^{-1}\left(\frac{e^x}{2}\right) + C$$

7.5.87

   a. Since $\frac{l}{10} = \csc(\theta)$, $\theta = \csc^{-1}\left(\frac{l}{10}\right)$, and $\frac{d\theta}{dl} = \frac{-1}{(l/10)\sqrt{(l/10)^2 - 1}} \cdot \frac{1}{10} = \frac{-10}{l\sqrt{l^2 - 100}}$.

   b. $\frac{d\theta}{dl}\big|_{l=50} = \frac{-10}{50\sqrt{2500-100}} \approx -0.0041$ radians per foot. $\frac{d\theta}{dl}\big|_{l=20} = \frac{-10}{20\sqrt{400-100}} \approx -0.029$ radians per foot.
$\frac{d\theta}{dl}\big|_{l=11} = \frac{-10}{11\sqrt{121-100}} \approx -0.198$ radians per foot.

   c. $\lim\limits_{l \to 10^+} \frac{-10}{l\sqrt{l^2 - 100}} = -\infty$. The angle changes very quickly as we approach the dock.

   d. $\frac{d\theta}{dl}$ is negative because this measures the change in θ as l increases – but when the boat is approaching the dock, l is decreasing.

7.5.89

   a. $\sin\theta = \frac{c}{R}$, so $\theta = \sin^{-1}\left(\frac{c}{R}\right)$. Thus $\frac{d\theta}{dc} = \frac{1/R}{\sqrt{1 - \left(\frac{c}{R}\right)^2}} = \frac{1}{\sqrt{R^2 - c^2}}$.

   b. $\frac{d\theta}{dc}\big|_{c=0} = \frac{1}{\sqrt{R^2}} = \frac{1}{R}$.

7.5.91 $(f^{-1})'(y_0) = \frac{1}{f'(x_0)}$ where $y_0 = f(x_0)$.

$$\frac{d}{dx}\sin^{-1}(x) = \frac{1}{\cos(\sin^{-1}(x))} = \frac{1}{\sqrt{1 - \sin^2(\sin^{-1}(x))}} = \frac{1}{\sqrt{1 - x^2}}.$$

7.5.93 Using the identity $\cot^{-1}(x) + \tan^{-1}(x) = \frac{\pi}{2}$, we have the $\frac{d}{dx}\cot^{-1}(x) + \frac{d}{dx}\tan^{-1}(x) = 0$, so $\frac{d}{dx}\cot^{-1}(x) = -\frac{d}{dx}\tan^{-1}(x)$. Likewise, since $\csc^{-1}(x) + \sec^{-1}(x) = \frac{\pi}{2}$, we have $\frac{d}{dx}\csc^{-1}(x) + \frac{d}{dx}\sec^{-1}(x) = 0$, so $\frac{d}{dx}\csc^{-1}(x) = -\frac{d}{dx}\sec^{-1}(x)$.

7.5.95 $\cos(2\sin^{-1}(x)) = \cos^2(\sin^{-1}(x)) - \sin^2(\sin^{-1}(x)) = 1 - \sin^2(\sin^{-1}(x)) - \sin^2(\sin^{-1}(x)) = 1 - 2x^2$ for $-1 \le x \le 1$.

7.5.97 $\sin(2\sin^{-1}(x)) = 2\sin(\sin^{-1}(x))\cos(\sin^{-1}(x)) = 2x\sqrt{1 - x^2}$ for $-1 \le x \le 1$.

7.6 L'Hôpital's Rule and Growth Rates of Functions

7.6.1 If $\lim\limits_{x \to a} f(x) = 1$ and $\lim\limits_{x \to a} g(x) = \infty$, then $f(x)^{g(x)} \to 1^\infty$ as $x \to a$, which is meaningless; so direct substitution does not work.

7.6.3 This means $\lim\limits_{x \to \infty} \frac{g(x)}{f(x)} = 0$.

7.6.5 By Theorem 4.15, we have $\ln x, x^3, 2^x, x^x$ in order of increasing growth rates.

7.6.7 Note that $\ln x^{2x} = 2x\ln x$, so we evaluate $L = \lim\limits_{x \to 0^+} 2x\ln x = 2\lim\limits_{x \to 0^+} \frac{\ln x}{1/x} = 2\lim\limits_{x \to 0^+} \frac{1/x}{-1/x^2} = 2\lim\limits_{x \to 0^+} -x = 0$ by l'Hôpital's rule. Therefore $\lim\limits_{x \to 0^+} x^{2x} = e^L = 1$.

7.6.9 Note that $\ln(\tan\theta)^{\cos\theta} = \cos\theta\ln\tan\theta$, so we evaluate $L = \lim\limits_{\theta\to\pi/2^-}\cos\theta\ln\tan\theta = \lim\limits_{\theta\to\pi/2^-}\dfrac{\ln\tan\theta}{\sec\theta}$.
L'Hôpital's rule gives $\lim\limits_{\theta\to\pi/2^-}\dfrac{\ln\tan\theta}{\sec\theta} = \lim\limits_{\theta\to\pi/2^-}\dfrac{\sec^2\theta/\tan\theta}{\sec\theta\tan\theta} = \lim\limits_{\theta\to\pi/2^-}\dfrac{\sec\theta}{\tan^2\theta} = \lim\limits_{\theta\to\pi/2^-}\dfrac{\cos\theta}{\sin^2\theta} = 0$, so $\lim\limits_{\theta\to\pi/2^-}(\tan\theta)^{\cos\theta} = e^L = 1$.

7.6.11 Note that $\ln(1+x)^{\cot x} = \cot x\ln(1+x)$, so we evaluate $L = \lim\limits_{x\to0+}\cot x\ln(1+x) = \lim\limits_{x\to0+}\dfrac{\ln(1+x)}{\tan x} = \lim\limits_{x\to0+}\dfrac{1/(1+x)}{\sec^2 x} = \lim\limits_{x\to0+}\dfrac{\cos^2 x}{1+x} = 1$ by l'Hôpital's rule. Therefore $\lim\limits_{x\to0+}(1+x)^{\cot x} = e^L = e$.

7.6.13 Note that $\ln(\tan x)^x = x\ln\tan x$, so we evaluate

$$L = \lim_{x\to0+}x\ln\tan x = \lim_{x\to0+}\frac{\ln\tan x}{1/x} = \lim_{x\to0+}\frac{\sec^2 x/\tan x}{-1/x^2} = -\lim_{x\to0+}\frac{x^2}{\sin x\cos x}$$

by l'Hôpital's rule. Next, observe that $\lim\limits_{x\to0+}\dfrac{x^2}{\sin x\cos x} = \lim\limits_{x\to0+}\dfrac{x}{\sin x}\cdot\lim\limits_{x\to0+}\dfrac{x}{\cos x} = 1\cdot0 = 0$. Therefore $L = 0$ and $\lim\limits_{x\to0+}(\tan x)^x = e^L = 1$.

7.6.15 Note that $\ln(x+\cos x)^{1/x} = (\ln(x+\cos x))/x$, so we evaluate

$$L = \lim_{x\to0}\frac{\ln(x+\cos x)}{x} = \lim_{x\to0}\frac{(x+\cos x)^{-1}(1-\sin x)}{1} = 1$$

by l'Hôpital's rule. Therefore $\lim\limits_{x\to0}(x+\cos x)^{1/x} = e^L = e$.

7.6.17 By Theorem 4.15, $e^{0.01x}$ grows faster than x^{10} as $x\to\infty$.

7.6.19 Note that $\ln x^{20} = 20\ln x$, so $\ln x^{20}$ and $\ln x$ have comparable growth rates as $x\to\infty$.

7.6.21 By Theorem 4.15, x^x grows faster than 100^x as $x\to\infty$.

7.6.23 By Theorem 4.15, 1.00001^x grows faster than x^{20} as $x\to\infty$.

7.6.25 Observe that $\lim\limits_{x\to\infty}\dfrac{(x/2)^x}{x^x} = \lim\limits_{x\to\infty}2^{-x} = 0$, so x^x grows faster than $(x/2)^x$ as $x\to\infty$.

7.6.27 Note that $\lim\limits_{x\to\infty}\dfrac{e^{x^2}}{e^{10x}} = \lim\limits_{x\to\infty}e^{x^2-10x} = \infty$, so e^{x^2} grows faster than e^{10x} as $x\to\infty$.

7.6.29

a. False; this limit has the form $0^\infty = 0$.

b. False; this limit has the indeterminate form 1^∞ which is not always 1.

c. True; $\ln x^{100} = 100\ln x$ so that $\lim\limits_{x\to\infty}\dfrac{\ln x^{100}}{\ln x} = 100 < \infty$.

d. True; note that $\lim\limits_{x\to\infty}\dfrac{e^x}{2^x} = \lim\limits_{x\to\infty}\left(\dfrac{e}{2}\right)^x = \infty$ since $e/2 > 1$.

7.6.31 Note that $\ln\left(\dfrac{\sin x}{x}\right)^{1/x^2} = \dfrac{\ln\sin x - \ln x}{x^2}$, so we evaluate

$$L = \lim_{x\to 0}\frac{\ln\sin x - \ln x}{x^2} = \lim_{x\to 0}\frac{\frac{\cos x}{\sin x} - \frac{1}{x}}{2x} = \lim_{x\to 0}\frac{x\cos x - \sin x}{2x^2\sin x}$$

by l'Hôpital's rule. Next, observe that

$$\lim_{x\to 0}\frac{x\cos x - \sin x}{2x^2\sin x} = \lim_{x\to 0}\frac{\cos x - x\sin x - \cos x}{4x\sin x + 2x^2\cos x} = \lim_{x\to 0}\frac{-\frac{\sin x}{x}}{\frac{4\sin x}{x} + 2\cos x} = -\frac{1}{6}$$

using l'Hôpital's rule and $\lim_{x\to 0}\sin x/x = 1$. Therefore $\lim_{x\to 0}\left(\dfrac{\sin x}{x}\right)^{1/x^2} = e^L = e^{-1/6}$.

7.6.33 $x^{1/x} = e^{\ln x/x}$. Now, $\lim_{x\to\infty}\ln x/x$ has the indeterminate form ∞/∞, so using L'Hôpital's rule,

$$\lim_{x\to\infty}\frac{\ln x}{x} = \lim_{x\to\infty}\frac{1}{x} = 0$$

so that $\lim_{x\to\infty} x^{1/x} = e^{\lim_{x\to\infty}\ln x/x} = e^0 = 1.$

7.6.35 $(\sqrt{x-1})^{\sin\pi x} = e^{(\sin\pi x)\ln\sqrt{x-1}} = e^{(\sin\pi x)\ln(x-1)/2}$. As $x\to 1^+$ the exponent has the indeterminate form $0\cdot\infty$. Using L'Hôpital's rule twice,

$$\lim_{x\to 1^+}\frac{(\sin\pi x)\ln(x-1)}{2} = \lim_{x\to 1^+}\frac{\ln(x-1)}{2/\sin\pi x} = \lim_{x\to 1^+}\frac{-\sin^2\pi x}{2(x-1)\pi\cos\pi x}$$

$$= -\lim_{x\to 1^+}\frac{2\pi\sin\pi x\cos\pi x}{2\pi(\cos\pi x - (x-1)\sin\pi x)} = -\lim_{x\to 1^+}\frac{\sin\pi x\cos\pi x}{\cos\pi x - (x-1)\sin\pi x} = 0$$

so that $\lim_{x\to 1^+}(\sqrt{x-1})^{\sin\pi x} = e^0 = 1.$

7.6.37

a. Approximately 3.44×10^{15}.

b. Approximately 3536.

c. We can explicitly solve for x in this case: $x^{x/100} = e^x \implies x^{1/100} = e \implies x = e^{100}$.

d. Approximately 163.

7.6.39 Note that $\ln(a^x - b^x)^x = x\ln(a^x - b^x)$, so we evaluate $L = \lim_{x\to 0^+} x\ln(a^x - b^x) = \lim_{x\to 0^+}\dfrac{\ln(a^x - b^x)}{1/x} =$

$\lim_{x\to 0^+} -x^2\left(\dfrac{(\ln a)a^x - (\ln b)b^x}{a^x - b^x}\right)$ by l'Hôpital's rule. We have

$$\lim_{x\to 0^+} -x^2\left(\frac{(\ln a)a^x - (\ln b)b^x}{a^x - b^x}\right) = -\lim_{x\to 0^+} x\left((\ln a)a^x - (\ln b)b^x\right)\frac{x}{a^x - b^x}$$

and one more application of l'Hôpital's rule gives $\lim_{x\to 0^+}\dfrac{x}{a^x - b^x} = \lim_{x\to 0^+}\dfrac{1}{(\ln a)a^x - (\ln b)b^x} = \dfrac{1}{\ln a - \ln b}$, so $L = 0$ and therefore $\lim_{x\to 0^+}(a^x - b^x)^x = e^L = 1.$

7.6.41 Apply l'Hôpital's rule: $\lim_{x\to 0}\dfrac{a^x - b^x}{x} = \lim_{x\to 0}\dfrac{(\ln a)a^x - (\ln b)b^x}{1} = \ln a - \ln b.$

7.6.43

a. After each year the balance increases by the factor $1 + r$; therefore the balance after t years is $B(t) = P(1 + r)^t$.

b. Observe that $\lim\limits_{m \to \infty} (1 + r/m)^m = \lim\limits_{m \to \infty} \left(1 + \dfrac{1}{m/r}\right)^{(m/r)r} = e^r$, since $\lim\limits_{n \to \infty} \left(1 + \dfrac{1}{n}\right)^n = e$. So with continuous compounding the balance after t years is $B(t) = Pe^{rt}$.

7.6.45 Let $t = b^x$, as in Example 8; then $x = \ln t / \ln b$ and we have $\lim\limits_{x \to \infty} \dfrac{x^p}{b^x} = \lim\limits_{t \to \infty} \dfrac{\ln^p t}{t \ln^p b} = 0$, by Theorem 4.15.

7.6.47 Note that $\log_a x = \ln x / \ln a$, so $\dfrac{\log_a x}{\log_b x} = \dfrac{\ln b}{\ln a}$, and therefore $\log_a x$ and $\log_b x$ grow at a comparable rate as $x \to \infty$.

7.6.49 Note that $\ln\left(1 + \frac{a}{x}\right)^x = \dfrac{\ln(1+a/x)}{1/x}$ so we evaluate $L = \lim\limits_{x \to \infty} \dfrac{\ln\left(1 + a/x\right)}{1/x} = \lim\limits_{x \to \infty} \dfrac{1}{1 + a/x} \cdot \dfrac{-a}{x^2} \cdot$ $\dfrac{1}{-1/x^2} = \lim\limits_{x \to \infty} \dfrac{a}{1 + a/x} = a$ by l'Hôpital's rule. Therefore $\lim\limits_{x \to \infty} \left(1 + \dfrac{a}{x}\right)^x = e^L = e^a$.

7.6.51

a. Observe that $\lim\limits_{x \to \infty} \dfrac{b^x}{e^x} = \lim\limits_{x \to \infty} \left(\dfrac{b}{e}\right)^x$. This limit is ∞ exactly when $b > e$.

b. Observe that $\lim\limits_{x \to \infty} \dfrac{e^{ax}}{e^x} = \lim\limits_{x \to \infty} e^{(a-1)x}$. This limit is ∞ exactly when $a > 1$.

7.7 Chapter 7 Review

7.7.1

a. True. $y = 2/x$; solving for x gives $x = 2/y$ so that $f^{-1}(x) = 2/x$.

b. False. $\ln(xy) = \ln x + \ln y$.

c. False, but it's close. $\dfrac{Ae^{0.1(t+1)}}{Ae^{0.1t}} = e^{0.1} \approx 1.1052 \neq 1.1$.

d. True. $\dfrac{d}{dx}(b^x) = b^x \ln b$, so $b^x = b^x \ln b$ only when $\ln b = 1$ so that $b = e$.

e. False. It is of the form 0^∞, which is not indeterminate. Looking at the limit, the base approaches zero while the exponent grows large, so clearly the power approaches zero.

f. False. That is the *range*. The domain of $\tan^{-1} x$ is $\mathbb{R}$.

g. False. $\pi^{3x} = e^{3x \ln \pi}$.

7.7.3 If $\log x^2 + 3 \log x = \log 32$, then $\log(x^2 \cdot x^3) = \log(32)$, so $x^5 = 32$ and $x = 2$. The answer does not depend on the base of the log.

7.7.5

By graphing, it is clear that this function is not one-to-one on its whole domain, but it is one-to-one on the interval $(-\infty, 0]$, on the interval $[0, 2]$, and on the interval $[2, \infty)$, so it would have an inverse if we restricted it to any of these particular intervals.

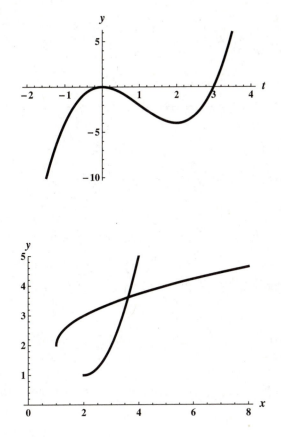

7.7.7

Completing the square gives $f(x) = x^2 - 4x + 4 + 1 = (x-2)^2 + 1$. Switching the x and y and solving for y yields $(y-2)^2 = x - 1$, so $|y - 2| = \sqrt{x-1}$, and thus $y = 2 + \sqrt{x-1}$ (we choose the "+" rather than the "−" because the domain of f is $x > 2$, so the range of f^{-1} must also consist of numbers greater than 2.)

7.7.9 $\left. (f^{-1}(x))' \right|_{x=f(\pi/4)} = \dfrac{1}{f'(\pi/4)} = \dfrac{1}{-\sin(\pi/4)} = -\sqrt{2}$

7.7.11 $\left. (f^{-1}(x))' \right|_{x=f(0)} = \dfrac{1}{f'(0)} = \dfrac{1}{4 \cdot 0^3 - 4 \cdot 0 - 1} = -1.$

7.7.13 If $f(x) = x^{-1/3}$, then $f^{-1}(x) = x^{-3}$. So $\left(f^{-1}\right)'(x) = -3x^{-4}$ for $x \neq 0$.

7.7.15 Let $f(y) = 3y - 4$. Then $f'(y) = 3$ for all y in the domain of f. Let $y = f^{-1}(x)$. $(f^{-1})'(x) = \frac{1}{f'(y)} = \frac{1}{3}$.

7.7.17 Let $x = f(y) = y^2 - 4$ for $y > 0$. Note that this means that $y = \sqrt{x+4}$. Then $f'(y) = 2y$. So $(f^{-1})'(x) = \frac{1}{f'(y)} = \frac{1}{2y} = \frac{1}{2\sqrt{x+4}}$.

7.7.19 For $y \geq -2$, let $x = \sqrt{y+2}$. Note that it then follows that $x \geq 0$. Then $1 = \frac{y'}{2\sqrt{y+2}}$, and $x^2 = y + 2$, so $y = x^2 - 2$. Thus we have $(f^{-1})'(x) = y' = 2\sqrt{x^2 - 2 + 2} = 2|x| = 2x$, since $x \geq 0$.

7.7.21 For $y > 0$, let $x = y^{-1/2}$. Then $1 = \frac{-1}{2}y^{-3/2}y'$, so $y' = -2y^{3/2} = -2(x^{-2})^{3/2} = -2x^{-3}$ where $x > 0$.

7.7.23

a. $\left(f^{-1}\right)'\left(\frac{1}{\sqrt{2}}\right) = \frac{1}{f'\left(\frac{\pi}{4}\right)} = \frac{1}{\cos\left(\frac{\pi}{4}\right)} = \sqrt{2}.$

b. $\frac{d}{dx}\sin^{-1}(x)\big|_{x=1/\sqrt{2}} = \frac{1}{\sqrt{1-(1/2)}} = \frac{1}{\sqrt{1/2}} = \sqrt{2}.$

7.7.25 $f'(x) = \ln^2 x + x \cdot 2 \ln x \cdot \left(\frac{1}{x}\right) = \ln x \cdot (\ln x + 2)$.

7.7.27 $f'(x) = 2^{x^2-x} \cdot \ln 2 \cdot (2x - 1)$.

7.7.29 $f'(x) = \dfrac{1}{\sqrt{1 - \left(\frac{1}{x}\right)^2}} \cdot \dfrac{-1}{x^2} = \dfrac{-1}{|x|\sqrt{x^2 - 1}}$.

7.7.31 $\frac{d}{dx} x^{1/x} = \frac{d}{dx} e^{\frac{\ln x}{x}} = e^{\frac{\ln x}{x}} \cdot \left(\frac{1 - \ln x}{x^2}\right) = x^{1/x} \left(\frac{1 - \ln x}{x^2}\right)$. So $\frac{d}{dx} x^{1/x}\big|_{x=1} = 1 \cdot \frac{1-0}{1^2} = 1$.

7.7.33 $f'(x) = \sec^{-1} x + \frac{1}{\sqrt{x^2-1}}$. So $f'(2/\sqrt{3}) = \frac{\pi}{6} + \sqrt{3}$.

7.7.35 Since $\cos(\pi/6) = \sqrt{3}/2$, $\cos^{-1}(\sqrt{3}/2) = \pi/6$.

7.7.37 Since $\sin(-\pi/2) = -1$, $\sin^{-1}(-1) = -\pi/2$.

7.7.39 $\sin(\sin^{-1}(x)) = x$, for all x in the domain of the inverse sine function.

7.7.41 If $\theta = \sin^{-1}(12/13)$, then $0 < \theta < \pi/2$, and $\sin \theta = 12/13$. Then (using the Pythagorean identity) we can deduce that $\cos \theta = 5/13$. It must follow that $\tan \theta = 12/5$, $\cot \theta = 5/12$, $\sec \theta = 13/5$, and $\csc \theta = 13/12$.

7.7.43

$$\sin(\cos^{-1}(x/2)) = \frac{\text{side opposite of } \cos^{-1}(x/2)}{\text{hypotenuse}} = \frac{\sqrt{4 - x^2}}{2}.$$

7.7.45

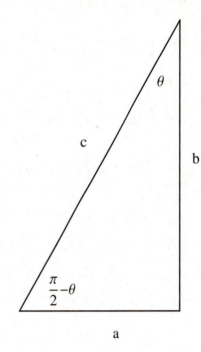

Note that

$$\tan\theta = \frac{a}{b} = \cot(\pi/2 - \theta).$$

Thus, $\cot^{-1}(\tan\theta) = \cot^{-1}(\cot(\pi/2 - \theta)) = \pi/2 - \theta$.

7.7.47 Let $\theta = \sin^{-1}(x)$. Then $\sin\theta = x$ and note that then $\sin(-\theta) = -\sin\theta = -x$, so $-\theta = \sin^{-1}(-x)$. Then $\sin^{-1}(x) + \sin^{-1}(-x) = \theta + -\theta = 0$.

7.7.49 Using the hint, we have $\cos(2\sin^{-1}(x)) = \cos^2(\sin^{-1}(x)) - \sin^2(\sin^{-1}(x)) = (\sqrt{1-x^2})^2 - x^2 = 1 - 2x^2$.

7.7.51 Let $u = \ln x$ so that $du = \frac{1}{x}\,dx$. Then we have $\int_2^8 \frac{1}{u}\,du = (\ln u)|_2^8 = \ln 4$.

7.7.53 Let $u = x^2 + 8x + 25$ so that $du = (2x+8)\,dx = 2(x+4)\,dx$. Then we have $\frac{1}{2}\int \frac{1}{u}\,du = \frac{1}{2}\ln|u| + C = \frac{1}{2}\ln|x^2 + 8x + 25| + C$.

7.7.55

$$\int \frac{3}{2x^2 + 1}\,dx = \frac{3}{2}\int \frac{1}{x^2 + 1/2}\,dx = \frac{3\sqrt{2}}{2}\tan^{-1}(\sqrt{2}x) + C$$

7.7.57

$$\int_{-1}^{2\sqrt{3}-1} \frac{dx}{x^2 + 2x + 5} = \int_{-1}^{2\sqrt{3}-1} \frac{dx}{(x+1)^2 + 4} = \frac{1}{2}\tan^{-1}\left(\frac{x+1}{2}\right)\Big|_{-1}^{2\sqrt{3}-1} = \frac{\pi}{6}$$

7.7.59

$$\int_{-2/3}^{2/\sqrt{3}} \frac{dx}{\sqrt{16 - 9x^2}} = \frac{1}{3}\int_{-2/3}^{2/\sqrt{3}} \frac{dx}{\sqrt{16/9 - x^2}} = \frac{1}{3}\sin^{-1}\left(\frac{3x}{4}\right)\Big|_{-2/3}^{2/\sqrt{3}}$$

$$= \frac{1}{3}\left(\sin^{-1}(\sqrt{3}/2) - \sin^{-1}(-1/2)\right) = \frac{\pi}{6}$$

7.7.61 $f'(x) = \dfrac{3}{x} - \dfrac{x}{12}$; then

$$1 + f'(x)^2 = 1 + \left(\frac{3}{x} - \frac{x}{12}\right)^2 = 1 + \frac{9}{x^2} + \frac{x^2}{144} - \frac{1}{2} = \left(\frac{3}{x} + \frac{x}{12}\right)^2$$

so that

$$\int_1^6 \sqrt{1 + \left(\frac{3}{x} - \frac{x}{12}\right)^2}\, dx = \int_1^6 \frac{3}{x} + \frac{x}{12}\, dx = \left(3\ln x + \frac{x^2}{24}\right)\bigg|_1^6 = 3\ln 6 + \frac{35}{24}$$

7.7.63 $f'(y) = 2\sqrt{2}e^{\sqrt{2}y} - \frac{\sqrt{2}}{16}e^{-\sqrt{2}y}$; then

$$1 + f'(y)^2 = 1 + 8e^{2\sqrt{2}y} - \frac{1}{2} + \frac{1}{128}e^{-2\sqrt{2}y} = \left(2\sqrt{2}e^{\sqrt{2}y} + \frac{\sqrt{2}}{16}e^{-\sqrt{2}y}\right)^2$$

so that

$$\int_0^{(\ln 2)/\sqrt{2}} \sqrt{1 + f'(y)^2}\, dy = \int_0^{(\ln 2)/\sqrt{2}} \left(2\sqrt{2}e^{\sqrt{2}y} + \frac{\sqrt{2}}{16}e^{-\sqrt{2}y}\right)\, dy$$

$$= 2e^{\sqrt{2}y} - \frac{1}{16}e^{-\sqrt{2}y}\bigg|_0^{(\ln 2)/\sqrt{2}} = \frac{65}{32}$$

7.7.65 Using the shell method, the volume is

$$\int_0^4 \frac{4\pi x}{x^2 + 1}\, dx = 2\pi \ln(x^2 + 1)\bigg|_0^4 = 2\pi \ln 17$$

7.7.67

a.

Tom's position function is given by

$$\int_0^t 20e^{-2x}\, dx = \left(-10e^{-2x}\right)\big|_0^t$$
$$= 10(1 - e^{-2t}).$$

Sue's position is given by

$$\int_0^t 15e^{-x}\, dx = \left(-15e^{-x}\right)\big|_0^t$$
$$= 15(1 - e^{-t}).$$

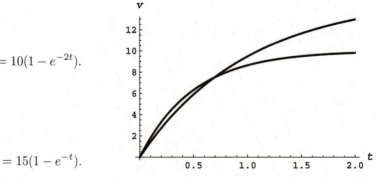

b. We are looking for t so that $10(1 - e^{-2t}) = 15(1 - e^{-t})$, which occurs when $10(e^{-t})^2 - 15e^{-t} + 5 = 0$, or $2u^2 - 3u + 1 = 0$ where $u = e^{-t}$. This quadratic factors as $(2u - 1)(u - 1) = 0$, so if $e^{-t} = 1$ or $e^{-t} = 1/2$, so $t = 0$ or $t = \ln 2$.

c. Sue takes the lead and doesn't relinquish it at $t = \ln 2$.

7.7.69 Growth is modeled by $p(t) = 150,000e^{kt}$ where $k = \ln(1.04)$. The population reaches 1,000,000 when $1,000,000 = 150,000e^{\ln(1.04)t}$, or when $\ln(20/3) = \ln(1.04)t$, so when $t = \frac{\ln(20/3)}{\ln(1.04)} \approx 48.37$ years.

7.7.71

The domain of f is the set of all real numbers. Note that $\lim_{x\to\infty} f(x) = \infty$ and $\lim_{x\to-\infty} f(x) = 0$. $f'(x) = e^x(2x-1) + (x^2-x)e^x = e^x(x^2+x-1)$, which is 0 for $x = \frac{-1\pm\sqrt{5}}{2}$. Note that $f'(x) > 0$ on the interval $(-\infty, (-1-\sqrt{5})/2)$ and on $((-1+\sqrt{5})/2, \infty)$, so f is increasing on those intervals, while $f'(x) < 0$ (and so f is decreasing) on $((-1-\sqrt{5})/2, (-1+\sqrt{5})/2)$. There is a local maximum at $(-1-\sqrt{5})/2$ and a local minimum at $(-1+\sqrt{5})/2$. Note that $f''(x) = e^x(2x+1) + (x^2+x-1)e^x = e^x(x^2+3x)$, which is 0 for $x = 0$ and $x = -3$. $f''(x) > 0$ on $(-\infty, -3)$ and on $(0, \infty)$, so f is concave up on those intervals, while $f''(x) < 0$ on $(-3, 0)$, so f is concave down there. There are inflection points at $x = -3$ and at $x = 0$.

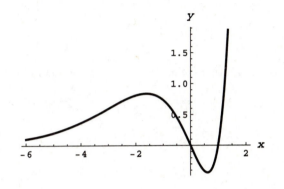

7.7.73 Note that $y' = 1/x$, so $1 + y'^2 = \frac{x^2+1}{x^2}$, and $\sqrt{1+y'^2} = \frac{\sqrt{x^2+1}}{x}$. So

$$L = \int_1^b \frac{\sqrt{x^2+1}}{x} \, dx$$

$$= \left(\sqrt{x^2+1} - \ln\left(\frac{1+\sqrt{x^2+1}}{x}\right)\right)\Bigg|_1^b = \sqrt{b^2+1} - \sqrt{2} + \ln\left(\frac{(\sqrt{b^2+1}-1)(1+\sqrt{2})}{b}\right).$$

Using a computer algebra system, we see that this has value 2 for $b \approx 2.715$.

7.7.75 Note that $\ln(\sin x)^{\tan x} = \tan x \ln \sin x$, so we evaluate $L = \lim_{x\to\pi/2^-} \tan x \ln \sin x = \lim_{x\to\pi/2^-} \frac{\ln \sin x}{\cot x} = \lim_{x\to\pi/2^-} \frac{\cot x}{-\csc^2 x} = \lim_{x\to\pi/2^-} -\cos x \sin x = 0$ by l'Hôpital's rule. Therefore $\lim_{x\to\pi/2^-} (\sin x)^{\tan x} = e^L = 1$.

7.7.77 Observe that $\lim_{x\to\infty} \frac{x+1}{x-1} = 1$, either by l'Hôpital's rule or Theorem 2.7. Hence $\lim_{x\to\infty} \ln\left(\frac{x+1}{x-1}\right) = \ln 1 = 0$.

7.7.79 Note that $\ln\left(1-\frac{3}{x}\right)^x = x\ln\left(1-\frac{3}{x}\right) = \frac{\ln(1-3/x)}{1/x}$, so we evaluate

$$L = \lim_{x\to\infty} \frac{\ln(1-3/x)}{1/x} = \lim_{x\to\infty} \frac{(1-3/x)^{-1}(3/x^2)}{-1/x^2} = -3$$

by l'Hôpital's rule. Therefore $\lim_{x\to\infty} \left(1-\frac{3}{x}\right)^x = e^L = e^{-3}$.

7.7.81 Observe that $\lim_{x\to\infty} \frac{x^{1/2}}{x^{1/3}} = \lim_{x\to\infty} x^{1/6} = \infty$, so $x^{1/2}$ grows faster than $x^{1/3}$ as $x \to \infty$.

7.7.83 By Theorem 4.15, $\sqrt{x}$ grows faster than $\ln^{10} x$ as $x \to \infty$.

7.7.85 Observe that $\lim_{x\to\infty} \frac{e^x}{3^x} = \lim_{x\to\infty} \left(\frac{e}{3}\right)^x = 0$ since $e/3 < 1$. Therefore 3^x grows faster than e^x as $x \to \infty$.

7.7.87 Observe that $4^{x/2} = (4^{1/2})^x = 2^x$, so these functions are identical and hence have comparable growth rates as $x \to \infty$.

7.7.89

For the second limit, let $y = x^2$:

$$\lim_{x \to 0} \frac{x^2}{1 - e^{-x^2}} = \lim_{y \to 0} \frac{y}{1 - e^{-y}}$$

$$= \lim_{y \to 0} \frac{1}{e^{-y}} = 1$$

by l'Hôpital's rule.   For the first, observe that for $x > 0$ $\dfrac{x}{\sqrt{1 - e^{-x^2}}} = \left(\dfrac{x^2}{1 - e^{-x^2}} \right)^{1/2}$, so $\lim_{x \to 0^+} \dfrac{x}{\sqrt{1 - e^{-x^2}}} = 1$ as well.

7.7.91 Note that $\lim_{x \to 0^+} x^x = 1$, as shown in Example 6 a. Therefore $\lim_{x \to 0^+} f(x) = \lim_{x \to 0^+} (x^x)^x = 1^0 = 1$, and $\lim_{x \to 0} g(x) = \lim_{x \to 0} x^{(x^x)} = 0^1 = 0$.

7.7.93 First, observe that $\lim_{x \to \infty} \left(1 + \dfrac{1}{x} \right)^{x+a} = \lim_{x \to \infty} \left(1 + \dfrac{1}{x} \right)^{x} \left(1 + \dfrac{1}{x} \right)^{a} = e \cdot 1 = e$. Thus $\lim_{x \to \infty} \ln g(x) = 1$. It suffices to determine whether $\ln g(x) - 1$ is positive or negative as $x \to \infty$. To do this, consider $\lim_{x \to \infty} x(\ln g(x) - 1) = \lim_{t \to 0} \dfrac{(1 + at) \ln(1 + t) - t}{t^2}$, where we make the change of variables $t = 1/x$. This limit can be evaluated by using l'Hôpital's rule twice: $\lim_{t \to 0} \dfrac{(1 + at) \ln(1 + t) - t}{t^2} = a - \dfrac{1}{2}$. Therefore when $a > 1/2$ we have $g(x) > e$ as $x \to \infty$, and when $0 < a < 1/2$, $g(x) < e$ as $x \to \infty$. In the case $a = 1/2$ we consider the limit $\lim_{x \to \infty} x^2(\ln g(x) - 1) = \lim_{t \to 0} \dfrac{(1 + at) \ln(1 + t) - t}{t^3}$, which can be evaluated by using l'Hôpital's three times: $\lim_{t \to 0} \dfrac{(1 + t/2) \ln(1 + t) - t}{t^3} = \dfrac{1}{12}$. Therefore $g(x) > e$ as $x \to \infty$ in this case as well.

7.7.95

a.

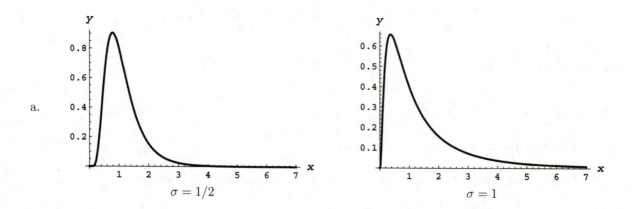

$\sigma = 1/2$ $\sigma = 1$

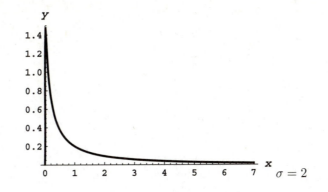

$\sigma = 2$

It appears that $\lim_{x \to 0} f(x) = 0$.

b. Let $x = e^y$, so that $y = \ln x$ and as $x \to 0$ we have $y \to -\infty$. Then $\lim_{x \to 0} f(x) = \lim_{y \to -\infty} \frac{e^{-y^2/2\sigma^2}}{\sigma\sqrt{2\pi}e^y} = \frac{1}{\sigma\sqrt{2\pi}} \lim_{y \to -\infty} \frac{1}{e^y \cdot e^{y^2/2\sigma^2}} = 0$.

c. Write f as $f(x) = \frac{1}{\sigma\sqrt{2\pi}} \left(\frac{1}{xe^{(\ln^2 x)/2\sigma^2}} \right)$. Then

$$f'(x) = \frac{-1}{\sigma\sqrt{2\pi}} \left(\frac{1}{x^2 e^{(\ln^2 x)/\sigma^2}} \right) \left(e^{(\ln^2 x)/2\sigma^2} + xe^{(\ln^2 x)/2\sigma^2} \cdot \frac{1}{\sigma^2} \ln x \cdot \frac{1}{x} \right) =$$

$$\frac{-1}{\sigma\sqrt{2\pi}} \left(\frac{1}{x^2} \right) \left(1 + \frac{\ln x}{\sigma^2} \right).$$

This quantity is 0 only when $1 + \frac{\ln x}{\sigma^2} = 0$, which occurs for $x^* = e^{-\sigma^2}$, and this critical number yields a maximum.

d. $f(e^{-\sigma^2}) = \frac{1}{\sigma\sqrt{2\pi}} \left(\frac{1}{(e^{-\sigma^2} e^{(\ln^2(e^{-\sigma^2}))/2\sigma^2}} \right) = \frac{1}{\sigma\sqrt{2\pi}} \left(\frac{1}{e^{-\sigma^2} e^{\sigma^2/2}} \right) = \frac{e^{\sigma^2/2}}{\sigma\sqrt{2\pi}}$.

e. Let $g(\sigma) = \frac{e^{\sigma^2/2}}{\sigma\sqrt{2\pi}}$. Then $g'(\sigma) = \frac{e^{\frac{\sigma^2}{2}}(\sigma^2-1)}{\sqrt{2\pi}\sigma^2}$, so there is a critical point at $\sigma = 1$. Note that g is decreasing on $(0, 1)$ and increasing on $(1, \infty)$, so g has a minimum at $\sigma = 1$.

7.7.97 The objective function to be minimized is the average number of tests required, given by $A(x) = N\left(1 - q^x + \frac{1}{x}\right)$ where x is the group size, $N = 10{,}000$ and $q = 0.95$. We may assume that $1 \leq x \leq 10{,}000$. The critical points of this function satisfy $A'(x) = N\left(-(\ln q)q^x - \frac{1}{x^2}\right) = 0$, which is equivalent to the equation $(0.95)^{-x} + \ln(0.95)x^2 = 0$. Using a numerical solver, we find that this equation has one root between 5 and 6, and one between 132 and 133. By the First Derivative Test, we see that the smaller of these roots gives a local minimum and the larger a local maximum. Therefore the minimum value of $A(x)$ for $1 \leq x \leq 10{,}000$ occurs either at the smaller root or at the endpoint 10,000. The group size x must be an integer, so the possible optimal choices are $x = 5, 6$ and 10,000; comparing the value of $A(x)$ at these points shows that $x = 5$ is the optimal group size.

Chapter 8

8.1 Integration by Parts

8.1.1 It is based on the product rule. In fact, the rule can be obtained by writing down the product rule, then integrating both sides and rearranging the terms in the result.

8.1.3 It is generally a good idea to let u be something easy to differentiate, keeping in mind that whatever is left for dv is something which you will need to be able to integrate. In this case, it would be prudent to let $u = x^n$ and $dv = \cos(ax)\,dx$. Note that differentiating x^n results in something simpler (lower degree,) while integrating it make it more complicated (higher degree.) However, differentiating or integrating $\cos(ax)$ yields essentially the same thing (a constant times the sine function).

8.1.5 Since integration by parts is related to the product rule, it is often useful when the integrand is the product of two or more functions.

8.1.7 Let $u = x$ and $dv = \cos x\,dx$. Then $du = dx$ and $v = \sin x$. Then $\int x \cos x\,dx = x \sin x - \int \sin x\,dx = x \sin x + \cos x + C$.

8.1.9 Let $u = t$ and $dv = e^t\,dt$. Then $du = dt$ and $v = e^t$. Then $\int t e^t\,dt = t e^t - \int e^t\,dt = t e^t - e^t + C$.

8.1.11 Let $u = x^2$ and $dv = \sin 2x\,dx$. Then $du = 2x\,dx$ and $v = \frac{-1}{2}\cos 2x$. Then $\int x^2 \sin 2x\,dx = \frac{-1}{2}x^2 \cos 2x + \int x \cos 2x\,dx$.

Now we consider computing this last term $\int x \cos 2x\,dx$ as a new problem. Let $u = x$ and $dv = \cos 2x\,dx$. Then $du = dx$ and $v = \frac{1}{2}\sin 2x$. So $\int x \cos 2x\,dx = \frac{1}{2}x \sin 2x - \frac{1}{2}\int \sin 2x\,dx = \frac{1}{2}x \sin 2x + \frac{1}{4}\cos 2x + C$.

Combining these results we have $\int x^2 \sin 2x\,dx = \frac{-1}{2}x^2 \cos 2x + \frac{1}{2}x \sin 2x + \frac{1}{4}\cos 2x + C$.

8.1.13 Let $u = x^2$ and $dv = e^{4x}\,dx$. Then $du = 2x\,dx$ and $v = \frac{e^{4x}}{4}$. Then $\int x^2 e^{4x}\,dx = \frac{1}{4}x^2 e^{4x} - \frac{1}{2}\int x e^{4x}\,dx$.

Now we consider computing this last integral $\int x e^{4x}\,dx$ as a new problem. Let $u = x$ and $dv = e^{4x}\,dx$. Then $du = dx$ and $v = \frac{e^{4x}}{4}$. Then $\int x e^{4x}\,dx = \frac{1}{4}x e^{4x} - \frac{1}{4}\int e^{4x}\,dx = \frac{1}{4}x e^{4x} - \frac{1}{16}e^{4x} + C$. Combining these results gives $\int x^2 e^{4x}\,dx = \frac{1}{4}x^2 e^{4x} - \frac{1}{8}x e^{4x} + \frac{1}{32}e^{4x} + C$.

8.1.15 Let $u = \ln x$ and $dv = x^2\,dx$. Then $du = \frac{1}{x}\,dx$ and $v = \frac{x^3}{3}$. Then $\int x^2 \ln x\,dx = \frac{x^3}{3}\ln x - \frac{1}{3}\int x^2\,dx = \frac{x^3}{3}\ln x - \frac{x^3}{9} + C$.

8.1.17 Let $u = \ln x$ and $dv = x^{-10}\,dx$. Then $du = \frac{1}{x}\,dx$ and $v = \frac{-1}{9}x^{-9}$. Then $\int \frac{\ln x}{x^{10}}\,dx = \frac{-1}{9x^9}\ln x + \frac{1}{9}\int x^{-10}\,dx = \frac{-1}{9x^9}\ln x + \frac{-1}{81x^9} + C$.

8.1.19 Let $u = \tan^{-1}(x)$ and $dv = dx$. Then $du = \frac{1}{1+x^2}\,dx$ and $v = x$. Then $\int \tan^{-1}(x)\,dx = x\tan^{-1}(x) - \int \frac{x}{1+x^2}\,dx = x\tan^{-1}(x) - \frac{1}{2}\ln(1+x^2) + C$. The fact that $-\int \frac{x}{1+x^2}\,dx = -\frac{1}{2}\ln(1+x^2) + C$ follows from the ordinary substitution $u = 1 + x^2$.

8.1.21 $\int x\sin x\cos x\,dx = \frac{1}{2}\int x\cdot(2\sin x\cos x)\,dx = \frac{1}{2}\int x\sin 2x\,dx$. Now using the result of problem 8, we have $\int x\sin x\cos x\,dx = \frac{1}{2}\cdot\left(\frac{-1}{2}x\cos 2x + \frac{1}{4}\sin 2x\right) + C = \frac{-1}{4}x\cos 2x + \frac{1}{8}\sin 2x + C$.

8.1.23 Let $u = \cos x$ and $dv = e^x\,dx$. Then $du = -\sin x\,dx$ and $v = e^x$. We have $\int e^x\cos x\,dx = e^x\cos x + \int e^x\sin x\,dx$.

Now in order to compute the integral which comprises this last term, we let $u = \sin x$ and $dv = e^x\,dx$. Then $du = \cos x\,dx$ and $v = e^x$. Thus, $\int e^x\sin x\,dx = e^x\sin x - \int e^x\cos x\,dx$.

Putting these results together gives

$$\int e^x\cos x\,dx = e^x\cos x + e^x\sin x - \int e^x\cos x\,dx$$

$$2\int e^x\cos x\,dx = e^x(\cos x + \sin x) + C$$

$$\int e^x\cos x\,dx = \frac{e^x}{2}(\cos x + \sin x) + C.$$

8.1.25 Let $u = \sin 4x$ and $dv = e^{-x}\,dx$. Then $du = 4\cos 4x\,dx$ and $v = -e^{-x}$. We have $\int e^{-x}\sin 4x\,dx = -e^{-x}\sin 4x + 4\int e^{-x}\cos 4x\,dx$.

Now in order to compute the integral which comprises this last term, we let $u = \cos 4x$ and $dv = e^{-x}\,dx$. Then $du = -4\sin 4x\,dx$ and $v = -e^{-x}$. Thus, $\int e^{-x}\cos 4x\,dx = -e^{-x}\cos 4x - 4\int e^{-x}\sin 4x\,dx$.

Putting these results together gives

$$\int e^{-x}\sin 4x\,dx = -e^{-x}\sin 4x - 4e^{-x}\cos 4x - 16\int e^{-x}\sin 4x\,dx$$

$$17\int e^{-x}\sin 4x\,dx = -e^{-x}\sin 4x - 4e^{-x}\cos 4x + C$$

$$\int e^{-x}\sin 4x\,dx = \frac{-e^{-x}}{17}(\sin 4x + 4\cos 4x) + C.$$

8.1.27 Let $u = t^3$ and $dv = e^{-t}\,dt$. Then $du = 3t^2\,dt$ and $v = -e^{-t}$. Then $\int t^3 e^{-t}\,dt = -t^3 e^{-t} + 3\int t^2 e^{-t}\,dt$. In order to compute the integral which comprises this last term, we let $u = t^2$ and $dv = e^{-t}\,dt$. Then $du = 2t\,dt$ and $v = -e^{-t}$. Then $\int t^2 e^{-t}\,dt = -t^2 e^{-t} + 2\int te^{-t}\,dt$. To compute this last integral, we let $u = t$ and $dv = e^{-t}\,dt$. Then $\int te^{-t}\,dt = -te^{-t} + \int e^{-t}\,dt = -te^{-t} - e^{-t} + C$.

Putting these results together, we obtain

$$\int t^3 e^{-t}\,dt = -t^3 e^{-t} - 3t^2 e^{-t} - 6te^{-t} - 6e^{-t} + C.$$

8.1.29 Let $u = x$ and $dv = \sin x \, dx$. Then $du = dx$ and $v = -\cos x$. Then $\displaystyle\int_0^\pi x \sin x \, dx = -x \cos x \big|_0^\pi + \displaystyle\int_0^\pi \cos x \, dx = \pi - 0 + \sin x \big|_0^\pi = \pi - 0 + 0 - 0 = \pi$.

8.1.31 Let $u = x$ and $dv = \cos 2x \, dx$. Then $du = dx$ and $v = \frac{1}{2} \sin 2x$. Then $\displaystyle\int_0^{\pi/2} x \cos 2x \, dx = \frac{1}{2} x \sin 2x \big|_0^{\pi/2} - \frac{1}{2} \int_0^{\pi/2} \sin 2x \, dx = 0 - \left(\frac{1}{2} \cdot \frac{(-\cos 2x)}{2} \right) \Big|_0^{\pi/2} = \frac{-1}{4} - \frac{1}{4} = \frac{-1}{2}$.

8.1.33 Let $u = \ln x$ and $dv = x^2 \, dx$. Then $du = \frac{1}{x} \, dx$ and $v = \frac{x^3}{3}$. Then $\displaystyle\int_1^{e^2} x^2 \ln x \, dx = \frac{1}{3} x^3 \ln x \big|_1^{e^2} - \frac{1}{3} \int_1^{e^2} x^2 \, dx = \frac{2}{3} e^6 - \left(\frac{1}{9} x^3 \right) \Big|_1^{e^2} = \frac{2}{3} e^6 - \frac{1}{9} \left(e^6 - 1 \right) = \frac{5}{9} e^6 + \frac{1}{9}$.

8.1.35 By problem 18, $\displaystyle\int \sin^{-1}(x) \, dx = x \sin^{-1}(x) - \int \frac{x}{\sqrt{1 - x^2}} \, dx = x \sin^{-1}(x) + \sqrt{1 - x^2}$. Thus,

$$\int_{1/2}^{\sqrt{3}/2} \sin^{-1}(x) \, dx = \left(x \sin^{-1}(x) + \sqrt{1 - x^2} \right) \Big|_{1/2}^{\sqrt{3}/2}$$

$$= \left(\frac{\sqrt{3}}{2} \cdot \frac{\pi}{3} + \frac{1}{2} \right) - \left(\frac{1}{2} \cdot \frac{\pi}{6} + \frac{\sqrt{3}}{2} \right)$$

$$= \frac{\pi}{6} \left(\sqrt{3} - \frac{1}{2} \right) + \frac{1}{2} \left(1 - \sqrt{3} \right).$$

8.1.37 Using shells, we have $\frac{V}{2\pi} = \displaystyle\int_0^{\ln 2} x e^{-x} \, dx$. Let $u = x$ and $dv = e^{-x} \, dx$, so that $du = dx$ and $v = -e^{-x}$. Then $\frac{V}{2\pi} = \left(-x e^{-x} \right) \big|_0^{\ln 2} + \displaystyle\int_0^{\ln 2} e^{-x} \, dx = \frac{-1}{2} \ln 2 - \left(e^{-x} \right) \big|_0^{\ln 2} = \frac{-\ln 2}{2} - \left(\frac{1}{2} - 1 \right) = \frac{1}{2} \left(1 - \ln 2 \right)$. Thus $V = \pi(1 - \ln 2)$.

8.1.39 Using disks, we have $\frac{V}{\pi} = \displaystyle\int_1^{e^2} x^2 \ln^2 x \, dx$. By problem 26, we have $\displaystyle\int x^2 \ln^2 x \, dx = \frac{1}{3} x^3 \ln^2 x - \frac{2}{9} x^3 \ln x + \frac{2}{27} x^3 + C$. Thus, $\frac{V}{\pi} = \left(\frac{1}{3} x^3 \ln^2 x - \frac{2}{9} x^3 \ln x + \frac{2}{27} x^3 \right) \Big|_1^{e^2} = \left(\frac{4}{3} e^6 - \frac{4}{9} e^6 + \frac{2}{27} e^6 \right) - \left(\frac{2}{27} \right) = \frac{26}{27} e^6 - \frac{2}{27}$. Thus, $V = \frac{\pi}{27} \left(26 e^6 - 2 \right)$.

8.1.41

a. False. For example, suppose $u = x$ and $dv = x \, dx$. Then $\displaystyle\int u v' \, dx = \int x^2 \, dx = \frac{x^3}{3} + C$, but $\displaystyle\int u \, dx \int v' \, dx = \left(\int x \, dx \right)^2 = \left(\frac{x^2}{2} + C \right)^2$.

b. True. This is one way to write the integration by parts formula.

8.1.43 Let $u = x^n$ and $dv = \cos ax \, dx$ Then $du = n x^{n-1} \, dx$ and $v = \frac{\sin ax}{a}$. Then $\displaystyle\int x^n \cos ax \, dx = \frac{x^n \sin ax}{a} - \frac{n}{a} \int x^{n-1} \sin ax \, dx$.

8.1.45 Let $u = \ln^n x$ and $dv = dx$ Then $du = \frac{n \ln^{n-1}(x)}{x} dx$ and $v = x$. Then $\int \ln^n(x)\, dx = x \ln^n x - n \int \ln^{n-1}(x)\, dx$.

8.1.47

$$\int x^2 \cos 5x\, dx = \frac{x^2 \sin 5x}{5} - \frac{2}{5} \int x \sin 5x\, dx$$

$$= \frac{x^2 \sin 5x}{5} - \frac{2}{5}\left(-\frac{x \cos 5x}{5} + \frac{1}{5} \int \cos 5x\, dx \right)$$

$$= \frac{1}{5}\left(x^2 \sin 5x + \frac{2}{5} x \cos 5x - \frac{2}{25} \sin 5x \right) + C.$$

8.1.49

$$\int \ln^4 x\, dx = x \ln^4 x - 4 \int \ln^3 x\, dx$$

$$= x \ln^4 x - 4\left(x \ln^3 x - 3 \int \ln^2 x\, dx \right)$$

$$= x \ln^4 x - 4x \ln^3 x + 12\left(x \ln^2 x - 2 \int \ln x\, dx \right)$$

$$= x \ln^4 x - 4x \ln^3 x + 12x \ln^2 x - 24\left(x \ln x - x \right) + C.$$

8.1.51 Let $u = \tan x + 2$, so that $du = \sec^2 x\, dx$. Then $\int \sec^2 x \ln(\tan x + 2)\, dx = \int \ln u\, du = u \ln u - u + C. = (\tan x + 2)\ln(\tan x + 2) - \tan x + C.$

8.1.53 Using the change of base formula, we have $\int \log_b x\, dx = \int \frac{\ln x}{\ln b}\, dx = \frac{1}{\ln b}\left(x \ln x - x \right) + C.$

8.1.55 Let $z = \sqrt{x}$, so that $dz = \frac{1}{2\sqrt{x}}\, dx$. Substituting yields $2 \int \frac{\sqrt{x} \cos \sqrt{x}}{2\sqrt{x}}\, dx = 2 \int z \cos z\, dz$. Now let $u = z$ and $dv = \cos z\, dz$. then $du = dz$ and $v = \sin z$. Then by Integration by Parts, we have $2 \int z \cos z\, dz = 2\left(z \sin z - \int \sin z\, dz \right) = 2z \sin z + 2 \cos z + C$. Thus, the original given integral is equal to $2(\sqrt{x} \sin \sqrt{x} + \cos \sqrt{x}) + C.$

8.1.57 By the Fundamental Theorem, $f'(x) = \sqrt{\ln^2 x - 1}$. So the arc length is $\int_e^{e^3} \sqrt{1 + (f'(x))^2}\, dx = \int_e^{e^3} |\ln x|\, dx = (x \ln x - x)\big|_e^{e^3} = 3e^3 - e^3 - (e - e) = 2e^3.$

8.1.59 Using shells, we have $\frac{V}{2\pi} = \int_0^{\pi/2} x \cos x\, dx$. Let $u = x$ and $dv = \cos x\, dx$, so that $du = dx$ and $v = \sin x$. We have $\frac{V}{2\pi} = x \sin x\big|_0^{\pi/2} - \int_0^{\pi/2} \sin x\, dx = \frac{\pi}{2} - 1$.
   Thus, $V = \pi(\pi - 2)$.

8.1.61 Let V_1 be the volume generated when R is revolved about the x-axis, and V_2 the volume generated when R is revolved about the y-axis.
   Using disks, we have

$$\frac{V_1}{\pi} = \int_0^\pi \sin^2 x\, dx = \frac{1}{2} \int_0^\pi (1 - \cos 2x)\, dx = \frac{1}{2}\left(\pi - \frac{1}{2} \int_0^\pi 2 \cos 2x\, dx \right) = \frac{1}{2}\left(\pi - \frac{1}{2} \int_0^{2\pi} \cos u\, du \right) = \frac{\pi}{2},$$

where the ordinary substitution $u = 2x$ was made. So $V_1 = \frac{\pi^2}{2}$.

Using shells to compute V_2, we have $\frac{V_2}{2\pi} = \int_0^\pi x \sin x \, dx$. Letting $u = x$ and $dv = \sin x \, dx$, we have $du = dx$ and $v = -\cos x$. Thus, $\frac{V_2}{2\pi} = -x \cos x \big|_0^\pi + \int_0^\pi \cos x \, dx = \pi$. Thus, $V_2 = 2\pi^2$, and $V_2 > V_1$.

8.1.63

a. Let $u = x$ and $dv = f'(x) \, dx$. Then $du = dx$ and $v = f(x)$. So $\int x f'(x) \, dx = x f(x) - \int f(x) \, dx$.

b. Letting $f'(x) = e^{3x}$ we have $\int x e^{3x} \, dx = \frac{1}{3} x e^{3x} - \frac{1}{9} e^{3x} + C$.

8.1.65 Let $u = \sec x$ and $dv = \sec^2 x \, dx$, so that $du = \sec x \tan x \, dx$ and $v = \tan x$. Then $\int \sec^3 x \, dx = \sec x \tan x - \int \sec x \tan^2 x \, dx = \sec x \tan x - \int \sec x \left(\sec^2 x - 1 \right) \, dx = \sec x \tan x - \int \sec^3 x \, dx + \int \sec x \, dx$. Thus $2 \int \sec^3 x \, dx = \sec x \tan x + \int \sec x \, dx$, so $\int \sec^3 x \, dx = \frac{1}{2} \sec x \tan x + \frac{1}{2} \int \sec x \, dx$.

8.1.67

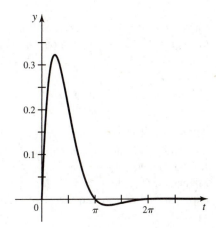

a. We have $s(t) = 0$ when $\sin t = 0$, which occurs for $t = k\pi$, where k is an integer.

b. This is given by $\frac{1}{\pi} \int_0^\pi e^{-t} \sin t \, dt$. Using the previous problem, this is equal to $\frac{1}{\pi} e^{-t} \cdot \frac{-\sin t - \cos t}{2} \Big|_0^\pi = \frac{1}{2\pi} (e^{-\pi} + 1)$.

c. This is given by $\frac{1}{\pi} \int_{n\pi}^{(n+1)\pi} e^{-t} \sin t \, dt$. Using the previous problem, this is equal to

$$\frac{1}{\pi} e^{-t} \cdot \frac{-\sin t - \cos t}{2} \bigg|_{n\pi}^{(n+1)\pi} = \frac{1}{2\pi} \left(e^{-(n+1)\pi} (-\sin(n+1)\pi - \cos(n+1)\pi) - e^{-n\pi} (-\sin n\pi - \cos n\pi) \right)$$

$$= \frac{1}{2\pi} \left(-e^{-(n+1)\pi} \cos(n+1)\pi + e^{-n\pi} \cos n\pi \right) = \frac{e^{-n\pi}}{2\pi} \left(\cos n\pi - e^{-\pi} \cos(n+1)\pi \right)$$

$$= \frac{e^{-n\pi}}{2\pi} \left((-1)^n - e^{-\pi}(-1)^{(n+1)} \right) = (-1)^n \frac{e^{-n\pi}}{2\pi} \left(1 + e^{-\pi} \right).$$

d. Each a_i is $e^{-\pi}$ times a_{i-1}.

8.1.69 We may think of A as $\int_a^b u\,dv$ and B as $\int_a^b v\,du$. Also, $pr = f(a)g(a)$ and $qs = f(b)g(b)$. So we have

$$\int_a^b u\,dv + \int_a^b v\,du = A + B = qs - pr = f(b)g(b) - f(a)g(a) = uv\big|_a^b.$$

Thus $\int_a^b u\,dv = uv\big|_a^b - \int_a^b v\,du$.

8.1.71

a. To compute I_1, we let $u = -x^2$, so that $du = -2x\,dx$. Then an ordinary substitution yield $\frac{-1}{2}\int e^u\,du = \frac{-1}{2}e^u + C = \frac{-1}{2}e^{-x^2} + C$.

b. To compute I_3, we let $u = x^2$ and $dv = xe^{-x^2}\,dx$, so that $du = 2x\,dx$ and $v = \frac{-1}{2}e^{-x^2}$ (by part (a)). Then $I_3 = \frac{-1}{2}x^2e^{-x^2} + \int xe^{-x^2}\,dx = \frac{-1}{2}x^2e^{-x^2} + \frac{-1}{2}e^{-x^2} + C = \frac{-1}{2}e^{-x^2}(x^2 + 1) + C$.

c. To compute I_5, we let $u = x^4$ and $dv = xe^{-x^2}\,dx$. Then $du = 4x^3\,dx$ and $v = \frac{-1}{2}e^{-x^2}$. Then $I_5 = \frac{-1}{2}x^4e^{-x^2} + 2\int x^3e^{-x^2}\,dx = \frac{-1}{2}x^4e^{-x^2} + 2I_3 = \frac{-1}{2}e^{-x^2}(x^4 + 2x^2 + 2)$.

d. Suppose that n is odd and that it is true that $I_n = \frac{-1}{2}e^{-x^2}p_{n-1}(x)$ where $p_{n-1}(x)$ is an even polynomial of degree $n-1$. We will show that I_{n+2} also has this property, so the result will follow by induction. Let $u = x^{n+1}$ and let $dv = xe^{-x^2}\,dx$, so that $du = (n+1)x^n\,dx$ and $v = \frac{-1}{2}e^{-x^2}$. Then $I_{n+2} = \frac{-1}{2}x^{n+1}e^{-x^2} + \frac{n+1}{2}I_n = \frac{-1}{2}e^{-x^2}(x^{n+1} + \frac{n+1}{2}p_{n-1}(x)) + C = \frac{-1}{2}e^{-x^2}p_{n+1}(x) + C$. Note that if $p_{n-1}(x)$ is an even polynomial of degree $n-1$ then $\frac{n+1}{2}p_{n-1}(x) + x^{n+1}$ is an even polynomial of degree $n+1$.

e. Note that $I_2 = \frac{-1}{2}xe^{-x^2} + \frac{1}{2}I_0$. (Using a similar technique to that above.) Now if I_2 were expressible in terms of elementary functions, then I_0 would be as well, but we are given that it isn't. Similarly, we can express I_{2k} in terms of I_{2k-2} using Integration by Parts, and if any of these were expressible in terms of elementary functions, then the even numbered one below it would be. So the inability to express I_0 that way implies the inability to express I_2 that way, which implies the inability to express I_4 that way, and so on.

8.2   Trigonometric Integrals

8.2.1 The half-angle identities for sine and cosine:

$$\sin^2 x = \frac{1 - \cos 2x}{2} \quad \text{and} \quad \cos^2 x = \frac{1 + \cos 2x}{2},$$

which are conjugates.

8.2.3 To integrate $\sin^3 x$, write $\sin^3 x = \sin x \sin^2 x = \sin x(1 - \cos^2 x)$, and let $u = \cos x$ so that $du = -\sin x\,dx$.

8.2.5 A reduction formula is a recursive formula involving integrals. Using it, one can rewrite an integral of a certain type in a simpler form – which can then perhaps be evaluated or further reduced.

8.2.7 One would compute this integral by letting $u = \tan x$, so that $du = \sec^2 x\,dx$. This substitution leads to the integral $\int u^{10}\,du$, which can easily be evaluated.

8.2.9 $\int \sin^2 x \, dx = \frac{1}{2} \int (1 - \cos 2x) \, dx = \frac{1}{2} \left(x - \frac{1}{2} \sin 2x \right) + C.$

8.2.11 $\int \sin^5 x \, dx = \int (\sin^2 x)^2 \sin x \, dx = \int (1 - \cos^2 x)^2 \sin x \, dx.$ Let $u = \cos x$ so that $du = -\sin x \, dx.$ Substituting yields $-\int (1 - u^2)^2 \, du = \int (-u^4 + 2u^2 - 1) \, du = \frac{-u^5}{5} + \frac{2u^3}{3} - u + C = \frac{-\cos^5 x}{5} + \frac{2\cos^3 x}{3} - \cos x + C.$

8.2.13 $\int \sin^2 x \cos^2 x \, dx = \int \left(\frac{1 - \cos 2x}{2} \right) \left(\frac{1 + \cos 2x}{2} \right) dx = \frac{1}{4} \int 1 - \cos^2 2x \, dx = \frac{1}{4} \int \left(1 - \frac{1 + \cos 4}{2} \right) dx$
$= \frac{1}{4} \int \left(\frac{1}{2} - \frac{1}{2} \cos 4x \right) dx = \frac{1}{4} \left(\frac{x}{2} - \frac{\sin 4x}{8} \right) + C.$

8.2.15 $\int \sin^5 x \cos^{-2} x \, dx = \int (\sin x) \left(\frac{(1 - \cos^2 x)^2}{\cos^2 x} \right) dx.$ Let $u = \cos x$ so that $du = -\sin x \, dx.$ Substituting yields $-\int \frac{(1 - u^2)^2}{u^2} \, du = \int -u^{-2} + 2 - u^2 \, du = \frac{1}{u} + 2u - \frac{u^3}{3} + C = \sec x + 2 \cos x - \frac{\cos^3 x}{3} + C.$

8.2.17 $\int \sin^2 x \cos^4 x \, dx = \int \left(\frac{1 - \cos 2x}{2} \right) \left(\frac{1 + \cos 2x}{2} \right)^2 dx = \frac{1}{8} \int (1 - \cos^2 2x)(1 + \cos 2x) \, dx = \frac{1}{8} \int 1 + \cos 2x - \cos^2 2x - \cos^3 2x \, dx = \frac{1}{8} \int 1 + \cos 2x - \frac{1 + \cos 4x}{2} - \cos^3 2x \, dx = \frac{1}{8} \int \frac{1}{2} + \cos 2x - \frac{1}{2} \cos 4x \, dx - \frac{1}{8} \int (1 - \sin^2 2x) \cos 2x \, dx = \frac{x}{16} + \frac{\sin 2x}{16} - \frac{\sin 4x}{64} - \frac{1}{8} \int (1 - \sin^2 2x) \cos 2x \, dx.$ To compute this last integral, we let $u = \sin 2x,$ so that $du = 2 \cos 2x \, dx.$ Then $\int (1 - \sin^2 2x) \cos 2x \, dx = \frac{1}{2} \int (1 - u^2) \, du = \frac{1}{2} \left(u - \frac{u^3}{3} \right) + C = \frac{1}{2} \left(\sin 2x - \frac{\sin^3 2x}{3} \right) + C.$

Thus, our original given integral is equal to $\frac{x}{16} + \frac{\sin 2x}{16} - \frac{\sin 4x}{64} - \frac{1}{8} \left(\frac{1}{2} \left(\sin 2x - \frac{\sin^3 2x}{3} \right) \right) + C = \frac{x}{16} - \frac{\sin 4x}{64} + \frac{\sin^3 2x}{48} + C.$

8.2.19 $\int \tan^2 x \, dx = \int (\sec^2 x - 1) \, dx = \tan x - x + C.$

8.2.21 $\int \tan^3 4x \, dx = \int (\tan 4x)(\sec^2 4x - 1) \, dx = \int (\tan 4x) \sec^2 4x \, dx - \int \tan 4x \, dx = \int (\tan 4x) \sec^2 4x \, dx + \frac{\ln|\cos 4x|}{4} + C.$ Let $u = \tan 4x$ so that $du = 4 \sec^2 4x \, dx.$ Substituting gives $\frac{1}{4} \int u \, du + \frac{\ln|\cos 4x|}{4} + C = \frac{u^2}{8} + \frac{\ln|cos4x|}{4} + C = \frac{\tan^2 4x}{8} + \frac{\ln|\cos 4x|}{4} + C.$

8.2.23

$$\int 20 \tan^6 x \, dx = 20 \int (\tan^4 x)(\sec^2 x - 1) \, dx = 20 \int ((\tan^4 x) \sec^2 x - (\tan^2 x)(\sec^2 x - 1)) \, dx$$
$$= 20 \int (\tan^4 x \sec^2 x - \tan^2 x \sec^2 x + \sec^2 x - 1) \, dx.$$

Let $u = \tan x$ so that $du = \sec^2 x \, dx.$ We have $20 \left(\int (u^4 - u^2) \, du + \tan x - x \right) + C = 4u^5 - \frac{20u^3}{3} + 20 \tan x - 20x + C = 4 \tan^5 x - \frac{20 \tan^3 x}{3} + 20 \tan x - 20x + C.$

8.2.25 Let $u = \tan x$ so that $du = \sec^2 x\, dx$. Then

$$\int \sec^2 x \tan^{1/2} x\, dx = \int u^{1/2}\, du = \frac{2}{3} u^{3/2} + C = \frac{2}{3} \tan^{3/2} x + C.$$

8.2.27 $\int \dfrac{\csc^4 x}{\cot^2 x}\, dx = \int (\csc^2 x)\left(\dfrac{\cot^2 x + 1}{\cot^2 x}\right)\, dx$. Let $u = \cot x$ so that $du = -\csc^2 x\, dx$. Substituting

gives $-\int \dfrac{u^2 + 1}{u^2}\, du = \int -1 - u^{-2}\, du = -u + \dfrac{1}{u} + C = -\cot x + \tan x + C$.

8.2.29 $\displaystyle\int_0^{\pi/4} \sec^4 \theta\, d\theta = \int_0^{\pi/4} (\sec^2 \theta)(1 + \tan^2 \theta)\, d\theta$. Let $u = \tan \theta$ so that $du = \sec^2 \theta\, d\theta$. Note that when

$\theta = 0$ we have $u = 0$ and when $\theta = \frac{\pi}{4}$ we have $u = 1$. So the original integral is equal to $\displaystyle\int_0^1 (1 + u^2)\, du =$

$u + \dfrac{u^3}{3}\Big|_0^1 = 1 + \dfrac{1}{3} = \dfrac{4}{3}$.

8.2.31 $\displaystyle\int_{\pi/6}^{\pi/3} \cot^3 \theta\, d\theta = \int_{\pi/6}^{\pi/3} (\cot \theta)(\csc^2 \theta - 1)\, d\theta = \int_{\pi/6}^{\pi/3} \cot \theta \csc^2 \theta\, d\theta - \int_{\pi/6}^{\pi/3} \dfrac{\cos \theta}{\sin \theta}\, d\theta$. For the first

integral, let $u = \cot \theta$ so that $du = -\csc^2 \theta\, d\theta$. For the second integral, let $w = \sin \theta$ so that $dw = \cos \theta\, d\theta$.
Substituting gives

$$-\int_{\sqrt{3}}^{1/\sqrt{3}} u\, du - \int_{1/2}^{\sqrt{3}/2} \frac{1}{w}\, dw = \frac{-u^2}{2}\Big|_{\sqrt{3}}^{1/\sqrt{3}} - \ln w\Big|_{1/2}^{\sqrt{3}/2} = \frac{-1}{2}\left(\frac{1}{3} - 3\right) - \left(\ln \sqrt{3} - \ln 2 - \ln 1 + \ln 2\right) = \frac{4}{3} - \frac{\ln 3}{2}.$$

8.2.33

a. True. We have $\displaystyle\int_0^\pi \cos^{2m+1} x\, dx = \int_0^\pi (\cos^2 x)^m \cos x\, dx = \int_0^\pi (1 - \sin^2 x)^m \cos x\, dx$. Let $u = \sin x$ so

that $du = \cos x\, dx$. Substituting yields $\displaystyle\int_0^0 (1 - u^2)^m\, du = 0$.

b. False. For example, suppose $m = 1$. Then $\displaystyle\int_0^\pi \sin x\, dx = -\cos x\Big|_0^\pi = -(-1 - 1) = 2 \neq 0$.

8.2.35 $\int \sec x\, dx = \int (\sec x)\left(\dfrac{\sec x + \tan x}{\sec x + \tan x}\right)\, dx = \int \dfrac{\sec^2 x + \sec x \tan x}{\sec x + \tan x}\, dx$. Let $u = \sec x + \tan x$ so that

$du = \sec^2 x + \sec x \tan x\, dx$. Substituting then yields $\int \dfrac{1}{u}\, du = \ln|u| + C = \ln|\sec x + \tan x| + C$.

8.2.37 Let $u = ax$ so that $du = a\, dx$. Then $\dfrac{1}{a}\int \tan u\, du = -\dfrac{1}{a} \ln|\cos ax| + C$ and $\dfrac{1}{a}\int \sec u\, du = \dfrac{1}{a} \ln|\sec ax + \tan ax| + C$.

8.2.39 $A = \displaystyle\int_0^{\pi/4} \sec x - \tan x\, dx = \ln|\sec x + \tan x| + \ln|\cos x|\Big|_0^{\pi/4} = \ln\left(\sqrt{2} + 1\right) + \ln\left(\dfrac{\sqrt{2}}{2}\right) - (0 + 0) =$
$\ln\left(1 + \sqrt{2}/2\right)$.

8.2.41 Let $u = \ln \theta$ so that $du = \dfrac{1}{\theta}\, d\theta$. Substituting yields $\int \sec^4 u\, du = \int (\sec^2 u)(1 + \tan^2 u)\, du$. Let

$w = \tan u$ so that $dw = \sec^2 u\, du$. Substituting again gives $\int (1 + w^2)\, dw = w + \dfrac{w^3}{3} + C = \tan(\ln(\theta)) + \dfrac{\tan^3(\ln(\theta))}{3} + C$.

8.2.43 $\displaystyle\int_{-\pi/3}^{\pi/3} \sqrt{\sec^2\theta - 1}\, d\theta = 2\int_0^{\pi/3} \sqrt{\sec^2\theta - 1}\, d\theta = 2\int_0^{\pi/3} \tan\theta\, d\theta = 2\left(-\ln|\cos\theta|\big|_0^{\pi/3}\right) = -2\ln(1/2)+$
$2\ln(1) = 2\ln(2)$.

8.2.45 $\displaystyle\int_0^{\pi} (1 - \cos 2x)^{3/2}\, dx = \int_0^{\pi} (2\sin^2 x)^{3/2}\, dx = 2\sqrt{2}\int_0^{\pi} \sin^3 x\, dx = 2\sqrt{2}\int_0^{\pi} (\sin x)(1 - \cos^2 x)\, dx$. Let

$u = \cos x$ so that $du = -\sin x\, dx$. Substituting yields $-2\sqrt{2}\displaystyle\int_1^{-1} (1 - u^2)\, du = 2\sqrt{2}\int_{-1}^{1} (1 - u^2)\, du =$

$4\sqrt{2}\displaystyle\int_0^1 (1 - u^2)\, du = 4\sqrt{2}\left(u - \frac{u^3}{3}\Big|_0^1\right) = \frac{8\sqrt{2}}{3}$.

8.2.47 $\displaystyle\int_0^{\pi/2} \sqrt{1 - \cos 2x}\, dx = \sqrt{2}\int_0^{\pi/2} \sin x\, dx = -\sqrt{2}\cos x\Big|_0^{\pi/2} = \sqrt{2}$.

8.2.49 $\displaystyle\int_0^{\pi/4} (1 + \cos 4x)^{3/2}\, dx = \int_0^{\pi/4} (2\cos^2 2x)^{3/2}\, dx = 2\sqrt{2}\int_0^{\pi/4} \cos^3 2x\, dx = 2\sqrt{2}\int_0^{\pi/4} (\cos 2x)(1 -$

$\sin^2 2x)\, dx$. Let $u = \sin 2x$ so that $du = 2\cos 2x\, dx$. Substituting gives $\sqrt{2}\displaystyle\int_0^1 (1 - u^2)\, du = \sqrt{2}\left(u - \frac{u^3}{3}\Big|_0^1\right) =$

$\dfrac{2\sqrt{2}}{3}$.

8.2.51 If $y = \ln(\cos x)$ then $\frac{dy}{dx} = -\tan x$. Thus, $L = \displaystyle\int_0^{\pi/4} \sqrt{1 + (-\tan x)^2}\, dx = \int_0^{\pi/4} \sec x\, dx =$
$\left(\ln|\sec x + \tan x|\big|_0^{\pi/4}\right) = \ln(\sqrt{2} + 1)$.

8.2.53 For $n \neq 1$, $\displaystyle\int \tan^n x\, dx = \int (\tan^{n-2} x)(\sec^2 x - 1)\, dx = \int \tan^{n-2} x \sec^2 x\, dx - \int \tan^{n-2} x\, dx$. Let

$u = \tan x$ so that $du = \sec^2 x\, dx$. Then substituting in the first of these last two integrals yields $\displaystyle\int u^{n-2}\, du -$

$\displaystyle\int \tan^{n-2} x\, dx = \frac{u^{n-1}}{n-1} - \int \tan^{n-2} x\, dx = \frac{\tan^{n-1} x}{n-1} - \int \tan^{n-2} x\, dx$.

Thus $\displaystyle\int_0^{\pi/4} \tan^3 x\, dx = \frac{\tan^2 x}{2}\Big|_0^{\pi/4} - \int_0^{\pi/4} \tan x\, dx = \frac{1}{2} + \ln|\cos x|\big|_0^{\pi/4} = \frac{1}{2} - \frac{\ln 2}{2}$.

8.2.55 $\displaystyle\int \sin 3x \cos 7x\, dx = \frac{1}{2}\left(\int \sin(-4x)\, dx + \int \sin 10x\, dx\right) = \frac{1}{2}\left(\frac{\cos(-4x)}{4} - \frac{\cos 10x}{10}\right) + C = \frac{\cos 4x}{8} -$
$\dfrac{\cos 10x}{20} + C$.

8.2.57 $\displaystyle\int \sin 3x \sin 2x\, dx = \frac{1}{2}\left(\int \cos x\, dx - \int \cos 5x\, dx\right) = \frac{\sin x}{2} - \frac{\sin 5x}{10} + C$.

8.2.59

  a. $\displaystyle\int_0^{\pi} \sin mx \sin nx\, dx = \frac{1}{2}\left(\int_0^{\pi} \cos(m-n)x\, dx - \int_0^{\pi} \cos(m+n)x\, dx\right) =$
  $\dfrac{1}{2}\left(\dfrac{1}{m-n}\displaystyle\int_0^{(m-n)\pi} \cos u\, du - \dfrac{1}{m+n}\int_0^{(m+n)\pi} \cos v\, dv\right)$ where $u = (m-n)x$ and $v = (m+n)x$. But
  this yields $\dfrac{1}{2}\left(\dfrac{1}{m-n}\sin u\Big|_0^{(m-n)\pi} - \dfrac{1}{m+n}\sin v\Big|_0^{(m+n)\pi}\right) = \dfrac{1}{2}(0 - 0) = 0$.

  b. $\displaystyle\int_0^{\pi} \cos mx \cos nx\, dx = \frac{1}{2}\left(\int_0^{\pi} \cos(m-n)x\, dx + \int_0^{\pi} \cos(m+n)x\, dx\right) = 0$ by the previous part of this
  problem,

c. $\displaystyle\int_0^\pi \sin mx \cos nx\, dx = \frac{1}{2}\left(\int_0^\pi \sin(m-n)x\, dx + \int_0^\pi \sin(m+n)x\, dx\right) =$

$\displaystyle\frac{1}{2}\left(\frac{1}{m-n}\int_0^{(m-n)\pi}\sin u\, du + \frac{1}{m+n}\int_0^{(m+n)\pi}\sin v\, dv\right)$ where $u=(m-n)x$ and $v=(m+n)x$. This

quantity is equal to $\displaystyle\frac{-1}{2}\left(\frac{1}{m-n}\cos u\bigg|_0^{(m-n)\pi} + \frac{1}{m+n}\cos v\bigg|_0^{(m+n)\pi}\right) =$

$\displaystyle\frac{-1}{2}\left(\frac{1}{m-n}\left(\cos(m-n)\pi - 1\right) + \frac{1}{m+n}\left(\cos(m+n)\pi - 1\right)\right) =$

$\begin{cases} 0 & \text{if } m \text{ and } n \text{ are both even or both odd;} \\ \frac{1}{m-n}+\frac{1}{m+n}=\frac{2m}{m^2-n^2} & \text{otherwise.} \end{cases}$

8.2.61

a.
$\displaystyle\int_0^\pi \sin^2 x\, dx \quad = \quad \frac{1}{2}\int_0^\pi (1 \; - \; \cos 2x)\, dx \quad =$
$\displaystyle\frac{1}{2}\left(x - \frac{\sin 2x}{2}\bigg|_0^\pi\right) = \frac{\pi}{2}.$
$\displaystyle\int_0^\pi \sin^2 2x\, dx \quad = \quad \frac{1}{2}\int_0^\pi (1 \; - \; \cos 4x)\, dx \quad =$
$\displaystyle\frac{1}{2}\left(x - \frac{\sin 4x}{4}\bigg|_0^\pi\right) = \frac{\pi}{2}.$

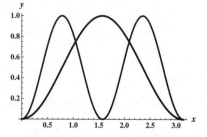

b.
$\displaystyle\int_0^\pi \sin^2 3x\, dx \quad = \quad \frac{1}{2}\int_0^\pi (1 \; - \; \cos 6x)\, dx \quad =$
$\displaystyle\frac{1}{2}\left(x - \frac{\sin 6x}{6}\bigg|_0^\pi\right) = \frac{\pi}{2}.$
$\displaystyle\int_0^\pi \sin^2 4x\, dx \quad = \quad \frac{1}{2}\int_0^\pi (1 \; - \; \cos 8x)\, dx \quad =$
$\displaystyle\frac{1}{2}\left(x - \frac{\sin 8x}{8}\bigg|_0^\pi\right) = \frac{\pi}{2}.$

c. $\displaystyle\int_0^\pi \sin^2 nx\, dx = \frac{1}{2}\int_0^\pi (1 - \cos 2nx)\, dx = \frac{1}{2}\left(x - \frac{\sin 2nx}{2n}\bigg|_0^\pi\right) = \frac{\pi}{2}.$

d. Yes. $\displaystyle\int_0^\pi \cos^2 nx\, dx = \frac{1}{2}\int_0^\pi (1 + \cos 2nx)\, dx = \frac{1}{2}\left(x + \frac{\sin 2nx}{2n}\bigg|_0^\pi\right) = \frac{\pi}{2}.$

e. Claim: The corresponding integrals are all equal to $\frac{3\pi}{8}$. Proof:

$\displaystyle\int_0^\pi \sin^4 nx\, dx = \int_0^\pi \left(\frac{1 - \cos 2nx}{2}\right)^2 dx$

$\displaystyle = \int_0^\pi \frac{1 - 2\cos 2nx + \cos^2 2nx}{4}\, dx = \int_0^\pi \frac{1}{4}\, dx - \frac{1}{2}\int_0^\pi \cos 2nx\, dx + \frac{1}{4}\int_0^\pi \cos^2 2nx\, dx$

$\displaystyle = \frac{\pi}{4} - \frac{1}{2}\left(\frac{\sin 2nx}{2n}\bigg|_0^\pi\right) + \frac{1}{4}\cdot\frac{\pi}{2} = \frac{\pi}{4} - 0 + \frac{\pi}{8} = \frac{3\pi}{8}.$

f. The proof is by strong induction. The base step ($m = 1$) follows by part (a) of this problem. Now suppose the result holds for all positive integer values less than m. Now

$$\int_0^\pi \sin^{2m} x \, dx = -\frac{\sin^{2m-1} x \cos x}{2m} \Big|_0^\pi + \frac{2m-1}{2m} \int_0^\pi \sin^{2m-2} x \, dx$$

$$= \frac{2m-1}{2m} \int_0^\pi \sin^{2(m-1)} x \, dx$$

$$= \frac{2m-1}{2m} \cdot \pi \cdot \frac{1 \cdot 3 \cdot 5 \cdots (2m-3)}{2 \cdot 4 \cdot 6 \cdots 2(m-1)} = \pi \cdot \frac{1 \cdot 3 \cdot 5 \cdots (2m-1)}{2 \cdot 4 \cdot 6 \cdots 2m}.$$

Thus the result holds for all positive numbers m. Note the use of the reduction formula from problem number 52.

To see that the same result holds for $\int_0^\pi \cos^{2m} x \, dx$ first note that $\int_{\pi/2}^\pi \sin^{2m} x \, dx = -\int_{\pi/2}^0 \sin^{2m}(\pi - u) \, du = \int_0^{\pi/2} \sin^{2m} u \, du$ (using the substitution $x = \pi - u$.)

Thus, $\int_0^\pi \cos^{2m} x \, dx = -\int_{\pi/2}^{-\pi/2} \cos^{2m}(\pi/2 - u) \, du$. (Via the substitution $x = \pi/2 - u$.) This last integral can be written as $\int_{-\pi/2}^{\pi/2} \sin^{2m} u \, du = 2 \int_0^{\pi/2} \sin^{2m} u \, du = \int_0^{\pi/2} \sin^{2m} u \, du + \int_{\pi/2}^\pi \sin^{2m} u \, du = \int_0^\pi \sin^{2m} x \, dx$.

8.3 Trigonometric Substitutions

8.3.1 This would suggest $x = 3 \sec \theta$, because then $\sqrt{x^2 - 9} = 3\sqrt{\sec^2 \theta - 1} = 3\sqrt{\tan^2 \theta} = 3 \tan \theta$, for $\theta \in [0, \pi/2) \cup (\pi/2, \pi]$.

8.3.3 This would suggest $x = 10 \sin \theta$, because then $\sqrt{100 - x^2} = 10\sqrt{1 - \sin^2 \theta} = 10\sqrt{\cos^2 \theta} = 10 \cos \theta$, for $|\theta| \leq \frac{\pi}{2}$.

8.3.5 If $x = 2 \sin \theta$ then $\frac{x^2}{4} = \sin^2 \theta = \frac{1}{\csc^2 \theta}$. Then $\cot^2 \theta = \csc^2 \theta - 1 = \frac{4}{x^2} - 1 = \frac{4-x^2}{x^2}$. So $\cot \theta = \frac{\sqrt{4-x^2}}{x}$ for $|\theta| \leq \frac{\pi}{2}$.

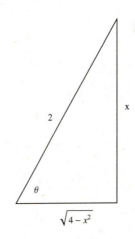

8.3.7 Let $x = 5 \sin \theta$, so that $dx = 5 \cos \theta \, d\theta$. Note that $\sqrt{25 - x^2} = 5 \cos \theta$. Then $\int_0^{5/2} \frac{1}{\sqrt{25 - x^2}} \, dx = \int_0^{\pi/6} \frac{5 \cos \theta}{5 \cos \theta} \, d\theta = \frac{\pi}{6}$.

8.3.9 Let $x = 10\sin\theta$ so that $dx = 10\cos\theta\,d\theta$. Note that $\sqrt{100 - x^2} = 10\cos\theta$. Then $\displaystyle\int_5^{10} \sqrt{100 - x^2}\,dx =$

$100\displaystyle\int_{\pi/6}^{\pi/2} \cos^2\theta\,d\theta = 50\int_{\pi/6}^{\pi/2} (1 + \cos 2\theta)\,d\theta = 50\left(\theta + \frac{\sin 2\theta}{2}\right)\Big|_{\pi/6}^{\pi/2} = 50\left(\frac{\pi}{2} + 0 - \left(\frac{\pi}{6} + \frac{\sqrt{3}}{4}\right)\right) = \frac{50\pi}{3} -$

$\dfrac{25\sqrt{3}}{2}$.

8.3.11 Let $x = 4\sin\theta$ so that $dx = 4\cos\theta\,d\theta$. Note that $\sqrt{16 - x^2} = 4\cos\theta$. Thus, $\displaystyle\int \frac{1}{\sqrt{16 - x^2}}\,dx =$

$\displaystyle\int \frac{4\cos\theta}{4\cos\theta}\,d\theta = \theta + C = \sin^{-1}\left(\frac{x}{4}\right) + C$.

8.3.13 Let $x = 3\sin\theta$ so that $dx = 3\cos\theta\,d\theta$ and $\sqrt{9 - x^2} = 3\cos\theta$. Then

$$\int \frac{\sqrt{9 - x^2}}{x}\,dx = \int \frac{3\cos\theta \cdot 3\cos\theta}{3\sin\theta}\,d\theta = 3\int \frac{1 - \sin^2\theta}{\sin\theta}\,d\theta$$

$$= 3\left(\int \csc\theta\,d\theta - \int \sin\theta\,d\theta\right) = 3\left(-\ln|\csc\theta + \cot\theta| + \cos\theta\right)$$

$$= -3\ln\left|\frac{3}{x} + \frac{\sqrt{9 - x^2}}{x}\right| + \sqrt{9 - x^2} + C.$$

8.3.15 Let $x = 8\sin\theta$ so that $dx = 8\cos\theta\,d\theta$ and $\sqrt{64 - x^2} = 8\cos\theta$. Then, $\displaystyle\int \sqrt{64 - x^2}\,dx = \int 64\cos^2\theta\,d\theta =$

$32\displaystyle\int (1 + \cos 2\theta)\,d\theta = 32\theta + 16\sin 2\theta + C = 32\theta + 32\sin\theta\cos\theta + C = 32\sin^{-1}\left(\frac{x}{8}\right) + \frac{x\sqrt{64 - x^2}}{2} + C$.

8.3.17 Le $x = 6\sin\theta$ so that $dx = 6\cos\theta\,d\theta$ and $\sqrt{36 - x^2} = 6\cos\theta$. Then $\displaystyle\int \frac{1}{\sqrt{36 - x^2}}\,dx = \int \frac{6\cos\theta}{6\cos\theta}\,d\theta =$

$\theta + C = \sin^{-1}\left(\frac{x}{6}\right) + C$.

8.3.19 Let $x = 9\sec\theta$ with $\theta \in (0, \pi/2)$. Then $dx = 9\sec\theta\tan\theta\,d\theta$ and $\sqrt{x^2 - 81} = 9\tan\theta$. Then

$\displaystyle\int \frac{1}{\sqrt{x^2 - 81}}\,dx = \int \frac{9\sec\theta\tan\theta}{9\tan\theta}\,d\theta = \int \sec\theta\,d\theta = \ln|\sec\theta + \tan\theta| + C = \ln\left|\frac{x}{9} + \frac{\sqrt{x^2 - 81}}{9}\right| + C$.

8.3.21 Let $x = \frac{\tan\theta}{2}$ so that $dx = \frac{\sec^2\theta}{2}\,d\theta$ and $\sqrt{1 + 4x^2} = \sec\theta$. Then $\displaystyle\int \frac{1}{(1 + 4x^2)^{3/2}}\,dx = \frac{1}{2}\int \frac{\sec^2\theta}{\sec^3\theta}\,d\theta =$

$\frac{1}{2}\displaystyle\int \cos\theta\,d\theta = \frac{\sin\theta}{2} + C = \frac{x}{\sqrt{1 + 4x^2}} + C$.

8.3.23 Let $x = 4\sin\theta$ so that $dx = 4\cos\theta\,d\theta$. Note that $\sqrt{16 - x^2} = 4\cos\theta$. Then $\displaystyle\int \frac{x^2}{\sqrt{16 - x^2}}\,dx =$

$\displaystyle\int \frac{16\sin^2\theta \cdot 4\cos\theta}{4\cos\theta}\,d\theta = 16\int \sin^2\theta\,d\theta = 8\int (1 - \cos 2\theta)\,d\theta = 8\left(\theta - \frac{\sin 2\theta}{2}\right) + C = 8\theta - 8\sin\theta\cos\theta + C =$

$8\sin^{-1}\left(\frac{x}{4}\right) - \frac{x\sqrt{16 - x^2}}{2} + C$.

8.3.25 Let $x = 3\sec\theta$ where $\theta \in (0, \pi/2)$. Then $dx = 3\sec\theta\tan\theta$ and $\sqrt{x^2 - 9} = 3\tan\theta$. Thus we

have $\displaystyle\int \frac{\sqrt{x^2 - 9}}{x}\,dx = \int \frac{3\sec\theta\tan\theta \cdot 3\tan\theta}{3\sec\theta}\,d\theta = 3\int \tan^2\theta\,d\theta = 3\int \sec^2\theta - 1\,d\theta = 3(\tan\theta - \theta) + C =$

$\sqrt{x^2 - 9} - 3\tan^{-1}\left(\frac{\sqrt{x^2 - 9}}{3}\right) + C$.

8.3.27 Let $x = 2\tan\theta$ so that $dx = 2\sec^2\theta\,d\theta$. Note that $\sqrt{4+x^2} = 2\sec\theta$. Then $\displaystyle\int \frac{x^2}{\sqrt{4+x^2}}\,dx =$

$\displaystyle\int \frac{4\tan^2\theta \cdot 2\sec^2\theta}{2\sec\theta}\,d\theta = 4\int \tan^2\theta\sec\theta\,d\theta = 4\int (\sec^2\theta - 1)\sec\theta\,d\theta = 4\left(\int \sec^3\theta\,d\theta - \int \sec\theta\,d\theta\right) =$

$\displaystyle 4\left(\frac{1}{2}\left(\sec\theta\tan\theta + \int \sec\theta\,d\theta\right) - \int \sec\theta\,d\theta\right) = 2\sec\theta\tan\theta - 2\int \sec\theta\,d\theta = 2\sec\theta\tan\theta - 2\ln|\sec\theta + \tan\theta| +$

$C = \dfrac{x\sqrt{4+x^2}}{2} - 2\ln\left|\dfrac{\sqrt{4+x^2}}{2} + \dfrac{x}{2}\right| + C.$

8.3.29 $\displaystyle\int \frac{1}{\sqrt{3 - 2x - x^2}}\,dx = \int \frac{1}{\sqrt{4 - (x+1)^2}}\,dx = \int \frac{1}{\sqrt{4 - u^2}}\,du$ where $u = x + 1$. Then let $u = 2\sin\theta$

so that $du = 2\cos\theta\,d\theta$. We have $\displaystyle\int \frac{2\cos\theta}{2\cos\theta}\,d\theta = \theta + C = \sin^{-1}\left(\frac{x+1}{2}\right) + C.$

8.3.31 Let $x = \frac{5}{3}\sec\theta$ where $\theta \in [0, \pi/2)$. Then $dx = \frac{5}{3}\sec\theta\tan\theta\,d\theta$ and $\sqrt{9x^2 - 25} = 5\tan\theta$. Thus,

$\displaystyle\int \frac{\sqrt{9x^2 - 25}}{x^3}\,dx = \int \frac{5\tan\theta \cdot \frac{5}{3}\cdot\sec\theta\tan\theta}{\frac{125}{27}\sec^3\theta}\,d\theta = \frac{9}{5}\int \frac{\tan^2\theta}{\sec^2\theta}\,d\theta = \frac{9}{5}\int \frac{\sec^2\theta - 1}{\sec^2\theta}\,d\theta = \frac{9}{5}\int 1 - \cos^2\theta\,d\theta =$

$\displaystyle\frac{9}{5}\int \sin^2\theta\,d\theta = \frac{9}{10}\int 1 - \cos 2\theta\,d\theta = \frac{9\theta}{10} - \frac{9\sin 2\theta}{20} + C = \frac{9\theta}{10} - \frac{9\sin\theta\cos\theta}{10} = \frac{9\cos^{-1}(5/3x)}{10} - \frac{\sqrt{9x^2 - 25}}{2x^2} + C.$

8.3.33 Let $x = 5\tan\theta$ so that $dx = 5\sec^2\theta\,d\theta$. Note that $25 + x^2 = 25\sec^2\theta$. Thus, $\displaystyle\int \frac{x^2}{(25 + x^2)^2}\,dx =$

$\displaystyle\int \frac{25\tan^2\theta \cdot 5\sec^2\theta}{25^2\sec^4\theta}\,d\theta = \frac{1}{5}\int \frac{\tan^2\theta}{\sec^2\theta}\,d\theta = \frac{1}{5}\int \frac{\sec^2\theta - 1}{\sec^2\theta}\,d\theta = \frac{1}{5}\int (1 - \cos^2\theta)\,d\theta = \frac{1}{5}\int \sin^2\theta\,d\theta =$

$\displaystyle\frac{1}{10}\int 1 - \cos 2\theta\,d\theta = \frac{1}{10}\left(\theta - \frac{\sin 2\theta}{2}\right) + C = \frac{1}{10}\left(\theta - \sin\theta\cos\theta\right) + C = \frac{1}{10}\left(\tan^{-1}(x/5) - \frac{5x}{25 + x^2}\right) + C.$

8.3.35 Let $x = 10\sin\theta$ so that $dx = 10\cos\theta\,d\theta$. Note that $\sqrt{100 - x^2} = 10\cos\theta$. Thus,

$$\int \frac{x^2}{(100 - x^2)^{3/2}}\,dx = \int \frac{100\sin^2\theta \cdot 10\cos\theta}{1000\cos^3\theta}\,d\theta = \int \tan^2\theta\,d\theta$$

$$= \int \sec^2\theta - 1\,d\theta = \tan\theta - \theta + C = \frac{x}{\sqrt{100 - x^2}} - \sin^{-1}(x/10) + C.$$

8.3.37 Let $x = 9\sin\theta$ so that $dx = 9\cos\theta\,d\theta$. Note that $81 - x^2 = 81\cos^2\theta$. Thus,

$$\int \frac{x^3}{(81 - x^2)^2}\,dx = \int \frac{9^3\sin^3\theta \cdot 9\cos\theta}{9^4\cos^4\theta}\,d\theta = \int \tan^3\theta\,d\theta$$

$$= \int (\tan\theta)(\sec^2\theta - 1)\,d\theta = \int \sec^2\theta\tan\theta\,d\theta - \int \tan\theta\,d\theta$$

$$= \int u\,du + \int (1/w)\,dw = \frac{u^2}{2} + \ln|w| + C$$

$$= \frac{\tan^2\theta}{2} + \ln|\cos\theta| + C = \frac{x^2}{2(81 - x^2)} + \ln\left|\frac{\sqrt{81 - x^2}}{9}\right| + C.$$

Note that we used the substitutions $u = \tan\theta$ and $w = \cos\theta$ along the way.

8.3.39 Let $x = \sec\theta$ where $\theta \in (0, \pi/2)$. Then $dx = \sec\theta\tan\theta\,d\theta$ and $\sqrt{x^2 - 1} = \tan\theta$. Then

$$\int \frac{1}{x(x^2 - 1)^{3/2}}\,dx = \int \frac{\sec\theta\tan\theta}{\sec\theta\tan^3\theta}\,d\theta = \int \cot^2\theta\,d\theta$$

$$= \int \csc^2\theta - 1\,d\theta = -\cot\theta - \theta + C = \frac{-1}{\sqrt{x^2 - 1}} - \sec^{-1}x + C.$$

8.3.41 Let $x = 4\tan\theta$ so that $dx = 4\sec^2\theta\,d\theta$. Note that $\sqrt{x^2 + 16} = 4\sec\theta$. Thus, $\int_0^1 \frac{1}{\sqrt{x^2 + 16}}\,dx =$
$\int_0^{\tan^{-1}(1/4)} \frac{4\sec^2\theta}{4\sec\theta}\,d\theta = \int_0^{\tan^{-1}(1/4)} \sec\theta\,d\theta = \ln|\sec\theta + \tan\theta||_0^{\tan^{-1}(1/4)} = \ln\left(\frac{\sqrt{17} + 1}{4}\right).$

8.3.43 Let $x = \frac{1}{3}\tan\theta$ so that $dx = \frac{1}{3}\sec^2\theta\,d\theta$. Note that $\sqrt{9x^2 + 1} = \sec\theta$. Thus $\int_0^{1/3} \frac{1}{(9x^2 + 1)^{3/2}}\,dx =$
$\int_0^{\pi/4} \frac{\frac{1}{3}\sec^2\theta}{\sec^3\theta}\,d\theta = \frac{1}{3}\int_0^{\pi/4} \cos\theta\,d\theta = \frac{1}{3}\sin\theta|_0^{\pi/4} = \frac{\sqrt{2}}{6}.$

8.3.45 Let $x = 2\sec\theta$ so that $dx = 2\sec\theta\tan\theta\,d\theta$ and $x^2 - 4 = 4\tan^2\theta$. Thus, $\int_{4/\sqrt{3}}^4 \frac{1}{x^2(x^2 - 4)}\,dx =$
$\int_{\pi/6}^{\pi/3} \frac{2\sec\theta\tan\theta}{4\sec^2\theta \cdot 4\tan^2\theta}\,d\theta = \frac{1}{8}\int_{\pi/6}^{\pi/3} \frac{\cos^2\theta}{\sin\theta}\,d\theta = \frac{1}{8}\int_{\pi/6}^{\pi/3} \frac{1 - \sin^2\theta}{\sin\theta}\,d\theta = \frac{1}{8}\int_{\pi/6}^{\pi/3} \csc\theta - \sin\theta\,d\theta =$
$\frac{1}{8}\left(-\ln|\csc\theta + \cot\theta| + \cos\theta\right)|_{\pi/6}^{\pi/3} = \frac{1}{8}\left(-\ln(\sqrt{3}(2 - \sqrt{3})) + \frac{1 - \sqrt{3}}{2}\right).$

8.3.47

a. False. In fact, we would have $\csc\theta = \frac{\sqrt{x^2 + 16}}{x}$.

b. True. Almost every number in the interval $[1, 2]$ is not in the domain of $\sqrt{1 - x^2}$, so this integral isn't defined.

c. False. It does represent a finite real number, since $\sqrt{x^2 - 1}$ is continuous on the interval $[1, 2]$.

d. False. It can be so evaluated. The integral is equivalent to $\int \frac{1}{(x + 2)^2 + 5}\,dx$, and this can be evaluated by the substitution $x + 2 = \sqrt{5}\tan\theta$.

8.3.49 Note that the given integral can be written $\int \frac{1}{(x + 3)^2 + 9}\,dx = \int \frac{1}{u^2 + 9}\,du$ where $u = x + 3$.
Now let $u = 3\tan\theta$ so that $du = 3\sec^2\theta\,d\theta$ and $u^2 + 9 = 9\sec^2\theta$. Thus we have $\int \frac{3\sec^2\theta}{9\sec^2\theta}\,d\theta = \frac{\theta}{3} + C =$
$\frac{\tan^{-1}((x + 3)/3)}{3} + C.$

8.3.51 Note that the given integral can be written as $\int \frac{(x - 1)^2}{\sqrt{(x - 1)^2 + 9}}\,dx = \int \frac{u^2}{\sqrt{u^2 + 9}}\,du$ where $u = x - 1$.
Now let $u = 3\tan\theta$ so that $du = 3\sec^2\theta\,d\theta$ and $u^2 + 9 = 9\sec^2\theta$. Thus we have

$$\int \frac{9\tan^2\theta \cdot 3\sec^2\theta}{3\sec\theta}\,d\theta = 9\int \tan^2\theta\sec\theta\,d\theta = 9\int (\sec^2\theta - 1)(\sec\theta)\,d\theta$$

$$= 9\int \sec^3\theta - \sec\theta\,d\theta = 9\left(\frac{1}{2}\sec\theta\tan\theta + \frac{1}{2}\int \sec\theta\,d\theta - \int \sec\theta\,d\theta\right)$$

$$= \frac{9}{2}\left(\sec\theta\tan\theta - \int \sec\theta\,d\theta\right)$$

$$= \frac{9}{2}\left(\sec\theta\tan\theta - \ln|\sec\theta + \tan\theta|\right) + C$$

$$= \frac{9}{2}\left(\frac{\sqrt{(x - 1)^2 + 9}(x - 1)}{9} - \ln\left|\frac{\sqrt{(x - 1)^2 + 9}}{3} + \frac{x - 1}{3}\right|\right) + C.$$

Note that in the middle of this derivation we used the reduction formula for $\int \sec^3\theta\,d\theta$ given in problem 54 in the previous section.

8.3.53 Note that the given integral can be written as $\int \dfrac{(x-4)^2}{(25-(x-4)^2)^{3/2}}\, dx$. Let $u = x - 4$, and note that

we have $\int \dfrac{u^2}{(25-u^2)^{3/2}}\, du$. Now let $u = 5 \sin\theta$ so that $du = 5\cos\theta\, d\theta$, and note that $\sqrt{25 - u^2} = 5\cos\theta$.

Thus we have $\int \dfrac{25 \sin^2\theta \cdot 5\cos\theta}{5^3 \cos^3\theta}\, d\theta = \int \tan^2\theta\, d\theta = \int \sec^2\theta - 1\, d\theta = \tan\theta - \theta + C = \dfrac{x-4}{\sqrt{25-(x-4)^2}} -$

$\sin^{-1}\left(\dfrac{x-4}{5}\right) + C.$

8.3.55 $\displaystyle\int_{1/2}^{(\sqrt{2}+3)/2\sqrt{2}} \dfrac{1}{8x^2 - 8x + 11}\, dx = \int_{1/2}^{(\sqrt{2}+3)/2\sqrt{2}} \dfrac{1}{8(x-1/2)^2 + 9}\, dx.$ Let $u = x - 1/2$, so that our

integral becomes $\displaystyle\int_0^{3/2\sqrt{2}} \dfrac{1}{8u^2 + 9}\, du.$ Now let $u = \frac{3}{\sqrt{8}}\tan\theta$ so that $du = \frac{3}{\sqrt{8}}\sec^2\theta\, d\theta.$ Substituting gives

$\displaystyle\int_0^{\pi/4} \dfrac{\frac{3}{\sqrt{8}}\sec^2\theta}{9\sec^2\theta}\, d\theta = \dfrac{1}{6\sqrt{2}}\int d\theta = \dfrac{1}{6\sqrt{2}}\theta\Big|_0^{\pi/4} = \dfrac{\pi\sqrt{2}}{48}.$

8.3.57

a. Recall that the area of a circular sector subtended by an angle θ is given by $\frac{\theta \cdot r^2}{2}$. So the area of the cap is this area minus the area of the isosceles triangle with two sides of length r and angle between them θ. So $A_{\text{cap}} = A - A_0 = \frac{\theta \cdot r^2}{2} - \frac{r^2 \sin\theta}{2} = \frac{r^2}{2}(\theta - \sin\theta).$

b. For a cap we have $0 \le \theta \le \pi$ so $0 \le \theta/2 \le \pi/2$. By symmetry, $\dfrac{A_{\text{cap}}}{2} = \displaystyle\int_{r\cos\theta/2}^{r} \sqrt{r^2 - x^2}\, dx.$ Let

$x = r\cos\alpha/2$ so that $dx = \frac{-r}{2}\sin\alpha/2\, d\alpha.$ Then we have $\dfrac{A_{\text{cap}}}{2} = \displaystyle\int_{\theta}^{0} r\sin(\alpha/2)\cdot\dfrac{-r}{2}\sin(\alpha/2)\, d\alpha =$

$\dfrac{r^2}{2}\displaystyle\int_0^\theta \sin^2(\alpha/2)\, d\alpha = \dfrac{r^2}{4}\displaystyle\int_0^\theta 1-\cos\alpha\, d\alpha = \dfrac{r^2}{4}(\alpha - \sin\alpha)\Big|_0^\theta = \dfrac{r^2}{4}(\theta - \sin\theta).$ Thus $A_{\text{cap}} = \frac{r^2}{2}(\theta - \sin\theta).$

8.3.59

a. The area is given by $\displaystyle\int_0^4 \dfrac{1}{\sqrt{9+x^2}}\, dx.$ Let $x = 3\tan\theta$, so that $dx = 3\sec^2\theta\, d\theta.$ Substituting yields

$\displaystyle\int_0^{\tan^{-1}(4/3)} \dfrac{3\sec^2\theta}{3\sec\theta}\, d\theta = \int_0^{\tan^{-1}(4/3)} \sec\theta\, d\theta = \ln|\sec\theta + \tan\theta|\Big|_0^{\tan^{-1}(4/3)} = \ln\left(\dfrac{5}{3} + \dfrac{4}{3}\right) = \ln 3.$

b. Using disks, we have $\dfrac{V}{\pi} = \displaystyle\int_0^4 \dfrac{1}{9+x^2}\, dx.$ Let $x = 3\tan\theta$ so that $dx = 3\sec^2\theta\, d\theta.$ Substituting yields

$\displaystyle\int_0^{\tan^{-1}(4/3)} \dfrac{3\sec^2\theta}{9\sec^2\theta}\, d\theta = \dfrac{1}{3}\int_0^{\tan^{-1}(4/3)} 1\, d\theta = \dfrac{1}{3}\theta\Big|_0^{\tan^{-1}(4/3)} = \dfrac{1}{3}\tan^{-1}(4/3).$ Thus $V = \frac{\pi}{3}\tan^{-1}(4/3).$

c. Using shells, we have $\dfrac{V}{2\pi} = \displaystyle\int_0^4 \dfrac{x}{(9+x^2)^{1/2}}\, dx.$ Let $u = 9 + x^2$, so that $du = 2x\, dx.$ Then we have

$\dfrac{1}{2}\displaystyle\int_9^{25} u^{-1/2}\, du = \sqrt{u}\Big|_9^{25} = 5 - 3 = 2.$ So $V = 4\pi.$

8.3.61 Since $y = ax^2$, we have $1 + \left(\dfrac{dy}{dx}\right)^2 = 1 + 4a^2x^2$, so the arc length is given by $\displaystyle\int_0^{10} \sqrt{1 + 4a^2x^2}\, dx.$ Let

$x = \frac{1}{2a}\tan\theta$ so that $dx = \frac{1}{2a}\sec^2\theta\, d\theta.$ Then we have $\dfrac{1}{2a}\displaystyle\int_0^{\tan^{-1}(20a)} \sec^2\theta\sec\theta\, d\theta = \dfrac{1}{2a}\int_0^{\tan^{-1}(20a)} \sec^3\theta\, d\theta =$

$\dfrac{1}{4a}(\sec\theta\tan\theta + \ln|\sec\theta + \tan\theta|)\Big|_0^{\tan^{-1}(20a)} = \dfrac{1}{4a}\left(\sqrt{1 + 400a^2}(20a) + \ln(\sqrt{1 + 400a^2} + 20a)\right).$

$\int_0^{3/2} \dfrac{1}{(9 - x^2)^2}\,dx.$ Let $x = 3\sin\theta$ so that $dx = 3\cos\theta\,d\theta.$ Then we have

8.3.63 $\displaystyle\int_0^{\pi/6} \dfrac{3\cos\theta}{9^2\cos^4\theta}\,d\theta = \dfrac{1}{27}\int_0^{\pi/6}\sec^3\theta\,d\theta =$

$\dfrac{1}{54}\left(\sec\theta\tan\theta + \ln|\sec\theta + \tan\theta|\right)\big|_0^{\pi/6} =$

$\dfrac{1}{54}\left(\dfrac{2}{\sqrt{3}}\cdot\dfrac{1}{\sqrt{3}} + \ln\left(\dfrac{2}{\sqrt{3}} + \dfrac{1}{\sqrt{3}}\right)\right) = \dfrac{1}{81} + \dfrac{\ln(\sqrt{3})}{54}.$

$\int_5^{10}\sqrt{x^2 - 25}\,dx.$ Let $x = 5\sec\theta$ so that $dx = 5\sec\theta\tan\theta\,d\theta.$ Then we have $\displaystyle\int_0^{\pi/3} 5\sec\theta\tan\theta \cdot 5\tan\theta\,d\theta =$

8.3.65 $25\displaystyle\int_0^{\pi/3}\sec\theta\tan^2\theta\,d\theta = 25\int_0^{\pi/3}\sec\theta(\sec^2\theta -$

$1)\,d\theta = 25\displaystyle\int_0^{\pi/3}\sec^3\theta - \sec\theta\,d\theta =$

$\dfrac{25}{2}\left(\sec\theta\tan\theta - \ln|\sec\theta + \tan\theta|\right)\big|_0^{\pi/3} =$

$\dfrac{25}{2}\left(2\sqrt{3} - \ln(2 + \sqrt{3})\right) = 25\sqrt{3} - \dfrac{25}{2}\cdot\ln(2 + \sqrt{3}).$

8.3.67 $\displaystyle\int_{2+\sqrt{2}}^4 \dfrac{1}{\sqrt{(x-1)(x-3)}}\,dx = \int_{2+\sqrt{2}}^4 \dfrac{1}{\sqrt{(x-2)^2 - 1}}\,dx = \int_{\sqrt{2}}^2 \dfrac{1}{\sqrt{u^2 - 1}}\,du$, where $u = x - 2$. Now let

$u = \sec\theta$, so that $du = \sec\theta\tan\theta\,d\theta$. Then $\displaystyle\int_{\pi/4}^{\pi/3} \dfrac{\sec\theta\tan\theta}{\tan\theta}\,d\theta = \int_{\pi/4}^{\pi/3}\sec\theta\,d\theta = \ln(\sec\theta + \tan\theta)\big|_{\pi/4}^{\pi/3} =$

$\ln(2 + \sqrt{3}) - \ln(\sqrt{2} + 1) = \ln\left(\dfrac{2 + \sqrt{3}}{\sqrt{2} + 1}\right).$

8.3.69 Using washers, $V = \displaystyle\int_{-4}^4 A(x)\,dx.$ Now $A(x) = \pi((f(x))^2 - (g(x))^2) = \pi((6 + \sqrt{16 - x^2})^2 -$

$(6 - \sqrt{16 - x^2})^2) = 24\pi\sqrt{16 - x^2}.$ Thus our integral is $\displaystyle\int_{-4}^4 24\pi\sqrt{16 - x^2}\,dx = 48\pi\int_0^4\sqrt{16 - x^2}\,dx =$

$48\pi\left(\dfrac{\pi\cdot 4^2}{4}\right) = 192\pi^2.$

8.3.71

a. $E_x(a) = \dfrac{kQa}{2L}\displaystyle\int_{-L}^L \dfrac{dy}{(a^2 + y^2)^{3/2}}.$ Let $y = a\tan\theta$ so that $dy = a\sec^2\theta\,d\theta.$ Then note that

$$\int_{-L}^L \dfrac{dy}{(a^2 + y^2)^{3/2}} = 2\int_0^L \dfrac{dy}{(a^2 + y^2)^{3/2}} = 2\int_0^{\tan^{-1}(L/a)} \dfrac{a\sec^2\theta}{a^3\sec^3\theta}\,d\theta = \dfrac{2}{a^2}\int_0^{\tan^{-1}(L/a)}\cos\theta\,d\theta$$

$$= \dfrac{2}{a^2}\sin\theta\Big|_0^{\tan^{-1}(L/a)} = \dfrac{2}{a^2}\cdot\dfrac{L}{\sqrt{a^2 + L^2}} = \dfrac{2L}{a^2\sqrt{a^2 + L^2}}.$$

Thus, $E_x(a) = \dfrac{kQ}{a\sqrt{a^2 + L^2}}.$

b. Set $\rho = Q/(2L)$. Then since $\displaystyle\lim_{L\to\infty}\dfrac{2L}{a^2\sqrt{a^2 + L^2}} = \dfrac{2}{a^2}$, we have $E_x(a, 0) \approx \dfrac{kQa}{2L}\displaystyle\lim_{L\to\infty}\int_{-L}^L \dfrac{dy}{(a^2 + y^2)^{3/2}} =$

$\dfrac{kQa}{2L}\displaystyle\lim_{L\to\infty}\dfrac{2L}{a^2\sqrt{a^2 + L^2}} = \dfrac{kQa}{2L}\left(\dfrac{2}{a^2}\right) = \dfrac{2kQ}{2aL} = \dfrac{2k\rho}{a}.$

8.3.73

a. $\int_a^b \sqrt{\dfrac{1-\cos t}{g(\cos a - \cos t)}}\, dt = \int_a^b \sqrt{\dfrac{(1-\cos t)(1+\cos t)}{g(1+\cos t)(\cos a - \cos t)}}\, dt = \int_a^b \sin t \sqrt{\dfrac{1}{g(1+\cos t)(\cos a - \cos t)}}\, dt$

since $t \in [0, \pi)$. Let $u = \cos t$ so that $du = -\sin t\, dt$. Then the given integral is equal to

$$\frac{-1}{\sqrt{g}} \int_{\cos a}^{\cos b} \sqrt{\frac{1}{(1+u)(\cos a - u)}}\, du.$$

Now we complete the square: $(1+u)(\cos a - u) = \cos a + (\cos a - 1)u - u^2 =$

$-\left(u^2 - (\cos a - 1)u + \left(\frac{\cos a - 1}{2}\right)^2 - \left(\frac{\cos a - 1}{2}\right)^2\right) + \cos a = \cos a + \left(\frac{\cos a - 1}{2}\right)^2 - \left(u - \frac{\cos a - 1}{2}\right)^2 =$

$\left(\frac{\cos a + 1}{2}\right)^2 - \left(u - \frac{\cos a - 1}{2}\right)^2$. Thus, setting $v = u - \frac{\cos a - 1}{2}$ we have that the original integral is equal to

$$\frac{-1}{\sqrt{g}} \int_{(\cos a + 1)/2}^{\cos b - \frac{\cos a - 1}{2}} \frac{1}{\sqrt{k^2 - v^2}}\, dv \text{ where } k = \frac{(\cos a + 1)}{2}.$$

Now, $\int \dfrac{1}{\sqrt{k^2 - v^2}}\, dv = \int \dfrac{k\cos\theta}{k\cos\theta}\, d\theta = \theta + C = \sin^{-1}(v/k) + C$ where $v = k\sin\theta$.

Therefore, the original integral is equal to $\frac{-1}{\sqrt{g}} \sin^{-1}\left(\frac{2v}{\cos a + 1}\right)\Big|_{(\cos a + 1)/2}^{\cos b - (\cos a - 1)/2} =$

$\frac{1}{\sqrt{g}}\left(\sin^{-1}\left(\frac{\cos a + 1}{\cos a + 1}\right) - \sin^{-1}\left(\frac{2\cos b - \cos a + 1}{\cos a + 1}\right)\right) = \frac{1}{\sqrt{g}}\left(\frac{\pi}{2} - \sin^{-1}\left(\frac{2\cos b - \cos a + 1}{\cos a + 1}\right)\right).$

b. Letting $b = \pi$, we have that the integral is equal to

$$\frac{1}{\sqrt{g}}\left(\frac{\pi}{2} - \sin^{-1}\left(\frac{-2 - \cos a + 1}{\cos a + 1}\right)\right)$$

$$= \frac{1}{\sqrt{g}}\left(\frac{\pi}{2} - \sin^{-1}(-1)\right) = \frac{1}{\sqrt{g}}\left(\frac{\pi}{2} - \left(\frac{-\pi}{2}\right)\right) = \frac{\pi}{\sqrt{g}}.$$

8.3.75 If $1 < x = \sec\theta$ then we need $\theta \in (0, \pi/2)$. Alternatively, if $-1 > x = \sec\theta$, then we need $\theta \in (\pi/2, \pi)$.

So, in the former, $\int \dfrac{1}{x\sqrt{x^2 - 1}}\, dx = \int \dfrac{\sec\theta\tan\theta}{\sec\theta\tan\theta}\, d\theta = \theta + C = \sec^{-1}x + C = \tan^{-1}\sqrt{x^2 - 1} + C$. In the

latter case, $\int \dfrac{1}{x\sqrt{x^2 - 1}}\, dx = \int \dfrac{\sec\theta}{\sec\theta(-\tan\theta)}\, d\theta = -\theta + C = -\sec^{-1}x + C = -\tan^{-1}\sqrt{x^2 - 1} + C$.

8.3.77

Let $x = 3\sec\theta$ for $x \in (\pi/2, \pi)$. Then $dx = 3\sec\theta\tan\theta\, d\theta$, and note that $|\tan\theta| = -\tan\theta$. We have $\int_{-6}^{-3} \dfrac{\sqrt{x^2 - 9}}{x}\, dx =$

$\int_{2\pi/3}^{\pi} \dfrac{3\sec\theta\tan\theta(-3\tan\theta)}{3\sec\theta}\, d\theta =$

$-3\int_{2\pi/3}^{\pi} \tan^2\theta\, d\theta = -3\left(\tan\theta - \theta\right)\big|_{2\pi/3}^{\pi} =$

$-3\left(-\pi - (-\sqrt{3} - 2\pi/3)\right) = \pi - 3\sqrt{3}$. Note that this number is less than zero, as suggested by the graph.

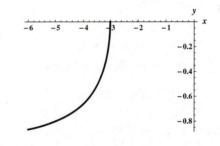

8.3.79

a. The area of sector OAB is given by the formula $\frac{\theta}{2}a^2$ where $\sin\theta = x/a$. The area of triangle OBC is $\frac{x}{2}\sqrt{a^2 - x^2}$. Thus, $F(x) = \dfrac{a^2\sin^{-1}(x/a)}{2} + \dfrac{x\sqrt{a^2 - x^2}}{2}$.

b. By the first fundamental theorem of calculus, $F(x)$ is an antiderivative of $\sqrt{a^2 - x^2}$, since $F(x) = \int_0^x \sqrt{a^2 - t^2}\, dt$. Thus, any other antiderivative differs from this by a constant, so $\int \sqrt{a^2 - x^2}\, dx = \dfrac{a^2\sin^{-1}(x/a)}{2} + \dfrac{x\sqrt{a^2 - x^2}}{2} + C$.

8.4 Partial Fractions

8.4.1 Proper rational functions can be integrated using partial fraction decomposition.

8.4.3

a. $\dfrac{A}{x-3}$.

b. $\dfrac{A_1}{x-4}, \dfrac{A_2}{(x-4)^2}, \dfrac{A_3}{(x-4)^3}$.

c $\dfrac{Ax+B}{x^2+2x+6}$.

8.4.5 $\dfrac{2}{x^2-2x-8} = \dfrac{2}{(x-4)(x+2)} = \dfrac{A}{x-4} + \dfrac{B}{x+2}$. Thus, $2 = A(x+2) + B(x-4)$. Letting $x = -2$ yields $B = \frac{-1}{3}$ and letting $x = 4$ yields $A = \frac{1}{3}$. Thus,

$$\frac{2}{x^2-2x-8} = \frac{1/3}{x-4} - \frac{1/3}{x+2}.$$

8.4.7 $\dfrac{x^2}{x^3-16x} = \dfrac{x}{(x-4)(x+4)} = \dfrac{A}{x-4} + \dfrac{B}{x+4}$. Thus, $x = A(x+4) + B(x-4)$. Letting $x = -4$ yields $B = 1/2$ and letting $x = 4$ yields $A = 1/2$. Thus,

$$\frac{x^2}{x^3-16x} = \frac{1/2}{x-4} + \frac{1/2}{x+4}.$$

8.4.9 If we write $\dfrac{1}{(x-1)(x+2)} = \dfrac{A}{x-1} + \dfrac{B}{x+2}$, we have $1 = A(x+2) + B(x-1)$. Letting $x = -2$ yields $B = -1/3$ and letting $x = 1$ yields $A = 1/3$. Thus, the original integral is equal to $\displaystyle\int \left(\frac{1/3}{x-1} - \frac{1/3}{x+2} \right) dx = \frac{1}{3}\ln|x-1| - \frac{1}{3}\ln|x+2| + C$.

8.4.11 If we write $\frac{3}{x^2-1} = \frac{3}{(x-1)(x+1)} = \frac{A}{x-1} + \frac{B}{x+1}$, then we have $3 = A(x+1) + B(x-1)$. Letting $x = -1$ yields $B = -3/2$ and letting $x = 1$ yields $A = 3/2$. Thus, the original integral is equal to $\displaystyle\int \left(\frac{3/2}{x-1} - \frac{3/2}{x+1} \right) dx = \frac{3}{2}\left(\ln|x-1| - \ln|x+1| \right) + C$.

8.4.13 If we write $\dfrac{2}{x^2-x-6} = \dfrac{A}{x-3} + \dfrac{B}{x+2}$, then we have $2 = A(x+2) + B(x-3)$. Letting $x = -2$ yields $B = -2/5$ and letting $x = 3$ yields $A = 2/5$. Thus the original integral is equal to $\displaystyle\int \left(\frac{2/5}{x-3} - \frac{2/5}{x+2} \right) dx = \frac{2}{5}\left(\ln|x-3| - \ln|x+2| \right) + C$.

8.4.15 If we write $\dfrac{1}{x^2-2x-24} = \dfrac{1}{(x-6)(x+4)} = \dfrac{A}{x-6} + \dfrac{B}{x+4}$, then we have $1 = A(x+4) + B(x-6)$. Letting $x = -4$ yields $B = -1/10$ and letting $x = 6$ yields $A = 1/10$. Thus the original integral is equal to $\displaystyle\int \left(\frac{1/10}{x-6} - \frac{1/10}{x+4} \right) dx = \frac{1}{10}\left(\ln|x-6| - \ln|x+4| \right) + C$.

8.4.17 If we write $\dfrac{1}{x^4-10x^2+9} = \dfrac{1}{(x-1)(x+1)(x-3)(x+3)} = \dfrac{A}{x-1} + \dfrac{B}{x+1} + \dfrac{C}{x-3} + \dfrac{D}{x+3}$ then $1 = A(x+1)(x-3)(x+3) + B(x-1)(x-3)(x+3) + C(x-1)(x+1)(x+3) + D(x-1)(x+1)(x-3)$. Letting $x = -1$ yields $B = 1/16$. Letting $x = 3$ yields $C = 1/48$. Letting $x = -3$ yields $D = -1/48$, and letting $x = 1$ yields $A = -1/16$. Thus the original integral is equal to $\displaystyle\int \left(\frac{-1/16}{x-1} + \frac{1/16}{x+1} + \frac{1/48}{x-3} - \frac{1/48}{x+3} \right) dx$ which is equal to $\dfrac{-1}{16}\ln|x-1| + \dfrac{1}{16}\ln|x+1| + \dfrac{1}{48}\ln|x-3| - \dfrac{1}{48}\ln|x+3| + C$.

8.4.19 If we write $\dfrac{3}{x^3 - 9x^2} = \dfrac{A}{x} + \dfrac{B}{x^2} + \dfrac{C}{x-9}$, then $3 = Ax(x-9) + B(x-9) + C(x^2)$. Letting $x = 0$ yields $B = -1/3$. Letting $x = 9$ yields $C = 1/27$. If we let $x = 10$, then we have $3 = 10A - \frac{1}{3} + \frac{100}{27}$, so $81 = 270A - 9 + 100$, so $A = -1/27$. Thus, the original integral is equal to $\displaystyle\int \left(\dfrac{-1/27}{x} - \dfrac{1/3}{x^2} + \dfrac{1/27}{x-9} \right) dx = \dfrac{1}{27} \left(\ln|x-9| - \ln|x| \right) + \dfrac{1}{3x} + C.$

8.4.21 If we write $\dfrac{x}{(x+3)^2} = \dfrac{A}{x+3} + \dfrac{B}{(x+3)^2}$, then we have $x = A(x+3) + B$. Letting $x = -3$ yields $B = -3$, and then letting $x = -2$ yields $A = 1$. Thus the original integral is equal to $\displaystyle\int \left(\dfrac{1}{x+3} - \dfrac{3}{(x+3)^2} \right) dx = \ln|x+3| + \dfrac{3}{x+3} + C.$

8.4.23 If we write $\dfrac{2}{x^3 + x^2} = \dfrac{A}{x} + \dfrac{B}{x^2} + \dfrac{C}{x+1}$, then $2 = Ax(x+1) + B(x+1) + Cx^2$. Letting $x = 0$ yields $B = 2$, and letting $x = -1$ yields $C = 2$. Then letting $x = 1$ yields $A = -2$. So the original integral is equal to $\displaystyle\int \left(\dfrac{-2}{x} + \dfrac{2}{x^2} + \dfrac{2}{x+1} \right) dx = 2 \left(\ln|x+1| - \ln|x| \right) - \dfrac{2}{x} + C.$

8.4.25 If we write $\dfrac{x-5}{x^2(x+1)} = \dfrac{A}{x} + \dfrac{B}{x^2} + \dfrac{C}{x+1}$, then we have $x - 5 = Ax(x+1) + B(x+1) + Cx^2$. Letting $x = 0$ yields $B = -5$, and letting $x = -1$ yields $C = -6$. Then letting $x = 1$ yields $-4 = 2A - 10 - 6$, so $A = 6$. The original integral is thus equal to $\displaystyle\int \left(\dfrac{6}{x} - \dfrac{5}{x^2} - \dfrac{6}{x+1} \right) dx = 6 \left(\ln|x| - \ln|x+1| \right) + \dfrac{5}{x} + C.$

8.4.27 $\dfrac{20x}{(x-1)^2(x^2+1)} = \dfrac{A}{x-1} + \dfrac{B}{(x-1)^2} + \dfrac{Cx+D}{x^2+1}.$

8.4.29 $\dfrac{2x^2 + 3}{(x^2 - 8x + 16)(x^2 + 3x + 4)} = \dfrac{2x^2 + 3}{(x-4)^2(x^2 + 3x + 4)} = \dfrac{A}{x-4} + \dfrac{B}{(x-4)^2} + \dfrac{Cx+D}{x^2 + 3x + 4}.$

8.4.31 If we write $\dfrac{2}{(x-4)(x^2 + 2x + 6)} = \dfrac{A}{x-4} + \dfrac{Bx+C}{x^2 + 2x + 6}$ then $2 = A(x^2 + 2x + 6) + (Bx + C)(x - 4)$. Letting $x = 4$ yields $A = 1/15$. Letting $x = 0$ yields $2 = 6/15 - 4C$, so $C = -2/5$. Then letting $x = 1$ results in the equation $2 = 9/15 + (B - 2/5)(-3)$, so $B = -1/15$. Thus, the original integral is equal to $\displaystyle\int \dfrac{1/15}{x-4}\, dx - \dfrac{1}{15} \int \dfrac{x+6}{x^2 + 2x + 6}\, dx$. This can be written as $\dfrac{1}{15} \ln|x-4| - \dfrac{1}{15} \int \dfrac{x+1}{x^2 + 2x + 6}\, dx - \dfrac{1}{15} \int \dfrac{5}{x^2 + 2x + 6}\, dx$. The middle term can be handled with the substitution $u = x^2 + 2x + 6$, but for the last term, we write $\dfrac{-1}{15} \int \dfrac{5}{(x+1)^2 + 5}\, dx$ which can be handled with the substitution $x + 1 = \sqrt{5} \tan\theta$. Putting this all together yields the result $\dfrac{1}{15} \left(\ln|x-4| - \ln\sqrt{x^2 + 2x + 6} \right) + \dfrac{-\sqrt{5}}{15} \tan^{-1}\left(\dfrac{x+1}{\sqrt{5}} \right) + C.$

8.4.33 If we write $\dfrac{x^2}{(x-1)(x^2 + 4x + 5)} = \dfrac{A}{x-1} + \dfrac{Bx+C}{x^2 + 4x + 5}$, then $x^2 = A(x^2 + 4x + 5) + (Bx + C)(x - 1)$. Letting $x = 1$ yields $A = 1/10$, and letting $x = 0$ yields $C = 1/2$. Also, the coefficient on x on the right-hand side is $4A - B + C = 9/10 - B$ and on the left-hand side this coefficient is zero, so $B = 9/10$. The original integral is therefore equal to $\displaystyle\int \left(\dfrac{1}{10(x-1)} + \dfrac{9x+5}{10(x^2 + 4x + 5)} \right) dx$. Now $\displaystyle\int \dfrac{9x+5}{10(x^2 + 4x + 5)}\, dx = \dfrac{9}{10} \int \dfrac{x+2}{x^2 + 4x + 5}\, dx - \dfrac{13}{10} \int \dfrac{1}{x^2 + 4x + 5}\, dx = \dfrac{9}{20} \ln|x^2 + 4x + 5| - \dfrac{13}{10} \int \dfrac{1}{(x+2)^2 + 1}\, dx$. This last term can be recognized as $\dfrac{-13}{10} \tan^{-1}(x+2) + C$. Putting this all together, we have that the original integral is equal to $\dfrac{1}{10} \ln|x-1| + \dfrac{9}{20} \ln|x^2 + 4x + 5| - \dfrac{13}{10} \tan^{-1}(x+2) + C.$

8.4.35 If we write $\dfrac{x^2}{x^3 - x^2 + 4x - 4} = \dfrac{x^2}{(x-1)(x^2+4)} = \dfrac{A}{x-1} + \dfrac{Bx+C}{x^2+4}$, then $x^2 = A(x^2+4) + (Bx + C)(x-1)$. Letting $x = 1$ yields $A = 1/5$. Letting $x = 0$ yields $C = 4/5$, and then letting $x = 2$ and solving for B yields $B = 4/5$. Thus the original integral is equal to $\displaystyle\int \dfrac{1/5}{x-1}\, dx + \dfrac{4}{5}\int \dfrac{x}{x^2+4}\, dx + \dfrac{4}{5}\int \dfrac{1}{x^2+4}\, dx$. Thus, the original integral is equal to $\frac{1}{5}\ln|x-1| + \frac{2}{5}\ln(x^2+4) + \frac{2}{5}\tan^{-1}(x/2) + C$.

8.4.37

a. False. Since the given integrand is improper, the first step would be to use long division to write the integrand as the sum of a polynomial and a proper rational function.

b. False. This is easy to evaluate via the substitution $u = 3x^2 + x$.

c. False. The discriminant of the denominator is $b^2 - 4ac = 169 - 168 = 1 > 0$, so the denominator factors into linear factors of the real numbers.

d. True. The discriminant of the denominator is $b^2 - 4ac = 169 - 172 = -3 < 0$, so the given quadratic expression is irreducible.

8.4.39

If we write $\dfrac{10}{x^2 - 2x - 24} = \dfrac{10}{(x-6)(x+4)} = \dfrac{A}{x-6} + \dfrac{B}{x+4}$, then $10 = A(x+4) + B(x-6)$. Letting $x = -4$ gives $B = -1$ and letting $x = 6$ gives $A = 1$. Thus the area in question is given by $-\displaystyle\int_{-2}^{2}\left(\dfrac{-1}{x+4} + \dfrac{1}{x-6}\right)dx = \ln(x+4) - \ln|x-6|\big|_{-2}^{2} = \ln 6 - \ln 4 - (\ln 2 - \ln 8) = \ln 6$.

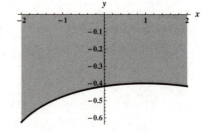

8.4.41

The curve intersects the x-axis at $x = 2 \pm 2\sqrt{2}$. We can write the integrand as $1 + \dfrac{1}{x^2 - 4x - 5} = 1 + \dfrac{1}{(x-5)(x+1)}$. If we write this second term in the form $\dfrac{A}{x-5} + \dfrac{B}{x+1}$, then $1 = A(x+1) + B(x-5)$. Letting $x = -1$ yields $B = -1/6$ and letting $x = 5$ yields $A = 1/6$. Thus the area in question is given by $\displaystyle\int_{2-2\sqrt{2}}^{2+2\sqrt{2}}\left(1 + \dfrac{1/6}{x-5} - \dfrac{1/6}{x+1}\right)dx = x + \dfrac{1}{6}\ln|x-5| - \dfrac{1}{6}\ln|x+1|\Big|_{2-2\sqrt{2}}^{2+2\sqrt{2}} = 4\sqrt{2} + \dfrac{1}{6}\ln\left(\dfrac{3-2\sqrt{2}}{3+2\sqrt{2}}\right)^2$.

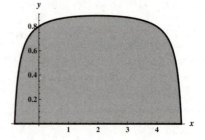

8.4.43 Using disks, we have $\dfrac{V}{\pi} = \displaystyle\int_{0}^{4}\dfrac{x^2}{(x+1)^2}\, dx$. Now we can perform long division to rewrite this integral as $\displaystyle\int_{0}^{4}\left(1 - \dfrac{2x+2}{x^2+2x+1} + \dfrac{1}{(x+1)^2}\right)dx$. Thus we have $V = \pi\left(x - \ln|x^2 + 2x + 1| - \dfrac{1}{x+1}\bigg|_{0}^{4}\right) = \pi\left(4 - \ln(25) - \dfrac{1}{5} - (-1)\right) = \pi\left(\dfrac{24}{5} - \ln(25)\right)$.

8.4.45 Using disks, we have $\dfrac{V}{\pi} = \displaystyle\int_1^2 \dfrac{1}{x(3-x)}\,dx$. Now if we write $\dfrac{1}{x(3-x)} = \dfrac{A}{x} + \dfrac{B}{3-x}$, then we have $1 = A(3-x) + Bx$. Letting $x = 0$ yields $A = 1/3$ and letting $x = 3$ yields $B = 1/3$. Thus we have $V = \dfrac{\pi}{3}\displaystyle\int_1^2 \left(\dfrac{1}{x} - \dfrac{1}{x-3}\right)dx = \dfrac{\pi}{3}\left(\ln x - \ln|x-3|\big|_1^2\right) = \dfrac{\pi}{3}\cdot 2\ln 2$.

8.4.47 Using shells, we have $\dfrac{V}{2\pi} = \displaystyle\int_0^3 \dfrac{x+1}{x+2}\,dx = \int_0^3 \left(1 - \dfrac{1}{x+2}\right)dx = x - \ln|x+2|\big|_0^3 = 3 + \ln(2/5)$. Thus, $V = 2\pi(3 + \ln(2/5))$.

8.4.49 Let $u = e^x$, so that $du = e^x\,dx$. Then $\displaystyle\int \dfrac{1}{1+e^x}\cdot\dfrac{e^x}{e^x}\,dx = \int \dfrac{1}{u(1+u)}\,du$. If we write $\dfrac{1}{u(1+u)} = \dfrac{A}{u} + \dfrac{B}{1+u}$, then $1 = A(1+u) + Bu$. Letting $u = 0$ yields $A = 1$ and letting $u = -1$ yields $B = -1$, so the original integral is equal to $\displaystyle\int \left(\dfrac{1}{u} - \dfrac{1}{1+u}\right)du = \ln|u| - \ln|1+u| + C = x - \ln(1+e^x) + C$.

8.4.51 After performing long division, we have that the original integrand is equal to $3 + \dfrac{13x - 12}{(x-1)(x-2)}$, and if we write $\dfrac{13x - 12}{(x-1)(x-2)} = \dfrac{A}{x-1} + \dfrac{B}{x-2}$, then $13x - 12 = A(x-2) + B(x-1)$. Letting $x = 1$ yields $A = -1$ and letting $x = 2$ yields $B = 14$. Thus the original integral is equal to $3x - \displaystyle\int\dfrac{1}{x-1}\,dx + 14\int\dfrac{1}{x-2}\,dx = 3x - \ln|x-1| + 14\ln|x-2| + C$.

8.4.53 $\displaystyle\int \dfrac{1}{2+e^{-t}}\,dt = \int \dfrac{e^t}{2e^t+1}\,dt$. Let $u = 2e^t + 1$, so that $du = 2e^t\,dt$. Then we have $\dfrac{1}{2}\displaystyle\int\dfrac{1}{u}\,du = \dfrac{1}{2}\ln|u| + C = \dfrac{1}{2}\ln(2e^t+1) + C$.

8.4.55 $\displaystyle\int\dfrac{\sec\theta}{1+\sin\theta}\,d\theta = \int\dfrac{\cos\theta}{\cos^2\theta(1+\sin\theta)}\,d\theta = \int\dfrac{\cos\theta}{(1-\sin\theta)(1+\sin\theta)^2}\,d\theta$. Let $u = \sin\theta$ so that $du = \cos\theta\,d\theta$. Then we have $\displaystyle\int\dfrac{1}{(1-u)(1+u)^2}\,du$. If we write this integrand as $\dfrac{A}{1-u} + \dfrac{B}{1+u} + \dfrac{C}{(1+u)^2}$, then $1 = A(1+u)^2 + B(1-u)(1+u) + C(1-u)$. Letting $u = -1$ yields $C = 1/2$ and letting $u = 1$ yields $A = 1/4$. Then letting $u = 0$ yields $B = 1/4$. Thus we have $\displaystyle\int\left(\dfrac{1/4}{u+1} - \dfrac{1/4}{u-1} + \dfrac{1}{2(u+1)^2}\right)du = \dfrac{1}{4}\left(\ln|u+1| - \ln|u-1|\right) - \dfrac{1}{2(u+1)} + C = \dfrac{1}{4}\left(\ln|\sin\theta+1| - \ln|\sin\theta-1|\right) - \dfrac{1}{2(\sin\theta+1)} + C$.

8.4.57 Let $u = e^x$ so that $du = e^x\,dx$. Then the original integral is equal to $\displaystyle\int\dfrac{1}{(u-1)(u+2)}\,du$. If we write $\dfrac{1}{(u-1)(u+2)} = \dfrac{A}{u-1} + \dfrac{B}{u+2}$, then $1 = A(u+2) + B(u-1)$. Letting $u = -2$ yields $B = -1/3$ and letting $u = 1$ yields $A = 1/3$. Thus we have $\displaystyle\int\left(\dfrac{1/3}{u-1} - \dfrac{1/3}{u+2}\right)du = \dfrac{1}{3}\left(\ln|u-1| - \ln|u+2|\right) + C = \dfrac{1}{3}\ln\left|\dfrac{e^x-1}{e^x+2}\right| + C$.

8.4.59 $\displaystyle\int\dfrac{1}{(e^x+e^{-x})^2}\cdot\dfrac{(e^x)^2}{(e^x)^2}\,dx = \int\dfrac{e^{2x}}{(e^{2x}+1)^2}\,dx$. Let $u = e^x$ so that $du = e^x\,dx$. Then we have $\displaystyle\int\dfrac{u}{(u^2+1)^2}\,du$. If we let $w = u^2 + 1$, then $dw = 2u\,du$, so we have $\dfrac{1}{2}\displaystyle\int w^{-2}\,dw = -\dfrac{1}{2w} + C = \dfrac{-1}{2u^2+2} + C = \dfrac{-1}{2e^{2x}+2} + C$.

8.4.61 If we let $u^4 = x + 2$, then $4u^3\, du = dx$. Substituting yields

$$\int \frac{4u^3}{u+1}\, du = \int \left(4u^2 - 4u + 4 - \frac{4}{u+1} \right) du = \frac{4}{3}u^3 - 2u^2 + 4u - 4\ln|u+1| + C$$

$$= \frac{4}{3}(x+2)^{3/4} - 2(x+2)^{1/2} + 4(x+2)^{1/4} - 4\ln((x+2)^{1/4}+1) + C.$$

8.4.63 If we let $u^6 = x$, then $6u^5\, du = dx$. Substituting yields

$$\int \frac{6u^5}{u^3 + u^2}\, du = 6\int \frac{u^3}{u+1}\, du = 6\int \left(u^2 - u + 1 - \frac{1}{u+1} \right) du$$

$$= 2u^3 - 3u^2 + 6u - 6\ln|u+1| + C = 2\sqrt{x} - 3\sqrt[3]{x} + 6\sqrt[6]{x} - \ln(\sqrt[6]{x}+1) + C.$$

8.4.65 If we let $(u^2 - 1)^2 = x$, then $2(u^2 - 1)\cdot 2u\, du = dx$. Substituting yields $\displaystyle\int \frac{4u(u^2-1)}{u}\, du = 4\int (u^2 - $

$1)\, du = \dfrac{4u^3}{3} - 4u + C = \dfrac{4}{3}u(u^2 - 3) + C = \dfrac{4}{3}\sqrt{1 + \sqrt{x}}(\sqrt{x} - 2) + C.$

8.4.67 If we write $\dfrac{2}{x(x^2+1)^2} = \dfrac{A}{x} + \dfrac{Bx+C}{x^2+1} + \dfrac{Dx+E}{(x^2+1)^2}$, then $2 = A(x^2+1)^2 + (Bx+C)x(x^2+1) + (Dx+E)x$.
Letting $x = 0$ yields $A = 2$. Expanding the right-hand side yields $2 = (2+B)x^4 + Cx^3 + (4+B+D)x^2 + (C+E)x + 2$. Equating coefficients gives us the equations $2 + B = 0$, $C = 0$, $4 + B + D = 0$, and $C + E = 0$, from which we can deduce that $B = -2$, $C = 0$, $D = -2$ and $E = 0$. The original integral is thus equal to
$\displaystyle\int \left(\frac{2}{x} - \frac{2x}{x^2+1} - \frac{2x}{(x^2+1)^2} \right) dx = 2\ln|x| - \ln|x^2+1| + \frac{1}{x^2+1} + C.$

8.4.69 If we write $\dfrac{x}{(x-1)(x^2+2x+2)^2} = \dfrac{A}{x-1} + \dfrac{Bx+C}{x^2+2x+2} + \dfrac{Dx+E}{(x^2+2x+2)^2}$, then $x = A(x^2 + 2x + 2)^2 + (Bx+C)(x-1)(x^2+2x+2) + (Dx+E)(x-1)$. Letting $x = 1$ yields $A = 1/25$. Then expanding the polynomial on the right-hand side gives $25x = (1 + 25B)x^4 + (4 + 25(B+C))x^3 + (8 + 25D + 25C)x^2 + (8 + 25E - 25D - 50B)x + (4 - 25E - 50C)$. Equating coefficients and then solving for the unknowns yields $B = -1/25$, $C = -3/25$, $D = -5/25$, and $E = 10/25$. The original integral is thus equal to

$$\frac{1}{25}\int \left(\frac{1}{x-1} - \frac{x+3}{x^2+2x+2} - \frac{5x-10}{(x^2+2x+2)^2} \right) dx$$

$$= \frac{1}{25}\int \left(\frac{1}{x-1} - \frac{x+1}{x^2+2x+2} - 5\frac{x+1}{(x^2+2x+2)^2} - 2\frac{1}{(x+1)^2+1} + 15\frac{1}{((x+1)^2+1)^2} \right) dx$$

$$= \frac{1}{25}\left(\ln|x-1| - \frac{1}{2}\ln|x^2+2x+2| + \frac{5}{2}\cdot\frac{1}{x^2+2x+2} - 2\tan^{-1}(x+1) \right) + \frac{3}{5}\int \frac{1}{(v^2+1)^2}\, dv,$$

where $v = x + 1$. To compute this last term, we let $v = \tan\theta$ so that $dv = \sec^2\theta\, d\theta$. Then the last term is equal to

$$\frac{3}{5}\int \frac{\sec^2\theta}{\sec^4\theta}\, d\theta = \frac{3}{5}\int \cos^2\theta\, d\theta = \frac{3}{10}\int (1 + \cos 2\theta)\, d\theta$$

$$= \frac{3}{10}\left(\theta + \frac{\sin 2\theta}{2} \right) + C = \frac{3}{10}\left(\tan^{-1}(x+1) + \frac{x+1}{(x+1)^2+1} \right) + C.$$

Putting this all together yields

$$\frac{1}{25}\left(\ln|x-1| - \frac{1}{2}\ln|x^2+2x+2| \right) + \frac{11}{50}\tan^{-1}(x+1) + \frac{1}{10}\left(\frac{3x+4}{(x+1)^2+1} \right) + C.$$

8.4.71 If we write $\dfrac{1}{x^2-1} = \dfrac{A}{x-1} + \dfrac{B}{x+1}$, then $1 = A(x+1) + B(x-1)$, so $A = 1/2$ and $B = -1/2$. Thus
we have $\dfrac{1}{2}\displaystyle\int \frac{1}{x-1} - \frac{1}{x+1}\, dx = \frac{1}{2}\left(\ln|x-1| - \ln|x+1| \right) + C.$

Now let $x = \sec\theta$, so that $dx = \sec\theta\tan\theta\,d\theta$. Then the original integral is equal to $\int \csc\theta\,d\theta =$

$-\ln|\csc\theta + \cot\theta| + C = -\ln\left|\dfrac{x}{\sqrt{x^2-1}} + \dfrac{1}{\sqrt{x^2-1}}\right| + C = -\ln\left(\dfrac{|x+1|}{\sqrt{x^2-1}}\right) + C = -\ln\left(\sqrt{\dfrac{x+1}{x-1}}\right) + C =$

$\ln\left(\sqrt{\dfrac{x-1}{x+1}}\right) + C$. The two answers are equivalent.

8.4.73 Using the substitution $x = 2\tan^{-1} u$ yields

$$\int \frac{1}{1+\sin x}\,dx = \int \frac{2}{1+u^2} \cdot \frac{1}{1 + \frac{2u}{1+u^2}}\,du = \int \frac{2}{u^2 + 2u + 1}\,du$$

$$= 2\int \frac{1}{(u+1)^2}\,du = \frac{-2}{u+1} + C = \frac{-2}{\tan(x/2)+1} + C.$$

8.4.75 Using the substitution $x = 2\tan^{-1} u$ yields $\displaystyle\int \frac{1}{1 - \frac{1-u^2}{1+u^2}} \cdot \frac{2}{1+u^2}\,du = \int u^{-2}\,du = \frac{-1}{u} + C = $ $-\cot(x/2) + C$.

8.4.77 Using the substitution $\theta = 2\tan^{-1} u$ yields $\displaystyle\int \frac{1}{\frac{1-u^2}{1+u^2} - \frac{2u}{1+u^2}} \cdot \frac{2}{1+u^2}\,du = -2\int \frac{1}{u^2 + 2u - 1}\,du = $

$-2\displaystyle\int \frac{1}{(u+1)^2 - 2}\,du$. Letting $u + 1 = \sqrt{2}\sec\alpha$ yields the integral $-\sqrt{2}\int \csc\alpha\,d\alpha = \sqrt{2}\ln|\csc\alpha + \cot\alpha| + $

$C = \sqrt{2}\ln\left|\dfrac{u+1}{\sqrt{(u+1)^2 - 2}} + \dfrac{\sqrt{2}}{\sqrt{(u+1)^2 - 2}}\right| + C = \sqrt{2}\ln\left|\dfrac{\tan(\theta/2) + 1 + \sqrt{2}}{\sqrt{(\tan(\theta/2)+1)^2 - 2}}\right| + C.$

8.4.79 $s_A(t) = \int v_A(t)\,dt = 88\int \frac{t}{t+1}\,dt = 88\int\left(1 - \frac{1}{1+t}\right)dt = 88t - 88\ln(t+1) + C$. Since $s_A(0) = 0$, we see that $C = 0$, so $s_A(t) = 88t - 88\ln(1+t)$.

$s_B(t) = \int v_B(t)\,dt = 88\int \frac{t^2}{(t+1)^2}\,dt = 88\int\left(1 - \frac{2t+2}{t^2+2t+1} + \frac{1}{(t+1)^2}\right)dt = 88(t - \ln(t^2+2t+1) - \frac{1}{t+1}) + D$. Since $s_B(0) = 0$, we see that $D = 88$, so $s_B(t) = 88(t - \ln(t^2+2t+1) - \frac{1}{t+1} + 1)$.

$s_C(t) = \int v_C(t)\,dt = 88\int \frac{t^2}{t^2+1}\,dt = 88\int\left(1 - \frac{1}{t^2+1}\right)dt = 88(t - \tan^{-1}(t)) + E$. Since $s_C(0) = 0$, we have that $E = 0$. Thus $s_C(t) = 88t - 88\tan^{-1}(t)$.

a. $s_A(1) = 88(1 - \ln(2)) \approx 27$. $s_B(1) = 88(1 - \ln 4 - (1/2) + 1) \approx 88(3/2 - \ln 4) \approx 10$. $s_C(1) = 88(1 - \tan^{-1}(1)) \approx 18.9$.

b. $s_A(5) = 88(5 - \ln 6) \approx 282$. $s_B(5) = 88(5 - \ln 36 - (1/6) + 1) \approx 198$. $s_C(5) = 88(5 - \tan^{-1}(5)) \approx 319$.

c. See the development above.

d. Ultimately car C gains the lead. This can be seen by the fact that car C's velocity function is greater than that of the other cars.

8.4.81 First note that the numerator of the given integrand can be written as $x^8 - 4x^7 + 6x^6 - 4x^5 + x^4$, and this quantity when divided by $x^2 + 1$ yields $x^6 - 4x^5 + 5x^4 - 4x^2 + 4 - \frac{4}{1+x^2}$. Thus the given integral is equal to $\left(\frac{x^7}{7} - \frac{2x^6}{3} + x^5 - \frac{4x^3}{3} + 4x\right)\Big|_0^1 - \left(4\tan^{-1}(x)\right)\Big|_0^1 = \frac{1}{7} - \frac{2}{3} + 1 - \frac{4}{3} + 4 - \pi = \frac{22}{7} - \pi$. Since the given integrand is positive on the interval $(0, 1)$, we know that this integral is positive. Thus,

$$0 < \int_0^2 \frac{x^4(1-x)^4}{1+x^2}\,dx = \frac{22}{7} - \pi.$$

Adding π to both sides of this inequality yields $\pi < \frac{22}{7}$.

8.5   Other Integration Strategies

8.5.1 The power rule, substitution, integration by parts, and partial fraction decomposition are examples of analytical methods.

8.5.3 The computer algebra system may use a different algorithm than whoever prepared the table, so the results may look different – however, they should differ by a constant.

8.5.5 Using the table entry for $\int \dfrac{1}{\sqrt{u^2 + a^2}}\, du$, we see that $\int \dfrac{1}{\sqrt{x^2 + 16}}\, dx = \ln(x + \sqrt{x^2 + 16}) + C$.

8.5.7 Using the table entry for $\int \dfrac{u}{bu + a}\, du$, we see that $\int \dfrac{3u}{2u + 7}\, du = \dfrac{3}{4}(7 + 2u - 7\ln|2u + 7|) + C$.

8.5.9 $\int \dfrac{1}{1 - \cos 4x} \cdot \dfrac{1 + \cos 4x}{1 + \cos 4x}\, dx = \int \dfrac{1 + \cos 4x}{\sin^2 4x}\, dx = \int \csc^2 4x\, dx + \int \cot 4x \csc 4x\, dx = \dfrac{-1}{4}\cot 4x +$
$\dfrac{-1}{4}\csc 4x + C$.

8.5.11 Letting $u = 4x + 1$, we substitute to obtain $\dfrac{1}{4}\int u^{-1/2}\, du = \dfrac{1}{2}\sqrt{u} + C = \dfrac{\sqrt{4x + 1}}{2} + C$.

8.5.13 Using the table entry for $\int \dfrac{1}{\sqrt{u^2 - a^2}}\, du$ we have $\int \dfrac{1}{\sqrt{9x^2 - 100}}\, du = \dfrac{1}{3}\int \dfrac{1}{\sqrt{u^2 - 100}}\, du$ where
$u = 3x$. This is then equal to $\frac{1}{3}\ln|u + \sqrt{u^2 - 100}| + C = \frac{1}{3}\ln|3x + \sqrt{9x^2 - 100}| + C$.

8.5.15 Using the table entry for $\int \dfrac{1}{(a^2 + u^2)^{3/2}}\, du$, we have $\int \dfrac{1}{(16 + 9x^2)^{3/2}}\, dx = \dfrac{1}{3}\int \dfrac{1}{(16 + u^2)^{3/2}}\, du =$
$\dfrac{1}{3}\cdot \dfrac{3x}{16\sqrt{16 + 9x^2}} + C = \dfrac{x}{16\sqrt{16 + 9x^2}} + C$.

8.5.17 Using the table entry for $\int \dfrac{1}{u\sqrt{a^2 - u^2}}\, du$, we have $\int \dfrac{1}{x\sqrt{(12)^2 - x^2}}\, dx = \dfrac{-1}{12}\ln\left|\dfrac{12 + \sqrt{144 - x^2}}{x}\right| +$
C.

8.5.19 $\int \dfrac{1}{x(x^{10} + 1)}\, dx = \int \dfrac{10x^9}{10x^{10}(x^{10} + 1)}\, dx$. Let $u = x^{10} + 1$ so that $du = 10x^9\, dx$. Substituting yields
$\int \dfrac{1}{10} \cdot \dfrac{1}{(u - 1)u}\, du$. Using the table entry for $\int \dfrac{du}{u(a + bu)}$ we have $\dfrac{1}{10}\ln\left|\dfrac{u - 1}{u}\right| + C = \dfrac{1}{10}\ln\left|\dfrac{x^{10}}{x^{10} + 1}\right| + C$.

8.5.21 $\int \dfrac{1}{x^2 + 2x + 10}\, dx = \int \dfrac{1}{(x + 1)^2 + 9}\, dx = \dfrac{1}{3}\tan^{-1}\left(\dfrac{x + 1}{3}\right) + C$.

8.5.23 $\int \dfrac{1}{\sqrt{x^2 - 6x}}\, dx = \int \dfrac{1}{\sqrt{(x - 3)^2 - 9}}\, dx = \ln|x - 3 + \sqrt{(x - 3)^2 - 9}| + C$.

8.5.25 $\int \dfrac{e^x}{\sqrt{e^{2x} + 4}}\, dx = \int \dfrac{1}{\sqrt{u^2 + 4}}\, du$ where $u = e^x$. Then we have $\ln(u + \sqrt{u^2 + 4}) + C = \ln(e^x +$
$\sqrt{e^{2x} + 4}) + C$.

8.5.27 $\int \dfrac{\cos x}{\sin^2 x + 2\sin x}\, dx = \int \dfrac{1}{u^2 + 2u}\, du = \int \dfrac{1}{(u + 1)^2 - 1}\, du$ where $u = \sin x$. Then we have

$$\dfrac{-1}{2}\ln\left|\dfrac{u + 2}{u}\right| + C = \dfrac{-1}{2}\ln\left|\dfrac{\sin x + 2}{\sin x}\right| + C.$$

8.5.29 Let $u = x^3$, so that $du = 3x^2\,dx$. Substituting yields

$$\frac{1}{3}\int \frac{\tan^{-1}(u)}{u^2}\,du = \frac{-1}{3}\left(\frac{1}{u}\tan^{-1}(u) - \int \frac{1}{u(1+u^2)}\,du\right) = \frac{-1}{3}\left(\frac{\tan^{-1}(x^3)}{x^3}\right) + \frac{1}{3}\int \frac{1}{u(1+u^2)}\,du.$$

Now let $w = u^2 + 1$ so that $dw = 2u\,du$. This last integral is thus equal to $\dfrac{1}{3}\displaystyle\int \dfrac{2u\,du}{2u^2(1+u^2)} = \dfrac{1}{6}\displaystyle\int \dfrac{dw}{(w-1)w} =$

$\dfrac{-1}{6}\ln\left|\dfrac{w}{w-1}\right| + C = \dfrac{-1}{6}\ln\left|\dfrac{u^2+1}{u^2}\right| + C = \dfrac{-1}{6}\ln\left|\dfrac{x^6+1}{x^6}\right| + C$. Thus the original integral is equal to

$\dfrac{-1}{3}\left(\dfrac{\tan^{-1}(x^3)}{x^3}\right) + \dfrac{-1}{6}\ln\left|\dfrac{x^2+1}{x^2}\right| + C$.

8.5.31 Let $u = \ln x$, so that $du = \frac{1}{x}\,dx$. Substituting yields $\displaystyle\int u\sin^{-1}(u)\,du = \dfrac{2u^2-1}{4}\sin^{-1}(u) + \dfrac{u\sqrt{1-u^2}}{4} +$

$C = \dfrac{2\ln^2 x - 1}{4}\sin^{-1}(\ln x) + \dfrac{\ln x\sqrt{1-\ln^2 x}}{4} + C$.

8.5.33 The integral which gives the length of the curve is $\dfrac{1}{2}\displaystyle\int_0^8 \sqrt{4+x^2}\,dx$. Using the table, we have

$\frac{1}{2}\left(\frac{x}{2}\sqrt{4+x^2} + \frac{4}{2}\ln(x + \sqrt{4+x^2})\right)\Big|_0^8 = 4\sqrt{17} + \ln(8 + 2\sqrt{17}) - \ln 2 \approx 18.59$.

8.5.35 The integral which gives the length of the curve is $\displaystyle\int_0^{\ln 2} \sqrt{1+e^{2x}}\,dx = \int_1^2 \dfrac{\sqrt{1+u^2}}{u}\,du$ where $u = e^x$.

This is equal to $\left(\sqrt{1+u^2} - \ln\left|\dfrac{1+\sqrt{1+u^2}}{u}\right|\right)\Big|_1^2 = \sqrt{5} - \sqrt{2} + \ln(1+\sqrt{2}) - \ln\left(\dfrac{1+\sqrt{5}}{2}\right) \approx 1.22$.

8.5.37 Using the method of shells, we have $\dfrac{V}{2\pi} = \displaystyle\int_0^{12} \dfrac{x}{\sqrt{x+4}}\,dx = \dfrac{2}{3}(x-8)\sqrt{x+4}\Big|_0^{12} = \dfrac{32}{3} + \dfrac{32}{3} = \dfrac{64}{3}$. Thus $V = \dfrac{128\pi}{3}$.

8.5.39 Using the method of disks, we have $\dfrac{V}{\pi} = \displaystyle\int_0^{\pi/2} \sin^2 y\,dy = \dfrac{y}{2} - \dfrac{\sin 2y}{4}\Big|_0^{\pi/2} = \dfrac{\pi}{4}$. Thus $V = \dfrac{\pi^2}{4}$.

8.5.41 $\displaystyle\int \dfrac{x}{\sqrt{2x+3}}\,dx = \dfrac{1}{3}(x-3)\sqrt{3+2x} + C$.

8.5.43 $\displaystyle\int \tan^2 3x\,dx = \dfrac{1}{3}\tan 3x - x + C$.

8.5.45 $\displaystyle\int \dfrac{(x^2-a^2)^{3/2}}{x}\,dx = \dfrac{1}{3}(x^2-a^2)^{3/2} - a^2\sqrt{x^2-a^2} + a^3\cos^{-1}(a/x) + C$.

8.5.47 $\displaystyle\int (a^2-x^2)^{3/2}\,dx = \dfrac{-x}{8}(2x^2 - 5a^2)\sqrt{a^2-x^2} + \dfrac{3a^4}{8}\sin^{-1}(x/a) + C$.

8.5.49 $\displaystyle\int_{2/3}^{4/5} x^8\,dx = \dfrac{(4/5)^9 - (2/3)^9}{9} \approx 0.01202283$.

8.5.51 $\displaystyle\int_0^4 (9+x^2)^{3/2}\,dx = \dfrac{1540 + 243\ln 3}{8} \approx 225.87035$.

8.5.53 $\displaystyle\int_0^{\pi/2} \dfrac{1}{1+\tan^2 x}\,dx = \dfrac{\pi}{4} \approx 0.78539816$.

8.5.55 $\displaystyle\int_0^1 \ln x \ln(1+x)\,dx = 2 - \frac{\pi^2}{12} - \ln 4 \approx -0.208761.$

8.5.57

a. Yes, it is possible, since these are equal for $x > 1$.

b. Yes, since $\frac{1}{9} = 0.\overline{11}$.

8.5.59 The two answers differ by a constant, namely the constant one. This can be seen as follows:

$$\frac{2\sin(x/2)}{\cos(x/2) + \sin(x/2)} = \frac{2\tan(x/2)}{1 + \tan(x/2)} = 2\frac{\sin x}{1 + \cos x} \cdot \frac{1}{1 + \frac{\sin x}{1 + \cos x}}$$

$$= \frac{2\sin x}{1 + \cos x + \sin x} = \frac{2\sin x}{1 + \cos x + \sin x} \cdot \frac{1 - \cos x - \sin x}{1 - \cos x - \sin x}$$

$$= \frac{2\sin x(1 - \cos x - \sin x)}{-2\sin x \cos x} = \frac{\sin x + \cos x - 1}{\cos x} = \frac{\sin x - 1}{\cos x} + 1.$$

8.5.61 $\displaystyle\int x^3 e^{2x}\,dx = \frac{1}{2}x^3 e^{2x} - \frac{3}{2}\int x^2 e^{2x}\,dx = \frac{1}{2}x^3 e^{2x} - \frac{3}{2}\left(\frac{1}{2}x^2 e^{2x} - \int x e^{2x}\,dx\right) = \frac{1}{2}x^3 e^{2x} - \frac{3}{4}x^2 e^{2x} +$

$\displaystyle\frac{3}{2}\left(\frac{1}{2}x e^{2x} - \frac{1}{2}\int e^{2x}\,dx\right) = \frac{1}{2}x^3 e^{2x} - \frac{3}{4}x^2 e^{2x} + \frac{3}{4}x e^{2x} - \frac{3}{8}e^{2x} + C.$

8.5.63 Let $u = 3y$. Then $\displaystyle\int \tan^4 3y\,dy = \frac{1}{3}\int \tan^4 u\,du = \frac{1}{3}\left(\frac{1}{3}\tan^3 u - \int \tan^2 u\,du\right) = \frac{1}{9}\tan^3 u -$

$\displaystyle\frac{1}{3}\left(\tan u - \int du\right) = \frac{1}{9}\tan^3 3y - \frac{1}{3}\tan 3y + y + C.$

8.5.65 Let $u = 2x$. Then $\displaystyle\int x \sin^{-1} 2x\,dx = \frac{1}{4}\int u \sin^{-1} u\,du = \frac{2u^2 - 1}{16}\sin^{-1} u + \frac{u\sqrt{1 - u^2}}{16} + C =$

$\displaystyle\frac{8x^2 - 1}{16}\sin^{-1} 2x + \frac{x\sqrt{1 - 4x^2}}{8} + C.$

8.5.67 $\displaystyle\int x^{-2} \tan^{-1} x\,dx = -(1/x)\tan^{-1} x + \int \frac{1}{x(1 + x^2)}\,dx = \frac{-\tan^{-1}(x)}{x} + \ln\left(\frac{|x|}{\sqrt{1 + x^2}}\right) + C.$

8.5.69 By a direct application of a table entry, $\displaystyle\int \frac{1}{\sqrt{2ax - x^2}}\,dx = \sin^{-1}\left(\frac{x - a}{a}\right) + C.$

8.5.71

a. Using a computer algebra system, we have

θ_0	T	Relative Error
0.1	6.27927	0.000623603
0.2	6.26762	0.0024778
0.3	6.24854	0.0051388
0.4	6.22253	0.00965413
0.5	6.19021	0.0147967
0.6	6.15236	0.0208215
0.7	6.10979	0.0275963
0.8	6.06338	0.0349831
0.9	6.01399	0.0428433
1.0	5.96247	0.0510427

b. All of these values are within 10 percent of 2π.

8.5.73 Let $u = ax + b$, so that $du = a\,dx$. Then we have $\dfrac{1}{a}\displaystyle\int \dfrac{ax + b - b}{ax + b}\,dx = \dfrac{1}{a}\displaystyle\int \left(1 - \dfrac{b}{ax + b}\right)dx =$

$\dfrac{1}{a}\left(x - \dfrac{b}{a}\displaystyle\int \dfrac{1}{u}\,du\right) = \dfrac{1}{a}\left(x - \dfrac{b}{a}\ln|u|\right) + C = \dfrac{1}{a}\left(x - \dfrac{b}{a}\ln|ax + b|\right) + C.$

8.5.75 Let $u = ax + b$, so that $du = a\,dx$. Then we have $\dfrac{1}{a}\displaystyle\int \dfrac{u - b}{a}u^n\,du = \dfrac{1}{a^2}\displaystyle\int (u^{n+1} - bu^n)\,du =$

$\dfrac{1}{a^2}\left(\dfrac{u^{n+2}}{n+2} - \dfrac{bu^{n+1}}{n+1}\right) + C = \dfrac{1}{a^2}\left(\dfrac{(ax + b)^{n+2}}{n + 2} - \dfrac{b(ax + b)^{n+1}}{n + 1}\right) + C.$

8.5.77

a. The result holds.

b. First note that $\displaystyle\int_0^{\pi/2} \cos^n x\,dx = -\int_{\pi/2}^0 \cos^n(\pi/2 - \theta)\,d\theta = \int_0^{\pi/2} \sin^n \theta\,d\theta.$ So we only need to show

the result for $\displaystyle\int_0^{\pi/2} \sin^{10} x\,dx.$ Repeatedly using the reduction formula we have $\displaystyle\int_0^{\pi/2} \sin^{10} x\,dx =$

$\dfrac{9}{10}\displaystyle\int_0^{\pi/2} \sin^8 x\,dx = \dfrac{9 \cdot 7}{10 \cdot 8}\displaystyle\int_0^{\pi/2} \sin^6 x\,dx = \dfrac{9 \cdot 7 \cdot 5}{10 \cdot 8 \cdot 6}\displaystyle\int_0^{\pi/2} \sin^4 x\,dx = \dfrac{9 \cdot 7 \cdot 5 \cdot 3}{10 \cdot 8 \cdot 6 \cdot 4}\displaystyle\int_0^{\pi/2} \sin^2 x\,dx =$

$\dfrac{9 \cdot 7 \cdot 5 \cdot 3}{10 \cdot 8 \cdot 6 \cdot 4 \cdot 2}\displaystyle\int_0^{\pi/2} dx = \dfrac{63\pi}{2^9}.$

c. The values decrease as n increases.

8.6 Numerical Integration

8.6.1 $\Delta x = \dfrac{18 - 4}{28} = \dfrac{1}{2}.$

8.6.3 The Trapezoidal Rule approximates the definite integral by using a trapezoid over each subinterval rather than a rectangle.

8.6.5 The endpoints of the subintervals are -1, 1, 3, 5, 7, and 9. The trapezoidal rule uses the value of f at each of these endpoints.

8.6.7 The absolute error is $|\pi - 3.14| \approx 0.0015926536$. The relative error is $\dfrac{|\pi - 3.14|}{\pi} \approx 5 \times 10^{-4}$.

8.6.9 The absolute error is $|e - 2.72| \approx 0.0017181715$. The relative error is $\dfrac{|e - 2.72|}{e} \approx 6.32 \times 10^{-4}$.

8.6.11
For $n = 1$, we have $f(6) \cdot 8 = 72 \cdot 8 = 576$.
For $n = 2$ we have $f(4) \cdot 4 + f(8) \cdot 4 = 32 \cdot 4 + 128 \cdot 4 = 640$.
For $n = 4$, we have $f(3) \cdot 2 + f(5) \cdot 2 + f(7) \cdot 2 + f(9) \cdot 2 = 18 \cdot 2 + 50 \cdot 2 + 98 \cdot 2 + 162 \cdot 2 = 656$.

8.6.13 We have

$$\dfrac{1}{6}\left(\sin(\pi/12) + \sin(\pi/4) + \sin(5\pi/12) + \sin(7\pi/12) + \sin(3\pi/4) + \sin(11\pi/12)\right) \approx 0.6439505509.$$

8.6.15 For $n = 2$ we have $T(2) = \dfrac{4}{2}(f(2) + 2f(6) + f(10)) = 2(8 + 2(72) + 200) = 704$.
Using the results of number 11: For $n = 4$, note that $T(4) = \dfrac{T(2) + M(2)}{2} = \dfrac{704 + 640}{2} = 672$.
Using the results of number 11: For $n = 8$ we have that $T(8) = \dfrac{T(4) + M(4)}{2} = \dfrac{672 + 656}{2} = 664$.

8.6.17 We have $T(6) = \dfrac{1}{12}(\sin 0 + 2\sin \pi/6 + 2\sin \pi/3 + 2\sin \pi/2 + 2\sin 2\pi/3 + 2\sin 5\pi/6 + \sin \pi) = \dfrac{1}{6}\left(\dfrac{1}{2} + \dfrac{\sqrt{3}}{2} + 1 + \dfrac{\sqrt{3}}{2} + \dfrac{1}{2}\right) = \dfrac{1}{6}(2 + \sqrt{3}).$

8.6.19 The width of each subinterval is 1/25, so $M(25) = \frac{1}{25}(\sin \pi/50 + \sin 3\pi/50 + \sin 5\pi/50 + \cdots + \sin 49\pi/50) \approx .6370388444$. Since $\int_0^1 \sin \pi x \, dx = \frac{2}{\pi}$, the absolute error is $|2/\pi - M(25)| \approx 4.19 \times 10^{-4}$ and the relative error is this number divided by $2/\pi$ which is approximately 6.58×10^{-4}. The Trapezoidal Rule yields approximately $.6357817937$, with a relative error of $\approx .0001316$.

8.6.21

n	$M(n)$	Absolute Error	$T(n)$	Absolute Error
4	99	1	102	2
8	99.75	0.250	100.5	0.5
16	99.9375	0.0625	100.125	0.125
32	99.984375	0.0156	100.03125	0.03125

8.6.23

n	$M(n)$	Absolute Error	$T(n)$	Absolute Error
4	1.50968181	9.68×10^{-3}	1.48067370	1.92×10^{-2}
8	1.50241228	2.41×10^{-3}	1.49517776	4.82×10^{-3}
16	1.50060256	6.03×10^{-4}	1.49879502	1.20×10^{-3}
32	1.50015061	1.51×10^{-4}	1.49969879	3.01×10^{-4}

8.6.25 Because the given function has odd symmetry about the midpoint of the interval $[0, \pi]$, the midpoint rule calculates to be zero for all even values of n, as does the trapezoidal rule.

8.6.27 Answers may vary.
$\overline{T} = \frac{1}{12}\int_0^{12} T(t)\,dt \approx \frac{1}{12}\text{Trapezoid}(12) \approx \frac{1}{24}(47+2(50+46+45+48+52+54+61+62+63+63+59)+55) = 54.5$.

8.6.29 Answers may vary.
$\overline{T} = \frac{1}{12}\int_0^{12} T(t)\,dt \approx \frac{1}{12}\text{Trapezoid}(12) \approx \frac{1}{24}(35 + 2(34 + 34 + 36 + 36 + 37 + 37 + 36 + 35 + 35 + 34 + 33) + 32)) \approx 35.042$.

8.6.31

n	$T(n)$	Absolute Error	$S(n)$	Absolute Error
25	3.19623162	—	—	—
50	3.19495398	4.26×10^{-4}	3.19452809	4.5×10^{-8}

8.6.33

n	$T(n)$	Absolute Error	$S(n)$	Absolute Error
50	1.00008509	—	—	—
100	1.00002127	2.13×10^{-5}	1.00000000	$< 10^{-8}$

8.6.35

n	$T(n)$	Absolute Error	$S(n)$	Absolute Error
4	1820	284	—	—
8	1607.75	71.8	1537	1
16	1553.9844	18	1536.0625	6.25×10^{-2}
32	1540.4990	4.5	1536.0037	3.98×10^{-3}

8.6.37

n	$T(n)$	Absolute Error	$S(n)$	Absolute Error
4	0.46911538	5.25×10^{-2}	—	—
8	0.50826998	1.33×10^{-2}	0.52132152	2.85×10^{-4}
16	0.51825968	3.35×10^{-3}	0.52158957	1.74×10^{-5}
32	0.52076933	8.38×10^{-4}	0.52160588	1.08×10^{-6}

8.6.39

a. True. In the case of a linear function, the region under the curve and over each subinterval is a trapezoid, so the trapezoidal rule gives the exact area.

b. False. Since $E_M(n) \leq \frac{k(b-a)^3}{24n^2}$, we have $E_M(3n) \leq \frac{k(b-a)^3}{24(9n^2)}$, so the error decreases by a factor of about 9.

c. True. Since $E_T(n) \leq \frac{k(b-a)^3}{12n^2}$, we have $E_T(4n) \leq \frac{k(b-a)^3}{24(16n^2)}$, so the error decreases by a factor of about 16.

8.6.41 $\displaystyle\int_0^{\pi/2} \cos^9 x \, dx = \frac{128}{315}.$

n	$M(n)$	Absolute Error	$T(n)$	Absolute Error
4	0.40635058	1.37×10^{-6}	0.40634783	1.38×10^{-6}
8	0.40634921	7.60×10^{-10}	0.40634921	7.62×10^{-9}
16	0.40634921	6.55×10^{-13}	0.40634921	6.56×10^{-13}
32	0.40634921	6.22×10^{-16}	0.40634921	6.22×10^{-16}

8.6.43 $\displaystyle\int_0^{\pi} \ln(5 + 3\cos x)x \, dx = \pi \ln(9/2).$

n	$M(n)$	Absolute Error	$T(n)$	Absolute Error
4	4.72531820	1.20×10^{-4}	4.72507878	1.20×10^{-4}
8	4.72519851	9.123×10^{-9}	4.72519849	9.12×10^{-9}
16	4.72519850	1.06×10^{-16}	4.72519850	1.06×10^{-16}
32	4.72519850	1×10^{-19}	4.72519850	3×10^{-19}

8.6.45 $\displaystyle\int_0^{\pi} \frac{\cos x}{(5/4) - \cos x} \, dx = \frac{2\pi}{3} \approx 2.094395102.$

n	$S(n)$	Absolute Error
4	1.916439963	.178
8	2.080919302	.0135
16	2.094341841	5.33×10^{-5}

8.6.47 $\displaystyle\int_0^{\pi} \sin 6x \cos 3x \, dx = \frac{4}{9} = 0.\overline{4}.$

n	$S(n)$	Absolute Error
8	0.0305049084	.475
16	0.4540112289	.0096
32	.44487	.00087

8.6.49 $\displaystyle\int_0^{2\pi} \sqrt{a^2 \cos^2 t + b^2 \sin^2 t}\, dt$, $a = 4$, $b = 8$.

n	$S(n)$
4	41.88790205
8	39.05860599

8.6.51 We are computing $\dfrac{1}{3\sqrt{2\pi}} \displaystyle\int_{66}^{72} e^{-(x-69)^2/18}\, dx$. If we use Simpson's rule we obtain

n	$S(n)$
4	68.3%
8	68.3%

8.6.53

a. For even n we have $S(n) = \frac{20}{n}(365)(f(a) + 4f(x_1) + 4f(x_2) + \ldots + 4f(x_{n-1}) + f(b))$. For $n = 6$ since there are 6 decades between 1940 and 2000, we have $S(6) \approx 157{,}440$ millions of barrels produced.

b. Following part (a) with $n = 6$ we have $S(n) \approx 66{,}000$ millions of barrels imported.

8.6.55

a. $T(40) = \frac{1}{80}(\sin 1 + 2\sum_{i=1}^{39} \sin e^{i/40} + \sin e) \approx .8748$.

b. $f(x) = \sin e^x$, so $f'(x) = e^x \cos e^x$. $f''(x) = -e^{2x} \sin e^x + e^x \cos e^x = e^x(\cos e^x - e^x \sin e^x)$.

c. $|f''(x)| = |e^x| \cdot |\cos e^x - e^x \sin e^x| \le |e^x|(|\cos e^x| + |e^x \sin e^x|) \le e(1 + e) < 10.11$. However, the graph of the absolute value of $f''(x)$ reveals that is is actually bounded by 5.75 on the interval $[0, 1]$.

d. Since $E_T(n) \le k(b-a)^3/12n^2 = 6/(12(40^2)) = .0003125$, $T(40)$ is accurate to at least 3 decimal places.

8.6.57 Partition $[a, b]$ into n subintervals of equal length so that the ith subinterval is $[x_{2i-2}, x_{2i}]$, which contains x_{2i-1} as its midpoint. The approximating quadratic (or linear function of the points are collinear) through the three points is the quadratic $Ax^2 + Bx + C$, since 3 points on a parabola determine the parabola. (When the 3 points are collinear, we have $A = 0$.) Thus we have $\int_{x_{2i-2}}^{x_{2i}} Ax^2 + Bx + C\, dx = S_{[x_{2i-2}, x_{2i}]}$. Thus $\int_a^b Ax^2 + Bx + C\, dx = \sum_{i=1}^n \int_{x_{2i-2}}^{x_{2i}} Ax^2 + Bx + C\, dx = \sum_{i=1}^n S_{[x_{2i-2}, x_{2i}]}(1) = S_{[a,b]}(n)$.

8.6.59 The trapezoidal rule will be an overestimate in this case. This is because of the fact that if the function is above the axis and concave up on the given interval, then each trapezoid on each subinterval lies over the area under the curve for that corresponding subinterval.

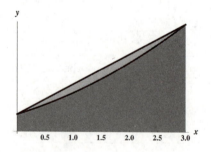

8.6.61 Using the previous results, we have

$$S(2n) = \frac{4T(2n) - T(n)}{3} = \frac{2(T(n) + M(n)) - T(n)}{3} = \frac{2M(n) + T(n)}{3}.$$

So $S(50) = \frac{1}{3}(2M(25) + T(25)) \approx .8298251909$.

8.7 Improper Integrals

8.7.1 The interval of integration is infinite or the integrand is unbounded on the interval of integration.

8.7.3 Compute $\int_0^1 \frac{1}{\sqrt{x}}\,dx = \lim_{b\to 0^+} \int_b^1 \frac{1}{\sqrt{x}}\,dx$.

8.7.5 $\int_1^\infty x^{-2}\,dx = \lim_{b\to\infty} \int_1^b x^{-2}\,dx = \lim_{b\to\infty} \left(-\frac{1}{x}\right)\big|_1^b = \lim_{b\to\infty} \left(1 - \frac{1}{b}\right) = 1$.

8.7.7 $\int_2^\infty \frac{dx}{\sqrt{x}} = \lim_{b\to\infty} \int_2^b \frac{dx}{\sqrt{x}} = \lim_{b\to\infty} 2\sqrt{x}\big|_2^b = \lim_{b\to\infty} 2(\sqrt{b} - \sqrt{2}) = \infty$, so the integral diverges.

8.7.9 $\int_0^\infty e^{-2x}\,dx = \lim_{b\to\infty} \int_0^b e^{-2x}\,dx = \lim_{b\to\infty} \left(-\frac{1}{2}e^{-2x}\right)\big|_0^b = \lim_{b\to\infty} \frac{1}{2}\left(1 - e^{-2b}\right) = \frac{1}{2}$.

8.7.11 $\int_{e^2}^\infty \frac{dx}{x\ln^p x} = \lim_{b\to\infty} \int_{e^2}^b \frac{dx}{x\ln^p x} = \lim_{b\to\infty} \left(\frac{1}{1-p}\ln^{1-p} x\right)\big|_{e^2}^b = \lim_{b\to\infty} \frac{1}{p-1}\left(2^{1-p} - \ln^{1-p} b\right) = \frac{1}{(p-1)2^{p-1}}$.

8.7.13 $\int_0^\infty xe^{-x^2}\,dx = \lim_{b\to\infty} \int_0^b xe^{-x^2}\,dx = \lim_{b\to\infty} \left(-\frac{1}{2}e^{-x^2}\right)\big|_0^b = \lim_{b\to\infty} \frac{1}{2}\left(1 - e^{-b^2}\right) = \frac{1}{2}$.

8.7.15 $\int_2^\infty \frac{\cos(\pi/x)}{x^2}\,dx = \lim_{b\to\infty} \int_2^b \frac{\cos(\pi/x)}{x^2}\,dx = \lim_{b\to\infty} \left(-\frac{1}{\pi}\sin(\pi/x)\right)\big|_2^b = \lim_{b\to\infty} \frac{1}{\pi}\left(1 - \sin(\pi/b)\right) = \frac{1}{\pi}$.

8.7.17 The indefinite integral $\int \frac{x}{\sqrt{x^4+1}}\,dx = \frac{1}{2}\ln(x^2 + \sqrt{x^4+1}) + C$ can be evaluated by the preliminary substitution $u = x^2$ and then the trig substitution $u = \tan\theta$. Thus we have $\int_0^\infty \frac{x}{\sqrt{x^4+1}}\,dx = \lim_{b\to\infty} \int_0^b \frac{x}{\sqrt{x^4+1}}\,dx = \lim_{b\to\infty} \left(\frac{1}{2}\ln(x^2 + \sqrt{x^4+1})\right)\big|_0^b = \lim_{b\to\infty} \frac{1}{2}\ln(b^2 + \sqrt{b^4+1}) = \infty$, so the integral diverges.

8.7.19 The method of partial fractions gives $\frac{x}{(x+2)^2} = \frac{1}{x+2} - \frac{2}{(x+2)^2}$, therefore

$$\int_2^\infty \frac{x}{(x+2)^2}\,dx = \lim_{b\to\infty} \int_2^b \frac{x}{(x+2)^2}\,dx = \lim_{b\to\infty} \left(\ln(x+2) + \frac{2}{x+1}\right)\big|_2^b$$

$$= \lim_{b\to\infty} \left(\ln(b+2) + \frac{2}{b+2} - \ln 4 - \frac{2}{3}\right) = \infty,$$

so the integral diverges.

8.7.21 Using the result from Example 2, we see that the volume is given by $V = \pi \int_1^\infty x^{-4}\,dx = \frac{\pi}{4-1} = \frac{\pi}{3}$.

8.7.23 Using the result from Example 2, we see that the volume is given by $V = \pi \int_1^\infty \left(\frac{1}{x^2} + \frac{1}{x^3}\right)\,dx = \frac{\pi}{2-1} + \frac{\pi}{3-1} = \frac{3\pi}{2}$.

8.7.25 $V = \pi \int_2^\infty \frac{dx}{x(\ln x)^2} = \pi \lim_{b\to\infty} \int_2^b \frac{dx}{x(\ln x)^2} = \pi \lim_{b\to\infty} \left(-\frac{1}{\ln x}\right)\big|_2^b = \pi \lim_{b\to\infty} \left(\frac{1}{\ln 2} - \frac{1}{\ln b}\right) = \frac{\pi}{\ln 2}$.

8.7.27 $\int_0^8 \frac{dx}{\sqrt[3]{x}} = \lim_{c\to 0^+} \int_c^8 x^{-1/3}\,dx = \lim_{c\to 0^+} \left(\frac{3}{2}x^{2/3}\right)\big|_c^8 = \frac{3}{2}\lim_{c\to 0^+}(4 - c^{2/3}) = 6$.

8.7.29 $\int_0^1 \frac{x^3}{x^4-1}\,dx = \lim_{c\to 1^-} \int_0^c \frac{x^3}{x^4-1}\,dx = \lim_{c\to 1^-} \left(\frac{1}{4}\ln|x^4-1|\right)\big|_0^c = \frac{1}{4}\lim_{c\to 1^-} \ln|c^4-1| = -\infty$, so the integral diverges.

8.7.31

$$\int_0^{10} \frac{dx}{\sqrt[4]{10-x}} = \lim_{c\to 10^-} \int_0^c (10-x)^{-1/4}\,dx = \lim_{c\to 10^-} \left(-\frac{4}{3}(10-x)^{3/4}\right)\big|_0^c$$

$$= \frac{4}{3}\lim_{c\to 10^-} \left(10^{3/4} - (10-c)^{3/4}\right) = \frac{4}{3}10^{3/4}.$$

8.7.33 $\int_0^1 \ln x^2\,dx = 2\lim_{c\to 0^+} \int_c^1 \ln x\,dx = 2\lim_{c\to 0^+} (x\ln x - x)\big|_c^1 = 2\lim_{c\to 0^+}(-1 - c\ln c + c) = -2$. (The fact that $\lim_{c\to 0^+} c\ln c = 0$ can be derived using L'Hôpital's rule, or from the result $\lim_{c\to 0^+} c^c = 1$.)

8.7.35 By symmetry,

$$\int_{-2}^{2} \frac{dx}{\sqrt{4-x^2}} = 2\int_{0}^{2} \frac{dx}{\sqrt{4-x^2}} = 2\lim_{c\to 2^-}\int_{0}^{c} \frac{dx}{\sqrt{4-x^2}} = 2\lim_{c\to 2^-}\left(\sin^{-1}(x/2)\right)\big|_0^c = 2(\sin^{-1}1 - \sin^{-1}0) = \pi.$$

8.7.37 $V = \pi\int_1^2 (x-1)^{-1/2}\,dx = \pi\lim_{c\to 1^+}\int_c^2 (x-1)^{-1/2}\,dx = \pi\lim_{c\to 1^+}\left(2(x-1)^{1/2}\right)\big|_c^2 = 2\pi\lim_{c\to 1^+}(1-\sqrt{c-1}) = 2\pi.$

8.7.39 $V = 2\pi\int_0^4 x(4-x)^{-1/3}\,dx = 2\pi\int_0^4 (4-u)u^{-1/3}\,du = 2\pi\int_0^4 (4u^{-1/3} - u^{2/3})\,du$ via the substitution $u = 4-x$. Therefore $V = 2\pi\lim_{c\to 0^+}\int_c^4 (4u^{-1/3} - u^{2/3})\,du = 2\pi\lim_{c\to 0^+}\left(6u^{2/3} - \frac{3}{5}u^{5/3}\right)\big|_c^4 = \frac{72\cdot 2^{1/3}\pi}{5}.$

8.7.41 Notice that this curve is symmetric in both the x and y-axes, so it suffices to find the length of the part of the curve in the first quadrant, which joins the points $(8,0)$ and $(0,8)$. Solving for y gives $y = (4 - x^{2/3})^{3/2}$, so $y' = (4-x^{2/3})^{1/2}x^{-1/3}$ and $1 + (y')^2 = \frac{4-x^{2/3}}{x^{2/3}} + 1 = \frac{4}{x^{2/3}}$, so the length of the part of the curve in the first quadrant is $L = 2\int_0^8 x^{-1/3}\,dx = 2\lim_{c\to 0^+}\int_c^8 x^{-1/3}\,dx = 2\lim_{c\to 0^+}\left(\frac{3}{2}x^{2/3}\right)\big|_c^8 = 12.$ Therefore the entire curve has length 48.

8.7.43 As in Example 6, we have $\text{AUC}_i = \int_0^\infty C_i(t)\,dt = 250\int_0^\infty e^{-0.08t}\,dt = \frac{250}{0.08} = 3125$ and $\text{AUC}_o = \int_0^\infty C_0(t)\,dt = 200\int_0^\infty (e^{-0.08t} - e^{-1.8t})\,dt = 200\left(\frac{1}{0.08} - \frac{1}{1.8}\right) = \frac{21{,}500}{9} \approx 2389$ (here we use the fact that $\int_0^\infty e^{-ax}\,dx = \frac{1}{a}$ for $a > 0$.) Therefore the bioavailability of the drug is $F = \frac{\text{AUC}_o}{\text{AUC}_i} = \frac{21{,}500}{9\cdot 3125} \approx 0.76.$

8.7.45 The maximum distance is

$$D = 10\int_0^\infty (t+1)^{-2}\,dt = 10\lim_{b\to\infty}\int_0^b (t+1)^{-2}\,dt = 10\lim_{b\to\infty}\left(-(t+1)^{-1}\right)\big|_0^b = 10\,\text{mi}.$$

8.7.47

a. True. The area under the curve $y = f(x)$ from 0 to ∞ is less than the area under $y = g(x)$ on this interval, which by assumption is finite.

b. False. For example, take $f(x) = 1$; then $\int_0^\infty f(x)\,dx = \infty$.

c. False. For example, take $p = 1/2$ and $q = 1$.

d. True. The area under the curve $y = x^{-q}$ from 1 to ∞ is less than the area under $y = x^{-p}$ on this interval, which by assumption is finite.

e. True. Using the result in Example 2, we see that this integral exists if and only if $3p + 2 > 1$, which is equivalent to $p > -1/3$.

8.7.49

a. The function $e^{-|x|}$ is even, so $\int_{-\infty}^\infty e^{-|x|}\,dx = 2\int_0^\infty e^{-x}\,dx = 2$ (Note that here we used the fact that $\int_0^\infty e^{-ax}\,dx = \frac{1}{a}$ for $a > 0$.)

b. The function $x^3/(1+x^8)$ is odd, so $\int_{-\infty}^\infty \frac{x^3}{1+x^8}\,dx = \int_0^\infty \frac{x^3}{1+x^8}\,dx + \int_{-\infty}^0 \frac{x^3}{1+x^8}\,dx = \int_0^\infty \frac{x^3}{1+x^8}\,dx - \int_0^\infty \frac{x^3}{1+x^8}\,dx = 0$ assuming $\int_0^\infty \frac{x^3}{1+x^8}\,dx$ exists, which is true because $0 < \frac{x^3}{1+x^8} < \frac{1}{x^5}$ on $[1,\infty)$ and $\int_1^\infty x^{-5}\,dx$ exists.

8.7.51

n	$T_2(n)$	$T_4(n)$	$T_8(n)$
4	.880619	.886319	1.036632
8	.881704	.886227	.886319
16	.881986	.886227	.886227
32	.882058	.886227	.886227

Based on these results, we conclude that $\int_0^\infty e^{-x^2}\,dx \approx 0.886227.$

8.7.53 Integration by parts gives $\int x \ln x \, dx = \frac{x^2}{4}(2 \ln x - 1) + C$, so $\int_0^1 x \ln x \, dx = \lim_{c \to 0+} \int_c^1 x \ln x \, dx = \lim_{c \to 0+} \left(\frac{x^2}{4}(2 \ln x - 1) \right) \Big|_c^1 = \frac{1}{4} \lim_{c \to 0+} (c^2(-2 \ln c + 1) - 1) = -\frac{1}{4}$. (The fact that $\lim_{c \to 0+} c^2 \ln c = 0$ can be derived using L'Hôpital's rule, or from the result $\lim_{c \to 0+} c^c = 1$.)

8.7.55 Let $u = x^2$ so that $2x \, dx = du$. The first integral is then equal to $\frac{1}{2} \int_0^\infty e^{-u} \, du = \frac{1}{2}$, using the fact that $\int_0^\infty e^{-ax} \, dx = \frac{1}{a}$ for $a > 0$. The second integral cannot be evaluated by finding an antiderivative for $x^2 e^{-x^2}$; however using more advanced methods it can be shown that $\int_0^\infty x^2 e^{-x^2} \, dx = \frac{\sqrt{\pi}}{4} \approx 0.443$.

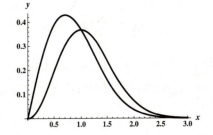

8.7.57 The region R has area $A = \int_0^\infty e^{-bx} \, dx - \int_0^\infty e^{-ax} \, dx = \frac{1}{b} - \frac{1}{a}$.

8.7.59

a. We have $A(a, b) = \int_b^\infty e^{-ax} \, dx = \lim_{c \to \infty} \int_b^c e^{-ax} \, dx = \lim_{c \to \infty} \left(-\frac{1}{a} e^{-ax} \right) \Big|_b^c = \frac{1}{a} \lim_{c \to \infty} (e^{-ab} - e^{-ac}) = \frac{e^{-ab}}{a}$.

b. Solving $e^{-ab} = 2a$ for b gives $b = g(a) = -\frac{1}{a} \ln(2a)$.

c. The function g has $g'(x) = \frac{1}{x^2} \ln(2x) - \frac{1}{x^2} = \frac{\ln(2x) - 1}{x^2}$, so g has a critical point at $x = e/2$, and the first derivative test shows that g takes a minimum at this point. Hence $b^* = g(e/2) = -\frac{2}{e}$.

8.7.61

a. The solid has volume $V = \pi \int_0^1 x^{-2p} \, dx$, which by the result in Problem 60 is finite if and only if $2p < 1$, or $p < 1/2$.

b. The solid has volume $V = 2\pi \int_0^1 x^{1-p} \, dx$, which by the result in Problem 60 is finite if and only if $p - 1 < 1$, or $p < 2$.

8.7.63 These integrals cannot be evaluated by finding an antiderivative of their integrands. However, if we make the substitution $u = \pi/2 - x$ we find that $\int_0^{\pi/2} \ln \sin x \, dx = -\int_{\pi/2}^0 \ln \cos u \, du = \int_0^{\pi/2} \ln \cos x \, dx$. We also have

$$\int_{\pi/2}^\pi \ln \sin x \, dx = \int_0^{\pi/2} \ln \sin(x + \pi/2) \, dx = \int_0^{\pi/2} \ln \cos x \, dx,$$

and therefore

$$\int_0^{\pi/2} \ln \sin x \, dx = \frac{1}{2} \int_0^\pi \ln \sin x \, dx = \int_0^{\pi/2} \ln \sin 2y \, dy = \int_0^{\pi/2} (\ln \sin y + \ln \cos y + \ln 2) \, dy.$$

This implies $\int_0^{\pi/2} \ln \cos x \, dx = -\int_0^{\pi/2} \ln 2 \, dx = -\frac{\pi \ln 2}{2}$.

8.7.65 This integral cannot be evaluated by finding an antiderivative of the integrand; however the result may be verified by numerical approximation.

8.7.67 We have the relation $B = I \int_0^\infty e^{-rt} \, dt = \frac{I}{r}$, using the result that $\int_0^\infty e^{-ax} \, dx = \frac{1}{a}$ for $a > 0$. Therefore $B = 5000/0.12 = \$41,666.67$.

8.7.69

a. We have $\int_0^\infty e^{-ax}\cos bx\, dx = \lim_{c\to\infty}\int_0^c e^{-ax}\cos bx\, dx = \lim_{c\to\infty}\left(\frac{e^{-ax}(b\sin bx - a\cos bx)}{a^2+b^2}\right)\Big|_0^c =$
$\lim_{c\to\infty}\frac{a+e^{-ac}(b\sin bc - a\cos bc)}{a^2+b^2} = \frac{a}{a^2+b^2}.$

b. We have $\int_0^\infty e^{-ax}\sin bx\, dx = \lim_{c\to\infty}\int_0^c e^{-ax}\sin bx\, dx = \lim_{c\to\infty}\left(-\frac{e^{-ax}(a\sin bx + b\cos bx)}{a^2+b^2}\right)\Big|_0^c =$
$\lim_{c\to\infty}\frac{b-e^{-ac}(a\sin bc + b\cos bc)}{a^2+b^2} = \frac{b}{a^2+b^2}.$

8.7.71 Evaluate the improper integral $\int_0^\infty te^{-at}\, dt = \lim_{b\to\infty}\int_0^b te^{-at}\, dt = \lim_{b\to\infty}\left(-\frac{e^{-at}(at+1)}{a^2}\right)\Big|_0^b = \frac{1}{a^2}\lim_{b\to\infty}\left(1-e^{-ab}(ab+1)\right) = \frac{1}{a^2}$, provided $a > 0$. Therefore $0.00005\int_0^\infty te^{-0.00005t}\, dt = \frac{0.00005}{0.00005^2} = 20{,}000$ hrs.

8.7.73

a. We have $W = GMm\int_R^\infty x^{-2}\, dx = GMm\lim_{b\to\infty}\left(-\frac{1}{x}\right)\Big|_R^b = \frac{GMm}{R} \approx 6.28\times 10^7$ m J.

b. Solve $\frac{1}{2}v_e^2 = 6.28\times 10^7$ to obtain $v_e \approx 11.2$ km/s.

c. We need $\frac{GM}{R} \geq \frac{1}{2}c^2 \iff R \leq \frac{2GM}{c^2} \approx 9$ mm.

8.7.75

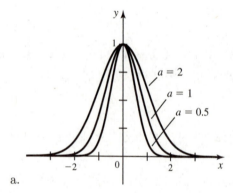

a.

b. The areas are $\sqrt{2\pi}, \sqrt{\pi}, \sqrt{\pi/2}$ respectively.

c. Completing the square gives

$$ax^2 + bx + c = a\left(x + \frac{b}{2a}\right)^2 + \frac{4ac - b^2}{4a},$$

and therefore

$$\int_{-\infty}^\infty e^{-(ax^2+bx+c)}\, dx = e^{(b^2-4ac)/(4a)}\int_{-\infty}^\infty e^{-a(x+(b/2a))^2}\, dx.$$

Make the substitution $y = x + b/(2a)$ to obtain

$$\int_{-\infty}^\infty e^{-(ax^2+bx+c)}\, dx = e^{(b^2-4ac)/(4a)}\int_{-\infty}^\infty e^{-ay^2}\, dy = e^{(b^2-4ac)/(4a)}\sqrt{\frac{\pi}{a}}.$$

8.7.77 The Laplace transform of $f(t) = e^{at}$ is given by $F(s) = \int_0^\infty e^{-st}e^{at}\, dt = \int_0^\infty e^{-(s-a)t}\, dt = \frac{1}{s-a}$, using the formula $\int_0^\infty e^{-cx}\, dx = \frac{1}{c}$ for $c > 0$.

8.7.79 The Laplace transform of $f(t) = \sin at$ is given by $F(s) = \int_0^\infty e^{-st}\sin at\, dt = \frac{a}{s^2+a^2}$ (this formula is derived in the solution to problem 69 b.)

8.7.81

a. Make the substitution $x = y + 2$; then

$$\int_1^3 \frac{dx}{\sqrt{(x-1)(3-x)}} = \int_{-1}^1 \frac{dy}{\sqrt{1-y^2}} = 2\int_0^1 \frac{dy}{\sqrt{1-y^2}} = 2\lim_{c\to 1^-}\left(\sin^{-1}x\right)\big|_0^c = \pi.$$

b. The substitution $y = e^x$ gives

$$\int_1^\infty \frac{dx}{e^{x+1}+e^{3-x}} = \frac{1}{e}\int_1^\infty \frac{e^x\,dx}{e^{2x}+e^2} = \frac{1}{e}\int_e^\infty \frac{dy}{y^2+e^2} = \frac{1}{e}\lim_{b\to\infty}\left(\frac{1}{e}\tan^{-1}\left(\frac{y}{e}\right)\right)\Big|_e^b = \frac{\pi}{4e^2}.$$

8.7.83 Assume $p > 0$, and observe that the functions $1/(x^p + x^{-p})$ and $1/x^p$ have the same growth rate as $x \to \infty$; therefore by the result in Example 2, we have $\int_0^\infty \frac{dx}{x^p+x^{-p}} < \infty \iff p > 1$.

8.7.85

a. Integrate by parts with $u = \sqrt{x}\ln x$ and $v = -1/(1+x)$: $\int_0^\infty \frac{\sqrt{x}\ln x}{(1+x)^2}\,dx = \frac{1}{2}\int_0^\infty \frac{\ln x + 2}{\sqrt{x}(x+1)}\,dx = \frac{1}{2}\int_0^\infty \frac{\ln x}{\sqrt{x}(x+1)}\,dx + \int_0^\infty \frac{dx}{\sqrt{x}(x+1)}$ (the integration by parts is legitimate for this improper integral because the product uv has limit 0 as $x \to \infty$ and as $x \to 0^+$).

b. Let $y = 1/x$. Then $dy = \frac{-1}{x^2}\,dx$.

c. We have $\int_0^1 \frac{\ln x}{\sqrt{x}(x+1)}\,dx = \int_\infty^1 \frac{-\ln y}{\frac{1}{\sqrt{y}}\left(\frac{1}{y}+1\right)}\left(-\frac{dy}{y^2}\right) = -\int_1^\infty \frac{\ln y}{\sqrt{y}(1+y)}\,dy$, and hence $\int_0^\infty \frac{\ln x}{\sqrt{x}(x+1)}\,dx = 0$.

d. The change of variables $z = \sqrt{x}$ gives $\int_0^\infty \frac{dx}{\sqrt{x}(x+1)} = 2\int_0^\infty \frac{dz}{z^2+1} = \pi$.

8.8 Differential Equations

8.8.1 Second-order, since the highest-order derivative appearing in the equation is second order.

8.8.3 The equation is second-order, so we expect two arbitrary constants in the general solution.

8.8.5 A separable first-order differential equation is one that can be written in the form $g(y)y'(t) = h(t)$.

8.8.7 Integrate both sides with respect to t and convert the integral on the left side to an integral with respect to y.

8.8.9 Integrate both sides with respect to t: $\int y'(t)\,dt = \int(3t^2 - 4t + 10)\,dt\, y(t) = t^3 - 2t^2 + 10t + C$; then substitute $y(0) = 20$ to obtain $y(0) = (t^3 - 2t^2 + 10t + C)\big|_{t=0} = C = 20$, so $y(t) = t^3 - 2t^2 + 10t + 20$.

8.8.11 Integrate both sides with respect to t: $\int y'(t)\,dt = \int \frac{2t^2+4}{t}\,dt = \int\left(2t + \frac{4}{t}\right)dt\, y(t) = t^2 + 4\ln t + C$; then substitute $y(1) = 2$ to obtain $y(1) = (t^2 + 4\ln t + C)\big|_{t=1} = 1 + C = 2$, so $C = 1$ and thus $y(t) = t^2 + 4\ln t + 1$.

8.8.13 The general solution is $y(t) = Ce^{3t} + \frac{4}{3}$.

8.8.15 The general solution is $y(x) = Ce^{-2x} - 2$.

8.8.17 The general solution is $y(t) = Ce^{3t} + 2$; substitute $y(0) = 9$ to obtain $C + 2 = 9$, so $C = 7$; hence $y(t) = 7e^{3t} + 2$.

8.8.19 The general solution is $y(t) = Ce^{-2t} - 2$; substitute $y(0) = 0$ to obtain $C - 2 = 0$, so $C = 2$; hence $y(t) = 2e^{-2t} - 2$.

8.8.21

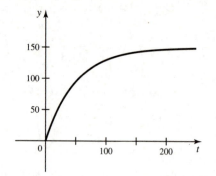

a. The general solution is $y(t) = Ce^{-0.02t} + 150$; substitute $y(0) = 0$ to obtain $C + 150 = 0$, so $C = -150$; hence $y(t) = 150(1 - e^{-0.02t})$.

b. The steady-state level is $\lim_{t \to \infty} 150(1 - e^{-0.02t}) = 150\,\text{mg}$.

c. We have $150(1 - e^{-0.02t}) = 0.9 \cdot 150$, so $e^{-0.02t} = 0.1$, and thus $t = \frac{\ln 10}{0.02} \approx 115\,\text{hrs}$.

8.8.23 The equation is separable, so we have $\int y\, dy = \int 3t^2\, dt$. So $\frac{y^2}{2} = t^3 + C$, and thus $y = \pm\sqrt{2t^3 + C}$.

8.8.25 The equation is separable, so we have $\int e^{-y/2}\, dy = \int \sin t\, dt$, and so $-2e^{-y/2} = -\cos t + C$. Thus, $y = -2\ln\left(\frac{1}{2}\cos t + C\right)$.

8.8.27 This equation is not separable.

8.8.29 This equation is separable, so we have $\int 2y\, dy = \int e^t\, dt$, so $y^2 = e^t + C$, and thus $y = \pm\sqrt{e^t + C}$. Substituting $y(\ln 2) = 1$ gives $1 = 2 + C$ so $C = -1$, and the solution to this initial value problem is $y = \sqrt{e^t - 1}$.

8.8.31 This equation is separable, so we have $\int e^y\, dy = \int e^x\, dx$, and thus $e^y = e^x + C$. Therefore, $y = \ln(e^x + C)$. Substituting $y(0) = \ln 3$ gives $\ln 3 = \ln(1 + C)$, so $C = 2$ and the solution to this initial value problem is $y = \ln(e^x + 2)$.

8.8.33

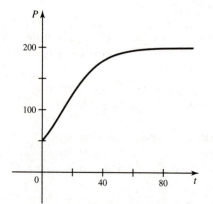

a. This equation is separable, so we have $\int \frac{200}{P(200-P)}\, dP = \int 0.08\, dt$, so $\int \left(\frac{1}{P} + \frac{1}{200-P}\right) dP = 0.08t + C$. Therefore, $\ln\left|\frac{P}{200-P}\right| = 0.08t + C$. Substituting $P(0) = 50$ gives $-\ln 3 = C$, and solving for P gives $P(t) = \frac{200}{3e^{-0.08t}+1}$.

b. The steady-state population is $\lim_{t \to \infty} P(t) = 200$.

8.8.35

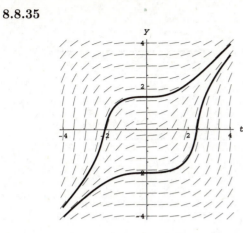

8.8.37

a. This matches with D.

b. This matches with B.

c. This matches with A.

d. This matches with C.

8.8.39

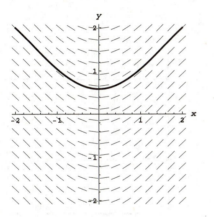

8.8.41

a. False. The general solution is $y = Ce^{20t}$.

b. False. They both do satisfy this differential equation.

c. False. Rewrite the equation as $y'(t) = (t + 2)(y + 2)$.

d. True. $y'(t) = 2(t + 1) = 2\sqrt{(t + 1)^2} = 2\sqrt{y}$.

8.8.43

a. $y = 0$ is an equilibrium solution.

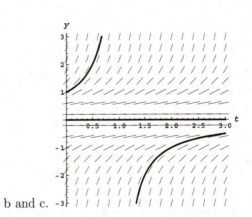

b and c.

8.8.45

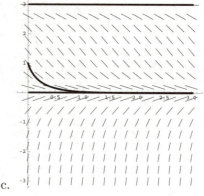

a. Solve $y(y - 3) = 0$ to get equilibrium solutions $y = 0$ and $y = 3$.

b and c.

8.8.47

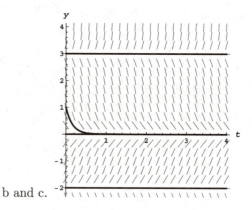

a. The equilibrium solutions are $y = 0$, $y = -2$ and $y = 3$.

b and c.

8.8.49 This equation is separable, so we have $\int \frac{dp}{p+1} = \int \frac{dt}{t^2}$, so $\ln|p+1| = -\frac{1}{t} + D$, and thus $p = Ce^{-1/t} - 1$. Substituting $p(1) = 3$ gives $3 = Ce^{-1} - 1$, so $C = 4e$ and the solution to this initial value problem is $p = 4e^{1-1/t} - 1$.

8.8.51 This equation is separable, so we have $\int \sec^2 w \, dw = \int 2t \, dt$, so $\tan w = t^2 + C$. Thus, $w = \tan^{-1}(t^2 + C)$. Substituting $w(0) = \pi/4$ gives $\pi/4 = \tan^{-1} C$, so $C = \tan(\pi/4) = 1$ and the solution to this initial value problem is $w = \tan^{-1}(t^2 + 1)$.

8.8.53

a. This equation is separable, so we have $\int \frac{1}{y(1-y)} \, dy = \int k \, dt$, so $\int \left(\frac{1}{y} + \frac{1}{1-y} \right) dy = kt + D$. Therefore, $\ln \left| \frac{y}{1-y} \right| = kt + D$, which is equivalent to $\frac{y}{1-y} = Ce^{kt}$. Substituting $y(0) = y_0$ gives $C = y_0/(1 - y_0)$, and thus $y = \frac{y_0}{(1-y_0)e^{-kt}+y_0}$.

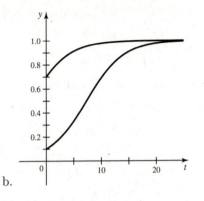

c. The denominator in $y(t)$ above is positive for all $t \geq 0$ when $0 < y_0 < 1$, so $y(t)$ is defined for all $t \geq 0$; we have $\lim_{t \to \infty} y(t) = 1$, which is the steady-state solution.

b.

8.8.55

a. We have $mv'(t) = mg - Rv$, so $v'(t) = g - bv$ with $b = R/m$.

b. Solve $bv = g$ to obtain terminal velocity $\tilde{v} = g/b = mg/R$.

c. The equation $v' = g - bv$ is first-order linear, with general solution $v = Ce^{-bv} + \tilde{v}$. The initial condition $v(0) = 0$ gives $C = -\tilde{v}$, which gives $v = \tilde{v}(1 - e^{-bt})$.

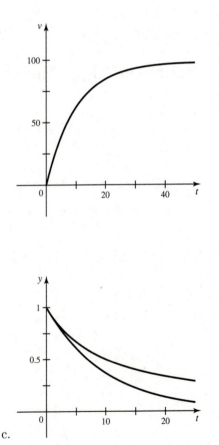

d. We have $b = 0.1$, $\tilde{v} = 98$ m/s.

8.8.57

a. The general solution to $y' = -ky$ is $y = Ce^{-kt}$.

b. The equation $y' = -ky^2$ is separable, so we have $-\int \frac{dy}{y^2} = \int k \, dt$, so $\frac{1}{y} = kt + C$. The initial condition $y(0) = y_0$ gives $C = 1/y_0$, and solving for y gives $y = \frac{1}{kt + 1/y_0} = \frac{y_0}{1 + ky_0 t}$.

c.

8.8.59

a. The equation $B' = aB - m$ is first-order linear, with general solution $B = Ce^{at} + m/a$; in this case $a = 0.05$, $m/a = 20{,}000$, and the initial condition $B_0 = 15{,}000$ gives $C = -5000$, so $B = 20{,}000 - 5000e^{0.05t}$. The balance decreases.

b. The steady-state (constant balance) solution is $B = m/a = \$50,000$, which gives $m = 0.05 \cdot 50,000 = \$2500$.

8.8.61

a. Solving $y' = 0$ gives the equilibrium solution $y = -b/a$, which is a horizontal line.

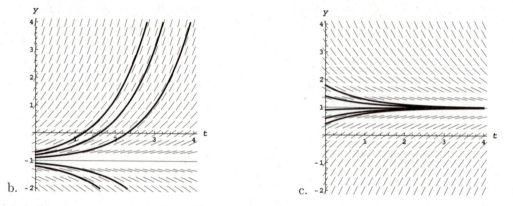

Note that the general solutions $y = (y_0 + \frac{b}{a})e^{at} - \frac{b}{a}$ increases without bound if $a > 0$ and $y_0 > \frac{-b}{a}$, and decreases without bound if $a > 0$ but $y_0 < \frac{-b}{a}$. But if $a < 0$, the general solutions have limit $\frac{-b}{a}$, but must increase to it if $y_0 < \frac{-b}{a}$ and decrease to it if $y_0 > \frac{-b}{a}$.

8.9 Chapter Eight Review

8.9.1

a. True. Two applications of integration by parts are needed to reduce to $\int e^{2x}\, dx$.

b. False. This integral can be done using a trigonometric substitution.

c. False. Both are correct, since $-\cos^2 x = \sin^2 x - 1$, so $\sin^2 x$ and $-\cos^2 x$ differ by a constant.

8.9.3 Two applications of integration by parts gives $\int e^x \sin x\, dx = e^x \sin x - \int e^x \cos x\, dx = e^x \sin x - e^x \cos x - \int e^x \sin x\, dx$. Thus, $2\int e^x \sin x\, dx = e^x \sin x - e^x \cos x$, so $\int e^x \sin x\, dx = \frac{1}{2}(e^x \sin x - e^x \cos x) + C$.

8.9.5 Let $u = 4\theta$, Then $\int \cos^2 \theta\, d\theta = \frac{1}{4}\int \cos^2 u\, du = \frac{1}{4}\int \frac{1+\cos 2u}{2}\, du = \frac{u}{8} + \frac{\sin 2u}{16} + C = \frac{\theta}{2} + \frac{\sin 8\theta}{16} + C$.

8.9.7 Let $u = \sec z$. Then $du = \sec z \tan z\, dz$ and $\int \sec^5 z \tan z\, dz = \int u^4\, du = \frac{u^5}{5} + C = \frac{\sec^5 z}{5} + C$.

8.9.9 Let $u = \cos x$ so that $du = -\sin x\, dx$. Then

$$\int_0^{\pi/6} \sin^5 \theta\, d\theta = \int_0^{\pi/6} \sin^4 \theta \cdot \sin \theta\, d\theta = -\int_1^{\sqrt{3}/2} (1-u^2)^2\, du = \int_{\sqrt{3}/2}^1 (1-u^2)^2\, du$$

$$= \int_{\sqrt{3}/2}^1 (1 - 2u^2 + u^4)\, du = \left(u - \frac{2}{3}u^3 + \frac{1}{5}u^5\right)\Big|_{\sqrt{3}/2}^1$$

$$= \frac{8}{15} - \frac{49\sqrt{3}}{160} = \frac{256 - 147\sqrt{3}}{480}.$$

8.9.11 Let $x = 2\sin\theta$ so that $dx = 2\cos\theta\, d\theta$. Then $\int \frac{dx}{\sqrt{4-x^2}} = \int \frac{2\cos\theta\, d\theta}{2\cos\theta} = \theta + C = \sin^{-1}(x/2) + C$.

8.9.13 Let $y = 3\sin\theta$ so that $dy = 3\cos\theta\, d\theta$. Then $\int \frac{dy}{y^2\sqrt{9-y^2}} = \int \frac{3\cos\theta\, d\theta}{9\sin^2\theta \cdot 3\cos\theta} = \frac{1}{9}\int \csc^2\theta\, d\theta = -\frac{1}{9}\cot\theta + C$. Now use $\sin\theta = y/3$, $\cos\theta = (1/3)\sqrt{9-y^2}$ to obtain $\int \frac{dy}{y^2\sqrt{9-y^2}} = -\frac{1}{9y}\sqrt{9-y^2} + C$.

8.9.15 Let $x = (3/2)\tan\theta$ so that $dx = (3/2)\sec^2\theta\,d\theta$, and note that $(3/2)\tan(\pi/6) = \sqrt{3}/2$. Then $\int_0^{\sqrt{3}/2} \frac{4}{9+4x^2}\,dx = \int_0^{\pi/6} \frac{4}{9\sec^2\theta} \cdot \frac{3}{2}\sec^2\theta\,d\theta = \frac{2}{3}\int_0^{\pi/6} d\theta = \frac{\pi}{9}$.

8.9.17 Using the method of partial fractions, we express $\frac{1}{x^2-2x-15} = \frac{1}{(x-5)(x+3)} = \frac{A}{x-5} + \frac{B}{x+3}$. Clearing denominators gives $1 = A(x+3) + B(x-5)$ and comparing coefficients gives $A + B = 0$, $3A - 5B = 1$ which has solution $A = 1/8$, $B = -1/8$. Hence $\int \frac{dx}{x^2-2x-15} = \frac{1}{8}\int\left(\frac{1}{x-5} - \frac{1}{x+3}\right)dx = \frac{1}{8}\ln\left|\frac{x-5}{x+3}\right| + C$.

8.9.19 Using the method of partial fractions, we express $\frac{1}{(y+1)(y^2+1)} = \frac{A}{y+1} + \frac{By+C}{y^2+1}$. Clearing denominators gives $1 = A(y^2+1) + (By+C)(y+1)$ and comparing coefficients gives $A + B = 0$, $B + C = 0$ and $A + C = 1$, which has solution $A = 1/2$, $B = -1/2$, $C = 1/2$. Hence $\int_0^1 \frac{dy}{(y+1)(y^2+1)} = \frac{1}{2}\int_0^1\left(\frac{1}{y+1} + \frac{1-y}{y^2+1}\right)dy$ $= \frac{1}{2}\left(\ln(y+1) + \tan^{-1}y - \frac{1}{2}\ln(y^2+1)\right)\big|_0^1 = \frac{1}{4}\ln 2 + \frac{\pi}{8}$.

8.9.21 Using a table of integrals, we find that $\int \frac{dx}{x\sqrt{ax-b}} = \frac{2}{\sqrt{b}}\tan^{-1}\left(\sqrt{\frac{ax-b}{b}}\right) + C$. Therefore $\int \frac{dx}{x\sqrt{4x-6}} = \frac{\sqrt{6}}{3}\tan^{-1}\left(\sqrt{\frac{2x-3}{3}}\right) + C$.

8.9.23 Using a CAS, we find that $\int_{-1}^1 e^{-2x^2}\,dx \approx 1.196288$.

8.9.25

a. Using a calculator program for the trapezoid and midpoint rules, we find that $T_6 = 9.125$, $M_6 = 8.9375$.

b. Similarly, $T_{12} = 9.03125$, $M_{12} = 8.984375$.

8.9.27 27 Use a calculator program for, say, Simpson's rule with $n = 100$, but replace the limits with 0.00000001 and 0.99999999 to avoid errors coming from trying to evaluate the function at $x = 0$ or $x = 1$; we obtain $\int_0^1 \frac{x^2-x}{\ln x}\,dx \approx 0.4054651$.

8.9.29 First evaluate
$$\int_0^b xe^{-x}\,dx = -e^{-x}(x+1)\big|_0^b = 1 - (b+1)e^{-b}.$$
Then $\int_0^\infty xe^{-x}\,dx = \lim_{b\to\infty}(1 - (b+1)e^{-b}) = 1$.

8.9.31 First take $0 < c < 3$ and evaluate
$$\int_0^c \frac{dx}{\sqrt{9-x^2}} = \sin^{-1}\left(\frac{x}{3}\right)\Big|_0^c = \sin^{-1}\left(\frac{c}{3}\right).$$
Then
$$\int_0^3 \frac{dx}{\sqrt{9-x^2}} = \lim_{c\to 3^-}\sin^{-1}\left(\frac{c}{3}\right) = \frac{\pi}{2}.$$

8.9.33 Factor $x^2 - x - 2 = (x-2)(x+1)$ and use the method of partial fractions: $\frac{1}{(x-2)(x+1)} = \frac{A}{x-2} + \frac{B}{x+1}$. Clearing denominators gives $1 = A(x+1) + B(x-2)$ and equating coefficients gives $A + B = 0$, $A - 2B = 1$, so $A = 1/3$, $B = -1/3$. Therefore $\int \frac{dx}{x^2-x-2} = \frac{1}{3}\int\left(\frac{1}{x-2} - \frac{1}{x+1}\right)dx = \frac{1}{3}\ln\left|\frac{x-2}{x+1}\right| + C$.

8.9.35 As a preliminary step, observe that $\frac{2x^2-4x}{x^2-4} = \frac{2x(x-2)}{(x-2)(x+2)} = \frac{2x}{x+2} = \frac{2(x+2-2)}{x+2} = 2 - \frac{4}{x+2}$. Hence $\int \frac{2x^2-4x}{x^2-4}\,dx = 2x - 4\ln|x+2| + C = 2(x - 2\ln|x+2|) + C$.

8.9.37 Make the preliminary substitution $x = e^{2t}$, $dx = 2e^{2t}$: $\int \frac{e^{2t}}{(1+e^{4t})^{3/2}}\,dt = \frac{1}{2}\int \frac{dx}{(1+x^2)^{3/2}}$. Now let $x = \tan\theta$, $dx = \sec^2\theta\,d\theta$ and $\frac{1}{2}\int \frac{dx}{(1+x^2)^{3/2}} = \frac{1}{2}\int \frac{\sec^2\theta}{\sec^3\theta}\,d\theta = \frac{1}{2}\int \cos\theta\,d\theta = \frac{1}{2}\sin\theta + C$. Now $\sin\theta = \frac{\tan\theta}{\sec\theta} = \frac{x}{\sqrt{1+x^2}} = \frac{e^{2t}}{\sqrt{1+e^{4t}}}$. Therefore $\int \frac{e^{2t}}{(1+e^{4t})^{3/2}}\,dt = \frac{1}{2}\frac{e^{2t}}{\sqrt{1+e^{4t}}} + C$.

8.9.39 The volume is $V = \pi \int_1^e (\ln x)^2\, dx = \pi x((\ln x)^2 - 2\ln x + 2)\big|_1^e = \pi(e-2)$.

8.9.41 The volume is $V = 2\pi \int_1^e (x-1)\ln x\, dx = \frac{\pi}{2} x \left(2(x-2)\ln x - x + 4\right)\big|_1^e = \frac{\pi}{2}\left(e^2 - 3\right)$.

8.9.43 The volume generated by revolving around the x-axis is $V_x = \pi \int_0^\pi \sin^2 x\, dx = \pi \left(\frac{x}{2} - \frac{\sin 2x}{4}\right)\big|_0^\pi = \frac{\pi^2}{2}$, and the volume generated by revolving around the y-axis is $V_y = 2\pi \int_0^\pi x \sin x\, dx = 2\pi(\sin x - x\cos x)\big|_0^\pi = 2\pi^2$, so the greater volume is obtained by revolving around the y-axis.

8.9.45

a. Observe that $\int_{1/2}^b \ln x\, dx = (x\ln x - x)\big|_{1/2}^b \approx b\ln b - b + 0.847$; solve $b\ln b - b + 0.847 = 0$ numerically to obtain $b \approx 1.603$.

b. Similarly, we have $\int_{1/3}^b \ln x\, dx = (x\ln x - x)\big|_{1/3}^b \approx b\ln b - b + 0.700$; solve $b\ln b - b + 0.700 = 0$ numerically to obtain $b \approx 1.870$.

c. In general, the pair (a, b) must satisfy the equation $\int_a^b \ln x\, dx = (b\ln b - b) - (a\ln a - a) = 0$, which gives $b\ln b - b = a\ln a - a$.

d. As a increases there is less negative area to the right of $x = 1$, so $b = g(a)$ is a decreasing function of a.

8.9.47 The average velocity is $\bar{v} = \frac{1}{\pi} \int_0^\pi 10\sin 3t\, dt = -\frac{10}{3\pi}\cos 3t\big|_0^\pi = \frac{20}{3\pi}$.

8.9.49 The number of cars is given by the integral $\int_0^4 800te^{-t/2}\, dt = -1600(t+2)e^{-t/2}\big|_0^4 = 3200(1 - 3e^{-2}) \approx 1901$.

8.9.51

a. Using integration by parts, we find that

$$I(p) = \int_1^e \frac{\ln x}{x^p}\, dx = -\frac{x^{1-p}}{(p-1)^2}\left((p-1)\ln x + 1\right)\Big|_1^e = \frac{1}{(p-1)^2}(1 - pe^{1-p})$$

for $p \neq 1$, and using the substitution $u = \ln x$ gives

$$I(1) = \int_1^e \frac{\ln x}{x}\, dx = \frac{(\ln x)^2}{2}\Big|_1^e = \frac{1}{2}.$$

b. We have

$$\lim_{p\to\infty} I(p) = \lim_{p\to\infty} \frac{1}{(p-1)^2}(1 - pe^{1-p}) = \lim_{p\to\infty} \left(\frac{1}{(p-1)^2} - \frac{pe}{(p-1)^2}e^{-p}\right) = 0,$$

and

$$\lim_{p\to-\infty} I(p) = \lim_{p\to-\infty} \left(\frac{1}{(p-1)^2} - \frac{pe}{(p-1)^2}e^{-p}\right) = \infty,$$

since e^{-p} grows much faster than $(p-1)^2/p$ as $p \to -\infty$.

c. By inspection we see that $I(0) = 1$.

8.9.53 Numerically approximate the integral to obtain $\int_0^1 \frac{\sin^{-1} x}{x}\, dx \approx 1.0889$; therefore $n = \pi \ln 2/1.0889 = 2$. (When approximating the integral, replace the lower limit 0 with a small positive number like 0.00001 to avoid an error from evaluating the integrand at $x = 0$.)

8.9.55 This first-order linear equation has general solution $y = Ce^{2t} - 2$; the initial condition gives $C - 2 = 8$, so $C = 10$ and hence $y = 10e^{2t} - 2$.

8.9.57 This equation is separable. We have $\int 2y\,dy = \int \left(1 + \frac{1}{t}\right) dt$, so $y^2 = t + \ln t + C$. The initial condition gives $16 = 1 + C$, so $C = 15$ and solving for y gives $y = \sqrt{t + \ln t + 15}$.

8.9.59 This equation is separable. We have $\int \cos y\,dy = \int \frac{dt}{t^2}$, so $\sin y = -\frac{1}{t} + C$. The initial condition gives $0 = -1 + C$, so $C = 1$ and solving for y gives $y = \sin^{-1}\left(1 - \frac{1}{t}\right)$. Therefore $\lim_{t \to \infty} y(t) = \lim_{t \to \infty} \sin^{-1}\left(1 - \frac{1}{t}\right) = \sin^{-1} 1 = \frac{\pi}{2}$.

8.9.61

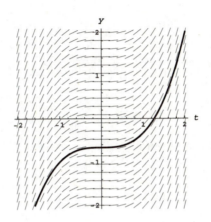

8.9.63 This equation is separable. We have $\int dt = -\frac{1}{10} \int \left(1 + \frac{5}{s}\right) ds \implies t = -\frac{s + 5\ln s}{10} + C$. The initial condition gives $s(0) = 50$ gives $C = (50 + 5\ln 50)/10$, so the solution is given implicitly by $t = \frac{-s - 5\ln s + 50 + 5\ln 50}{10}$. As $t \to \infty$ we must have $s + 5\ln s \to -\infty$, which implies that $\lim_{t \to \infty} s(t) = 0$.

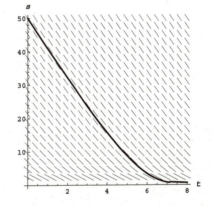

8.9.65

a. Using integration by parts or a CAS, we find that the volume is $V_1(a) = \pi \int_1^a (\ln x)^2\,dx = \pi x((\ln x)^2 - 2\ln x + 2)\big|_1^a = \pi[(a\ln^2 a - 2a\ln a + 2(a-1)]$.

b. Using integration by parts or a CAS, we find that the volume is $V_2(a) = 2\pi \int_1^a x\ln x\,dx = \frac{\pi}{2}x^2(2\ln x - 1)\big|_1^a = \frac{\pi}{2}(2a^2\ln a - a^2 + 1)$.

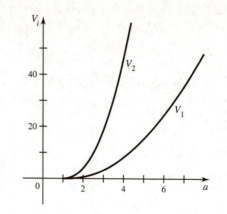

c. As shown by the graph, $V_2(a) > V_1(a)$ for all $a > 1$.

8.9.67 We have

$$V_1 = \pi \int_0^b e^{-2ax}\,dx = -\frac{\pi}{2a}e^{-2ax}\Big|_0^b = \frac{\pi}{2a}\left(1 - e^{-2ab}\right),$$

and

$$V_2 = \pi \int_b^\infty e^{-2ax}\,dx = \lim_{c\to\infty}\left(-\frac{\pi}{2a}e^{-2ax}\Big|_b^c\right) = \frac{\pi}{2a}e^{-2ab}.$$

Equating and solving gives $e^{-2ab} = 1/2$, which is equivalent to $ab = (1/2)\ln 2$.

Chapter 9

9.1 An Overview

9.1.1 A *sequence* is a list of numbers $a_1, a_2, a_3, \ldots$, often written $\{a_1, a_2, \ldots\}$ or $\{a_n\}$. For example, the natural numbers $\{1, 2, 3, \ldots\}$ are a sequence where $a_n = n$ for every n.

9.1.3 $a_1 = 1$ (given); $a_2 = 1 \cdot a_1 = 1$; $a_3 = 2 \cdot a_2 = 2$; $a_4 = 3 \cdot a_3 = 6$; $a_5 = 4 \cdot a_4 = 24$.

9.1.5 An *infinite series* is an infinite sum of numbers. Thus if $\{a_n\}$ is a sequence, then $a_1 + a_2 + \cdots = \sum_{i=1}^{\infty} a_i$, is an infinite series. For example, if $a_i = \frac{1}{i}$, then $\sum_{i=1}^{\infty} a_i = \sum_{i=1}^{\infty} \frac{1}{i}$ is an infinite series.

9.1.7 $S_1 = \sum_{k=1}^{1} k^2 = 1$; $S_2 = \sum_{k=1}^{2} k^2 = 1 + 4 = 5$; $S_3 = \sum_{k=1}^{3} k^2 = 1 + 4 + 9 = 14$; $S_4 = \sum_{k=1}^{4} k^2 = 1 + 4 + 9 + 16 = 30$.

9.1.9 $a_1 = \dfrac{1}{10}$; $a_2 = \dfrac{1}{100}$; $a_3 = \dfrac{1}{1000}$; $a_4 = \dfrac{1}{10000}$.

9.1.11 $a_1 = 1 + \sin(\pi/2) = 2$; $a_2 = 1 + \sin(2\pi/2) = 1 + \sin(\pi) = 1$; $a_3 = 1 + \sin(3\pi/2) = 0$; $a_4 = 1 + \sin(4\pi/2) = 1 + \sin(2\pi) = 1$.

9.1.13 $a_1 = 10$ (given); $a_2 = 3 \cdot a_1 - 12 = 30 - 12 = 18$; $a_3 = 3 \cdot a_2 - 12 = 54 - 12 = 42$; $a_4 = 3 \cdot a_3 - 12 = 126 - 12 = 114$.

9.1.15 $a_1 = 0$ (given); $a_2 = 3 \cdot a_1^2 + 1 + 1 = 2$; $a_3 = 3 \cdot a_2^2 + 2 + 1 = 15$; $a_4 = 3 \cdot a_3^2 + 3 + 1 = 679$.

9.1.17

a. $\frac{1}{32}, \frac{1}{64}$.

b. $a_1 = 1$; $a_{n+1} = \frac{a_n}{2}$.

c. $a_n = \frac{1}{2^{n-1}}$.

9.1.19

a. 32, 64.

b. $a_1 = 1$; $a_{n+1} = 2 \cdot a_n$.

c. $a_n = 2^{n-1}$.

9.1.21

a. 243, 729.

b. $a_1 = 1$; $a_{n+1} = 3 \cdot a_n$.

c. $a_n = 3^{n-1}$.

9.1.23 $a_1 = 9$, $a_2 = 99$, $a_3 = 999$, $a_4 = 9999$. This sequence diverges, since the terms get larger without bound.

9.1.25 $a_1 = -1$, $a_2 = \frac{1}{2}$, $a_3 = -\frac{1}{3}$, $a_4 = \frac{1}{4}$. This sequence converges to 0 since each term is smaller in absolute value than the preceding term and they get arbitrarily close to zero.

9.1.27 Rewrite the recurrence as $a_{n+1} = 10^{-1}a_n^2$. Then $a_0 = 1$, $a_1 = 10^{-1} = \frac{1}{10}$, $a_2 = 10^{-1}(10^{-1})^2 = 10^{-3} = \frac{1}{1000}$, $a_3 = 10^{-1}(10^{-3})^2 = 10^{-7} = \frac{1}{10000000}$, $a_4 = 10^{-1}(10^{-7})^2 = 10^{-15} = \frac{1}{1000000000000000}$. This sequence converges to 0.

9.1.29 $a_0 = 100$, $a_1 = 0.5 \cdot 100 + 50 = 100$, $a_2 = 0.5 \cdot 100 + 50 = 100$, $a_3 = 0.5 \cdot 100 + 50 = 100$, $a_4 = 0.5 \cdot 100 + 50 = 100$. This sequence obviously converges to 100.

9.1.31

a. 1, 2, 3, 4.

b. 1, 2, 3, 4, 5, 6, 7, 8, 9, 10, 11, ... This sequence diverges.

9.1.33

a. 0, 2, 6, 12.

b.

n	1	2	3	4	5	6	7	8	9	10
a_n	0	2	6	12	20	30	42	56	72	90

This sequence appears to diverge.

9.1.35

a. 1/3, 1/2, 3/5, 2/3.

b.

n	2	3	4	5	6	7	8	9	10	11
a_n	1/3	1/2	3/5	2/3	5/7	3/4	7/9	4/5	9/11	5/6

This sequence appears to converge to 1.

9.1.37

a. 5/2, 9/4, 17/8, 33/16.

b. The limit is 2.

9.1.39

a. $a_0 = 3$, $a_1 = 5$, $a_2 = 7$, $a_3 = 9$.

b. $a_n = 2n + 3$.

c.

n	0	1	2	3	4	5	6	7	8	9	10
a_n	3	5	7	9	11	13	15	17	19	21	23

This sequence diverges.

9.1.41

a. $a_0 = 0$, $a_1 = 1$, $a_2 = 3$, $a_3 = 7$, $a_4 = 15$.

b. $a_n = 2^n - 1$.

c.

n	0	1	2	3	4	5	6	7	8	9	10
a_n	0	1	3	7	15	31	63	127	255	511	1023

This sequence diverges.

9.1.43

a. $a_0 = 1$, $a_1 = 3/2$, $a_2 = 7/4$, $a_3 = 15/8$, $a_4 = 31/16$.

b. $a_n = \frac{2^{n+1}-1}{2^n} = 2 - \frac{1}{2^n}$.

c.

n	0	1	2	3	4	5	6	7	8	9	10
a_n	1	3/2	7/4	15/8	31/16	63/32	127/64	255/128	511/256	1023/512	2047/1024

This sequence converges to 2.

9.1.45

a. 20, 10, 5, 2.5.

b. $h_n = 20 \cdot (0.5)^n$.

9.1.47

a. 30, 7.5, 1.875, 0.46875.

b. $h_n = 30 \cdot (0.25)^n$.

9.1.49 $S_1 = 0.3$, $S_2 = 0.33$, $S_3 = 0.333$, $S_4 = 0.3333$. It appears that the infinite series has a value of $0.3333\ldots = \frac{1}{3}$.

9.1.51 $S_1 = 4$, $S_2 = 4.9$, $S_3 = 4.99$, $S_4 = 4.999$. The infinite series has a value of $4.999\cdots = 5$.

9.1.53

a. $S_1 = \frac{2}{3}$, $S_2 = \frac{4}{5}$, $S_3 = \frac{6}{7}$, $S_4 = \frac{8}{9}$.

b. It appears that $S_n = \frac{2n}{2n+1}$.

c. The series has a value of 1 (the partial sums converge to 1).

9.1.55

a. $S_1 = \frac{1}{3}$, $S_2 = \frac{2}{5}$, $S_3 = \frac{3}{7}$, $S_4 = \frac{4}{9}$.

b. $S_n = \frac{n}{2n+1}$.

c. The partial sums converge to $\frac{1}{2}$, which is the value of the series.

9.1.57

a. True. For example, $S_2 = 1 + 2 = 3$, and $S_4 = a_1 + a_2 + a_3 + a_4 = 1 + 2 + 3 + 4 = 10$.

b. False. For example, $\frac{1}{2}$, $\frac{3}{4}$, $\frac{7}{8}$, $\cdots$ where $a_n = 1 - \frac{1}{2^n}$ converges to 1, but each term is greater than the previous one.

c. True. In order for the partial sums to converge, they must get closer and closer together. In order for this to happen, the difference between successive partial sums, which is just the value of a_n, must approach zero.

9.1.59 Using the work from the previous problem:

a. Here $h_0 = 20$, $r = 0.75$, so $S_0 = 40$, $S_1 = 40 + 40 \cdot 0.75 = 70$, $S_2 = S_1 + 40 \cdot (0.75)^2 = 92.5$, $S_3 = S_2 + 40 \cdot (0.75)^3 = 109.375$, $S_4 = S_3 + 40 \cdot (0.75)^4 = 122.03125$

b.

n	0	1	2	3	4	5
a_n	40	70	92.5	109.375	122.0313	131.5234
n	6	7	8	9	10	11
a_n	138.6426	143.9819	147.9865	150.9898	153.2424	154.9318
n	12	13	14	15	16	17
a_n	156.1988	157.1491	157.8618	158.3964	158.7973	159.0980
n	18	19	20	21	22	23
a_n	159.3235	159.4926	159.6195	159.715	159.786	159.839

The sequence converges to 160.

9.1.61

  a. 0.5, 0.75, 0.875, .9375.

  b. The limit is 1.

9.1.63

  a. $\frac{1}{3}, \frac{4}{9}, \frac{13}{27}, \frac{40}{81}$.

  b. The limit is 1/2.

9.1.65

  a. $-1, 0, -1, 0$.

  b. The limit does not exist.

9.1.67

  a. $\frac{3}{10} = 0.3, \frac{33}{100} = 0.33, \frac{333}{1000} = 0.333, \frac{3333}{10000} = 0.3333$.

  b. The limit is 1/3.

9.1.69

  a. $M_0 = 20, \quad M_1 = 20 \cdot 0.5 = 10, \quad M_2 = 20 \cdot 0.5^2 = 5, \ M_3 = 20 \cdot 0.5^3 = 2.5, \quad M_4 = 20 \cdot 0.5^4 = 1.25$

  b. $M_n = 20 \cdot 0.5^n$.

  c. The initial mass is M_0. We are given that 50% of the mass is gone after each decade, so that $M_{n+1} = 0.5 \cdot M_n, n \geq 0$.

  d. The amount of material goes to 0.

9.1.71

  a. $d_0 = 200, \quad d_1 = 200 \cdot .95 = 190, \ d_2 = 200 \cdot .95^2 = 180.5, d_3 = 200 \cdot .95^3 = 171.475, \ d_4 = 200 \cdot .95^4 = 162.90125$.

  b. $d_n = 200 \cdot (0.95)^n, n \geq 0$.

  c. We are given $d_0 = 200$; since 5% of the drug is washed out every hour, that means that 95% of the preceding amount is left every hour, so that $d_{n+1} = 0.95 \cdot d_n$.

  d. The sequence converges to 0.

9.1.73

  a. $0.333\ldots = \sum_{i=1}^{\infty} 3(0.1)^i$.

  b. The limit of the sequence of partial sums is 1/3.

9.1.75

  a. $0.111\ldots = \sum_{i=1}^{\infty} (0.1)^i$.

  b. The limit of the sequence of partial sums is 1/9.

9.1.77

  a. $0.0909\ldots = \sum_{i=1}^{\infty} 9(0.01)^i$.

  b. The limit of the sequence of partial sums is 1/11.

9.1.79

  a. $0.037037037\ldots = \sum_{i=1}^{\infty} 37(0.001)^i$.

  b. The limit of the sequence of partial sums is $37/999 = 1/27$.

9.2   Sequences

9.2.1 There are many examples; one is $a_n = \frac{1}{n}$. This sequence is nonincreasing (in fact, it is strictly decreasing) and has a limit of 0.

9.2.3 There are many examples; one is $a_n = \frac{1}{n}$. This sequence is nonincreasing (in fact, it is strictly decreasing), is bounded above by 1 and below by 0, and has a limit of 0.

9.2.5 $\{r^n\}$ converges for $|r| < 1$ and for $r = 1$. It diverges for all other values of r (see Theorem 8.3).

9.2.7 A sequence a_n converges to l if, given any $\epsilon > 0$, there exists a positive integer N, such that whenever $n > N$, $|a_n - L| < \varepsilon$.

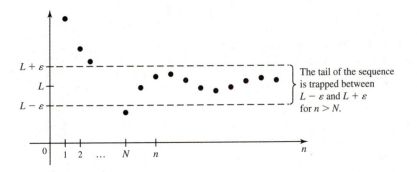

The tail of the sequence is trapped between $L - \varepsilon$ and $L + \varepsilon$ for $n > N$.

9.2.9 Divide numerator and denominator by n^4 to get $\lim\limits_{n\to\infty} \frac{1/n}{1+\frac{1}{n^4}} = 0$.

9.2.11 Divide numerator and denominator by n^3 to get $\lim\limits_{n\to\infty} \frac{3-n^{-3}}{2+n^{-3}} = \frac{3}{2}$.

9.2.13 As $n \to \infty$, $\tan^{-1} n \to \pi/2$, so $\frac{\tan^{-1} n}{n} \to 0$.

9.2.15 Find the limit of the logarithm of the expression, which is $n \ln\left(1 + \frac{2}{n}\right)$. Using L'Hôpital's rule:
$\lim\limits_{n\to\infty} n \ln\left(1 + \frac{2}{n}\right) = \lim\limits_{n\to\infty} \frac{\ln\left(1+\frac{2}{n}\right)}{1/n} = \lim\limits_{n\to\infty} \frac{\frac{1}{1+(2/n)}\left(\frac{-2}{n^2}\right)}{-1/n^2} = \lim\limits_{n\to\infty} \frac{2}{1+(2/n)} = 2$. Thus the limit of the original expression is e^2.

9.2.17 Take the logarithm of the expression and use L'Hôpital's rule:

$$\lim\limits_{n\to\infty} \frac{n}{2} \ln\left(1 + \frac{1}{2n}\right) = \lim\limits_{n\to\infty} \frac{\ln(1 + (1/2n))}{2/n} = \lim\limits_{n\to\infty} \frac{\frac{1}{1+(1/2n)} \cdot \frac{-1}{2n^2}}{-2/n^2} = \lim\limits_{n\to\infty} \frac{1}{4(1 + (1/2n))} = \frac{1}{4}.$$

Thus the original limit is $e^{1/4}$.

9.2.19 Taking logs, we have $\lim\limits_{n\to\infty} \frac{1}{n} \ln(1/n) = \lim\limits_{n\to\infty} -\frac{\ln n}{n} = \lim\limits_{n\to\infty} \frac{-1}{n} = 0$ by L'Hôpital's rule. Thus the original sequence has limit $e^0 = 1$.

9.2.21 Except for a finite number of terms, this sequence is just $a_n = ne^{-n}$, so it has the same limit as this sequence. Note that $\lim\limits_{n\to\infty} \frac{n}{e^n} = \lim\limits_{n\to\infty} \frac{1}{e^n} = 0$, by L'Hôpital's rule.

9.2.23 $\ln(\sin(1/n)) + \ln n = \ln(n \sin(1/n)) = \ln\left(\frac{\sin(1/n)}{1/n}\right)$. As $n \to \infty$, $\sin(1/n)/(1/n) \to 1$, so the limit of the original sequence is $\ln 1 = 0$.

9.2.25 $\lim\limits_{n\to\infty} n \sin(6/n) = \lim\limits_{n\to\infty} \frac{\sin(6/n)}{1/n} = \lim\limits_{n\to\infty} \frac{\frac{-6\cos(6/n)}{n^2}}{(-1/n^2)} = \lim\limits_{n\to\infty} 6\cos(6/n) = 6$.

9.2.27 When n is an integer, $\sin\left(\frac{n\pi}{2}\right)$ oscillates between the values ± 1 and 0, so this sequence does not converge.

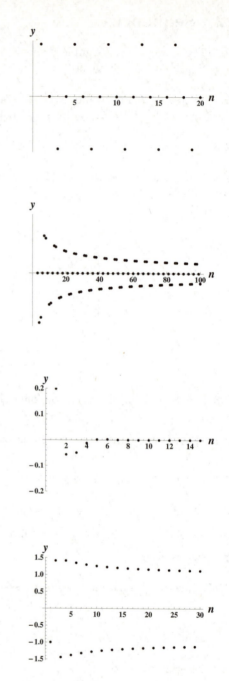

9.2.29 The numerator is bounded in absolute value by 1, while the denominator goes to ∞, so the limit of this sequence is 0.

9.2.31 This is the sequence $\frac{\cos n}{e^n}$; the numerator is bounded in absolute value by 1 and the denominator increases without bound, so the limit is zero.

9.2.33 Ignoring the factor of $(-1)^n$ for the moment, we see, taking logs, that $\lim\limits_{n\to\infty}\frac{\ln n}{n}=0$, so that $\lim\limits_{n\to\infty}\sqrt[n]{n}=e^0=1$. Taking the sign into account, the odd terms converge to -1 while the even terms converge to 1. Thus the sequence does not converge.

9.2.35 Since $0.2 < 1$, this sequence converges to 0. Since $0.2 > 0$, the convergence is monotone.

9.2.37 Since $|-0.7| < 1$, the sequence converges to 0; since $-0.7 < 0$, it does not do so monotonically.

9.2.39 Since $1.00001 > 1$, the sequence diverges; since $1.00001 > 0$, the divergence is monotone.

9.2.41 Since $|-2.5| > 1$, the sequence diverges; since $-2.5 < 0$, the divergence is not monotone.

9.2.43 Since $-1 \leq \sin n \leq 1$ for all n, the given sequence satisfies $\frac{-1}{2^n} \leq \frac{\sin n}{2^n} \leq \frac{1}{2^n}$, and since both $\pm\frac{1}{2^n} \to 0$ as $n \to \infty$, the given sequence converges to zero as well by the Squeeze Theorem.

9.2.45 $\tan^{-1}$ takes values between $-\pi/2$ and $\pi/2$, so the numerator is always between $-\pi$ and π. Thus $\frac{-\pi}{n^3+4} \leq \frac{2\tan^{-1}n}{n^3+4} \leq \frac{\pi}{n^3+4}$, and by the Squeeze Theorem, the given sequence converges to zero.

9.2.47

a. After the nth dose is given, the amount of drug in the bloodstream is $d_n = 0.5 \cdot d_{n-1} + 80$, since the half-life is one day. The initial condition is $d_1 = 80$.

b. The limit of this sequence is 160 mg.

c. Let $L = \lim_{n\to\infty} d_n$. Then from the recurrence relation, we have $d_n = 0.5 \cdot d_{n-1} + 80$, and thus $\lim_{n\to\infty} d_n = 0.5 \cdot \lim_{n\to\infty} d_{n-1} + 80$, so $L = 0.5 \cdot L + 80$, and therefore $L = 160$.

9.2.49

a.

$$B_0 = 0$$
$$B_1 = 1.0075 \cdot B_0 + \$100 = \$100$$
$$B_2 = 1.0075 \cdot B_1 + \$100 = \$200.75$$
$$B_3 = 1.0075 \cdot B_2 + \$100 = \$302.26$$
$$B_4 = 1.0075 \cdot B_3 + \$100 = \$404.52$$
$$B_5 = 1.0075 \cdot B_4 + \$100 = \$507.56$$

b. $B_n = 1.0075 \cdot B_{n-1} + \$100$.

c. Using a calculator or computer program, $B_n > \$5,000$ during the 43rd month.

9.2.51 $\{n^2\} \ll \{n^2 \ln n\}$ from the second inequality in Theorem 8.6: $\{n^p\} \ll \{n^p \ln^r n\}$.

9.2.53 $\{100n!\} \ll \{3n^n\}$ since $\{n!\} \ll \{n^n\}$ from Theorem 8.6, and the sequences given are constant multiples of those sequences, so the rankings do not change.

9.2.55 Since $\frac{1}{2} > \frac{1}{10}$, Theorem 8.6 tells us that $\{n^{1/10}\} \ll \{n^{1/2}\}$.

9.2.57 Let $\varepsilon > 0$ be given and let N be an integer with $N > \frac{1}{\varepsilon}$. Then if $n > N$, we have $\left|\frac{1}{n} - 0\right| = \frac{1}{n} < \frac{1}{N} < \varepsilon$.

9.2.59 Let $\varepsilon > 0$ be given. We wish to find N such that for $n > N$, $\left|\frac{3n^2}{4n^2+1} - \frac{3}{4}\right| = \left|\frac{-3}{4(4n^2+1)}\right| = \frac{3}{4(4n^2+1)} < \varepsilon$.
But this means that $3 < 4\varepsilon(4n^2 + 1)$, or $16\varepsilon n^2 + (4\varepsilon - 3) > 0$. Solving the quadratic, we get $n > \frac{1}{4}\sqrt{\frac{3}{\varepsilon} - 4}$, provided $\varepsilon < 3/4$. So let $N = \frac{1}{4}\sqrt{\frac{3}{\varepsilon} - 4}$ if $\epsilon < 3/4$ and let $N = 1$ otherwise.

9.2.61 Let $\varepsilon > 0$ be given. We wish to find N such that for $n > N$, $\left|\frac{cn}{bn+1} - \frac{c}{b}\right| = \left|\frac{-c}{b(bn+1)}\right| = \frac{c}{b(bn+1)} < \varepsilon$.
But this means that $\varepsilon b^2 n + (b\varepsilon - c) > 0$, so that $N > \frac{c - b\varepsilon}{b^2 \varepsilon}$ will work.

9.2.63

a. True. See Theorem 8.2 part 4.

b. False. For example, if $a_n = e^n$ and $b_n = 1/n$, then $\lim_{n\to\infty} a_n b_n = \infty$.

c. True. The definition of the limit of a sequence involves only the behavior of the nth term of a sequence as n gets large (see the Definition of Limit of a Sequence). Thus suppose a_n, b_n differ in only finitely many terms, and that M is large enough so that $a_n = b_n$ for $n > M$. Suppose a_n has limit L. Then for $\varepsilon > 0$, if N is such that $|a_n - L| < \varepsilon$ for $n > N$, first increase N if required so that $N > M$ as well. Then we also have $|b_n - L| < \varepsilon$ for $n > N$. Thus a_n and b_n have the same limit. A similar argument applies if a_n has no limit.

d. True. Note that a_n converges to zero. Intuitively, the nonzero terms of b_n are those of a_n, which converge to zero. More formally, given ϵ, choose N_1 such that for $n > N_1$, $a_n < \epsilon$. Let $N = 2N_1 + 1$. Then for $n > N$, consider b_n. If n is even, then $b_n = 0$ so certainly $b_n < \epsilon$. If n is odd, then $b_n = a_{(n-1)/2}$, and $(n-1)/2 > ((2N_1 + 1) - 1)/2 = N_1$ so that $a_{(n-1)/2} < \epsilon$. Thus b_n converges to zero as well.

e. False. If $\{a_n\}$ happens to converge to zero, the statement is true. But consider for example $a_n = 2 + \frac{1}{n}$. Then $\lim_{n \to \infty} a_n = 2$, but $(-1)^n a_n$ does not converge (it oscillates between positive and negative values increasingly close to ± 2).

f. True. Suppose $\{0.000001 a_n\}$ converged to L, and let $\epsilon > 0$ be given. Choose N such that for $n > N$, $|0.000001 a_n - L| < \epsilon \cdot 0.000001$. Dividing through by 0.000001, we get that for $n > N$, $|a_n - 1000000L| < \epsilon$, so that a_n converges as well (to $1000000L$).

9.2.65 $\{(n-2)^2 + 6(n-2) - 9\}_{n=3}^{\infty} = \{n^2 + 2n - 17\}_{n=3}^{\infty}$.

9.2.67 Evaluate the limit of each term separately: $\lim_{n \to \infty} \frac{75^{n-1}}{99^n} = \frac{1}{99} \lim_{n \to \infty} \left(\frac{75}{99}\right)^{n-1} = 0$, while $\frac{-5^n}{8^n} \leq \frac{5^n \sin n}{8^n} \leq \frac{5^n}{8^n}$, so by the Squeeze Theorem, this second term converges to 0 as well. Thus the sum of the terms converges to zero.

9.2.69 Since $\lim_{n \to \infty} 0.99^n = 0$, and since cosine is continuous, the first term converges to $\cos 0 = 1$. The limit of the second term is $\lim_{n \to \infty} \frac{7^n + 9^n}{63^n} = \lim_{n \to \infty} \left(\frac{7}{63}\right)^n + \lim_{n \to \infty} \left(\frac{9}{63}\right)^n = 0$. Thus the sum converges to 1.

9.2.71 A graph shows that the sequence appears to converge. Let its supposed limit be L, then $\lim_{n \to \infty} a_{n+1} = \lim_{n \to \infty} (2a_n(1 - a_n)) = 2(\lim_{n \to \infty} a_n)(1 - \lim_{n \to \infty} a_n)$, so $L = 2L(1-L) = 2L - 2L^2$, and thus $2L^2 - L = 0$, so $L = 0, \frac{1}{2}$. Thus the limit appears to be either 0 or 1/2; with the given initial condition, doing a few iterations by hand confirms that the sequence converges to 1/2: $a_0 = 0.3$; $a_1 = 2 \cdot 0.3 \cdot 0.7 = .42$; $a_2 = 2 \cdot 0.42 \cdot 0.58 = 0.4872$.

9.2.73 Computing three terms gives $a_0 = 0.5, a_1 = 4 \cdot .5 \cdot 0.5 = 1, a_2 = 4 \cdot 1 \cdot (1-1) = 0$. All successive terms are obviously zero, so the sequence converges to 0.

9.2.75 For $b = 2$, $2^3 > 3!$ but $16 = 2^4 < 4! = 24$. For e, $e^5 \approx 148.41 > 5! = 120$ while $e^6 \approx 403.4 < 6! = 720$. For 10, $24! \approx 6.2 \times 10^{23} < 10^{24}$, while $25! \approx 1.55 \times 10^{25} > 10^{25}$.

9.2.77

a. The profits for each of the first ten days, in dollars are:

n	0	1	2	3	4	5	6	7	8	9	10
h_n	130.00	130.75	131.40	131.95	132.40	132.75	133.00	133.15	133.20	133.15	133.00

b. The profit on an item is revenue minus cost. The total cost of keeping the hippo for n days is $.45n$, and the revenue for selling the hippo on the nth day is $(200 + 5n) \cdot (.65 - .01n)$, since the hippo gains 5 pounds per day but is worth a penny less per pound each day. Thus the total profit on the nth day is $h_n = (200 + 5n) \cdot (.65 - .01n) - .45n = 130 + 0.8n - 0.05n^2$. The maximum profit occurs when $-.1n + .8 = 0$, which occurs when $n = 8$. The maximum profit is achieved by selling the hippo on the 8th day.

9.2.79 The approximate first few values of this sequence are:

n	0	1	2	3	4	5	6
c_n	.7071	.6325	.6136	.6088	.6076	.6074	.6073

The value of the constant appears to be around 0.607.

9.2.81

a. If we "cut off" the expression after n square roots, we get a_n from the recurrence given. We can thus *define* the infinite expression to be the limit of a_n as $n \to \infty$.

b. $a_0 = 1$, $a_1 = \sqrt{2}$, $a_2 = \sqrt{1 + \sqrt{2}} \approx 1.5538,$, $a_3 \approx 1.598$, $a_4 \approx 1.6118$, and $a_5 \approx 1.6161$.

c. $a_{10} \approx 1.618286$, which differs from $\frac{1 + \sqrt{5}}{2} \approx 1.61803394$ by less than .001.

d. Assume $\lim_{n \to \infty} a_n = L$. Then $\lim_{n \to \infty} a_{n+1} = \lim_{n \to \infty} \sqrt{1 + a_n} = \sqrt{1 + \lim_{n \to \infty} a_n}$, so $L = \sqrt{1 + L}$, and thus $L^2 = 1 + L$. Therefore we have $L^2 - L - 1 = 0$, so $L = \frac{1 \pm \sqrt{5}}{2}$.

Since clearly the limit is positive, it must be the positive square root.

e. Letting $a_{n+1} = \sqrt{p + \sqrt{a_n}}$ with $a_0 = p$ and assuming a limit exists we have $\lim_{n \to \infty} a_{n+1} = \lim_{n \to \infty} \sqrt{p + a_n} = \sqrt{p + \lim_{n \to \infty} a_n}$, so $L = \sqrt{p + L}$, and thus $L^2 = p + L$. Therefore, $L^2 - L - p = 0$, so $L = \frac{1 \pm \sqrt{1 + 4p}}{2}$, and since we know that L is positive, we have $L = \frac{1 + \sqrt{4p+1}}{2}$. The limit exists for all positive p.

9.2.83

a. Define a_n as given in the problem statement. Then we can *define* the value of the continued fraction to be $\lim_{n \to \infty} a_n$.

b. $a_0 = 1$, $a_1 = 1 + \frac{1}{a_0} = 2$, $a_2 = 1 + \frac{1}{a_1} = \frac{3}{2} = 1.5$, $a_3 = 1 + \frac{1}{a_2} = \frac{5}{3} \approx 1.67$, $a_4 = 1 + \frac{1}{a_3} = \frac{8}{5} \approx 1.6$, $a_5 = 1 + \frac{1}{a_4} = \frac{13}{8} \approx 1.625$.

c. From the list above, the values of the sequence alternately decrease and increase, so we would expect that the limit is somewhere between 1.6 and 1.625.

d. Assume that the limit is equal to L. Then from $a_{n+1} = 1 + \frac{1}{a_n}$, we have $\lim_{n \to \infty} a_{n+1} = 1 + \frac{1}{\lim_{n \to \infty} a_n}$, so $L = 1 + \frac{1}{L}$, and thus $L^2 - L - 1 = 0$. Therefore, $L = \frac{1 \pm \sqrt{5}}{2}$, and since L is clearly positive, it must be equal to $\frac{1 + \sqrt{5}}{2}$.

e. Here $a_0 = a$ and $a_{n+1} = a + \frac{b}{a_n}$. Assuming that $\lim_{n \to \infty} a_n = L$ we have $L = a + \frac{b}{L}$, so $L^2 = aL + b$, and thus $L^2 - aL - b = 0$. Therefore, $L = \frac{a \pm \sqrt{a^2 + 4b}}{2}$, and since $L > 0$ we have $L = \frac{a + \sqrt{a^2 + 4b}}{2}$.

9.2.85

a. $f_0 = f_1 = 1$, $f_2 = 2$, $f_3 = 3$, $f_4 = 5$, $f_5 = 8$, $f_6 = 13$, $f_7 = 21$, $f_8 = 34$, $f_9 = 55$, $f_{10} = 89$.

b. The sequence is clearly not bounded.

c. $\frac{f_{10}}{f_9} \approx 1.61818$

d. We use induction. Note that $\frac{1}{\sqrt{5}} \left(\varphi + \frac{1}{\varphi} \right) = \frac{1}{\sqrt{5}} \left(\frac{1 + \sqrt{5}}{2} + \frac{2}{1 + \sqrt{5}} \right) = \frac{1}{\sqrt{5}} \left(\frac{1 + 2\sqrt{5} + 5 + 4}{2(1 + \sqrt{5})} \right) = 1 = f_1$. Also note that $\frac{1}{\sqrt{5}} \left(\varphi^2 - \frac{1}{\varphi^2} \right) = \frac{1}{\sqrt{5}} \left(\frac{3 + \sqrt{5}}{2} - \frac{2}{3 + \sqrt{5}} \right) = \frac{1}{\sqrt{5}} \left(\frac{9 + 6\sqrt{5} + 5 - 4}{2(3 + \sqrt{5})} \right) = 1 = f_2$. Now note that

$$f_{n-1} + f_{n-2} = \frac{1}{\sqrt{5}} (\varphi^{n-1} - (-1)^{n-1} \varphi^{1-n} + \varphi^{n-2} - (-1)^{n-2} \varphi^{2-n})$$

$$= \frac{1}{\sqrt{5}} ((\varphi^{n-1} + \varphi^{n-2}) - (-1)^n (\varphi^{2-n} - \varphi^{1-n})).$$

Now, note that $\varphi - 1 = \frac{1}{\varphi}$, so that

$$\varphi^{n-1} + \varphi^{n-2} = \varphi^{n-1} \left(1 + \frac{1}{\varphi} \right) = \varphi^{n-1}(\varphi) = \varphi^n$$

and

$$\varphi^{2-n} - \varphi^{1-n} = \varphi^{-n}(\varphi^2 - \varphi) = \varphi^{-n}(\varphi(\varphi - 1)) = \varphi^{-n}$$

Making these substitutions, we get

$$f_{n-1} + f_{n-2} = \frac{1}{\sqrt{5}}(\varphi^n - (-1)^n \varphi^{-n}) = f_n$$

9.2.87

a.

$$
\begin{aligned}
2: \quad & 1 \\
3: \quad & 10,\ 5,\ 16,\ 8,\ 4,\ 2,\ 1 \\
4: \quad & 2,\ 1 \\
5: \quad & 16,\ 8,\ 4,\ 2,\ 1 \\
6: \quad & 3,\ 10,\ 5,\ 16,\ 8,\ 4,\ 2,\ 1 \\
7: \quad & 22,\ 11,\ 34,\ 17,\ 52,\ 26,\ 13,\ 40,\ 20,\ 10,\ 5,\ 16,\ 8,\ 4,\ 2,\ 1 \\
8: \quad & 4,\ 2,\ 1 \\
9: \quad & 28,\ 14,\ 7,\ 22,\ 11,\ 34,\ 17,\ 52,\ 26,\ 13,\ 40,\ 20,\ 10,\ 5,\ 16,\ 8,\ 4,\ 2,\ 1 \\
10: \quad & 5,\ 16,\ 8,\ 4,\ 2,\ 1
\end{aligned}
$$

b. From the above, $H_2 = 1, H_3 = 7$, and $H_4 = 2$.

c. This plot is for $1 \le n \le 100$. Like hailstones, the numbers in the sequence a_n rise and fall but eventually crash to the earth. The conjecture appears to be true.

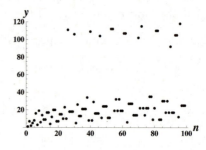

9.3   Infinite Series

9.3.1 A geometric series is a series in which the ratio of successive terms in the underlying sequence is a constant. Thus a geometric series has the form $\sum ar^k$ where r is the constant. One example is $3 + 6 + 12 + 24 + 48 + \cdots$ in which $a = 3$ and $r = 2$.

9.3.3 The ratio is the common ratio between successive terms in the sum.

9.3.5 No. For example, the geometric series with $a_n = 3 \cdot 2^n$ does not have a finite sum.

9.3.7 $S = 1 \cdot \dfrac{1 - 3^9}{1 - 3} = \dfrac{19682}{2} = 9841.$

9.3.9 $S = 1 \cdot \dfrac{1 - (4/25)^{21}}{1 - 4/25} = \dfrac{25^{21} - 4^{21}}{25^{21} - 4 \cdot 25^{20}} \approx 1.1905.$

9.3.11 $S = 1 \cdot \dfrac{1 - (-3/4)^{10}}{1 + 3/4} = \dfrac{4^{10} - 3^{10}}{4^{10} + 3 \cdot 4^9} = \dfrac{141361}{262144} \approx 0.5392.$

9.3.13 $S = 1 \cdot \dfrac{1 - \pi^7}{1 - \pi} = \dfrac{\pi^7 - 1}{\pi - 1} \approx 1409.84.$

9.3.15 $S = 1 \cdot \dfrac{1 - (-1)^{21}}{2} = 1.$

9.3.17 $\dfrac{1093}{2916}.$

9.3.19 $\dfrac{1}{1 - 1/4} = \dfrac{4}{3}.$

9.3.21 $\dfrac{1}{1 - 0.9} = 10.$

9.3.23 Divergent, since $r > 1$.

9.3.25 $\dfrac{e^{-2}}{1 - e^{-2}} = \dfrac{1}{e^2 - 1}.$

9.3.27 $\dfrac{2^{-3}}{1 - 2^{-3}} = \dfrac{1}{7}.$

9.3.29 $\dfrac{1/625}{1 - 1/5} = \dfrac{1}{500}.$

9.3.31 $\dfrac{1}{1 - e/\pi} = \dfrac{\pi}{\pi - e}.$ (Note that $e < \pi$, so $r < 1$ for this series.)

9.3.33 $\displaystyle\sum_{k=0}^{\infty} \left(\dfrac{1}{4}\right)^k 5^{6-k} = 5^6 \sum_{k=0}^{\infty} \left(\dfrac{1}{20}\right)^k = 5^6 \cdot \dfrac{1}{1 - 1/20} = \dfrac{5^6 \cdot 20}{19} = \dfrac{312500}{19}.$

9.3.35 $\dfrac{1}{1 + 9/10} = \dfrac{10}{19}.$

9.3.37 $3 \cdot \dfrac{1}{1 + 1/\pi} = \dfrac{3\pi}{\pi + 1}.$

9.3.39 $\dfrac{0.15^2}{1.15} = \dfrac{9}{460} \approx 0.0196.$

9.3.41 $0.121212\ldots = \displaystyle\sum_{i=0}^{\infty} .12 \cdot 10^{-2i} = \dfrac{.12}{1 - 1/100} = \dfrac{12}{99} = \dfrac{4}{33}.$

9.3.43 $0.456456456\ldots = \displaystyle\sum_{i=0}^{\infty} .456 \cdot 10^{-3i} = \dfrac{.456}{1 - 1/1000} = \dfrac{456}{999} = \dfrac{152}{333}.$

9.3.45 $0.00952952\ldots = \displaystyle\sum_{i=0}^{\infty} .00952 \cdot 10^{-3i} = \dfrac{.00952}{1 - 1/1000} = \dfrac{9.52}{999} = \dfrac{952}{99900} = \dfrac{238}{24975}.$

9.3.47 The second term of each summand cancels with the first term of the succeeding summand, so $S_n = \frac{1}{1+1} - \frac{1}{n+2} = \frac{n}{2n+4}$, and $\lim_{n \to \infty} \frac{n}{2n+4} = \frac{1}{2}$.

9.3.49 $\dfrac{1}{(k+1)(k+2)} = \dfrac{1}{k+1} - \dfrac{1}{k+2}$, so the series given is the same as $\sum_{k=1}^{\infty} \left(\frac{1}{k+1} - \frac{1}{k+2}\right)$. In that series, the second term of each summand cancels with the first term of the succeeding summand, so $S_n = \frac{1}{1+1} - \frac{1}{n+2} = \frac{n}{2n+4}$ and $\lim_{n \to \infty} \frac{n}{2n+4} = \frac{1}{2}$

9.3.51 $\ln\left(\dfrac{k+1}{k}\right) = \ln(k+1) - \ln k$, so the series given is the same as $\sum_{k=1}^{\infty}(\ln(k+1) - \ln k)$, in which the first term of each summand cancels with the second term of the next summand, so we get $S_{n-1} = \ln n - \ln 1 = \ln n$, and thus the series diverges.

9.3.53 $\dfrac{1}{(k+p)(k+p+1)} = \dfrac{1}{k+p} - \dfrac{1}{k+p+1}$, so that $\displaystyle\sum_{k=1}^{\infty}\dfrac{1}{(k+p)(k+p+1)} = \sum_{k=1}^{\infty}\left(\dfrac{1}{k+p} - \dfrac{1}{k+p+1}\right)$ and this series telescopes to give $S_n = \frac{1}{p+1} - \frac{1}{n+p+1} = \frac{n}{n(p+1)+(p+1)^2}$ so that $\lim_{n\to\infty} S_n = \frac{1}{p+1}$.

9.3.55 Let $a_n = \dfrac{1}{\sqrt{n+1}} - \dfrac{1}{\sqrt{n+3}}$. Then the second term of a_n cancels with the first term of a_{n+2}, so the series telescopes and $S_n = \frac{1}{\sqrt{2}} + \frac{1}{\sqrt{3}} - \frac{1}{\sqrt{n-1+3}} - \frac{1}{\sqrt{n+3}}$ and thus the sum of the series is the limit of S_n, which is $\dfrac{1}{\sqrt{2}} + \dfrac{1}{\sqrt{3}}$.

9.3.57 $16k^2 + 8k - 3 = (4k+3)(4k-1)$, so $\frac{1}{16k^2+8k-3} = \frac{1}{(4k+3)(4k-1)} = \frac{1}{4}\left(\frac{1}{4k-1} - \frac{1}{4k+3}\right)$. Thus the series given is equal to $\dfrac{1}{4}\displaystyle\sum_{k=0}^{\infty}\left(\dfrac{1}{4k-1} - \dfrac{1}{4k+3}\right)$. This series telescopes, so $S_n = \frac{1}{4}\left(-1 - \frac{1}{4n+3}\right)$ and so the sum of the series is equal to $\lim_{n\to\infty} S_n = -\frac{1}{4}$.

9.3.59

a. True. $\left(\dfrac{\pi}{e}\right)^{-k} = \left(\dfrac{e}{\pi}\right)^{k}$; since $e < \pi$, this is a geometric series with ratio less than 1.

b. True. If $\displaystyle\sum_{k=12}^{\infty} a^k = L$, then $\displaystyle\sum_{k=0}^{\infty} a^k = \left(\sum_{k=0}^{11} a^k\right) + L$.

c. False. For example, let $0 < a < 1$ and $b > 1$.

9.3.61 At the nth stage, there are 2^{n-1} triangles of area $A_n = \frac{1}{8}A_{n-1} = \frac{1}{8^{n-1}}A_1$, so the total area of the triangles formed at the nth stage is $\dfrac{2^{n-1}}{8^{n-1}}A_1 = \left(\dfrac{1}{4}\right)^{n-1} A_1$. Thus the total area under the parabola is

$$\sum_{n=1}^{\infty}\left(\frac{1}{4}\right)^{n-1} A_1 = A_1 \sum_{n=1}^{\infty}\left(\frac{1}{4}\right)^{n-1} = A_1 \frac{1}{1-1/4} = \frac{4}{3}A_1.$$

9.3.63 It appears that the loan is paid off after about 470 months. Let B_n be the loan balance after n months. Then $B_0 = 180000$ and $B_n = 1.005 \cdot B_{n-1} - 1000$. Then $B_n = 1.005 \cdot B_{n-1} - 1000 = 1.005(1.005 \cdot B_{n-2} - 1000) - 1000 = (1.005)^2 \cdot B_{n-2} - 1000(1+1.005) = (1.005)^2 \cdot (1.005 \cdot B_{n-3} - 1000) - 1000(1+1.005) = (1.005)^3 \cdot B_{n-3} - 1000(1 + 1.005 + (1.005)^2) = \cdots = (1.005)^n B_0 - 1000(1 + 1.005 + (1.005)^2 + \cdots + (1.005)^{n-1}) = (1.005)^n \cdot 180000 - 1000\left(\frac{(1.005)^n - 1}{1.005 - 1}\right)$. Solving this equation for $B_n = 0$ gives $n \approx 461.66$ months, so the loan is paid off after 462 months.

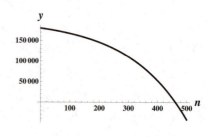

9.3.65 $F_n = 1.015F_{n-1} - 120$ with $F_0 = 4000$. Assume F_n reaches a limit L; then $L = 1.015L - 120$, so $.015L = 120$, and thus $L = 8000$. However, starting with only 4000 fish, the replacement rate of .015% only adds 60 fish in the first month, which is not enough to make up for the harvest of 120 fish. Thus the population in fact decreases to zero. If the initial population were above 8000, it would level off at 8000.

9.3.67 Under the one-child policy, each couple will have one child. Under the one-son policy, we compute the expected number of children as follows: with probability $1/2$ the first child will be a son; with probability $(1/2)^2$, the first child will be a daughter and the second child will be a son; in general, with probability $(1/2)^n$, the first $n-1$ children will be girls and the nth a boy. Thus the expected number of children is the sum $\sum_{i=1}^{\infty} i \cdot \left(\frac{1}{2}\right)^i$. To evaluate this series, use the following "trick": Let $f(x) = \sum_{i=1}^{\infty} i x^i$. Then $f(x) + \sum_{i=1}^{\infty} x^i = \sum_{i=1}^{\infty} (i+1) x^i$. Now, let

$$g(x) = \sum_{i=1}^{\infty} x^{i+1} = -1 - x + \sum_{i=0}^{\infty} x^i = -1 - x + \frac{1}{1-x}$$

and

$$g'(x) = f(x) + \sum_{i=1}^{\infty} x^i = f(x) - 1 + \sum_{i=0}^{\infty} x^i = f(x) - 1 + \frac{1}{1-x}.$$

Evaluate $g'(x) = -1 - \frac{1}{(1-x)^2}$; then

$$f(x) = 1 - \frac{1}{1-x} - 1 - \frac{1}{(1-x)^2} = \frac{-1 + x + 1}{(1-x)^2} = \frac{x}{(1-x)^2}$$

Finally, evaluate at $x = \frac{1}{2}$ to get $f\left(\frac{1}{2}\right) = \sum_{i=1}^{\infty} i \cdot \left(\frac{1}{2}\right)^i = \frac{1/2}{(1-1/2)^2} = 2$. There will thus be twice as many children under the one-son policy as under the one-child policy.

9.3.69 Ignoring the initial drop for the moment, the height after the nth bounce is $10p^n$, so the total time spent in that bounce is $2 \cdot \sqrt{2 \cdot 10p^n/g}$ seconds. The total time before the ball comes to rest (now including the time for the initial drop) is then $\sqrt{20/g} + \sum_{i=1}^{\infty} 2 \cdot \sqrt{2 \cdot 10p^n/g} = \sqrt{\frac{20}{g}} + 2\sqrt{\frac{20}{g}} \sum_{i=1}^{\infty} (\sqrt{p})^n = \sqrt{\frac{20}{g}} + 2\sqrt{\frac{20}{g}} \frac{\sqrt{p}}{1-\sqrt{p}} = \sqrt{\frac{20}{g}} \left(1 + \frac{2\sqrt{p}}{1-\sqrt{p}}\right) = \sqrt{\frac{20}{g}} \left(\frac{1+\sqrt{p}}{1-\sqrt{p}}\right)$ seconds.

9.3.71

a. I_{n+1} is obtained by I_n by dividing each edge into three equal parts, removing the middle part, and adding two parts equal to it. Thus 3 equal parts turn into 4, so $L_{n+1} = \frac{4}{3}L_n$. This is a geometric sequence with a ratio greater than 1, so the nth term grows without bound.

b. As the result of part (a), I_n has $3 \cdot 4^n$ sides of length $\frac{1}{3^n}$; each of those sides turns into an added triangle in I_{n+1} of side length 3^{-n-1}. Thus the added area in I_{n+1} consists of $3 \cdot 4^n$ equilateral triangles with side 3^{-n-1}. The area of an equilateral triangle with side x is $\frac{x^2 \sqrt{3}}{4}$. Thus $A_{n+1} = A_n + 3 \cdot 4^n \cdot \frac{3^{-2n-2}\sqrt{3}}{4} = A_n + \frac{\sqrt{3}}{12} \cdot \left(\frac{4}{9}\right)^n$, and $A_0 = \frac{\sqrt{3}}{4}$. Thus $A_{n+1} = A_0 + \sum_{i=0}^{n} \frac{\sqrt{3}}{12} \cdot \left(\frac{4}{9}\right)^i$, so that

$$A_\infty = A_0 + \frac{\sqrt{3}}{12} \sum_{i=0}^{\infty} \left(\frac{4}{9}\right)^i = \frac{\sqrt{3}}{4} + \frac{\sqrt{3}}{12} \frac{1}{1-4/9} = \frac{\sqrt{3}}{4}\left(1 + \frac{3}{5}\right) = \frac{2}{5}\sqrt{3}$$

9.3.73 $|S - S_n| = \left|\sum_{i=n}^{\infty} r^k\right| = \left|\frac{r^n}{1-r}\right|$ since the latter sum is simply a geometric series with first term r^n and ratio r.

9.3.75

a. Solve $\left|\frac{(-0.8)^n}{1.8}\right| = \frac{0.8^n}{1.8} < 10^{-6}$ for n to get $n = 60$.

b. Solve $\frac{0.2^n}{0.8} < 10^{-6}$ for n to get $n = 9$.

9.3.77

a. Solve $\frac{1/\pi^n}{1-1/\pi} < 10^{-6}$ for n to get $n = 13$.

b. Solve $\frac{1/e^n}{1-1/e} < 10^{-6}$ for n to get $n = 15$.

9.3.79 $f(x) = \sum_{k=0}^{\infty} (-1)^k x^k = \frac{1}{1+x}$; since f is a geometric series, $f(x)$ exists only when the ratio, $-x$, is such that $|-x| = |x| < 1$. Then $f(0) = 1$, $f(0.2) = \frac{1}{1.2} = \frac{5}{6}$, $f(0.5) = \frac{1}{1+.05} = \frac{2}{3}$. Neither $f(1)$ nor $f(1.5)$ exists. The domain of f is $|x| < 1$.

9.3.81 $f(x)$ is a geometric series with ratio $\frac{1}{1+x}$; thus $f(x)$ converges when $\left|\frac{1}{1+x}\right| < 1$. For $x > -1$, $\left|\frac{1}{1+x}\right| = \frac{1}{1+x}$ and $\frac{1}{1+x} < 1$ when $1 < 1+x$, $x > 0$. For $x < -1$, $\left|\frac{1}{1+x}\right| = \frac{1}{-1-x}$, and this is less than 1 when $1 < -1 - x$, i.e. $x < -2$. So $f(x)$ converges for $x > 0$ and for $x < -2$. When $f(x)$ converges, its value is $\frac{1}{1-\frac{1}{1+x}} = \frac{1+x}{x}$, so $f(x) = 3$ when $1 + x = 3x$, $x = \frac{1}{2}$.

9.4 The Divergence and Integral Tests

9.4.1 A series may diverge so slowly that no reasonable number of terms may definitively show that it does so.

9.4.3 Yes. Either the series and the integral both converge, or both diverge.

9.4.5 For the same values of p as in the previous problem – it converges for $p > 1$, and diverges for all other values of p.

9.4.7 The remainder of an infinite series is the error in approximating a convergent infinite series by a finite number of terms.

9.4.9 $\sum_{k=0}^{\infty} \left(3\left(\frac{2}{5}\right)^k - 2\left(\frac{5}{7}\right)^k \right) = 3\sum_{k=0}^{\infty}\left(\frac{2}{5}\right)^k - 2\sum_{k=0}^{\infty}\left(\frac{5}{7}\right)^k = 3\left(\frac{1}{3/5}\right) - 2\left(\frac{1}{2/7}\right) = 5 - 7 = -2$.

9.4.11 $\sum_{k=1}^{\infty} \left(\frac{1}{3}\left(\frac{5}{6}\right)^k + \frac{3}{5}\left(\frac{7}{9}\right)^k \right) = \frac{1}{3}\sum_{k=1}^{\infty}\left(\frac{5}{6}\right)^k + \frac{3}{5}\sum_{k=1}^{\infty}\left(\frac{7}{9}\right)^k = \frac{1}{3}\left(\frac{5/6}{1/6}\right) + \frac{3}{5}\left(\frac{7/9}{2/9}\right) = \frac{5}{3} + \frac{21}{10} = \frac{113}{30}$.

9.4.13 $\sum_{k=1}^{\infty} \left(\left(\frac{1}{6}\right)^k + \left(\frac{1}{3}\right)^{k-1} \right) = \sum_{k=1}^{\infty}\left(\frac{1}{6}\right)^k + \sum_{k=1}^{\infty}\left(\frac{1}{3}\right)^{k-1} = \frac{1/6}{5/6} + \frac{1}{2/3} = \frac{17}{10}$.

9.4.15 $a_k = \frac{k}{2k+1}$ and $\lim_{k\to\infty} a_k = \frac{1}{2}$, so the series diverges.

9.4.17 $a_k = \frac{k}{\ln k}$ and $\lim_{k\to\infty} a_k = \infty$, so the series diverges.

9.4.19 $a_k = \frac{1}{1000+k}$ and $\lim_{k\to\infty} a_k = 0$, so the divergence test is inconclusive.

9.4.21 $a_k = \frac{\sqrt{k}}{\ln^{10} k}$ and $\lim_{k\to\infty} a_k = \infty$, so the series diverges.

9.4.23 Let $f(x) = \frac{1}{x \ln x}$. Then $f(x)$ is continuous and decreasing on $(1, \infty)$, since $x \ln x$ is increasing there. Since $\int_1^{\infty} f(x)\,dx = \infty$, the series diverges.

9.4.25 Let $f(x) = x \cdot e^{-2x^2}$. This function is continuous for $x \geq 1$. Its derivative is $e^{-2x^2}(1 - 4x^2) < 0$ for $x \geq 1$, so $f(x)$ is decreasing. Since $\int_1^{\infty} x \cdot e^{-2x^2}\,dx = \frac{1}{4e^2}$, the series converges.

9.4.27 Let $f(x) = \frac{1}{\sqrt{x+8}}$. $f(x)$ is obviously continuous and decreasing for $x \geq 1$. Since $\int_1^\infty \frac{1}{\sqrt{x+8}}\, dx = \infty$, the series diverges.

9.4.29 Let $f(x) = \frac{x}{e^x}$. $f(x)$ is clearly continuous for $x > 1$, and its derivative, $f'(x) = \frac{e^x - xe^x}{e^{2x}} = (1-x)\frac{e^x}{e^{2x}}$, is negative for $x > 1$ so that $f(x)$ is decreasing. Since $\int_1^\infty f(x)\, dx = 2e^{-1}$, the series converges.

9.4.31 This is a p-series with $p = 10$, so this series converges.

9.4.33 $\sum_{k=3}^\infty \frac{1}{(k-2)^4} = \sum_{k=1}^\infty \frac{1}{k^4}$, which is a p-series with $p = 4$, thus convergent.

9.4.35

a. The remainder R_n is bounded by $\int_n^\infty \frac{1}{x^6}\, dx = \frac{1}{5n^5}$.

b. We solve $\frac{1}{5n^5} < 10^{-3}$ to get $n = 3$.

c. $L_n = S_n + \int_{n+1}^\infty \frac{1}{x^6}\, dx = S_n + \frac{1}{5(n+1)^5}$, and $U_n = S_n + \int_n^\infty \frac{1}{x^6}\, dx = S_n + \frac{1}{5n^5}$.

d. $S_{10} \approx 1.017341512$, so $L_{10} \approx 1.017341512 + \frac{1}{5 \cdot 11^5} \approx 1.017342754$, and $U_{10} \approx 1.017341512 + \frac{1}{5 \cdot 10^5} \approx 1.017343512$.

9.4.37

a. The remainder R_n is bounded by $\int_n^\infty \frac{1}{3^x}\, dx = \frac{1}{3^n \ln(3)}$.

b. We solve $\frac{1}{3^n \ln(3)} < 10^{-3}$ to obtain $n = 7$.

c. $L_n = S_n + \int_{n+1}^\infty \frac{1}{3^x}\, dx = S_n + \frac{1}{3^{n+1} \ln(3)}$, and $U_n = S_n + \int_n^\infty \frac{1}{3^x}\, dx = S_n + \frac{1}{3^n \ln(3)}$.

d. $S_{10} \approx 0.4999915325$, so $L_{10} \approx 0.4999915325 + \frac{1}{3^{11} \ln 3} \approx 0.4999966708$, and $U_{10} \approx 0.4999915325 + \frac{1}{3^{10} \ln 3} \approx 0.5000069475$.

9.4.39

a. The remainder R_n is bounded by $\int_n^\infty \frac{1}{x^{3/2}}\, dx = 2n^{-1/2}$.

b. We solve $2n^{-1/2} < 10^{-3}$ to get $n > 4 \times 10^6$, so let $n = 4 \times 10^6 + 1$.

c. $L_n = S_n + \int_{n+1}^\infty \frac{1}{x^{3/2}}\, dx = S_n + 2(n+1)^{-1/2}$, and $U_n = S_n + \int_n^\infty \frac{1}{x^{3/2}}\, dx = S_n + 2n^{-1/2}$.

d. $S_{10} = \sum_{k=1}^{10} \frac{1}{k^{3/2}} \approx 1.995336494$, so $L_{10} \approx 1.995336494 + 2 \cdot 11^{-1/2} \approx 2.598359183$, and $U_{10} \approx 1.995336494 + 2 \cdot 10^{-1/2} \approx 2.627792026$.

9.4.41

a. The remainder R_n is bounded by $\int_n^\infty \frac{1}{x^3}\, dx = \frac{1}{2n^2}$.

b. We solve $\frac{1}{2n^2} < 10^{-3}$ to get $n = 23$.

c. $L_n = S_n + \int_{n+1}^\infty \frac{1}{x^3}\, dx = S_n + \frac{1}{2(n+1)^2}$, and $U_n = S_n + \int_n^\infty \frac{1}{x^3}\, dx = S_n + \frac{1}{2n^2}$.

d. $S_{10} \approx 1.197531986$, so $L_{10} \approx 1.197531986 + \frac{1}{2 \cdot 11^2} \approx 1.201664217$, and $U_{10} \approx 1.197531986 + \frac{1}{2 \cdot 10^2} \approx 1.202531986$.

9.4.43

a. True. The two series differ by a finite amount ($\sum_{k=1}^9 a_k$), so if one converges, so does the other.

b. True. The same argument applies as in part (a).

c. False. If $\sum a_k$ converges, then $a_k \to 0$ as $k \to \infty$, so that $a_k + 0.0001 \to 0.0001$ as $k \to \infty$, so that $\sum(a_k + 0.0001)$ cannot converge.

d. False. Suppose $p = .9999$. Then $\sum p^k$ converges, but $p + 0.001 = 1.009$ so that $\sum(p + 0.001)^k$ diverges.

e. False. Let $p = 1.0005$; then $-p + .001 = -(p - .001) = -.9995$, so that $\sum k^{-p}$ converges (p-series) but $\sum k^{-p+.001}$ diverges.

f. False. Let $a_k = \frac{1}{k}$, the harmonic series.

9.4.45 Converges by the Integral Test since $\displaystyle\int_1^\infty \frac{1}{(3x+1)(3x+4)}\, dx = \int_1^\infty \frac{1}{3(3x+1)} - \frac{1}{3(3x+4)}\, dx =$
$\displaystyle\lim_{b\to\infty} \int_1^b \left(\frac{1}{3(3x+1)} - \frac{1}{3(3x+4)}\right) dx = \lim_{b\to\infty} \frac{1}{9}\left(\ln\left(\frac{3x+1}{3x+4}\right)\right)\Big|_1^b = \lim_{b\to\infty} = \frac{-1}{9}\cdot\ln(4/7) \approx 0.06217 < \infty.$

9.4.47 Diverges by the Divergence Test since $\displaystyle\lim_{k\to\infty} a_k = \lim_{k\to\infty} \frac{k}{\sqrt{k^2+4}} = 1 \ne 0.$

9.4.49 Converges by the Integral Test since $\displaystyle\int_2^\infty \frac{4}{x\ln^2 x}\, dx = \lim_{b\to\infty}\left(\frac{-4}{\ln x}\Big|_2^b\right) = \frac{4}{\ln 2} < \infty.$

9.4.51

a. We must be somewhat careful here, as the function $\dfrac{1}{x\ln x(\ln\ln x)^p}$ has a discontinuity at $x = e$. So instead we evaluate the convergence of the series starting with a lower bound of 3. In this case, $\int \frac{1}{x\ln x(\ln\ln x)^p}\, dx = \frac{1}{1-p}(\ln\ln x)^{1-p}$, and thus the improper integral with bounds n and ∞ exists only if $p > 1$ since $\ln\ln x > 0$ for $x > e$. So this series converges for $p > 1$.

b. For large values of z, clearly $\sqrt{z} > \ln z$, so that $z > (\ln z)^2$. Write $z = \ln x$; then for large x, $\ln x > (\ln\ln x)^2$; multiplying both sides by $x\ln x$ we get that $x\ln^2 x > x\ln x(\ln\ln x)^2$, so that the first series converges faster since the terms get smaller faster.

9.4.53 Let $S_n = \sum_{k=1}^n \frac{1}{\sqrt{k}}$. Then this looks like a left Riemann sum for the function $y = \frac{1}{\sqrt{x}}$ on $[1, n+1]$. Since each rectangle lies above the curve itself, we see that S_n is bounded below by the integral of $\frac{1}{\sqrt{x}}$ on $[1, n+1]$. Now,

$$\int_1^{n+1} \frac{1}{\sqrt{x}}\, dx = \int_1^{n+1} x^{-1/2}\, dx = 2\sqrt{x}\Big|_1^{n+1} = 2\sqrt{n+1} - 2$$

This integral diverges as $n \to \infty$, so the series does as well by the bound above.

9.4.55 $\sum_{k=1}^\infty ca_k = \lim_{n\to\infty} \sum_{k=1}^n ca_k = \lim_{n\to\infty} c\sum_{k=1}^n a_k = c\lim_{n\to\infty} \sum_{k=1}^n a_k$, so that one sum converges if and only if the other one does.

9.4.57 To approximate the sequence for $\zeta(m)$, note that the remainder R_n after n terms is bounded by

$$\int_n^\infty \frac{1}{x^m}\, dx = \frac{1}{m-1}n^{1-m}.$$

For $m = 3$, if we wish to approximate the value to within 10^{-3}, we must solve $\frac{1}{2}n^{-2} < 10^{-3}$, so that $n = 23$,

and $\displaystyle\sum_{i=1}^{23} \frac{1}{i^3} \approx 1.201151926$. The true value is ≈ 1.202056903.

For $m = 5$, if we wish to approximate the value to within 10^{-3}, we must solve $\frac{1}{4}n^{-4} < 10^{-3}$, so that $n = 4$,

and $\displaystyle\sum_{i=1}^{4} \frac{1}{i^5} \approx 1.036341789$. The true value is ≈ 1.036927755.

For $m = 7$, if we wish to approximate the value to within 10^{-3}, we must solve $\frac{1}{6}n^{-6} < 10^{-3}$, so that $n = 3$,

and $\sum_{i=1}^{3} \frac{1}{i^7} \approx 1.008269747$. The true value is ≈ 1.008349277.

9.4.59 $\sum_{k=1}^{\infty} \frac{1}{k^2} = \sum_{k=1}^{\infty} \frac{1}{(2k)^2} + \sum_{k=1}^{\infty} \frac{1}{(2k-1)^2}$, splitting the series into even and odd terms. But $\sum_{k=1}^{\infty} \frac{1}{(2k)^2} =$

$\frac{1}{4}\sum_{k=1}^{\infty} \frac{1}{k^2}$. Thus $\frac{\pi^2}{6} = \frac{1}{4}\frac{\pi^2}{6} + \sum_{k=1}^{\infty} \frac{1}{(2k-1)^2}$, so that the sum in question is $\frac{3\pi^2}{24} = \frac{\pi^2}{8}$.

9.4.61

a. $x_1 = \sum_{k=2}^{2} \frac{1}{k} = \frac{1}{2}$, $x_2 = \sum_{k=3}^{4} \frac{1}{k} = \frac{1}{3} + \frac{1}{4} = \frac{7}{12}$, $x_3 = \sum_{k=4}^{6} \frac{1}{k} = \frac{1}{4} + \frac{1}{5} + \frac{1}{6} = \frac{37}{60}$.

b. x_n has n terms. Each term is bounded below by $\frac{1}{2n}$ and bounded above by $\frac{1}{n+1}$. Thus $x_n \geq n \cdot \frac{1}{2n} = \frac{1}{2}$, and $x_n \leq n \cdot \frac{1}{n+1} < n \cdot \frac{1}{n} = 1$.

c. The right Riemann sum for $\int_1^2 \frac{dx}{x}$ using n subintervals has n rectangles of width $\frac{1}{n}$; the right edges of those rectangles are at $1 + \frac{i}{n} = \frac{n+i}{n}$ for $i = 1, 2, \ldots, n$. The height of such a rectangle is the value of $\frac{1}{x}$ at the right endpoint, which is $\frac{n}{n+i}$. Thus the area of the rectangle is $\frac{1}{n} \cdot \frac{n}{n+i} = \frac{1}{n+i}$. Adding up over all the rectangles gives x_n.

d. The limit $\lim_{n\to\infty} x_n$ is the limit of the right Riemann sum as the width of the rectangles approaches zero. This is precisely $\int_1^2 \frac{dx}{x} = \ln x \Big|_1^2 = \ln 2$.

9.4.63

a. Note that the center of gravity of any stack of dominoes is the average of the locations of their centers. Define the midpoint of the zeroth (top) domino to be $x = 0$, and stack additional dominoes down and to its right (to increasingly positive x-coordinates.) Let $m(n)$ be the x-coordinate of the midpoint of the nth domino. Then in order for the stack not to fall over, the left edge of the nth domino must be placed directly under the center of gravity of dominos 0 through $n - 1$, which is $\frac{1}{n}\sum_{i=0}^{n-1} m(i)$, so that $m(n) = 1 + \frac{1}{n}\sum_{i=0}^{n-1} m(i)$. Claim that in fact $m(n) = \sum_{k=1}^{n} \frac{1}{k}$. Use induction. This is certainly true for $n = 1$. Note first that $m(0) = 0$, so we can start the sum at 1 rather than at 0. Now, $m(n) = 1 + \frac{1}{n}\sum_{i=1}^{n-1} m(i) = 1 + \frac{1}{n}\sum_{i=1}^{n-1}\sum_{j=1}^{i} \frac{1}{j}$. Now, 1 appears $n - 1$ times in the double sum, 2 appears $n - 2$ times, and so forth, so we can rewrite this sum as $m(n) = 1 + \frac{1}{n}\sum_{i=1}^{n-1} \frac{n-i}{i} = 1 + \frac{1}{n}\sum_{i=1}^{n-1}\left(\frac{n}{i} - 1\right) = 1 + \frac{1}{n}\left(n\sum_{i=1}^{n-1} \frac{1}{i} - (n-1)\right) = \sum_{i=1}^{n-1} \frac{1}{i} + 1 - \frac{n-1}{n} = \sum_{i=1}^{n} \frac{1}{i}$, and we are done by induction (noting that the statement is clearly true for $n = 0$, $n = 1$). Thus the maximum overhang is $\sum_{k=2}^{n} \frac{1}{k}$.

b. For an infinite number of dominos, since the overhang is the harmonic series, the distance is potentially infinite.

9.5 The Ratio, Root, and Comparison Tests

9.5.1 Given a series $\sum a_k$ of positive terms, compute $\lim_{k\to\infty} \frac{a_{k+1}}{a_k}$ and call it r. If $0 \leq r < 1$, the given series converges. If $r > 1$ (including $r = \infty$), the given series diverges. If $r = 1$, the test is inconclusive.

9.5.3 Given a series of positive terms $\sum a_k$ that you suspect converges, find a series $\sum b_k$ that you know converges, for which $\lim_{k\to\infty} \frac{a_k}{b_k} = L$ where $L \geq 0$ is a finite number. If you are successful, you will have shown that the series $\sum a_k$ converges.

Given a series of positive terms $\sum a_k$ that you suspect diverges, find a series $\sum b_k$ that you know diverges, for which $\lim_{k\to\infty} \frac{a_k}{b_k} = L$ where $L > 0$ (including the case $L = \infty$). If you are successful, you will have shown that $\sum a_k$ diverges.

9.5.5 The Ratio Test.

9.5.7 The difference between successive partial sums is a term in the sequence. Since the terms are positive, differences between successive partial sums are as well, so the sequence of partial sums is increasing.

9.5.9 The ratio between successive terms is $\frac{a_{k+1}}{a_k} = \frac{1}{(k+1)!} \cdot \frac{(k)!}{1} = \frac{1}{k+1}$, which goes to zero as $k \to \infty$, so the given series converges by the Ratio Test.

9.5.11 The ratio between successive terms is $\frac{a_{k+1}}{a_k} = \frac{(k+1)^2}{4(k+1)} \cdot \frac{4^k}{(k)^2} = \frac{1}{4}\left(\frac{k+1}{k}\right)^2$. The limit is $1/4$ as $k \to \infty$, so the given series converges by the Ratio Test.

9.5.13 The ratio between successive terms is $\frac{a_{k+1}}{a_k} = \frac{(k+1)e^{-(k+1)}}{(k)e^{-(k)}} = \frac{k+1}{(k)e}$. The limit of this ratio as $k \to \infty$ is $1/e < 1$, so the given series converges by the Ratio Test.

9.5.15 The ratio between successive terms is $\frac{2^{k+1}}{(k+1)^{99}} \cdot \frac{(k)^{99}}{2^k} = 2\left(\frac{k}{k+1}\right)^{99}$; the limit as $k \to \infty$ is 2, so the given series diverges by the Ratio Test.

9.5.17 The ratio between successive terms is $\frac{((k+1)!)^2}{(2(k+1))!} \cdot \frac{(2k)!}{((k)!)^2} = \frac{(k+1)^2}{(2k+2)(2k+1)}$; the limit as $k \to \infty$ is $1/4$, so the given series converges by the Ratio Test.

9.5.19 $\lim\limits_{k\to\infty} \sqrt[k]{a_k} = \lim\limits_{k\to\infty} \frac{4k^3+k}{9k^3+k+1} = \frac{4}{9} < 1$, so the given series converges by the Root Test.

9.5.21 $\lim\limits_{k\to\infty} \sqrt[k]{a_k} = \lim\limits_{k\to\infty} \frac{k^{2/k}}{2} = \frac{1}{2} < 1$, so the given series converges by the Root Test.

9.5.23 $\lim\limits_{k\to\infty} \sqrt[k]{a_k} = \lim\limits_{k\to\infty} \left(\frac{k}{k+1}\right)^{2k} = e^{-2} < 1$, so the given series converges by the Root Test.

9.5.25 $\lim\limits_{k\to\infty} \sqrt[k]{a_k} = \lim\limits_{k\to\infty} \sqrt[k]{\left(\frac{1}{k^k}\right)} = \lim\limits_{k\to\infty} \frac{1}{k} = 0$, so the given series converges by the Root Test.

9.5.27 $\frac{1}{k^2+4} < \frac{1}{k^2}$, and $\sum_{k=1}^{\infty} \frac{1}{k^2}$ converges, so $\sum_{k=1}^{\infty} \frac{1}{k^2+4}$ converges as well, by the Comparison Test.

9.5.29 Use the Limit Comparison Test with $\left\{\frac{1}{k}\right\}$. The ratio of the terms of the two series is $\frac{k^3-k}{k^3+4}$ which has limit 1 as $k \to \infty$. Since the comparison series diverges, the given series does as well.

9.5.31 For all k, $\frac{1}{k^{3/2}+1} < \frac{1}{k^{3/2}}$. The series whose terms are $\frac{1}{k^{3/2}}$ is a p-series which converges, so the given series converges as well by the Comparison Test.

9.5.33 $\sin(1/k) > 0$ for $k \geq 1$, so we can apply the Comparison Test with $1/k^2$. $\sin(1/k) < 1$, so $\frac{\sin(1/k)}{k^2} < \frac{1}{k^2}$. Since the comparison series converges, the given series converges as well.

9.5.35 Use the Limit Comparison Test with $\{1/k\}$. The ratio of the terms of the two series is $\frac{k}{2k-\sqrt{k}} = \frac{1}{2-1/\sqrt{k}}$, which has limit $1/2$ as $k \to \infty$. Since the comparison series diverges, the given series does as well.

9.5.37 Use the Limit Comparison Test with $\frac{k^{2/3}}{k^{3/2}}$. The ratio of corresponding terms of the two series is $\frac{\sqrt[3]{k^2+1}}{\sqrt{k^3+1}} \cdot \frac{k^{3/2}}{k^{2/3}} = \frac{\sqrt[3]{k^2+1}}{\sqrt[3]{k^2}} \cdot \frac{\sqrt{k^3}}{\sqrt{k^3+1}}$, which has limit 1 as $k \to \infty$. The comparison series is the series whose terms are $k^{2/3-3/2} = k^{-5/6}$, which is a p-series with $p < 1$, so it, and the given series, both diverge.

9.5.39

a. False. For example, let $\{a_k\}$ be all zeros, and $\{b_k\}$ be all 1's.

b. True. This is a result of the Comparison Test.

c. True. Both of these statements follow from the Comparison Test.

9.5.41 Use the Comparison Test. Each term $\frac{1}{k} + 2^{-k} > \frac{1}{k}$. Since the harmonic series diverges, so does this series.

9.5.43 Use the Ratio Test. $\frac{a_{k+1}}{a_k} = \frac{2^{k+1}(k+1)!}{(k+1)^{k+1}} \cdot \frac{(k)^k}{2^k(k)!} = 2\left(\frac{k}{k+1}\right)^k$, which has limit $\frac{2}{e}$ as $k \to \infty$, so the given series converges.

9.5.45 Use the Limit Comparison Test with $\{1/k^3\}$. The ratio of corresponding terms is $\frac{k^{11}}{k^{11}+3}$, which has limit 1 as $k \to \infty$. Since the comparison series converges, so does the given series.

9.5.47 This is a p-series with exponent greater than 1, so it converges.

9.5.49 $\ln\left(\frac{k+2}{k+1}\right) = \ln(k+2) - \ln(k+1)$, so this series telescopes. We get $\sum_{k=1}^{n} \ln\left(\frac{k+2}{k+1}\right) = \ln(n+2) - \ln 2$. Since $\lim_{n \to \infty} \ln(n+2) - \ln(2) = \infty$, the sequence of partial sums diverges, so the given series is divergent.

9.5.51 For $k > 10$, $\ln k > 2$ so note that $\frac{1}{k^{\ln k}} < \frac{1}{k^2}$. Since $\sum_{k=1}^{\infty} \frac{1}{k^2}$ converges, the given series converges as well.

9.5.53 Use the Limit Comparison Test with the harmonic series. $\frac{\tan(1/k)}{1/k}$ has limit 1 as $k \to \infty$ since $\lim_{x \to 0} \frac{\tan(x)}{x} = 1$. Thus the original series diverges.

9.5.55 Note that $\frac{1}{(2k+1) \cdot (2k+3)} = \frac{1}{2}\left(\frac{1}{2k+1} - \frac{1}{2k+3}\right)$. Thus this series telescopes.

$$\sum_{k=0}^{n} \frac{1}{(2k+1)(2k+3)} = \frac{1}{2}\sum_{k=0}^{n}\left(\frac{1}{2k+1} - \frac{1}{2k+3}\right) = \frac{1}{2}\left(-\frac{1}{2n+3} + 1\right),$$

so the given series converges to $1/2$, since that is the limit of the sequence of partial sums.

9.5.57 This series is $\sum_{k=1}^{\infty} \frac{k^2}{k!}$. By the Ratio Test, $\frac{a_{k+1}}{a_k} = \frac{(k+1)^2}{(k+1)!} \cdot \frac{k!}{k^2} = \frac{1}{k+1}\left(\frac{k+1}{k}\right)^2$, which has limit 0 as $k \to \infty$, so the given series converges.

9.5.59 For $p \leq 1$ and $k > e$, $\frac{\ln k}{k^p} > \frac{1}{k^p}$. The series $\sum_{k=1}^{\infty} \frac{1}{k^p}$ diverges, so the given series diverges. For $p > 1$, let $q < p - 1$; then for sufficiently large k, $\ln k < k^q$, so that by the Comparison Test, $\frac{\ln k}{k^p} < \frac{k^q}{k^p} = \frac{1}{k^{p-q}}$. But $p - q > 1$, so that $\sum_{k=1}^{\infty} \frac{1}{k^{p-q}}$ is a convergent p-series. Thus the original series is convergent precisely when $p > 1$.

9.5.61 For $p \leq 1$, $\frac{(\ln k)^p}{k^p} > \frac{1}{k^p}$, and $\sum_{k=1}^{\infty} \frac{1}{k^p}$ diverges for $p \leq 1$, so the original series diverges. For $p > 1$, let $q < p - 1$; then for sufficiently large k, $(\ln k)^p < k^q$. Note that $\frac{(\ln k)^p}{k^p} < \frac{k^q}{k^p} = \frac{1}{k^{p-q}}$. But $p - q > 1$, so $\sum_{k=1}^{\infty} \frac{1}{k^{p-q}}$ converges, so the given series converges. Thus, the given series converges exactly for $p > 1$.

9.5.63 Use the Ratio Test:

$$\frac{a_{k+1}}{a_k} = \frac{1 \cdot 3 \cdot 5 \cdots (2k+1)}{(k+1)p^{k+2}(k+1)!} \cdot \frac{(k)p^{k+1}(k)!}{1 \cdot 3 \cdot 5 \cdots (2k-1)} = \frac{(2k+1)(k)}{(k+1)^2 p}$$

and this expression has limit $\frac{2}{p}$ as $k \to \infty$. Thus the series converges for $p > 2$.

9.5.65 $\lim_{k \to \infty} a_k = \lim_{k \to \infty}\left(1 - \frac{p}{k}\right)^k = e^{-p} \neq 0$, so this sequence diverges for all p by the Divergence Test.

9.5.67 These tests apply only for series with positive terms, so assume $r > 0$. Clearly the series do not converge for $r = 1$, so we assume $r \neq 1$ in what follows. Using the Integral Test, $\sum r^k$ converges if and only if $\int_1^\infty r^x dx$ converges. This improper integral has value $\lim_{b \to \infty} \frac{r^x}{\ln r}\Big|_1^b$, which converges only when $\lim_{b \to \infty} r^b$ exists, which occurs only for $r < 1$. Using the Ratio Test, $\frac{a_{k+1}}{a_k} = \frac{r^{k+1}}{r^k} = r$, so by the Ratio Test, the series converges if and only if $r < 1$. Using the Root Test, $\lim_{k \to \infty} \sqrt[k]{a_k} = \lim_{k \to \infty} \sqrt[k]{r^k} = \lim_{k \to \infty} r = r$, so again we have convergence if and only if $r < 1$.

9.5.69 To prove case (2), assume $L = 0$ and that $\sum b_k$ converges. Since $L = 0$, for every $\varepsilon > 0$, there is some N such that for all $n > N$, $\left|\frac{a_k}{b_k}\right| < \varepsilon$. Take $\varepsilon = 1$; this then says that there is some N such that for all $n > N$, $0 < a_k < b_k$. By the Comparison Test, since $\sum b_k$ converges, so does $\sum a_k$. To prove case (3), since $L = \infty$, then $\lim\limits_{k \to \infty} \frac{b_k}{a_k} = 0$, so by the argument above, we have $0 < b_k < a_k$ for sufficient large k. But $\sum b_k$ diverges, so by the Comparison Test, $\sum a_k$ does as well.

9.5.71 $\dfrac{a_{k+1}}{a_k} = \dfrac{x^{k+1}}{x^k} = x$. This has limit x as $k \to \infty$, so the series converges for $x < 1$. It clearly does not converge for $x = 1$.

9.5.73 $\dfrac{a_{k+1}}{a_k} = \dfrac{x^{k+1}}{(k+1)^2} \cdot \dfrac{k^2}{x^k} = x\left(\dfrac{k}{k+1}\right)^2$, which has limit x as $k \to \infty$. Thus the series converges for $x < 1$. When $x = 1$, the series is $\frac{1}{k^2}$, which converges. Thus the original series converges for $x \le 1$.

9.5.75 $\dfrac{a_{k+1}}{a_k} = \dfrac{x^{k+1}}{2^{k+1}} \cdot \dfrac{2^k}{x^k} = \dfrac{x}{2}$, which has limit $x/2$ as $k \to \infty$. Thus the series converges for $x < 2$. For $x = 2$, it is obviously divergent.

9.5.77

a. $\ln \prod_{k=0}^{\infty} e^{1/2^k} = \sum_{k=0}^{\infty} \frac{1}{2^k} = 2$, so that the original product converges to e^2.

b. $\ln \prod_{k=2}^{\infty} \left(1 - \frac{1}{k}\right) = \ln \prod_{k=2}^{\infty} \frac{k-1}{k} = \sum_{k=2}^{\infty} \ln \frac{k-1}{k} = \sum_{k=2}^{\infty} (\ln(k-1) - \ln(k))$. This series telescopes to give $S_n = -\ln(n)$, so the original series has limit $\lim\limits_{n \to \infty} P_n = \lim\limits_{n \to \infty} e^{-\ln(n)} = 0$.

9.6   Alternating Series

9.6.1 If $a_n < 0$ (which happens for every other value of n), then $S_n = S_{n-1} + a_n < S_{n-1}$.

9.6.3 Since we are far enough out that the terms are nonincreasing in magnitude, we may assume the series is $\sum_{k=0}^{\infty} (-1)^k a_k$, where $a_k > 0$ for all k, and such that the a_k are nonincreasing in magnitude starting with a_1. Then

$$S = S_{2n+1} + (a_{2n} - a_{2n+1}) + (a_{2n+2} - a_{2n+3}) + \cdots$$

and each term of the form $a_{2k} - a_{2k+1} > 0$, so that $S_{2n+1} < S$. Also

$$S = S_{2n} + (-a_{2n+1} + a_{2n+2}) + (-a_{2n+3} + a_{2n+4}) + \cdots$$

and each term of the form $-a_{2k+1} + a_{2k+2} < 0$, so that $S < S_{2n}$. Thus the sum of the series is trapped between the odd partial sums and the even partial sums.

9.6.5 The remainder is less than the first neglected term since

$$L - S_n = (-1)^{n+1}(a_{n+1} + (-a_{n+2} + a_{n+3}) + \cdots)$$

so that the sum of the series *after* the first disregarded term has the opposite sign from the first disregarded term.

9.6.7 No. If the terms are positive, then the absolute value of each term is the term itself, so convergence and absolute convergence would mean the same thing in this context.

9.6.9 Yes. For example, $\sum \frac{(-1)^k}{k^3}$ converges absolutely and thus not conditionally (see the definition).

9.6.11 The terms of the series decrease in magnitude, and $\lim\limits_{k \to \infty} \frac{1}{k^3} = 0$, so the given series converges.

9.6.13 The terms of the series decrease in magnitude, and

$$\lim_{k \to \infty} \frac{k^2}{k^3 + 1} = \lim_{k \to \infty} \frac{1}{k + 1/k^2} = 0,$$

so the given series converges.

9.6.15 $\lim_{k \to \infty} \frac{k^2 - 1}{k^2 + 3} = 1$, so the terms of the series do not tend to zero and thus the given series diverges.

9.6.17 $\lim_{k \to \infty} \left(1 + \frac{1}{k}\right) = 1$, so the given series diverges.

9.6.19 The derivative of $f(k) = \frac{k^{10} + 2k^5 + 1}{k(k^{10} + 1)}$ is $f'(k) = \frac{-(k^{20} + 2k^{10} + 12k^{15} - 8k^5 + 1)}{k^2(k^{10} + 1)^2}$. The numerator is negative for large enough values of k, and the denominator is always positive, so the derivative is negative for large enough k. Also, $\lim_{k \to \infty} \frac{k^{10} + 2k^5 + 1}{k(k^{10} + 1)} = \lim_{k \to \infty} \frac{1 + 2k^{-5} + k^{-10}}{k + k^{-9}} = 0$. Thus the given series converges.

9.6.21 $\lim_{k \to \infty} k^{1/k} = 1$ (for example, take logs and apply L'Hôpital's rule), so the given series diverges by the Divergence Test.

9.6.23 $\frac{1}{\sqrt{k^2 + 4}}$ is decreasing and tends to zero as $k \to \infty$, so the given series converges.

9.6.25 We want $\frac{1}{n+1} < 10^{-4}$, or $n + 1 > 10^4$, so $n = 10^4$.

9.6.27 The series starts with $k = 0$, so we want $\frac{1}{2n+1} < 10^{-4}$, or $2n + 1 > 10^4$, $n = 5000$.

9.6.29 We want $\frac{1}{(n+1)^4} < 10^{-4}$, or $(n+1)^4 > 10^4$, so $n = 10$.

9.6.31 The series starts with $k = 0$, so we want $\frac{1}{3n+1} < 10^{-4}$, or $3n + 1 > 10^4$, $n = 3334$.

9.6.33 The series starts with $k = 0$, so we want $\frac{1}{4^n} \left(\frac{2}{4n+1} + \frac{2}{4n+2} + \frac{1}{4n+3}\right) < 10^{-4}$, or $\frac{4^n(4n+1)(4n+2)(4n+3)}{4(20n^2 + 21n + 5)} > 10000$, which occurs first for $n = 6$.

9.6.35 To figure out how many terms we need to sum, we must find n such that $\frac{1}{(n+1)^5} < 10^{-3}$, so that $(n+1)^5 > 1000$; this occurs first for $n = 3$. Thus $\frac{-1}{1} + \frac{1}{2^5} - \frac{1}{3^5} \approx -0.972865$.

9.6.37 To figure how many terms we need to sum, we must find n such that $\frac{1}{(n+1)^{n+1}} < 10^{-3}$, or $(n+1)^{n+1} > 1000$, so $n = 4$ ($5^5 = 3125$). Thus the approximation is $\sum_{k=1}^{4} \frac{(-1)^n}{n^n} \approx -.7831307870$.

9.6.39 The series of absolute values is a p-series with $p = 3/2$, so it converges absolutely.

9.6.41 The series of absolute values is $\sum \frac{|\cos(k)|}{k^3}$, which converges by the Comparison Test since $\frac{|\cos(k)|}{k^3} \leq \frac{1}{k^3}$. Thus the series converges absolutely.

9.6.43 The series of absolute values is $\sum \frac{k}{2k+1}$, but $\lim_{k \to \infty} \frac{k}{2k+1} = \frac{1}{2}$, so by the Divergence Test, this series diverges. The original series does not converge conditionally, either, since $\lim_{k \to \infty} a_k = \frac{1}{2} \neq 0$.

9.6.45 The series of absolute values is $\sum \frac{\tan^{-1}(k)}{k^3}$, which converges by the Comparison Test since $\frac{\tan^{-1}(k)}{k^3} < \frac{\pi}{2} \frac{1}{k^3}$, and $\sum \frac{\pi}{2} \frac{1}{k^3}$ converges since it is a constant multiple of a convergent p-series. So the original series converges absolutely.

9.6.47

a. False. For example, consider the alternating harmonic series.

b. True. This is part of Theorem 8.21.

c. True. This statement is simply saying that a convergent series converges.

d. True. This is part of Theorem 8.21.

e. False. Let $a_k = \frac{1}{k}$.

f. True. Use the Comparison Test: $\lim\limits_{k \to \infty} \frac{a_k^2}{a_k} = \lim\limits_{k \to \infty} a_k = 0$ since $\sum a_k$ converges, so $\sum a_k^2$ and $\sum a_k$ converge or diverge together. Since the latter converges, so does the former.

g. True, by definition. If $\sum |a_k|$ converged, the original series would converge absolutely, not conditionally.

9.6.49 $\sum_{k=1}^{\infty} \frac{1}{k^2} - \sum_{k=1}^{\infty} \frac{(-1)^{k+1}}{k^2} = 2 \sum_{k=1}^{\infty} \frac{1}{(2k)^2} = 2 \cdot \frac{1}{4} \sum_{k=1}^{\infty} \frac{1}{k^2}$, and thus $\sum_{k=1}^{\infty} \frac{(-1)^{k+1}}{k^2} = \frac{\pi^2}{6} - \frac{1}{2} \cdot \frac{\pi^2}{6} = \frac{\pi^2}{12}$.

9.6.51 Write $r = -s$; then $0 < s < 1$ and $\sum r^k = \sum (-1)^k s^k$. Since $|s| < 1$, the terms s^k are nonincreasing and tend to zero, so by the Alternating Series Test, the series $\sum s^k$ converges, so $\sum (-1)^k s^k = \sum r^k$ does too.

9.6.53 Let $S = 1 - \frac{1}{2} + \frac{1}{3} - \cdots$. Then

$$S = \left(1 - \tfrac{1}{2}\right) + \left(\tfrac{1}{3} - \tfrac{1}{4}\right) + \left(\tfrac{1}{5} - \tfrac{1}{6}\right) + \left(\tfrac{1}{7} - \tfrac{1}{8}\right) + \cdots$$

$$\tfrac{1}{2}S = \quad \tfrac{1}{2} \quad - \quad \tfrac{1}{4} \quad + \quad \tfrac{1}{6} \quad - \quad \tfrac{1}{8} \quad + \cdots$$

Add these two series together to get

$$\frac{3}{2}S = \frac{3}{2}\ln 2 = 1 + \frac{1}{3} - \frac{1}{2} + \frac{1}{5} + \cdots$$

To see that the results are as desired, consider a collection of four terms:

$$\cdots + \left(\frac{1}{4k+1} - \frac{1}{4k+2}\right) + \left(\frac{1}{4k+3} - \frac{1}{4k+4}\right) + \cdots$$

$$\cdots + \frac{1}{4k+2} - \frac{1}{4k+4} + \cdots$$

Adding these results in the desired sign pattern. This repeats for each group of four elements.

9.6.55 Both series diverge, so comparisons of their values are not meaningful.

9.7   Chapter Nine Review

9.7.1

a. False. Let $a_n = 1 - \frac{1}{n}$. This sequence has limit 1.

b. False. The terms of a sequence tending to zero is necessary but not sufficient for convergence of the series.

c. True. This is the definition of convergence of a series.

d. False. If a series converges absolutely, the definition says that it does not converge conditionally.

9.7.3 $\lim\limits_{n \to \infty} \frac{8^n}{n!} = 0$ since exponentials grow more slowly than factorials.

9.7.5 Take logs and compute $\lim\limits_{n \to \infty} (1/n) \ln n = \lim\limits_{n \to \infty} (\ln n)/n = \lim\limits_{n \to \infty} \frac{1}{n} = 0$ by L'Hôpital's rule. Thus the original limit is $e^0 = 1$.

9.7.7 Take logs, and then evaluate $\lim\limits_{n\to\infty}\frac{1}{\ln n}\ln(1/n) = \lim\limits_{n\to\infty}(-1) = -1$, so the original limit is e^{-1}.

9.7.9 $a_n = (-1/0.9)^n = (-10/9)^n$. The terms grow without bound so the sequence does not converge.

9.7.11

a. $S_1 = \frac{1}{3}$, $S_2 = \frac{11}{24}$, $S_3 = \frac{21}{40}$, $S_4 = \frac{17}{30}$.

b. $S_n = \frac{1}{2}\left(\frac{1}{1} + \frac{1}{2} - \frac{1}{n+1} - \frac{1}{n+2}\right)$, since the series telescopes.

c. From part (b), $\lim\limits_{n\to\infty} S_n = \frac{3}{4}$, which is the sum of the series.

9.7.13 $\sum_{k=1}^{\infty} 3(1.001)^k = 3\sum_{k=1}^{\infty}(1.001)^k$. This is a geometric series with ratio greater than 1, so it diverges.

9.7.15 $\frac{1}{k(k+1)} = \frac{1}{k} - \frac{1}{k+1}$, so the series telescopes, and $S_n = 1 - \frac{1}{n+1}$. Thus $\lim\limits_{n\to\infty} S_n = 1$, which is the value of the series.

9.7.17 This series telescopes. $S_n = 3 - \frac{3}{3n+1}$, so that $\lim\limits_{n\to\infty} S_n = 3$, which is the value of the series.

9.7.19 $\sum\limits_{k=1}^{\infty} \frac{2^k}{3^{k+2}} = \frac{1}{9}\sum\limits_{k=1}^{\infty}\left(\frac{2}{3}\right)^k = \frac{1}{9} \cdot \frac{2/3}{1-2/3} = \frac{2}{9}$.

9.7.21

a. It appears that the series converges, since the sequence of partial sums appears to converge to 1.

b. It appears that series B converges to about 4.

c. This series clearly appears to diverge, since the partial sums seem to be growing without bound.

9.7.23 The series can be written $\sum \frac{1}{k^{2/3}}$, which is a p-series with $p = 2/3 < 1$, so this series diverges.

9.7.25 This is a geometric series with ratio $2/e < 1$, so the sum is $\frac{2/e}{1-2/e} = \frac{2}{e-2}$.

9.7.27 Applying the Ratio Test:

$$\lim_{k\to\infty} \frac{a_{k+1}}{a_k} = \lim_{k\to\infty} \frac{2^{k+1}(k+1)!}{(k+1)^{k+1}} \cdot \frac{k^k}{2^k k!} = \lim_{k\to\infty} 2\left(\frac{k}{k+1}\right)^k = \frac{2}{e} < 1,$$

so the given series converges.

9.7.29 Use the Comparison Test: $\frac{3}{2+e^k} < \frac{3}{e^k}$, but $\sum \frac{3}{e^k}$ converges since it is a geometric series with ratio $\frac{1}{e} < 1$. Thus the original series converges as well.

9.7.31 $a_k = \frac{k^{1/k}}{k^3} = \frac{1}{k^{3-1/k}}$. For $k \geq 2$, then, $a_k < \frac{1}{k^2}$. Since $\sum \frac{1}{k^2}$ converges, the given series also converges, by the Comparison Test.

9.7.33 Use the Ratio Test: $\frac{a_{k+1}}{a_k} = \frac{(k+1)^5}{e^{k+1}} \cdot \frac{e^k}{k^5} = \frac{1}{e} \cdot \left(\frac{k+1}{k}\right)^5$, which has limit $1/e < 1$ as $k \to \infty$. Thus the given series converges.

9.7.35 Use the Comparison Test. Since $\lim\limits_{k\to\infty} \frac{\ln k}{k^{1/2}} = 0$, we have that for sufficiently large k, $\ln k < k^{1/2}$, so that $a_k = \frac{2\ln k}{k^2} < \frac{2k^{1/2}}{k^2} = \frac{2}{k^{3/2}}$. Now $\sum \frac{2}{k^{3/2}}$ is convergent, since it is a p-series with $p = 3/2 > 1$. Thus the original series is convergent.

9.7.37 $|a_k| = \frac{1}{k^2-1}$. Use the Limit Comparison Test with the convergent series $\sum \frac{1}{k^2}$. Since $\lim_{k\to\infty} \frac{1}{k^2-1} \Big/ \frac{1}{k^2} = \lim_{k\to\infty} \frac{k^2}{k^2-1} = 1$, the given series converges absolutely.

9.7.39 Use the Ratio Test on the absolute values of the sequence of terms: $\lim_{k\to\infty} \left| \frac{a_{k+1}}{a_k} \right| = \lim_{k\to\infty} \frac{k+1}{e^{k+1}} \cdot \frac{e^k}{k} = \lim_{k\to\infty} \frac{1}{e} \cdot \frac{k+1}{k} = \frac{1}{e} < 1$. Thus, the original series is absolutely convergent.

9.7.41 Use the Ratio Test on the absolute values of the sequence of terms: $\lim_{k\to\infty} \left| \frac{a_{k+1}}{a_k} \right| = \lim_{k\to\infty} \frac{10}{k+1} = 0$, so the series converges absolutely.

9.7.43

    a. For $|x| < 1$, $\lim_{k\to\infty} x^k = 0$, so this limit is zero.

    b. This is a geometric series with ratio $-4/5$, so the sum is $\frac{1}{1+4/5} = \frac{5}{9}$.

9.7.45 Since the series converges, we must have $\lim_{k\to\infty} a_k = 0$. Since it converges to 8, the partial sums converge to 8, so that $\lim_{k\to\infty} S_k = 8$.

9.7.47 The series converges absolutely for $p > 1$, conditionally for $0 < p \le 1$ in which case $\{k^{-p}\}$ is decreasing to zero.

9.7.49 The sum is 0.2500000000 to ten decimal places. The maximum error is

$$\int_{20}^{\infty} \frac{1}{5^x}\, dx = \lim_{b\to\infty} \frac{-1}{\ln(5)5^x} \Big|_{20}^{b} = \frac{1}{\ln(5)\cdot 5^{20}} \approx 6.5 \times 10^{-15}.$$

9.7.51 The maximum error is a_{n+1}, so we want $a_{n+1} = \frac{1}{(k+1)^4} < 10^{-8}$, or $(k+1)^4 > 10^8$, so $k = 100$.

9.7.53

    a. Let T_n be the amount of additional tunnel dug during week n. Then $T_0 = 100$ and $T_n = .95 \cdot T_{n-1} = (.95)^n T_0 = 100(0.95)^n$, so the total distance dug in N weeks is

$$S_N = 100 \sum_{k=0}^{N-1} (0.95)^k = 100 \left(\frac{1 - (0.95)^N}{1 - 0.95} \right) = 2000(1 - 0.95^N).$$

    Then $S_{10} \approx 802.5$ meters and $S_{20} \approx 1283.03$ meters.

    b. The longest possible tunnel is $S_\infty = 100 \sum_{k=0}^{\infty} (0.95)^k = \frac{100}{1-.95} = 2000$ meters.

9.7.55

    a. The area of a circle of radius r is πr^2. For $r = 2^{1-n}$, this is $\pi 2^{2-2n}$. There are 2^{n-1} circles on the nth page, so the total area of circles on the nth page is $2^{n-1} \cdot \pi 2^{2-2n} = 2^{1-n}\pi$.

    b. The sum of the areas on all pages is $\sum_{k=1}^{\infty} 2^{1-k}\pi = 2\pi \sum_{k=1}^{\infty} 2^{-k} = 2\pi \cdot \frac{1/2}{1/2} = 2\pi$.

9.7.57

    a. $B_n = 1.0025 \cdot B_{n-1} + 100$ and $B_0 = 100$.

    b. $B_n = 100 \cdot 1.0025^n + 100 \cdot \frac{1-1.0025^n}{1-1.0025} = 100 \cdot 1.0025^n - 40000(1 - 1.0025^n)$.

9.7.59

a. $T_1 = \frac{\sqrt{3}}{16}$ and $T_2 = \frac{7\sqrt{3}}{64}$.

b. At stage n, 3^{n-1} triangles of side length $1/2^n$ are removed. Each of those triangles has an area of $\frac{\sqrt{3}}{4 \cdot 4^n} = \frac{\sqrt{3}}{4^{n+1}}$, so a total of

$$3^{n-1} \cdot \frac{\sqrt{3}}{4^{n+1}} = \frac{\sqrt{3}}{16} \cdot \left(\frac{3}{4}\right)^{n-1}$$

is removed at each stage. Thus

$$T_n = \frac{\sqrt{3}}{16} \sum_{k=1}^{n} \left(\frac{3}{4}\right)^{k-1} = \frac{\sqrt{3}}{16} \sum_{k=0}^{n-1} \left(\frac{3}{4}\right)^{k} = \frac{\sqrt{3}}{4}\left(1 - \left(\frac{3}{4}\right)^{n}\right)$$

c. $\lim_{n \to \infty} T_n = \frac{\sqrt{3}}{4}$ since $\left(\frac{3}{4}\right)^n \to 0$ as $n \to \infty$.

d. The area of the triangle was originally $\frac{\sqrt{3}}{4}$, so none of the original area is left.

Chapter 10

10.1   Approximating Functions With Polynomials

10.1.1 Let the polynomial be $p(x)$. Then $p(0) = f(0)$, $p'(0) = f'(0)$, and $p''(0) = f''(0)$.

10.1.3 The approximations are $p_0(0.1) = 1$, $p_1(0.1) = 1 + \frac{0.1}{2} = 1.05$, and $p_2(0.1) = 1 + \frac{0.1}{2} - \frac{.01}{8} = 1.04875$.

10.1.5 The remainder is the difference between the value of the Taylor polynomial at a point and the true value of the function at that point, $R_n(x) = f(x) - p_n(x)$.

10.1.7

   a. $f'(x) = -e^{-x}$, so $p_1(x) = f(0) + f'(0)(x) = 1 - x$.

   b. $f''(x) = e^{-x}$, so $p_2(x) = f(0) + f'(0)(x) + \frac{1}{2}f''(0)(x^2) = 1 - x + \frac{1}{2}x^2$.

   c. $p_1(0.2) = 0.8$, and $p_2(0.2) = 1 - 0.2 + \frac{1}{2}(0.04) = 0.82$.

10.1.9

   a. $f'(x) = -\frac{1}{(x+1)^2}$, so $p_1(x) = f(0) + f'(0)(x) = 1 - x$.

   b. $f''(x) = \frac{2}{(x+1)^3}$, so $p_2(x) = f(0) + f'(0)(x) + \frac{1}{2}f''(0)x^2 = 1 - x + x^2$.

   c. $p_1(.05) = .95$, and $p_2(.05) = 1 - .05 + .0025 = .9525$.

10.1.11

   a. $f'(x) = (1/3)x^{-2/3}$, so $p_1(x) = f(8) + f'(8)(x - 8) = 2 + \frac{1}{12}(x - 8)$.

   b. $f''(x) = (-2/9)x^{-5/3}$, so $p_2(x) = f(8) + f'(8)(x - 8) + \frac{1}{2}f''(8)(x - 8)^2 = 2 + \frac{1}{12}(x - 8) - \frac{1}{288}(x - 8)^2$.

   c. $p_1(7.5) \approx 1.958333333$, $p_2(7.5) \approx 1.957465278$.

10.1.13 $f(0) = 1, f'(0) = -\sin 0 = 0, f''(0) = -\cos 0 = -1$, so that $p_0(x) = 1$, $p_1(x) = 1$, $p_2(x) = 1 - \frac{1}{2}x^2$.

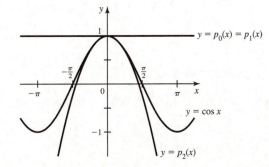

10.1.15 $f(0) = 0$, $f'(0) = \frac{-1}{1-0} = -1$, $f''(0) = \frac{1}{(1-0)^2} = -1$, so that $p_0(x) = 0$, $p_1(x) = -x$, $p_2(x) = -x - \frac{1}{2}x^2$.

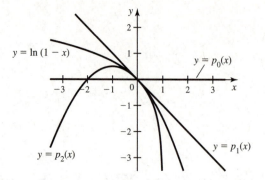

10.1.17 $f(0) = 0$. $f'(x) = 1 + \tan^2 x$, $f''(x) = 2\tan x \sec^2 x$, so that $f'(0) = 1$, $f''(0) = 0$. Thus $p_0(x) = 0$, $p_1(x) = x$, $p_2(x) = x$.

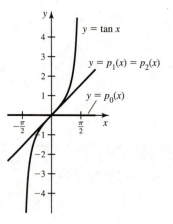

10.1.19 $f(0) = 1$, $f'(0) = -3(1+0)^{-4} = -3$, $f''(0) = 12(1+0)^{-5} = 12$, so that $p_0(x) = 1$, $p_1(x) = 1 - 3x$, $p_2(x) = 1 - 3x + 6x^2$.

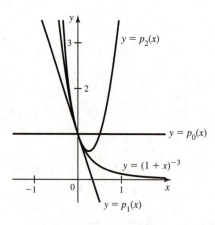

10.1.21

  a. $p_2(.05) = 1.0246875$.

  b. The absolute error is $\sqrt{1.05} - p_2(1.05) \approx 1.024695077 - 1.0246875 = 7.58 \times 10^{-6}$.

10.1.23

  a. $p_2(.08) = 0.9624$.

b. The absolute error is $p_2(.08) - 1/\sqrt{1.08} \approx 0.9624 - 0.9622504482 \approx 1.50 \times 10^{-4}$.

10.1.25

a. $p_2(0.15) = 0.86125$.

b. The absolute error is $p_2(0.15) - e^{-0.15} \approx 0.86125 - 0.8607079764 \approx 5.42 \times 10^{-4}$.

10.1.27 $p_0(x) = \frac{\sqrt{2}}{2}$, $p_1(x) = \frac{\sqrt{2}}{2} + \frac{\sqrt{2}}{2}(x - \frac{\pi}{4})$, $p_2(x) = \frac{\sqrt{2}}{2} + \frac{\sqrt{2}}{2}(x - \frac{\pi}{4}) - \frac{\sqrt{2}}{4}(x - \frac{\pi}{4})^2$.

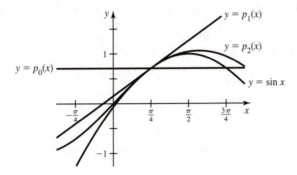

10.1.29 $p_0(x) = 3$, $p_1(x) = 3 + \frac{1}{6}(x - 9)$, $p_2(x) = 3 + \frac{1}{6}(x - 9) - \frac{1}{216}(x - 9)^2$.

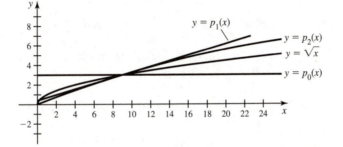

10.1.31 $p_0(x) = 1$, $p_1(x) = 1 + \frac{1}{e}(x - e)$, $p_2(x) = 1 + \frac{1}{e}(x - e) - \frac{1}{2e^2}(x - e)^2$.

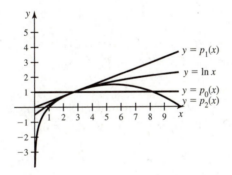

10.1.33

a. $f(x) = e^x$. $p_3(x) = 1 + x + \frac{1}{2}x^2 + \frac{1}{6}x^3$. $p_3(0.12) \approx 1.127488$.

b. $f(0.12) - p_3(0.12) \approx 1.127496852 - 1.127488 = .000008852$.

10.1.35

a. $f(x) = \tan(x)$. $p_3(x) = x + \frac{1}{3}x^3$. $p_3(-0.1) \approx -0.1003333333$.

b. $p_3(-0.1) - f(-0.1) \approx -0.1003333333 + 0.1003346721 = 0.0000013388$.

10.1.37

   a. $f(x) = \sqrt{1+x}$. $p_3(x) = 1 + \frac{1}{2}x - \frac{1}{8}x^2 + \frac{1}{16}x^3$. $p_3(0.06) \approx 1.029563500$.

   b. $f(0.06) - p_3(0.06) \approx 1.029563014 - 1.029563500 \approx 4.86 \times 10^{-7}$.

10.1.39

   a. $f(x) = \sqrt{x}$. $p_3(x) = 10 + \frac{1}{20}(x - 100) - \frac{1}{8000}(x - 100)^2 + \frac{1}{1600000}(x - 100)^3$. $p_3(101) \approx 10.04987563$.

   b. $p_3(101) - f(101) \approx 10.04987563 - 10.04987562 \approx 1 \times 10^{-8}$.

10.1.41 $R_n(x) = \dfrac{\sin^{(n+1)}(c)}{(n+1)!}x^{n+1}$ for some c between 0 and x.

10.1.43 $R_n(x) = \dfrac{(-1)^n e^{-c}}{(n+1)!}x^{n+1}$ for some c between 0 and x.

10.1.45 $R_n(x) = \dfrac{\sin^{(n+1)}(c)}{(n+1)!}\left(x - \dfrac{\pi}{2}\right)^{n+1}$ for some c between $\frac{\pi}{2}$ and x.

10.1.47 $f(x) = \sin x$, so $f^{(4)}(x) = \cos x$. Since $\cos x$ is bounded in magnitude by 1, the remainder is bounded by $|R_4(x)| \leq \frac{0.3^5}{5!} \approx 2.03 \times 10^{-5}$.

10.1.49 $f(x) = e^x$, so $f^{(5)}(x) = e^x$. Since $e^{0.25}$ is bounded by 2, $|R_4(x)| \leq 2 \cdot \frac{0.25^5}{5!} \approx 1.63 \times 10^{-5}$.

10.1.51 $f(x) = e^{-x}$, so $f^{(5)}(x) = -e^{-x}$. Since $f^{(5)}$ achieves its maximum magnitude in the range at $x = -0.5$, which is $\sqrt{e} < 2$, $|R_4(x)| \leq 2 \cdot \frac{0.5^5}{5!} \approx 5.2 \times 10^{-4}$.

10.1.53 Here $n = 3$ or 4, so use $n = 4$, and $M = 1$ since $f^{(5)}(x) = \cos x$, so that $R_4(x) \leq \frac{(\pi/4)^5}{5!} \approx 2.49 \times 10^{-3}$.

10.1.55 $n = 2$ and $M = e^{1/2} < 2$, so $R_2(x) \leq 2 \cdot \frac{(1/2)^3}{3!} \approx 4.17 \times 10^{-2}$.

10.1.57 $n = 2$; $f^{(2)}(x) = \frac{2}{(1+x)^3}$, which achieves its maximum at $x = -0.2$: $|f^{(3)}(x)| = \frac{2}{0.8^3} < 4$. Then $R_2(x) \leq 4 \cdot \frac{0.2^3}{3!} \approx 5.3 \times 10^{-3}$.

10.1.59 Use the Taylor series for e^x at $x = 0$. The derivatives of e^x are e^x. On $[-0.5, 0]$, the maximum magnitude of any derivative is thus 1 and $x = 0$, so $|R_n(-0.5)| \leq \frac{0.5^{n+1}}{(n+1)!}$, so for $R_n(-0.5) < 10^{-3}$ we need $n = 4$.

10.1.61 Use the Taylor series for $\cos x$ at $x = 0$. The magnitude of any derivative of $\cos x$ is bounded by 1, so $|R_n(-0.25)| \leq \frac{0.25^{n+1}}{(n+1)!}$, so for $|R_n(-0.25)| < 10^{-3}$ we need $n = 3$.

10.1.63 Use the Taylor series for $f(x) = \sqrt{x}$ at $x = 1$. Then $|f^{(n+1)}(x)| = \frac{1 \cdot 3 \cdots (2n-1)}{2^{n+1}}x^{-(2n+1)/2}$, which achieves its maximum on $[1, 1.06]$ at $x = 1$. Then

$$|R_n(1.06)| \leq \frac{1 \cdot 3 \cdots (2n-1)}{2^{n+1}} \cdot \frac{(1.06-1)^{n+1}}{(n+1)!},$$

and for $|R_n(0.06)| < 10^{-3}$ we need $n = 1$.

10.1.65

   a. False. If $f(x) = e^{-2x}$, then $f^{(n)}(x) = (-1)^n 2^n e^{-2x}$, so that $f^{(n)}(0) \neq 0$ and all powers of x are present in the Taylor series.

b. True. The constant term of the Taylor series is $f(0) = 1$. Higher-order terms all involve derivatives of $f(x) = x^5 - 1$ evaluated at $x = 0$; clearly for $n < 5$, $f^{(n)}(0) = 0$, and for $n > 5$, the derivative itself vanishes. Only for $n = 5$, where $f^{(5)}(x) = 5!$, is the derivative nonzero, so the coefficient of x^5 in the Taylor series is $f^{(5)}(0)/5! = 1$ and the Taylor polynomial of order 10 is in fact $x^5 - 1$.

c. True. The odd derivatives of $\sqrt{1 + x^2}$ vanish at $x = 0$, while the even ones do not.

10.1.67

a. This matches (C) since for $f(x) = (1 + 2x)^{1/2}$, $f''(x) = -(1 + 2x)^{-3/2}$ so $\frac{f''(0)}{2!} = \frac{-1}{2}$.

b. This matches (E) since for $f(x) = (1 + 2x)^{-1/2}$, $f''(x) = 3(1 + 2x)^{-5/2}$, so $\frac{f''(0)}{2!} = \frac{3}{2}$.

c. This matches (A) since $f^{(n)}(x) = 2^n e^{2x}$, so that $f^{(n)}(0) = 2^n$, which is (A)'s pattern.

d. This matches (D) since $f''(x) = 8(1 + 2x)^{-3}$ and $f''(0) = 8$, so that $f''(0)/2! = 4$

e. This matches (B) since $f'(x) = -6(1 + 2x)^{-4}$ so that $f'(0) = -6$.

f. This matches (F) since $f^{(n)}(x) = (-2)^n e^{-2x}$, so $f^{(n)}(0) = (-2)^n$, which is (F)'s pattern.

10.1.69

a. $p_2(0.1) = 0.1$. The maximum error in the approximation is $1 \cdot \frac{0.1^3}{3!} \approx 1.67 \times 10^{-4}$.

b. $p_2(0.2) = 0.2$. The maximum error in the approximation is $1 \cdot \frac{0.2^3}{3!} \approx 1.33 \times 10^{-3}$.

10.1.71

a. $p_3(0.1) = 1 - .01/2 = 0.995$. The maximum error is $1 \cdot \frac{0.1^4}{4!} \approx 4.17 \times 10^{-6}$.

b. $p_3(0.2) = 1 - .04/2 = 0.98$. The maximum error is $1 \cdot \frac{0.2^4}{4!} \approx 6.67 \times 10^{-5}$.

10.1.73

a. $p_1(0.1) = 1.05$. Since $|f''(x)| = \frac{1}{4}(1 + x)^{-3/2}$ has a maximum value of $1/4$ at $x = 0$, the maximum error is $\frac{1}{4} \cdot \frac{0.1^2}{2} = 1.25 \times 10^{-3}$.

b. $p_1(0.2) = 1.1$. The maximum error is $\frac{1}{4} \cdot \frac{0.2^2}{2} = 5 \times 10^{-3}$.

10.1.75

a. $p_1(0.1) = 1.1$. Since $f''(x) = e^x$ is less than 2 on $[0, 0.2]$, the maximum error is less than $2 \cdot \frac{0.1^3}{3!} \approx 3.33 \times 10^{-4}$.

b. $p_1(0.2) = 1.2$. The maximum error is less than $2 \cdot \frac{0.2^3}{3!} \approx 2.67 \times 10^{-3}$.

10.1.77

a.

	$|\sin x - p_3(x)|$	$|\sin x - p_5(x)|$
-0.2	2.66×10^{-6}	2.54×10^{-9}
-0.1	8.33×10^{-8}	1.98×10^{-11}
0.0	0	0
0.1	8.33×10^{-8}	1.98×10^{-11}
0.2	2.66×10^{-6}	2.54×10^{-9}

b. The errors are equal for positive and negative x. This makes sense, since $\sin(-x) = -\sin(x)$ and $p_n(-x) = -p_n(x)$ for $n = 3, 5$. The errors appear to get larger as x gets farther from zero.

10.1.79

	$\left\|e^{-x}-p_1(x)\right\|$	$\left\|e^{-x}-p_2(x)\right\|$
-0.2	2.14×10^{-2}	1.40×10^{-3}
-0.1	5.17×10^{-3}	1.71×10^{-4}
0.0	0	0
0.1	4.84×10^{-3}	1.63×10^{-4}
0.2	1.87×10^{-2}	1.27×10^{-3}

a.

b. The errors are different for positive and negative displacements from zero, and appear to get larger as x gets farther from zero.

10.1.81

	$\left\|\tan(x)-p_1(x)\right\|$	$\left\|\tan(x)-p_3(x)\right\|$
-0.2	2.71×10^{-3}	4.34×10^{-5}
-0.1	3.35×10^{-4}	1.34×10^{-6}
0.0	0	0
0.1	3.35×10^{-4}	1.34×10^{-6}
0.2	2.71×10^{-3}	4.34×10^{-5}

a.

b. The errors are equal for positive and negative x. This makes sense, since $\tan(-x) = -\tan(x)$ and $p_n(-x) = -p_n(x)$ for $n = 1, 3$. The errors appear to get larger as x gets farther from zero.

10.1.83 The true value of $e^{0.35} \approx 1.419067549$. The 6th-order Taylor polynomial for e^x centered at $x = 0$ is

$$p_6(x) \quad = \quad 1 + x + \frac{x^2}{2} + \frac{x^3}{6} + \frac{x^4}{24} + \frac{x^5}{120} + \frac{x^6}{720}.$$

Evaluating the polynomials at $x = 0.35$ produces the following table:

n	$p_n(0.35)$	$\left\|p_n(0.35) - e^{0.35}\right\|$
1	1.350000000	6.91×10^{-2}
2	1.411250000	7.82×10^{-3}
3	1.418395833	6.72×10^{-4}
4	1.419021094	4.65×10^{-5}
5	1.419064862	2.69×10^{-6}
6	1.419067415	1.33×10^{-7}

The 6th-order Taylor polynomial for e^x centered at $x = \ln 2$ is

$$p_6(x) = \qquad 2 + 2(x - \ln 2) + (x - \ln 2)^2 + \frac{1}{3}(x - \ln 2)^3 + \frac{1}{12}(x - \ln 2)^4$$

$$+ \frac{1}{60}(x - \ln 2)^5 + \frac{1}{360}(x - \ln 2)^6.$$

Evaluating the polynomials at $x = 0.35$ produces the following table:

n	$p_n(0.35)$	$\left\|p_n(0.35) - e^{0.35}\right\|$
1	1.313705639	1.05×10^{-1}
2	1.431455626	1.24×10^{-2}
3	1.417987101	1.08×10^{-3}
4	1.419142523	7.50×10^{-5}
5	1.419063227	4.32×10^{-6}
6	1.419067762	2.13×10^{-7}

Comparing the tables shows that using the polynomial centered at $x = 0$ is more accurate for all n. To see why, consider the remainder. Let $f(x) = e^x$. By Theorem 9.2, the magnitude of the remainder when approximating $f(0.35)$ by the polynomial p_n centered at 0 is:

$$|R_n(0.35)| = \frac{|f^{(n+1)}(c)|}{(n+1)!}(0.35)^{n+1} = \frac{e^c}{(n+1)!}(0.35)^{n+1}$$

for some c with $0 < c < 0.35$ while the magnitude of the remainder when approximating $f(0.35)$ by the polynomial p_n centered at $\ln 2$ is:

$$|R_n(0.35)| = \frac{|f^{(n+1)}(c)|}{(n+1)!}|0.35 - \ln 2|^{n+1} = \frac{e^c}{(n+1)!}(\ln 2 - 0.35)^{n+1}$$

for some c with $0.35 < c < \ln 2$. Since $\ln 2 - 0.35 \approx 0.35$, the relative size of the magnitudes of the remainders is determined by e^c in each remainder. Since e^x is an increasing function, the remainder in using the polynomial centered at 0 will be less than the remainder in using the polynomial centered at $\ln 2$, and the former polynomial will be more accurate.

10.1.85

a. The slope of the tangent line to $f(x)$ at $x = a$ is by definition $f'(a)$; by the point-slope form for the equation of a line, we have $y - f(a) = f'(a)(x - a)$, or $y = f(a) + f'(a)(x - a)$.

b. The Taylor polynomial centered at a is $p_1(x) = f(a) + f'(a)(x - a)$, which is the tangent line at a.

10.2   Properties of Power Series

10.2.1 $c_0 + c_1 x + c_2 x^2 + c_3 x^3$.

10.2.3 Generally the Ratio Test or Root Test is used.

10.2.5 The radius of convergence does not change, but the interval may change at the endpoints.

10.2.7 $|x| < \frac{1}{4}$.

10.2.9 Using the Root Test: $\lim_{k \to \infty} \sqrt[k]{a_k} = \lim_{k \to \infty} \frac{x}{3} = \frac{x}{3}$, so the radius of convergence is 3. At -3, the series is $\sum (-1)^k$, which diverges. At 3, the series is $\sum 1$, which diverges. So the interval of convergence is $(-3, 3)$.

10.2.11 Using the Root Test: $\lim_{k \to \infty} \sqrt[k]{a_k} = \lim_{k \to \infty} \frac{x}{k} = 0$, so the radius of convergence is infinite and the interval of convergence is $(-\infty, \infty)$.

10.2.13 Using the Ratio Test: $\lim_{k \to \infty} \left| \frac{(k+1)^2 x^{2k+2}}{(k+1)!} \cdot \frac{k!}{k^2 x^{2k}} \right| = \lim_{k \to \infty} \frac{k+1}{k^2} x^2 = 0$, so the radius of convergence is infinite, and the interval of convergence is $(-\infty, \infty)$.

10.2.15 Using the Ratio Test: $\lim_{k \to \infty} \left| \frac{a_{k+1}}{a_k} \right| = \left| \frac{x^{2k+3}}{3^k} \cdot \frac{3^{k-1}}{x^{2k+1}} \right| = \frac{x^2}{3}$ so that the radius of convergence is $\sqrt{3}$. At $x = \sqrt{3}$, the series is $\sum 3\sqrt{3}$, which diverges. At $x = -\sqrt{3}$, the series is $\sum(-3\sqrt{3})$, which also diverges, so the interval of convergence is $(-\sqrt{3}, \sqrt{3})$.

10.2.17 Using the Root Test: $\lim_{k \to \infty} \sqrt[k]{|a_k|} = \lim_{k \to \infty} \frac{(|x-1|)k}{k+1} = |x - 1|$, so the series converges when $|x - 1| < 1$, so for $0 < x < 2$. The radius of convergence is 1. At $x = 2$, the series diverges by the Divergence Test. At $x = 0$, the series diverges as well by the Divergence Test. Thus the interval of convergence is $(0, 2)$.

10.2.19 Using the Ratio Test: $\lim_{k \to \infty} \left| \frac{a_{k+1}}{a_k} \right| = \left| \frac{(k+1)^{20} x^{k+1}}{(2k+3)!} \cdot \frac{(2k+1)!}{x^k k^{20}} \right| = \lim_{k \to \infty} \left(\frac{k+1}{k} \right)^{20} \frac{|x|}{(2k+2)(2k+3)} = 0$, so the radius of convergence is infinite, and the interval of convergence is $(-\infty, \infty)$.

10.2.21 $f(3x) = \frac{1}{1-3x} = \sum_{k=0}^{\infty} 3^k x^k$, which converges for $|x| < 1/3$, and diverges at the endpoints.

10.2.23 $h(x) = \frac{2x^3}{1-x} = \sum_{k=0}^{\infty} 2x^{k+3}$, which converges for $|x| < 1$ and is divergent at the endpoints.

10.2.25 $p(x) = \frac{4x^{12}}{1-x} = \sum_{k=0}^{\infty} 4x^{k+12} = 4\sum_{k=0}^{\infty} x^{k+12}$, which converges for $|x| < 1$. It is divergent at the endpoints.

10.2.27 $f(3x) = \ln(1-3x) = -\sum_{k=1}^{\infty} \frac{(3x)^k}{k} = -\sum_{k=1}^{\infty} \frac{3^k}{k} x^k$. Using the Ratio Test: $\lim_{k \to \infty} \left| \frac{a_{k+1}}{a_k} \right| = \lim_{k \to \infty} \frac{3k}{k+1} |x| = 3|x|$, so the radius of convergence is $1/3$. The series diverges at $1/3$ (harmonic series), and converges at $-1/3$ (alternating harmonic series).

10.2.29 $h(x) = x\ln(1-x) = -\sum_{k=1}^{\infty} \frac{x^{k+1}}{k}$. Using the Ratio Test: $\lim_{k \to \infty} \left| \frac{a_{k+1}}{a_k} \right| = \lim_{k \to \infty} \frac{k}{k+1} |x| = |x|$, so the radius of convergence is 1, and the series diverges at 1 (harmonic series) but converges at -1 (alternating harmonic series).

10.2.31 $p(x) = 2x^6 \ln(1-x) = -2\sum_{k=1}^{\infty} \frac{x^{k+6}}{k}$. Using the Ratio Test: $\lim_{k \to \infty} \left| \frac{a_{k+1}}{a_k} \right| = \lim_{k \to \infty} \frac{k}{k+1} |x| = |x|$, so the radius of convergence is 1. The series diverges at 1 (harmonic series) but converges at -1 (alternating harmonic series).

10.2.33 The power series for $f(x)$ is $\sum_{k=0}^{\infty} x^k$, convergent for $-1 < x < 1$, so the power series for $g(x) = f'(x)$ is $\sum_{k=1}^{\infty} kx^{k-1} = \sum_{k=0}^{\infty} (k+1)x^k$, also convergent on $|x| < 1$.

10.2.35 The power series for $f(x)$ is $\sum_{k=0}^{\infty} x^k$, convergent for $-1 < x < 1$, so the power series for $g(x) = \frac{1}{6}f'''(x)$ is $\frac{1}{6}\sum_{k=3}^{\infty} k(k-1)(k-2)x^{k-3} = \frac{1}{6}\sum_{k=0}^{\infty} (k+1)(k+2)(k+3)x^k$, also convergent on $|x| < 1$.

10.2.37 The power series for $\frac{1}{1-3x}$ is $\sum_{k=0}^{\infty} (3x)^k$, convergent on $|x| < 1/3$. Since $g(x) = \ln(1-3x) = -3\int \frac{1}{1-3x}\,dx$, the power series for $g(x)$ is $-3\sum_{k=0}^{\infty} 3^k \frac{1}{k+1} x^{k+1} = -\sum_{k=1}^{\infty} \frac{3^k}{k} x^k$, also convergent on $[-1/3, 1/3)$.

10.2.39 Start with $g(x) = \frac{1}{1+x}$. The power series for $g(x)$ is $\sum_{k=0}^{\infty} (-1)^k x^k$. Since $f(x) = g(x^2)$, its power series is $\sum_{k=0}^{\infty} (-1)^k x^{2k}$. The radius of convergence is still 1, and the series is divergent at both endpoints. The interval of convergence is $(-1, 1)$.

10.2.41 Note that $f(x) = \frac{3}{3+x} = \frac{1}{1+(1/3)x}$. Let $g(x) = \frac{1}{1+x}$. The power series for $g(x)$ is $\sum_{k=0}^{\infty} (-1)^k x^k$, so the power series for $f(x) = g((1/3)x)$ is $\sum_{k=0}^{\infty} (-1)^k 3^{-k} x^k = \sum_{k=0}^{\infty} \left(\frac{-x}{3} \right)^k$. Using the Ratio Test: $\lim_{k \to \infty} \left| \frac{a_{k+1}}{a_k} \right| = \lim_{k \to \infty} \left| \frac{3^{-(k+1)} x^{k+1}}{3^{-k} x^k} \right| = \frac{|x|}{3}$, so the radius of convergence is 3. The series diverges at both endpoints. The interval of convergence is $(-3, 3)$.

10.2.43 Note that $f(x) = \ln\sqrt{4-x^2} = \frac{1}{2}\ln(4-x^2) = \frac{1}{2}\left(\ln 4 + \ln\left(1 - \frac{x^2}{4} \right) \right) = \ln 2 + \frac{1}{2}\ln\left(1 - \frac{x^2}{4} \right)$. Now, the power series for $g(x) = \ln(1-x)$ is $-\sum_{k=1}^{\infty} \frac{1}{k} x^k$, so the power series for $f(x)$ is $\ln 2 - \frac{1}{2}\sum_{k=1}^{\infty} \frac{1}{k} \frac{x^{2k}}{4^k} = \ln 2 - \sum_{k=1}^{\infty} \frac{x^{2k}}{k2^{2k+1}}$. Now, $\lim_{k \to \infty} \left| \frac{a_{k+1}}{a_k} \right| = \lim_{k \to \infty} \left| \frac{x^{2k+2}}{(k+1)2^{2k+3}} \cdot \frac{k2^{2k+1}}{x^{2k}} \right| = \lim_{k \to \infty} \frac{k}{4(k+1)} x^2 = \frac{x^2}{4}$, so that the radius of convergence is 2. The series diverges at both endpoints, so its interval of convergence is $(-2, 2)$.

10.2.45

  a. True. This power series is centered at $x = 3$, so its interval of convergence will be symmetric about 3.

  b. True. Use the Root Test.

  c. True. Use the Root Test.

  d. True. Since the power series is zero on the interval, all its derivatives are as well, which implies (differentiating the power series) that all the c_k are zero.

10.2.47 $\sum_{k=0}^{\infty}(-1)^k\frac{1}{k+1}x^k$

10.2.49 $\sum_{k=1}^{\infty}(-1)^k\frac{x^{2k}}{k!}$

10.2.51 The power series for $f(x-a)$ is $\sum c_k(x-a)^k$. Then $\sum c_k(x-a)^k$ converges if and only if $|x-a| < R$, which happens if and only if $a - R < x < a + R$, so the radius of convergence is the same.

10.2.53 This is a geometric series with ratio $\sqrt{x} - 2$, so its sum is $\frac{1}{1-(\sqrt{x}-2)} = \frac{1}{3-\sqrt{x}}$. Again using the Root Test, $\lim_{k\to\infty}\sqrt[k]{|a_k|} = \sqrt{x} - 2$, so the interval of convergence is given by $|\sqrt{x} - 2| < 1$, so $1 < \sqrt{x} < 3$ and $1 < x < 9$. The series diverges at both endpoints.

10.2.55 This is a geometric series with ratio e^{-x}, so its sum is $\frac{1}{1-e^{-x}}$. By the Root Test, $\lim_{k\to\infty}\sqrt[k]{|a_k|} = e^{-x}$, so the power series converges for $x > 0$.

10.2.57 This is a geometric series with ratio $(x^2 - 1)/3$, so its sum is $\frac{1}{1-\frac{x^2-1}{3}} = \frac{3}{3-(x^2-1)} = \frac{3}{4-x^2}$. Using the Root Test, the series converges for $|x^2 - 1| < 3$, so that $-2 < x^2 < 4$ or $-2 < x < 2$. It diverges at both endpoints.

10.2.59 The power series for e^x is $\sum_{k=0}^{\infty}\frac{x^k}{k!}$. Substitute $-x$ for x to get $e^{-x} = \sum_{k=0}^{\infty}(-1)^k\frac{x^k}{k!}$. The series converges for all x.

10.2.61 Substitute $-3x$ for x in the power series for e^x to get $e^{-3x} = \sum_{k=0}^{\infty}\frac{(-3x)^k}{k!} = \sum_{k=0}^{\infty}(-1)^k\frac{3^k}{k!}x^k$. The series converges for all x.

10.2.63 The power series for $x^m f(x)$ is $\sum c_k x^{k+m}$. The ratio used in the Ratio Test is: $\frac{a_{k+1}}{a_k} = \frac{c_{k+1}x^{k+m+1}}{c_k x^{k+m}} = \frac{c_{k+1}}{c_k}x$, which is the same ratio that results when applying the Ratio Test to the original series, so the two power series have the same radius of convergence. Now suppose that the power series for $f(x)$ converges at $x = b$, an endpoint of the interval of convergence. Then $\sum c_k b^k$ converges, and since $\lim_{k\to\infty}\frac{c_k b^{k+m}}{c_k b^k} = b^m < \infty$, the power series for $x^m f(x)$ converges at $x = b$ as well by the Limit Comparison Test.

10.2.65

a. $f(x)g(x) = c_0d_0 + (c_0d_1 + c_1d_0)x + (c_0d_2 + c_1d_1 + c_2d_0)x^2 + \ldots$

b. The coefficient of x^n in $f(x)g(x)$ is $\sum_{i=0}^{n}c_i d_{n-i}$.

10.2.67

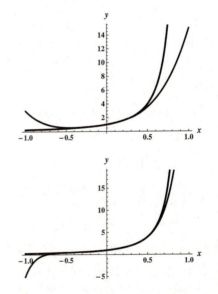

a. For both graphs, the difference between the true value and the estimate is greatest at the two ends of the range; the difference at 0.9 is greater than that at -0.9.

b. The difference between $f(x)$ and $S_n(x)$ is greatest for $x = 0.9$; at that point, $f(x) = \frac{1}{(1-0.9)^2} = 100$, so we want to find n such that $S_n(x)$ is within 0.01 of 100. But $S_{111} \approx 99.98991435$ and $S_{112} \approx 99.99084790$, so $n = 112$.

10.3 Taylor Series

10.3.1 The Taylor series is in a sense the "limit" of the Taylor polynomials. It is the infinite sum of terms of the form appearing in the Taylor polynomials.

10.3.3 The nth coefficient is $\frac{f^{(n)}(a)}{n!}(x-a)^n$.

10.3.5 Substitute x^2 for x in the Taylor series. By theorems proved in the previous section about power series, the interval of convergence does not change except perhaps at the endpoints of the interval.

10.3.7 It means that the limit of the remainder term is zero.

10.3.9

a. Note that $f(0) = 1$, $f'(0) = -1$, $f''(0) = 1$, and $f'''(0) = -1$. So the Maclaurin series is $1 - x + x^2/2 - x^3/6 + \ldots$.

b. $\sum_{k=0}^{\infty} (-1)^k \frac{x^k}{k!}$.

c. The series converges on $(-\infty, \infty)$, as can be seen from the Ratio Test.

10.3.11

a. Since the series for $\frac{1}{1+x}$ is $1 - x + x^2 - x^3 + \ldots$, the series for $\frac{1}{1+x^2}$ is $1 - x^2 + x^4 - x^6 + \ldots$.

b. $\sum_{k=0}^{\infty} (-1)^k x^{2k}$.

c. The absolute value of the ratio of consecutive terms is x^2, so by the Ratio Test, the radius of convergence is 1. The series diverges at the endpoints by the Divergence Test, so the interval of convergence is $(-1, 1)$.

10.3.13

a. Note that $f(0) = 1$, and that $f^{(n)}(0) = 2^n$. Thus, the series is given by $1 + 2x + \frac{4x^2}{2} + \frac{8x^3}{6} + \ldots$.

b. $\sum_{k=0}^{\infty} \frac{(2x)^k}{k!}$.

c. The absolute value of the ratio of consecutive terms is $\frac{2|x|}{n}$, which has limit 0 as $n \to \infty$. So by the Ratio Test, the interval of convergence is $(-\infty, \infty)$.

10.3.15

a. By integrating the Taylor series for $\frac{1}{1+x^2}$ (which is the derivative of $\tan^{-1}(x)$), we obtain the series $x - \frac{x^3}{3} + \frac{x^5}{5} - \frac{x^7}{7} + \ldots$.

b. $\sum_{k=0}^{\infty} (-1)^k \frac{1}{2k+1} x^{2k+1}$.

c. By the Ratio Test (the ratio of consecutive terms has limit x^2), the radius of convergence is $|x| < 1$. Also, at the endpoints we have convergence by the Alternating Series Test, so the interval of convergence is $[-1, 1]$.

10.3.17

a. Note that $f(\pi/2) = 1$, $f'(\pi/2) = \cos(\pi/2) = 0$, $f''(\pi/2) = -\sin(\pi/2) = -1$, $f'''(\pi/2) = -\cos(\pi/2) = 0$, and so on. Thus the series is given by $1 - \frac{1}{2}\left(x - \frac{\pi}{2}\right)^2 + \frac{1}{24}\left(x - \frac{\pi}{2}\right)^4 - \frac{1}{720}\left(x - \frac{\pi}{2}\right)^6 + \ldots$.

b. $\sum_{k=0}^{\infty}(-1)^k \frac{1}{(2k)!}\left(x - \frac{\pi}{2}\right)^{2k}$.

10.3.19

a. Note that $f^{(k)}(1) = (-1)^k \frac{k!}{1^{k+1}} = (-1)^k \cdot k!$. Thus the series is given by $1-(x-1)+(x-1)^2-(x-1)^3+\cdots$.

b. $\sum_{k=0}^{\infty}(-1)^k(x-1)^k$.

10.3.21

a. Note that $f^{(k)}(3) = (-1)^{k-1}\frac{(k-1)!}{3^k}$. Thus the series is given by $\ln(3)+\frac{x-3}{3}-\frac{1}{18}(x-3)^2+\frac{1}{81}(x-3)^3+\cdots$.

b. $\ln(3) + \sum_{k=1}^{\infty}(-1)^{k+1}\frac{1}{k\cdot 3^k}(x-3)^k$.

10.3.23 Since the Taylor series for $\ln(1+x)$ is $x - \frac{x^2}{2} + \frac{x^3}{3} - \frac{x^4}{4} + \cdots$, the first four terms of the Taylor series for $\ln(1+x^2)$ are $x^2 - \frac{x^4}{2} + \frac{x^6}{3} - \frac{x^8}{4}$, obtained by substituting x^2 for x.

10.3.25 The Taylor series for $e^x - 1$ is the Taylor series for e^x, less the constant term of 1, so it is $x + \frac{x^2}{2} + \frac{x^3}{3!} + \frac{x^4}{4!} + \cdots$. Thus, the first four terms of the Taylor series for $\frac{e^x-1}{x}$ are $1 + \frac{x}{2!} + \frac{x^2}{3!} + \frac{x^3}{4!}$, obtained by dividing the terms of the first series by x.

10.3.27 Since the Taylor series for $(1+x)^{-1}$ is $1 - x + x^2 - x^3 + \cdots$, if we substitute x^4 for x, we obtain $1 - x^4 + x^8 - x^{12} + \cdots$.

10.3.29

a. The binomial coefficients are $\binom{-2}{0} = 1$, $\binom{-2}{1} = \frac{-2}{1!} = -2$, $\binom{-2}{2} = \frac{(-2)(-3)}{2!} = 3$, $\binom{-2}{3} = \frac{(-2)(-3)(-4)}{3!} = -4$.

Thus the first four terms of the series are $1 - 2x + 3x^2 - 4x^3$.

b. $1 - 2\cdot 0.1 + 3\cdot 0.01 - 4\cdot 0.001 = 0.826$

10.3.31

a. The binomial coefficients are $\binom{1/4}{0} = 1$, $\binom{1/4}{1} = \frac{1/4}{1} = \frac{1}{4}$, $\binom{1/4}{2} = \frac{(1/4)(-3/4)}{2!} = -\frac{3}{32}$, $\binom{1/4}{3} = \frac{(1/4)(-3/4)(-7/4)}{3!} = \frac{7}{128}$, so the first four terms of the series are $1 + \frac{1}{4}x - \frac{3}{32}x^2 + \frac{7}{128}x^3$.

b. Substitute $x = 0.12$ to get 1.0287445.

10.3.33

a. The binomial coefficients are $\binom{-2/3}{0} = 1$, $\binom{-2/3}{1} = -\frac{2}{3}$, $\binom{-2/3}{2} = \frac{(-2/3)(-5/3)}{2!} = \frac{5}{9}$, $\binom{-2/3}{3} = \frac{(-2/3)(-5/3)(-8/3)}{3!} = -\frac{40}{81}$, so the first four terms of the series are $1 - \frac{2}{3}x + \frac{5}{9}x^2 - \frac{40}{81}x^3$.

b. Substitute $x = 0.18$ to get 0.89512.

10.3.35 $\sqrt{1+x^2} = 1+\frac{x^2}{2}-\frac{x^4}{8}+\frac{x^6}{16}-\cdots$. By the Ratio Test, the radius of convergence is 1. At the endpoints, the series obtained are convergent by the Alternating Series Test. Thus, the interval of convergence is $[-1,1]$.

10.3.37 $\sqrt{9-9x} = 3\sqrt{1-x} = 3 - \frac{3}{2}x - \frac{3}{8}x^2 - \frac{3}{16}x^3 - \cdots$. The interval of convergence is $[-1,1)$.

10.3.39 $\sqrt{a^2+x^2} = a\sqrt{1+\frac{x^2}{a^2}} = a + \frac{x^2}{2a} - \frac{x^4}{8a^3} + \frac{x^6}{16a^5} - \cdots$. The series converges when $\frac{x^2}{a^2}$ is less than 1 in magnitude, so the radius of convergence is a. The series given by the endpoints is convergent by the Alternating Series Test, so the interval of convergence is $[-a,a]$.

10.3.41 $(1+4x)^{-2} = 1 - 2(4x) + 3(4x)^2 - 4(4x)^3 + \cdots = 1 - 8x + 48x^2 - 256x^3 + \cdots$.

10.3.43 $\frac{1}{(4+x^2)^2} = (4+x^2)^{-2} = \frac{1}{16}(1+(x^2/4))^{-2} = \frac{1}{16}\left(1 - 2\cdot\frac{x^2}{4} + 3\cdot\frac{x^4}{16} - 4\cdot\frac{x^6}{64} + \dots\right) = \frac{1}{16} - \frac{1}{32}x^2 + \frac{3}{256}x^4 - \frac{1}{256}x^6 + \dots$

10.3.45 $(3+4x)^{-2} = \frac{1}{9}\left(1+\frac{4x}{3}\right)^{-2} = \frac{1}{9}\left(1 - 2\frac{4}{3}x + 3\frac{16}{9}x^2 - 4\frac{64}{27}x^3 + \dots\right) = \frac{1}{9} - \frac{8}{27}x + \frac{16}{27}x^2 - \frac{256}{243}x^3 + \dots$

10.3.47 The interval of convergence for the Taylor series for $f(x) = \sin x$ is $(-\infty, \infty)$. The remainder is $R_n(x) = \frac{f^{(n+1)}(c)}{(n+1)!}x^{n+1}$ for some c. Since $f^{(n+1)}(x)$ is $\pm\sin x$ or $\pm\cos x$, we have

$$\lim_{n\to\infty}|R_n(x)| \le \lim_{n\to\infty}\frac{1}{(n+1)!}\left|x^{n+1}\right| = 0$$

for any x.

10.3.49 The interval of convergence for the Taylor series for e^{-x} is $(-\infty, \infty)$. The remainder is $R_n(x) = \frac{(-1)^{n+1}e^{-c}}{(n+1)!}x^{n+1}$ for some c. Thus $\lim_{n\to\infty}|R_n(x)| = 0$ for any x.

10.3.51

a. False. Not all of its derivatives are defined at zero - in fact, none of them are.

b. True. The derivatives of $\csc x$ involve positive powers of $\csc x$ and $\cot x$, both of which are defined at $\pi/2$, so that $\csc x$ has continuous derivatives at $\pi/2$.

c. False. For example, the Taylor series for $f(x^2)$ doesn't converge at $x = 1.9$, since the Taylor series for $f(x)$ doesn't converge at $1.9^2 = 3.61$.

d. False. The Taylor series centered at 1 involves derivatives of f evaluated at 1, not at 0.

e. True. The follows because the Taylor series must itself be an even function.

10.3.53

a. The relevant Taylor series are: $e^x = 1+x+\frac{x^2}{2!}+\frac{x^3}{3!}+\frac{x^4}{4!}+\frac{x^5}{5!}+\frac{x^6}{6!}+\cdots$ and $e^{-x} = 1-x+\frac{x^2}{2!}-\frac{x^3}{3!}+\frac{x^4}{4!}-\frac{x^5}{5!}+\frac{x^6}{6!}+\cdots$. Thus the first four terms of the resulting series are $\frac{1}{2}(e^x+e^{-x}) = 1+\frac{x^2}{2!}+\frac{x^4}{4!}+\frac{x^6}{6!}+\cdots$.

b. Since each series converges (absolutely) on $(-\infty, \infty)$, so does their sum. The radius of convergence is ∞.

10.3.55

a. Use the binomial theorem. The binomial coefficients are $\binom{-2/3}{0} = 1$, $\binom{-2/3}{1} = -\frac{2}{3}$, $\binom{-2/3}{2} = \frac{(-2/3)(-5/3)}{2!} = \frac{5}{9}$, $\binom{-2/3}{3} = \frac{(-2/3)(-5/3)(-8/3)}{3!} = -\frac{40}{81}$ and then, substituting x^2 for x, we obtain $1 - \frac{2}{3}x^2 + \frac{5}{9}x^4 - \frac{40}{81}x^6 + \dots$

b. From Theorem 9.6 the radius of convergence is determined from $\left|x^2\right| < 1$, so it is 1.

10.3.57

a. From the binomial formula, the Taylor series for $(1-x)^p$ is $\sum \binom{p}{k}(-1)^k x^k$, so the Taylor series for $(1-x^2)^p$ is $\sum \binom{p}{k}(-1)^k x^{2k}$. Here $p = 1/2$, and the binomial coefficients are $\binom{1/2}{0} = 1$, $\binom{1/2}{1} = \frac{1/2}{1!} = \frac{1}{2}$, $\binom{1/2}{2} = \frac{(1/2)(-1/2)}{2!} = -\frac{1}{8}$, $\binom{1/2}{3} = \frac{(1/2)(-1/2)(-3/2)}{3!} = \frac{1}{16}$ so that $(1-x^2)^{1/2} = 1 - \frac{1}{2}x^2 - \frac{1}{8}x^4 - \frac{1}{16}x^6 + \dots$

b. From Theorem 9.6 the radius of convergence is determined from $\left|x^2\right| < 1$, so it is 1.

10.3.59

a. $f(x) = (1+x^2)^{-2}$; using the binomial series and substituting x^2 for x we obtain $1 - 2x^2 + 3x^4 - 4x^6 + \dots$.

b. From Theorem 9.6 the radius of convergence is determined from $\left|x^2\right| < 1$, so it is 1.

10.3.61 Since $f(64) = 4$, and $f'(x) = \frac{1}{3}x^{-2/3}$, $f'(64) = \frac{1}{48}$, $f''(x) = -\frac{2}{9}x^{-5/3}$, $f''(64) = -\frac{1}{4608}$, $f'''(x) = \frac{10}{27}x^{8/3}$, and $f'''(64) = \frac{10}{1769472} = \frac{5}{884736}$, the first four terms of the Taylor series are $4 + \frac{1}{48}(x-64) - \frac{1}{4608 \cdot 2!}(x - 64)^2 + \frac{5}{884736 \cdot 3!}(x - 64)^3$. Evaluating at $x = 60$, we get 3.914870274.

10.3.63 Since $f(16) = 2$, and $f'(x) = \frac{1}{4}x^{-3/4}$, $f'(16) = \frac{1}{32}$, $f''(x) = -\frac{3}{16}x^{-7/4}$, $f''(16) = -\frac{3}{2048}$, $f'''(x) = \frac{21}{64}x^{-11/4}$, and $f'''(16) = \frac{21}{131072}$, the first four terms of the Taylor series are $2 + \frac{1}{32}(x-16) - \frac{3}{2048 \cdot 2!}(x - 16)^2 + \frac{21}{131072 \cdot 3!}(x - 16)^3$. Evaluating at $x = 13$, we get 1.898937225.

10.3.65 Evaluate the binomial coefficient $\binom{1/2}{k} = \frac{(1/2)(-1/2)(-3/2)\ldots(1/2-k+1)}{k!} = \frac{(1/2)(-1/2)\ldots((3-2k)/2)}{k!} = (-1)^{k-1}2^{-k}\frac{1 \cdot 3 \cdots (2k-3)}{k!} = (-1)^{k-1}2^{-k}\frac{(2k-2)!}{2^{k-1} \cdot (k-1)! \cdot k!} = (-1)^{k-1}2^{1-2k} \cdot \frac{1}{k}\binom{2k-2}{k-1}$. This is the coefficient of x^k in the Taylor series for $\sqrt{1+x}$. Substituting $4x$ for x, the Taylor series becomes $\sum_{k=0}^{\infty}(-1)^{k-1}2^{1-2k} \cdot \frac{1}{k}\binom{2k-2}{k-1}(4x)^k = \sum_{k=0}^{\infty}(-1)^{k-1}\frac{2}{k}\binom{2k-2}{k-1}x^k$. If we can show that k divides $\binom{2k-2}{k-1}$, we will be done, for then the coefficient of x^k will be an integer. But $\binom{2k-2}{k-1} - \binom{2k-2}{k-2} = \frac{(2k-2)!}{(k-1)!(k-1)!} - \frac{(2k-2)!}{(k-2)!k!} = \frac{(2k-2)!}{(k-1)!(k-1)!} - \frac{(2k-2)!(k-1)}{(k-1)!(k-1)!k} = \frac{k(2k-2)!-(k-1)(2k-2)!}{k(k-1)!(k-1)!} = \frac{1}{k}\frac{(2k-2)!}{(k-1)!(k-1)!} = \frac{1}{k}\binom{2k-2}{k-1}$ and thus we have shown that k divides $\binom{2k-2}{k-1}$.

10.3.67 The Maclaurin series for $\sin x$ is $x - \frac{1}{3!}x^3 + \frac{1}{5!}x^5 - \frac{1}{7!}x^7 + \ldots$. Squaring the first four terms yields

$$(x - \frac{1}{3!}x^3 + \frac{1}{5!}x^5 - \frac{1}{7!}x^7)^2$$
$$= x^2 - \frac{2}{3!}x^4 + (\frac{2}{5!} + \frac{1}{3!3!})x^6 + (-2\frac{1}{7!} - 2\frac{1}{3!5!})x^8$$
$$= x^2 - \frac{1}{3}x^4 + \frac{2}{45}x^6 - \frac{1}{315}x^8.$$

The Maclaurin series for $\cos x$ is $1 - \frac{1}{2}x^2 + \frac{1}{4!}x^4 - \frac{1}{6!}x^6 + \frac{1}{8!}x^8 - \ldots$. Substituting $2x$ for x in the Maclaurin series for $\cos x$ and then computing $(1 - \cos 2x)/2$, we obtain

$$(1 - (1 - \frac{1}{2}(2x)^2 + \frac{1}{4!}(2x)^4 - \frac{1}{6!}(2x)^6) + \frac{1}{8!}(2x)^8)/2$$
$$= (2x^2 - \frac{2}{3}x^4 + \frac{4}{45}x^6 - \frac{2}{315}x^8)/2$$
$$= x^2 - \frac{1}{3}x^4 + \frac{2}{45}x^6 - \frac{1}{315}x^8,$$

and the two are the same.

10.3.69 There are many solutions. For example, first find a series that has $(-1, 1)$ as an interval of convergence, say $\frac{1}{1-x} = \sum_{k=0}^{\infty}x^k$. Then the series $\frac{1}{1-x/2} = \sum_{k=0}^{\infty}\left(\frac{x}{2}\right)^k$ has $(-2, 2)$ as its interval of convergence. Now shift the series up so that it is centered at 4: $\sum_{k=0}^{\infty}\left(\frac{x-4}{2}\right)^k$, and the interval of convergence is $(2, 6)$.

10.3.71 $\frac{1 \cdot 3 \cdot 5 \cdot 7}{2 \cdot 4 \cdot 6 \cdot 8}x^4 - \frac{1 \cdot 3 \cdot 5 \cdot 7 \cdot 9}{2 \cdot 4 \cdot 6 \cdot 8 \cdot 10}x^5$.

10.3.73 Use the Taylor series for $\cos x$ centered at $\pi/4$: $\frac{\sqrt{2}}{2}(1 - (x - \pi/4) - \frac{1}{2}(x - \pi/4)^2 + \frac{1}{6}(x - \pi/4)^3 + \ldots)$. The remainder after n terms (since the derivatives of $\cos x$ are bounded by 1 in magnitude) is $|R_n(x)| \leq \frac{1}{(n+1)!} \cdot \left(\frac{\pi}{4} - \frac{2\pi}{9}\right)^{n+1}$.

Solving for $|R_n(x)| < 10^{-4}$, we obtain $n = 3$. Evaluating the first four terms (through $n = 3$) of the series we get 0.7660427050. The true value is ≈ 0.7660444431.

10.3.75 Use the Taylor series for $f(x) = x^{1/3}$ centered at 64: $4 + \frac{1}{48}(x - 64) - \frac{1}{9216}(x - 64)^2 + \cdots$. Since we wish to evaluate this series at $x = 83$, $|R_n(x)| = \frac{|f^{(n+1)}(c)|}{(n+1)!}(83 - 64)^{n+1}$. We compute that $|f^{(n+1)}(c)| = \frac{2 \cdot 5 \cdots (3n-1)}{3^{n+1}c^{(3n+2)/3}}$, which is maximized at $c = 64$. Thus

$$|R_n(x)| \leq \frac{2 \cdot 5 \cdots (3n-1)}{3^{n+1}64^{(3n+2)/3}(n+1)!}19^{n+1}$$

Solving for $|R_n(x)| < 10^{-4}$, we obtain $n = 5$. Evaluating the terms of the series through $n = 5$ gives 4.362122553. The true value is ≈ 4.362070671.

10.3.77

   a. Use the Taylor series for $(125 + x)^{1/3}$ centered at $x = 0$. Using the first four terms and evaluating at $x = 3$ gives a result (5.03968) accurate to within 10^{-4}.

   b. Use the Taylor series for $x^{1/3}$ centered at $x = 125$. Note that this gives the identical Taylor series except that the exponential terms are $(x - 125)^n$ rather than x^n. Thus we need terms up through $(x - 125)^3$, just as before, evaluated at $x = 128$, and we obtain the identical result.

   c. Since the two Taylor series are the same except for the shifting, the results are equivalent.

10.3.79 Consider the remainder after the first term of the Taylor series. Taylor's Theorem indicates that $R_1(x) = \frac{f''(c)}{2}(x - a)^2$ for some c between x and a, so that $f(x) = f(a) + f'(a)(x - a) + \frac{f''(c)}{2}(x - a)^2$. But $f'(a) = 0$, so that for every x in an interval containing a, there is a c between x and a such that $f(x) = f(a) + \frac{f''(c)}{2}(x - a)^2$.

   a. If $f''(x) > 0$ on the interval containing a, then for every x in that interval, we have $f(x) = f(a) + \frac{f''(c)}{2}(x - a)^2$ for some c between x and a. But $f''(c) > 0$ and $(x - a)^2 > 0$, so that $f(x) > f(a)$ and a is a local minimum.

   b. If $f''(x) < 0$ on the interval containing a, then for every x in that interval, we have $f(x) = f(a) + \frac{f''(c)}{2}(x - a)^2$ for some c between x and a. But $f''(c) < 0$ and $(x - a)^2 > 0$, so that $f(x) < f(a)$ and a is a local maximum.

10.4 Working with Taylor Series

10.4.1 Replace f and g by their Taylor series centered at a, and evaluate the limit.

10.4.3 Substitute -0.6 for x in the Taylor series for e^x centered at 0. Note that this series is an alternating series, so the error can easily be estimated by looking at the magnitude of the first neglected term.

10.4.5 The series is $f'(x) = \sum_{k=1}^{\infty} kc_k x^{k-1}$, which converges for $|x| < b$.

10.4.7 We compute that

$$\frac{e^x - e^{-x}}{x} = \frac{1}{x}\left(\left(1 + x + \frac{x^2}{2} + \frac{x^3}{6} + \cdots\right) - \left(1 - x + \frac{x^2}{2} - \frac{x^3}{6} + \cdots\right)\right)$$

$$= \frac{1}{x}\left(2x + \frac{x^3}{3} + \cdots\right) = 2 + \frac{x^2}{3} + \cdots$$

so the limit of $\dfrac{e^x - e^{-x}}{x}$ as $x \to 0$ is 2.

10.4.9 We compute that

$$\frac{2\cos 2x - 2 + 4x^2}{2x^4} = \frac{1}{2x^4}\left(2\left(1 - \frac{(2x)^2}{2} + \frac{(2x)^4}{24} - \frac{(2x)^6}{720} + \cdots\right) - 2 + 4x^2\right)$$

$$= \frac{1}{2x^4}\left(\frac{(2x)^4}{12} - \frac{(2x)^6}{360} + \cdots\right) = \frac{2}{3} - \frac{4x^2}{45} + \cdots$$

so the limit of $\dfrac{2\cos 2x - 2 + 4x^2}{2x^4}$ as $x \to 0$ is $\dfrac{2}{3}$.

10.4.11 We compute that

$$
\begin{aligned}
\frac{3\tan x - 3x - x^3}{x^5} &= \frac{1}{x^5}\left(3\left(x + \frac{x^3}{3} + \frac{2x^5}{15} + \frac{17x^7}{315} + \cdots\right) - 3x - x^3\right) \\
&= \frac{1}{x^5}\left(\frac{2x^5}{5} + \frac{17x^7}{105} + \cdots\right) = \frac{2}{5} + \frac{17x^2}{105} + \cdots
\end{aligned}
$$

so the limit of $\dfrac{3\tan x - 3x - x^3}{x^5}$ as $x \to 0$ is $\dfrac{2}{5}$.

10.4.13 We compute that

$$
\begin{aligned}
\frac{3\tan^{-1} x - 3x + x^3}{x^5} &= \frac{1}{x^5}\left(3\left(x - \frac{x^3}{3} + \frac{x^5}{5} - \frac{x^7}{7} + \cdots\right) - 3x + x^3\right) \\
&= \frac{1}{x^5}\left(\frac{3x^5}{5} + \frac{3x^7}{7} + \cdots\right) == \frac{3}{5} + \frac{3x^2}{7} + \cdots
\end{aligned}
$$

so the limit of $\dfrac{3\tan^{-1} x - 3x + x^3}{x^5}$ as $x \to 0$ is $\dfrac{3}{5}$.

10.4.15 We compute that

$$
\begin{aligned}
\frac{\sin x - \tan x}{3x^3 \cos x} &= \frac{\left(x - \frac{x^3}{6} + \frac{x^5}{120} - \cdots\right) - \left(x + \frac{x^3}{3} + \frac{2x^5}{15} + \cdots\right)}{3x^3\left(1 - \frac{x^2}{2} + \cdots\right)} \\
&= \frac{-\frac{x^3}{2} - \frac{x^5}{8} + \cdots}{3x^3 - \frac{3x^5}{2} + \cdots} = \frac{-\frac{1}{2} - \frac{x^2}{8} + \cdots}{3 - \frac{3x^2}{2} + \cdots}
\end{aligned}
$$

so the limit of $\dfrac{\sin x - \tan x}{3x^3 \cos x}$ as $x \to 0$ is $\dfrac{-1/2}{3} = -\dfrac{1}{6}$.

10.4.17 The Taylor series for $\ln(x - 1)$ centered at 2 is

$$
\ln(x - 1) = (x - 2) + \frac{1}{2}(x - 2)^2 + \cdots.
$$

We compute that

$$
\frac{x - 2}{\ln(x - 1)} = \frac{x - 2}{(x - 2) + \frac{1}{2}(x - 2)^2 + \cdots} = \frac{1}{1 + \frac{1}{2}(x - 2) + \cdots}
$$

so the limit of $\dfrac{x - 2}{\ln(x - 1)}$ as $x \to 2$ is 1.

10.4.19 The Taylor series for $(1 + x)^{-2}$ centered at 0 is

$$
(1 + x)^{-2} = 1 - 2x + 3x^2 - 4x^3 + \cdots.
$$

and the Taylor series for $\cos\sqrt{x}$ centered at 0 is

$$
\cos\sqrt{x} = 1 - \frac{x}{2} + \frac{x^2}{24} - \frac{x^3}{720} + \cdots
$$

We compute that

$$
\begin{aligned}
&\frac{(1 + x)^{-2} - 4\cos\sqrt{x} + 3}{2x^2} \\
&= \frac{1}{2x^2}\left(\left(1 - 2x + 3x^2 - 4x^3 + \cdots\right) - 4\left(1 - \frac{x}{2} + \frac{x^2}{24} - \frac{x^3}{720} + \cdots\right) + 3\right) \\
&= \frac{1}{2x^2}\left(\frac{17x^2}{6} - \frac{719x^3}{180} + \cdots\right) = \frac{17}{12} - \frac{719x}{360} + \cdots
\end{aligned}
$$

so the limit of $\dfrac{(1 + x)^{-2} - 4\cos\sqrt{x} + 3}{2x^2}$ as $x \to 0$ is $\dfrac{17}{12}$.

10.4.21

a. $f'(x) = \frac{d}{dx}\left(\sum_{k=0}^{\infty} \frac{x^k}{k!}\right) = \sum_{k=1}^{\infty} k\frac{x^{k-1}}{k!} = \sum_{k=0}^{\infty} \frac{x^k}{k!} = f(x)$.

b. $f'(x) = e^x$ as well.

c. The series converges on $(-\infty, \infty)$.

10.4.23

a. $f'(x) = \frac{d}{dx}(\ln(1+x)) = \frac{d}{dx}\left(\sum_{k=1}^{\infty}(-1)^{k+1}\frac{1}{k}x^k\right) = \sum_{k=1}^{\infty}(-1)^{k+1}x^{k-1}$.

b. This is the power series for $\frac{1}{1+x}$.

c. The Taylor series for $\ln(1+x)$ converges on $(-1, 1)$, as does the Taylor series for $\frac{1}{1+x}$.

10.4.25

a. $f'(x) = \frac{d}{dx}(e^{-2x}) = \frac{d}{dx}\left(\sum_{k=0}^{\infty}\frac{(-2x)^k}{k!}\right) = \left(\sum_{k=0}^{\infty}(-2)^k\frac{x^k}{k!}\right)' = -2\sum_{k=1}^{\infty}(-2)^{k-1}\frac{x^{k-1}}{k-1} = -2\sum_{k=0}^{\infty}\frac{(-2x)^k}{k!}$.

b. This is the Taylor series for $-2e^{-2x}$.

c. Since the Taylor series for e^{-2x} converges on $(-\infty, \infty)$, so does this one.

10.4.27

a. Since $y(0) = 2$, we have $0 = y'(0) - y(0) = y'(0) - 2$ so that $y'(0) = 2$. Differentiating the equation gives $y''(0) = y'(0)$, so that $y''(0) = 2$. Successive derivatives also have the value 2 at 0, so the Taylor series is $2\sum_{k=0}^{\infty}\frac{t^k}{k!}$.

b. $2\sum_{k=0}^{\infty}\frac{t^k}{k!} = 2e^t$.

10.4.29

a. $y(0) = 2$, so that $y'(0) = 16$. Differentiating, $y''(t) - 3y'(t) = 0$, so that $y''(0) = 48$, and in general $y^{(k)}(0) = 3y^{(k-1)}(0) = 3^{k-1} \cdot 16$. Thus the power series is $2 + \frac{16}{3}\sum_{k=1}^{\infty}\frac{(3t)^k}{k!}$.

b. $2 + \frac{16}{3}\sum_{k=1}^{\infty}\frac{(3t)^k}{k!} = 2 + \frac{16}{3}(e^{3t} - 1)$.

10.4.31 The Taylor series for e^{-x^2} is $\sum_{k=0}^{\infty}(-1)^k\frac{x^{2k}}{k!}$. Thus, the desired integral is $\int_0^{0.25}\sum_{k=0}^{\infty}(-1)^k\frac{x^{2k}}{k!}\,dx =$ $\sum_{k=0}^{\infty}(-1)^k\frac{x^{2k+1}}{(2k+1)k!}\Big|_0^{0.25} = \sum_{k=0}^{\infty}(-1)^k\frac{1}{(2k+1)k!4^{2k+1}}$. Since this is an alternating series, to approximate it to within 10^{-4}, we must find n such that $a_{n+1} < 10^{-4}$, or $\frac{1}{(2n+3)(n+1)!\cdot 4^{2n+3}} < 10^{-4}$. This occurs for $n = 1$, so $\sum_{k=0}^{1}(-1)^k\frac{1}{(2k+1)\cdot k!\cdot 4^{2k+1}} = \frac{1}{4} - \frac{1}{192} \approx 0.2447916667$.

10.4.33 The Taylor series for $\cos 2x^2$ is $\sum_{k=0}^{\infty}(-1)^k\frac{(2x^2)^{2k}}{(2k)!} = \sum_{k=0}^{\infty}(-1)^k\frac{4^k x^{4k}}{(2k)!}$. Note that $\cos x$ is an even function, so we compute the integral from 0 to 0.35 and double it: $2\int_0^{0.35}\sum_{k=0}^{\infty}(-1)^k\frac{4^k x^{4k}}{(2k)!}\,dx =$ $2\left(\sum_{k=0}^{\infty}(-1)^k\frac{4^k x^{4k+1}}{(4k+1)(2k)!}\right)\Big|_0^{0.35} = 2\left(\sum_{k=0}^{\infty}(-1)^k\frac{4^k 0.35^{4k+1}}{(4k+1)(2k)!}\right)$. Since this is an alternating series, to approximate it to within $\frac{1}{2}\cdot 10^{-4}$, we must find n such that $a_{n+1} < \frac{1}{2}\cdot 10^{-4}$, or $\frac{4^{n+1}0.35^{4n+3}}{(4n+3)(2n+2)!} < \frac{1}{2}\cdot 10^{-4}$. This occurs first for $n = 1$, and we have $2\left(.35 - \frac{4\cdot 0.35^5}{5\cdot 2!}\right) \approx 0.69579825$.

10.4.35 The Taylor series for $\frac{\sin x}{x}$ is $\sum_{k=0}^{\infty}(-1)^k\frac{x^{2k}}{(2k+1)!}$, so the desired integral is $\int_0^{0.15}\sum_{k=0}^{\infty}(-1)^k\frac{x^{2k}}{(2k+1)!}\,dx =$ $\sum_{k=0}^{\infty}(-1)^k\frac{x^{2k+1}}{(2k+1)(2k+1)!}\Big|_0^{.15} = \sum_{k=0}^{\infty}(-1)^k\frac{0.1^{2k+1}}{(2k+1)(2k+1)!}$. This is an alternating series, so to approximate it to within 10^{-4}, we must find n such that $a_{n+1} < 10^{-4}$, or $\frac{0.15^{2n+3}}{(2n+3)(2n+3)!} < 10^{-4}$. This occurs first for $n = 1$, and we have $0.15 - \frac{0.15^3}{3\cdot 3!} = 0.1498125$.

10.4.37 The Taylor series for $(1+x^6)^{-1/2}$ is $\sum_{k=0}^{\infty} \binom{-1/2}{k} x^{6k}$, so the desired integral is $\int_0^{0.5} \sum_{k=0}^{\infty} \binom{-1/2}{k} x^{6k}\, dx = \sum_{k=0}^{\infty} \frac{1}{6k+1} \binom{-1/2}{k} x^{6k+1} \Big|_0^{0.5} = \sum_{k=0}^{\infty} \frac{1}{6k+1} \binom{-1/2}{k} 0.5^{6k+1}$. This is an alternating series since the binomial coefficients alternate in sign, so to approximate it to within 10^{-4}, we must find n such that $a_{n+1} < 10^{-4}$, or $\left| \frac{1}{6n+7} \binom{-1/2}{n+1} 0.5^{6n+7} \right| < 10^{-4}$. This occurs first for $n = 1$, so we have $\binom{-1/2}{0} 0.5 + \frac{1}{7} \binom{-1/2}{1}(0.5)^7 \approx 0.4994419643$.

10.4.39 Use the Taylor series for e^x at 0: $1 + \frac{2}{1!} + \frac{2^2}{2!} + \frac{2^3}{3!}$.

10.4.41 Use the Taylor series for $\cos x$ at 0: $1 - \frac{2^2}{2!} + \frac{2^4}{4!} - \frac{2^8}{8!}$.

10.4.43 Use the Taylor series for $\ln(1+x)$ evaluated at $x = 1/2$: $\frac{1}{2} - \frac{1}{2} \cdot \frac{1}{4} + \frac{1}{3} \cdot \frac{1}{8} - \frac{1}{4} \cdot \frac{1}{16}$.

10.4.45 The Taylor series for f centered at 0 is $\frac{-1 + \sum_{k=0}^{\infty} \frac{x^k}{k!}}{x} = \frac{\sum_{k=1}^{\infty} \frac{x^k}{k!}}{x} = \sum_{k=1}^{\infty} \frac{x^{k-1}}{k!} = \sum_{k=0}^{\infty} \frac{x^k}{(k+1)!}$. Evaluating both sides at $x = 1$, we have $e - 1 = \sum_{k=0}^{\infty} \frac{1}{(k+1)!}$.

10.4.47 The Taylor series for $\ln(1+x)$ centered at 0 is $x - \frac{1}{2}x^2 + \frac{1}{3}x^3 - \frac{1}{4}x^4 + \cdots = \sum_{k=1}^{\infty} (-1)^{k+1} \frac{x^k}{k}$. By the Ratio Test, $\lim_{k \to \infty} \left| \frac{a_{k+1}}{a_k} \right| = \lim_{k \to \infty} \left| \frac{x^{k+1} k}{x^k (k+1)} \right| = |x|$, so the radius of convergence is 1. The series diverges at -1 and converges at 1, so the interval of convergence is $(-1, 1]$. Evaluating at 1 gives $\ln 2 = \sum_{k=1}^{\infty} (-1)^{k+1} \frac{1}{k} = 1 - \frac{1}{2} + \frac{1}{3} - \frac{1}{4} + \cdots$.

10.4.49 $\sum_{k=0}^{\infty} \frac{x^k}{2^k} = \sum_{k=0}^{\infty} \left(\frac{x}{2} \right)^k = \frac{1}{1 - \frac{x}{2}} = \frac{2}{2-x}$.

10.4.51 $\sum_{k=0}^{\infty} (-1)^k \frac{x^{2k}}{4^k} = \sum_{k=0}^{\infty} \left(\frac{-x^2}{4} \right)^k = \frac{1}{1 + \frac{x^2}{4}} = \frac{4}{4+x^2}$.

10.4.53 $\ln(1+x) = -\sum_{k=1}^{\infty} (-1)^k \frac{x^k}{k}$, so $\ln(1-x) = -\sum_{k=1}^{\infty} \frac{x^k}{k}$, and finally $-\ln(1-x) = \sum_{k=1}^{\infty} \frac{x^k}{k}$.

10.4.55

$$
\begin{aligned}
\sum_{k=1}^{\infty} (-1)^k \frac{k x^{k+1}}{3^k} &= \sum_{k=1}^{\infty} (-1)^k \frac{k}{3^k} x^{k+1} = \sum_{k=1}^{\infty} k \left(-\frac{1}{3} \right)^k x^{k+1} \\
&= x^2 \sum_{k=1}^{\infty} \left(-\frac{1}{3} \right)^k k x^{k-1} = x^2 \sum_{k=1}^{\infty} \left(-\frac{1}{3} \right)^k \frac{d}{dx}(x^k) \\
&= x^2 \frac{d}{dx} \left(\sum_{k=1}^{\infty} \left(-\frac{x}{3} \right)^k \right) = x^2 \frac{d}{dx} \left(\frac{1}{1 + \frac{x}{3}} \right) = -\frac{3x^2}{(x+3)^2}.
\end{aligned}
$$

10.4.57 $\sum_{k=2}^{\infty} \frac{k(k-1)x^k}{3^k} = x^2 \sum_{k=2}^{\infty} \frac{k(k-1)x^{k-2}}{3^k} = x^2 \frac{d^2}{dx^2} \left(\sum_{k=2}^{\infty} \frac{x^k}{3^k} \right)$
$= x^2 \frac{d^2}{dx^2} \left(\sum_{k=2}^{\infty} \left(\frac{x}{3} \right)^k \right) = x^2 \frac{d^2}{dx^2} \left(\frac{x^2}{9} \cdot \frac{1}{1 - \frac{x}{3}} \right) = x^2 \frac{d^2}{dx^2} \left(\frac{x^2}{9 - 3x} \right) = x^2 \frac{-6}{(x-3)^3} = \frac{-6x^2}{(x-3)^3}$.

10.4.59

a. False, since $\frac{1}{1-x}$ is not continuous at 1, which is in the interval of integration.

b. False, since the Ratio Test shows that the radius of convergence for the Taylor series for $\tan^{-1}(x)$ centered at 0 is 1.

c. True, since $\sum_{k=0}^{\infty} \frac{x^k}{k!} = e^x$. Substitute $x = \ln 2$.

10.4.61 The Taylor series for $\sin x$ centered at 0 is

$$\sin x = x - \frac{x^3}{6} + \frac{x^5}{120} - \cdots .$$

We compute that

$$\frac{\sin ax}{\sin bx} = \frac{ax - \frac{(ax)^3}{6} + \frac{(ax)^5}{120} - \cdots}{bx - \frac{(bx)^3}{6} + \frac{(bx)^5}{120} - \cdots}$$

$$= \frac{a - \frac{a^3 x^2}{6} + \frac{a^5 x^4}{120} - \cdots}{b - \frac{b^3 x^2}{6} + \frac{b^5 x^4}{120} - \cdots}$$

so the limit of $\dfrac{\sin ax}{\sin bx}$ as $x \to 0$ is $\dfrac{a}{b}$.

10.4.63 Compute instead the limit of the log of this expression, $\lim\limits_{x\to 0} \frac{\ln(\sin x/x)}{x^2}$. If the Taylor expansion of $\ln(\sin x/x)$ is $\sum_{k=0}^{\infty} c_k x^k$, then $\lim\limits_{x\to 0} \frac{\ln(\sin x/x)}{x^2} = \lim\limits_{x\to 0} \sum_{k=0}^{\infty} c_k x^{k-2} = \lim\limits_{x\to 0} c_0 x^{-2} + c_1 x^{-1} + c_2$, since the higher-order terms have positive powers of x and thus approach zero as x does. So compute the terms of the Taylor series of $\ln\left(\frac{\sin x}{x}\right)$ up through the quadratic term. The relevant Taylor series are: $\frac{\sin x}{x} = 1 - \frac{1}{6}x^2 + \frac{1}{120}x^4 - \cdots$, $\ln(1+x) = x - \frac{1}{2}x^2 + \frac{1}{3}x^3 - \cdots$ and we substitute the Taylor series for $\frac{\sin x}{x} - 1$ for x in the Taylor series for $\ln(1+x)$. Since the lowest power of x in the first Taylor series is 2, it follows that only the linear term in the series for $\ln(1+x)$ will give any powers of x that are at most quadratic. The only term that results is $-\frac{1}{6}x^2$. Thus $c_0 = c_1 = 0$ in the above, and $c_2 = -\frac{1}{6}$, so that $\lim\limits_{x\to 0} \frac{\ln(\sin x/x)}{x^2} = -\frac{1}{6}$ and thus $\lim\limits_{x\to 0} \left(\frac{\sin x}{x}\right)^{1/x^2} = e^{-1/6}$.

10.4.65 The Taylor series we need are $\cos x = 1 - \frac{1}{2}x^2 + \frac{1}{24}x^4 + \ldots$, $e^t = 1 + t + \frac{1}{2!}t^2 + \frac{1}{3!}t^3 + \frac{1}{4!}t^4 + \cdots$. We are looking for powers of x^3 and x^4 that occur when the first series is substituted for t in the second series. Clearly there will be no odd powers of x, since $\cos x$ has only even powers. Thus the coefficient of x^3 is zero, so that $f^{(3)}(0) = 0$. The coefficient of x^4 comes from the expansion of $1 - \frac{1}{2}x^2 + \frac{1}{24}x^4$ in each term of e^t. Higher powers of x clearly cannot contribute to the coefficient of x^4. Thus consider $\left(1 - \frac{1}{2}x^2 + \frac{1}{24}x^4\right)^k$. The term $-\frac{1}{2}x^2$ generates $\binom{k}{2}$ terms of value $\frac{1}{4}x^4$ for $k \geq 2$, while the other term generates k terms of value $\frac{1}{24}x^4$ for $k \geq 1$. These terms all have to be divided by the $k!$ appearing in the series for e^t. So the total coefficient of x^4 is $\frac{1}{24}\sum_{k=1}^{\infty}\frac{k}{k!} + \frac{1}{4}\sum_{k=2}^{\infty}\binom{k}{2}\frac{1}{k!}$, $= \frac{1}{24}\sum_{k=1}^{\infty}\frac{1}{(k-1)!} + \frac{1}{4}\sum_{k=2}^{\infty}\frac{1}{2\cdot(k-2)!}$, $= \frac{1}{24}\sum_{k=0}^{\infty}\frac{1}{k!} + \frac{1}{8}\sum_{k=0}^{\infty}\frac{1}{k!}$, $= \frac{1}{24}e + \frac{1}{8}e = \frac{e}{6}$ Thus $f^{(4)}(0) = \frac{e}{6}\cdot 4! = 4e$.

10.4.67 The Taylor series for $\sin t^2$ is $\sin t^2 = t^2 - \frac{1}{3!}t^6 + \frac{1}{5!}t^{10} - \ldots$, so that $\int_0^x \sin t^2\, dt = \frac{1}{3}t^3 - \frac{1}{7\cdot 3!}t^7 + \ldots \Big|_0^x = $ $\frac{1}{3}x^3 - \frac{1}{7\cdot 3!}x^6 + \ldots$. Thus $f^{(3)}(0) = \frac{3!}{3} = 2$ and $f^{(4)}(0) = 0$.

10.4.69 Consider the series $\sum_{k=1}^{\infty} x^k = \frac{x}{1-x}$. Differentiating both sides gives $\frac{1}{(1-x)^2} = \sum_{k=0}^{\infty} kx^{k-1} = \frac{1}{x}\sum_{k=0}^{\infty} kx^k$ so that $\frac{x}{(1-x)^2} = \sum_{k=0}^{\infty} kx^k$. Evaluate both sides at $x = 1/2$ to see that the sum of the series is $\frac{1/2}{(1-1/2)^2} = 2$. Thus the expected number of tosses is 2.

10.4.71

a. We look first for a Taylor series for $(1 - k^2\sin^2(\theta))^{-1/2}$. Since $(1 - k^2 x^2)^{-1/2} = (1 - (kx)^2)^{-1/2} = \sum_{i=0}^{\infty}\binom{-1/2}{i}(kx)^{2i}$, and $\sin\theta = \theta - \frac{1}{3!}\theta^3 + \frac{1}{5!}\theta^5 - \cdots$, substituting the second series into the first gives $\frac{1}{\sqrt{1-k^2\sin^2\theta}} = 1 + \frac{1}{2}k^2\theta^2 + \left(-\frac{1}{6}k^2 + \frac{3}{8}k^4\right)\theta^4 + \left(\frac{1}{45}k^2 - \frac{1}{4}k^4 + \frac{5}{16}k^6\right)\theta^6 + \left(\frac{-1}{630}k^2 + \frac{3}{40}k^4 - \frac{5}{16}k^6 + \frac{35}{128}k^8\right)\theta^8 + \cdots$.

Integrating with respect to θ and evaluating at $\pi/2$ (the value of the antiderivative is 0 at 0) gives $\frac{1}{2}\pi + \frac{1}{48}k^2\pi^3 + \frac{1}{160}\left(-\frac{1}{6}k^2 + \frac{3}{8}k^4\right)\pi^5 + \frac{1}{896}\left(\frac{1}{45}k^2 - \frac{1}{4}k^4 + \frac{5}{16}k^6\right)\pi^7 + \frac{1}{4608}\left(-\frac{1}{630}k^2 + \frac{3}{40}k^4 - \frac{5}{16}k^6 + \frac{35}{128}k^8\right)\pi^9$. Evaluating these terms for $k = 0.1$ gives $F(0.1) \approx 1.574749680$. (The true value is approximately 1.574745562.)

 b. The terms above, with coefficients of k^n converted to decimal approximations, is $1.5707 + .3918 \cdot k^2 + .3597 \cdot k^4 - .9682 \cdot k^6 + 1.7689 \cdot k^8$. The coefficients are all less than 2 and do not appear to be increasing very much if at all, so if we want the result to be accurate to within 10^{-3} we should probably take n such that $k^n < \frac{1}{2} \times 10^{-3} = .0005$, so $n = 4$ for this value of k.

 c. By the above analysis, we would need a larger n since $0.2^n > 0.1^n$ for a given value of n.

10.4.73

 a. By the Fundamental Theorem, $S'(x) = \sin x^2$, $C'(x) = \cos x^2$.

 b. The relevant Taylor series are $\sin t^2 = t^2 - \frac{1}{3!}t^6 + \frac{1}{5!}t^{10} - \frac{1}{7!}t^{14} + \dots$, and $\cos t^2 = 1 - \frac{1}{2!}t^4 + \frac{1}{4!}t^8 - \frac{1}{6!}t^{12} + \dots$. Integrating, we have $S(x) = \frac{1}{3}x^3 - \frac{1}{7 \cdot 3!}x^7 + \frac{1}{11 \cdot 5!}x^{11} - \frac{1}{15 \cdot 7!}x^{15} + \dots$, and $C(x) = x - \frac{1}{5 \cdot 2!}x^5 + \frac{1}{9 \cdot 4!}x^9 - \frac{1}{13 \cdot 6!}x^{13} + \dots$.

 c. $S(0.05) \approx \frac{1}{3}(0.05)^3 - \frac{1}{42}(0.05)^5 + \frac{1}{1320}(0.05)^{11} - \frac{1}{75600}(0.05)^{15} \approx 4.166664807 \times 10^{-5}$. $C(-0.25) \approx (-0.25) - \frac{1}{10}(-0.25)^5 + \frac{1}{216}(-0.25)^9 - \frac{1}{9360}(-0.25)^{13} \approx -.2499023616$.

 d. The series is alternating. Since $a_{n+1} = \frac{1}{(4n+7)(2n+3)!}(0.05)^{4n+7}$, and this is less than 10^{-4} for $n = 0$, only one term is required.

 e. The series is alternating. Since $a_{n+1} = \frac{1}{(4n+5)(2n+2)!}(0.25)^{4n+5}$, and this is less than 10^{-6} for $n = 1$, two terms are required.

10.4.75

 a. $J_0(x) = 1 - \frac{1}{4}x^2 + \frac{1}{16 \cdot 2!^2}x^4 - \frac{1}{2^6 \cdot 3!^2}x^6 + \dots$.

 b. Using the Ratio Test: $\left| \frac{a_{k+1}}{a_k} \right| = \frac{x^{2k+2}}{2^{2k+2}((k+1)!)^2} \cdot \frac{2^{2k}(k!)^2}{x^{2k}} = \frac{x^2}{4(k+1)^2}$, which has limit 0 as $k \to \infty$ for any x. Thus the radius of convergence is infinite and the interval of convergence is $(-\infty, \infty)$.

 c. Starting only with terms up through x^8, we have $J_0(x) = 1 - \frac{1}{4}x^2 + \frac{1}{64}x^4 - \frac{1}{2304}x^6 + \frac{1}{147456}x^8 + \dots$, $J_0'(x) = -\frac{1}{2}x + \frac{1}{16}x^3 - \frac{1}{384}x^5 + \frac{1}{18432}x^7 + \dots$, $J_0''(x) = -\frac{1}{2} + \frac{3}{16}x^2 - \frac{5}{384}x^4 + \frac{7}{18432}x^6 + \dots$ so that $x^2 J_0(x) = x^2 - \frac{1}{4}x^4 + \frac{1}{64}x^6 - \frac{1}{2304}x^8 + \frac{1}{147456}x^{10} + \dots$, $x J_0'(x) = -\frac{1}{2}x^2 + \frac{1}{16}x^4 - \frac{1}{384}x^6 + \frac{1}{18432}x^8 + \dots$, $x^2 J_0''(x) = -\frac{1}{2}x^2 + \frac{3}{16}x^4 - \frac{5}{384}x^6 + \frac{7}{18432}x^8 + \dots$, and $x^2 J_0''(x) + x J_0'(x) + x^2 J_0(x) = 0$.

10.4.77

 a. The power series for $\cos x$ has only even powers of x, so that the power series has the same value evaluated at $-x$ as it does at x.

 b. The power series for $\sin x$ has only odd powers of x, so that evaluating it at $-x$ gives the opposite of its value at x.

10.4.79

 a. Since $f(a) = g(a) = 0$, we use the Taylor series for $f(x)$ and $g(x)$ centered at a to compute that

$$
\begin{aligned}
\lim_{x \to a} \frac{f(x)}{g(x)} &= \lim_{x \to a} \frac{f(a) + f'(a)(x-a) + \frac{1}{2}f''(a)(x-a)^2 + \cdots}{g(a) + g'(a)(x-a) + \frac{1}{2}g''(a)(x-a)^2 + \cdots} \\
&= \lim_{x \to a} \frac{f'(a)(x-a) + \frac{1}{2}f''(a)(x-a)^2 + \cdots}{g'(a)(x-a) + \frac{1}{2}g''(a)(x-a)^2 + \cdots} \\
&= \lim_{x \to a} \frac{f'(a) + \frac{1}{2}f''(a)(x-a) + \cdots}{g'(a) + \frac{1}{2}g''(a)(x-a) + \cdots} = \frac{f'(a)}{g'(a)}.
\end{aligned}
$$

Since $f'(x)$ and $g'(x)$ are assumed to be continuous at a and $g'(a) \neq 0$,

$$
\frac{f'(a)}{g'(a)} = \lim_{x \to a} \frac{f'(x)}{g'(x)}
$$

we have that

$$\lim_{x \to a} \frac{f(x)}{g(x)} = \lim_{x \to a} \frac{f'(x)}{g'(x)}$$

which is one form of L'Hôpital's Rule.

b. Since $f(a) = g(a) = f'(a) = g'(a) = 0$, we use the Taylor series for $f(x)$ and $g(x)$ centered at a to compute that

$$\begin{aligned}
\lim_{x \to a} \frac{f(x)}{g(x)} &= \lim_{x \to a} \frac{f(a) + f'(a)(x - a) + \frac{1}{2}f''(a)(x - a)^2 + \frac{1}{6}f'''(a)(x - a)^3 + \cdots}{g(a) + g'(a)(x - a) + \frac{1}{2}g''(a)(x - a)^2 + \frac{1}{6}g'''(a)(x - a)^3 + \cdots} \\
&= \lim_{x \to a} \frac{\frac{1}{2}f''(a)(x - a)^2 + \frac{1}{6}f'''(a)(x - a)^3 + \cdots}{\frac{1}{2}g''(a)(x - a)^2 + \frac{1}{6}g'''(a)(x - a)^3 + \cdots} \\
&= \lim_{x \to a} \frac{\frac{1}{2}f''(a) + \frac{1}{6}f'''(a)(x - a) + \cdots}{\frac{1}{2}g''(a) + \frac{1}{6}g'''(a)(x - a) + \cdots} = \frac{f''(a)}{g''(a)}.
\end{aligned}$$

Since $f''(x)$ and $g''(x)$ are assumed to be continuous at a and $g''(a) \neq 0$,

$$\frac{f''(a)}{g''(a)} = \lim_{x \to a} \frac{f''(x)}{g''(x)}$$

we have that

$$\lim_{x \to a} \frac{f(x)}{g(x)} = \lim_{x \to a} \frac{f''(x)}{g''(x)}$$

which is consistent with two applications of L'Hôpital's Rule.

10.5   Chapter Ten Review

10.5.1

a. True. The approximations tend to get better as n increases in size, and also when the value being approximated is closer to the center of the series. Since 2.1 is closer to 2 than 2.2 is, and since $3 > 2$, we should have $|p_3(2.1) - f(2.1)| < |p_2(2.2) - f(2.2)|$.

b. False. The interval of convergence may or may not include the endpoints.

c. True. The interval of convergence is an interval centered at 0, and the endpoints may or may not be included.

d. True. Since $f(x)$ is a polynomial, all its derivatives vanish after a certain point (in this case, $f^{(12)}(x)$ is the last nonzero derivative).

10.5.3 $p_2(x) = 1$.

10.5.5 $p_3(x) = x - \frac{x^2}{2} + \frac{x^3}{3}$.

10.5.7 $p_2(x) = x - 1 - \frac{1}{2}(x - 1)^2$.

10.5.9

a. $p_0(x) = 1$, $p_1(x) = 1 + x$, and $p_2(x) = 1 + x + \frac{x^2}{2}$.

b.

n	$p_n(-0.08)$	$\left\vert p_n(-0.08) - e^{-0.08} \right\vert$
0	1	7.69×10^{-2}
1	0.92	3.12×10^{-3}
2	0.9232	8.37×10^{-5}

10.5.11

a. $p_0(x) = \frac{\sqrt{2}}{2}$, $p_1(x) = \frac{\sqrt{2}}{2}(1 + (x - \pi/4))$, and $p_2(x) = \frac{\sqrt{2}}{2}\left(1 + (x - \pi/4) - \frac{1}{2}(x - \pi/4)^2\right)$.

b.

n	$p_n(\pi/5)$	$|p_n(\pi/5) - \sin(\pi/5)|$
0	0.7071	0.119
1	0.5960	8.25×10^{-3}
2	0.5873	4.74×10^{-4}

10.5.13 The derivatives of $\sin x$ are bounded in magnitude by 1, so $|R_n(x)| \le M\frac{|x|^{n+1}}{(n+1)!} \le \frac{|x|^{n+1}}{(n+1)!}$. But $|x| < \pi$, so $|R_3(x)| \le \frac{\pi^4}{24}$.

10.5.15 Using the Ratio Test, $\lim_{k\to\infty}\left|\frac{a_{k+1}}{a_k}\right| = \lim_{k\to\infty}\left|\frac{(k+1)^2 x^{k+1}}{(k+1)!} \cdot \frac{k!}{k^2 x^k}\right| = \lim_{k\to\infty}\left(\frac{k+1}{k}\right)^2 \frac{|x|}{k+1} = 0$, so the interval of convergence is $(-\infty, \infty)$.

10.5.17 Using the Ratio Test, $\lim_{k\to\infty}\frac{a_{k+1}}{a_k} = \lim_{k\to\infty}\left|\frac{(x+1)^{2k+2}}{(k+1)!} \cdot \frac{k!}{(x+1)^{2k}}\right| = \lim_{k\to\infty}\frac{1}{k+1}(x+1)^2 = 0$, so the interval of convergence is $(-\infty, \infty)$.

10.5.19 By the Root Test, $\lim_{k\to\infty}\left(\frac{|x|}{9}\right)^3 = \frac{|x^3|}{729}$, so the series converges for $|x| < 9$. The series given by letting $x = \pm 9$ are both divergent by the Divergence Test. Thus, $(-9, 9)$ is the interval of convergence.

10.5.21 The Maclaurin series for $f(x)$ is $\sum_{k=0}^{\infty} x^{2k}$. By the Root Test, this converges for $|x^2| < 1$, so $-1 < x < 1$. It diverges at both endpoints, so the interval of convergence is $(-1, 1)$.

10.5.23 The Maclaurin series for $f(x)$ is $\sum_{k=0}^{\infty}(3x)^k = \sum_{k=0}^{\infty} 3^k x^k$. By the Root Test, this has radius of convergence $1/3$. Checking the endpoints, we obtain an interval of convergence of $(-1/3, 1/3)$.

10.5.25 Taking the derivative of $\frac{1}{1-x}$ gives $f(x)$. Thus, the Maclaurin series for $f(x)$ is $\sum_{k=1}^{\infty} kx^{k-1}$. Using the Ratio Test, $\lim_{k\to\infty}\left|\frac{a_{k+1}}{a_k}\right| = \lim_{k\to\infty}\left|\frac{(k+1)x^k}{kx^{k-1}}\right| = \lim_{k\to\infty}\frac{k+1}{k}|x| = |x|$, so the radius of convergence is 1. Checking the endpoints, we obtain $(-1, 1)$ for the interval of convergence.

10.5.27 The first three terms are $1 + 3x + \frac{9x^2}{2}$. The series is $\sum_{k=0}^{\infty} \frac{(3x)^k}{k!}$.

10.5.29 The first three terms are $-(x - \pi/2) + \frac{1}{6}(x - \pi/2)^3 - \frac{1}{120}(x - \pi/2)^5$. The series is

$$\sum_{k=0}^{\infty}(-1)^{k+1}\frac{1}{(2k+1)!}\left(x - \frac{\pi}{2}\right)^{2k+1}.$$

10.5.31 The first three terms are $x - \frac{1}{3}x^3 + \frac{1}{5}x^5$. The series is $\sum_{k=0}^{\infty}(-1)^k\frac{x^{2k+1}}{2k+1}$.

10.5.33 $f(x) = \binom{1/3}{0} + \binom{1/3}{1}x + \binom{1/3}{2}x^2 + \cdots = 1 + \frac{1}{3}x - \frac{1}{9}x^2 + \cdots$.

10.5.35 $f(x) = \binom{-3}{0} + \binom{-3}{1}\frac{x}{2} + \binom{-3}{2}\frac{x^2}{4} + \cdots = 1 - \frac{3}{2}x + \frac{3}{2}x^2 + \cdots$.

10.5.37 $R_n(x) = \frac{(-1)^{n+1}e^{-c}}{(n+1)!}x^{n+1}$ for some c between 0 and x, and $\lim_{n\to\infty}|R_n(x)| = e^{-c}\lim_{n\to\infty}\frac{|x|^{n+1}}{(n+1)!} = 0$, since $n!$ grows faster than $|x|^n$ as $n \to \infty$ for all x.

10.5.39 $R_n(x) = \frac{f^{(n+1)}(c)}{(n+1)!}x^{n+1}$ for some c in $(-1/2, 1/2)$. Now, $|f^{(n+1)}(c)| = \frac{n!}{(1+c)^{n+1}}$, so $\lim_{n\to\infty}|R_n(x)| = \lim_{n\to\infty}\left(\frac{|x|}{1+c}\right)^{n+1} \cdot \frac{1}{n+1} < \lim_{n\to\infty} 1^{n+1}\frac{1}{n+1} = 0$.

10.5.41 The Taylor series for $\cos x$ centered at 0 is

$$\cos x = 1 - \frac{x^2}{2} + \frac{x^4}{24} - \frac{x^6}{720} + \cdots.$$

We compute that

$$\frac{x^2/2 - 1 + \cos x}{x^4} = \frac{1}{x^4}\left(x^2/2 - 1 + \left(1 - \frac{x^2}{2} + \frac{x^4}{24} - \frac{x^6}{720} + \cdots\right)\right)$$

$$= \frac{1}{x^4}\left(\frac{x^4}{24} - \frac{x^6}{720} + \cdots\right) = \frac{1}{24} - \frac{x^2}{720} + \cdots$$

so the limit of $\dfrac{x^2/2 - 1 + \cos x}{x^4}$ as $x \to 0$ is $\dfrac{1}{24}$.

10.5.43 The Taylor series for $\ln(x - 3)$ centered at 4 is

$$\ln(x - 3) = (x - 4) - \frac{1}{2}(x - 4)^2 + \frac{1}{3}(x - 4)^3 - \cdots.$$

We compute that

$$\frac{\ln(x - 3)}{x^2 - 16} = \frac{1}{(x - 4)(x + 4)}\left((x - 4) - \frac{1}{2}(x - 4)^2 + \frac{1}{3}(x - 4)^3 - \cdots\right)$$

$$= \frac{1}{(x - 4)(x + 4)}\left((x - 4)\left(1 - \frac{1}{2}(x - 4) + \frac{1}{3}(x - 4)^2 - \cdots\right)\right)$$

$$= \frac{1}{x + 4}\left(1 - \frac{1}{2}(x - 4) + \frac{1}{3}(x - 4)^2 - \cdots\right)$$

so the limit of $\dfrac{\ln(x - 3)}{x^2 - 16}$ as $x \to 4$ is $\dfrac{1}{8}$.

10.5.45 The Taylor series for $\sec x$ centered at 0 is

$$\sec x = 1 + \frac{x^2}{2} + \frac{5x^4}{24} + \frac{61x^6}{720} + \cdots$$

and the Taylor series for $\cos x$ centered at 0 is

$$\cos x = 1 - \frac{x^2}{2} + \frac{x^4}{24} - \frac{x^6}{720} + \cdots.$$

We compute that

$$\frac{\sec x - \cos x - x^2}{x^4}$$

$$= \frac{1}{x^4}\left(\left(1 + \frac{x^2}{2} + \frac{5x^4}{24} + \frac{61x^6}{720} + \cdots\right) - \left(1 - \frac{x^2}{2} + \frac{x^4}{24} - \frac{x^6}{720} + \cdots\right) - x^2\right)$$

$$= \frac{1}{x^4}\left(\frac{x^4}{6} + \frac{31x^6}{360} + \cdots\right) = \frac{1}{6} + \frac{31x^2}{360} + \cdots$$

so the limit of $\dfrac{\sec x - \cos x - x^2}{x^4}$ as $x \to 0$ is $\dfrac{1}{6}$.

10.5.47 Since $y(0) = 4$, we have $y'(0) - 16 + 12 = 0$, so $y'(0) = 4$. Differentiating the equation $n - 1$ times and evaluating at 0 we obtain $y^{(n)}(0) = 4y^{(n-1)}(0)$, so that $y^{(n)}(0) = 4^n$. The Taylor series for $y(x)$ is thus $y(x) = 4 + 4x + \frac{4^2 x^2}{2!} + \frac{4^3 x^3}{3!} + \ldots$, or $y(x) = 3 + e^{4x}$.

10.5.49

a. The Taylor series for $\ln(1+x)$ is $\sum_{k=1}^{\infty}(-1)^{k+1}\frac{x^k}{k}$. Evaluating at $x=1$ gives $\ln 2 = \sum_{k=1}^{\infty}(-1)^{k+1}\frac{1}{k}$.

b. The Taylor series for $\ln(1-x)$ is $-\sum_{k=1}^{\infty}\frac{x^k}{k}$. Evaluating at $x=1/2$ gives $\ln(1/2) = -\sum_{k=1}^{\infty}\frac{1}{k2^k}$, so that $\ln 2 = \sum_{k=1}^{\infty}\frac{1}{k2^k}$.

c. $f(x) = \ln\left(\frac{1+x}{1-x}\right) = \ln(1+x) - \ln(1-x)$. Using the two Taylor series above we have $f(x) = \sum_{k=1}^{\infty}(-1)^{k+1}\frac{x^k}{k} - \left(-\sum_{k=1}^{\infty}\frac{x^k}{k}\right) = \sum_{k=1}^{\infty}(1+(-1)^{k+1})\frac{x^k}{k} = 2\sum_{k=0}^{\infty}\frac{x^{2k+1}}{2k+1}$.

d. Since $\frac{1+x}{1-x} = 2$ when $x = \frac{1}{3}$, the resulting infinite series for $\ln 2$ is $2\sum_{k=0}^{\infty}\frac{1}{3^{2k+1}(2k+1)}$.

e. The first four terms of each series are: $1 - \frac{1}{2} + \frac{1}{3} - \frac{1}{4} \approx 0.5833333333$, $\frac{1}{2} + \frac{1}{8} + \frac{1}{24} + \frac{1}{64} \approx 0.6822916667$, $\frac{2}{3} + \frac{2}{81} + \frac{2}{1215} + \frac{2}{15309} \approx 0.6931347573$ The true value is $\ln 2 \approx 0.6931471806$. The third series converges the fastest, since it has 3^{k+1} in the denominator as opposed to 2^k, so its terms get small faster.

Chapter 11

11.1 Parametric Equations

11.1.1 Given an input value of t, the point $(x(t), y(t))$ can be plotted in the xy-plane, generating a curve.

11.1.3 Let $x = R\cos(\pi t/5)$ and $y = R\sin(\pi t/5)$. Note that as t ranges from 0 to 10, $\pi t/5$ ranges from 0 to 2π. Since $x^2 + y^2 = R^2$, this curve represents a circle of radius R.

11.1.5 Let $x = t$ and $y = t^2$ for $t \in (-\infty, \infty)$.

11.1.7

a.

t	x	y
-10	-20	-34
-5	-10	-19
0	0	-4
5	10	11
10	20	26

b.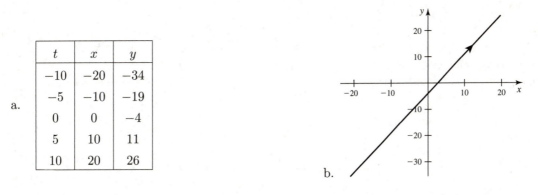

c. Solving $x = 2t$ for t yields $t = x/2$, so $y = 3t - 4 = 3x/2 - 4$.

d. The curve is the line segment from $(-20, -34)$ to $(20, 26)$.

11.1.9

a.

t	x	y
-5	11	-18
-3	9	-12
0	6	-3
3	3	6
5	1	12

b.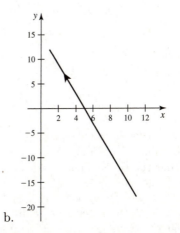

c. Solving $x = -t + 6$ for t yields $t = 6 - x$, so $y = 3t - 3 = 18 - 3x - 3 = 15 - 3x$.

d. The curve is the line segment from $(11, -18)$ to $(1, 12)$.

11.1.11

a. Solving $x = \sqrt{t} + 4$ for t yields $t = (x - 4)^2$. Thus, $y = 3\sqrt{t} = 3(x - 4)$, where x ranges from 4 to 8. Note that all $t > 0$, $x > 0$, and $y > 0$.

b. The curve is the line segment from $(4, 0)$ to $(8, 12)$.

11.1.13

a. Solving $x = t - 1$ for t yields $t = x + 1$. Thus, $y = t^3 = (x + 1)^3$, where $-5 \leq x \leq 3$.

b. The curve is the part of the standard cubic curve, shifted one unit to the left, from $(-5, -64)$ to $(3, 64)$.

11.1.15 Note that $x^2 + y^2 = 9\cos^2 t + 9\sin^2 t = 9$, so this represents an arc of the circle of radius 3 centered at the origin from $(-3, 0)$ to $(3, 0)$ traversed counterclockwise.

11.1.17 Note that $x^2 + y^2 = 49\cos^2 2t + 49\sin^2 2t = 49$, so this represents an arc of the circle of radius 7 centered at the origin from $(-7, 0)$ to $(-7, 0)$ traversed counterclockwise. (So the whole circle is represented.)

11.1.19

Let $x = 4\cos t$ and $y = 4\sin t$ for $0 \leq t \leq 2\pi$. Then $x^2 + y^2 = 16\cos^2 t + 16\sin^2 t = 16$.

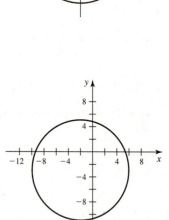

11.1.21

Let $x = -2 + 8\sin t$ and $y = -3 + 8\cos t$ for $0 \leq t \leq 2\pi$. Then $(x+2)^2 + (y+3)^2 = 64\sin^2 t + 64\cos^2 t = 64$.

11.1.23 Let t be time in minutes, so $0 \leq t \leq 1.5$ Let $x = 400\cos(4\pi/3)t$ and $y = 400\sin(4\pi/3)t$. Then since $x^2 + y^2 = 400^2$, the path is a circle of radius 400. Note that the values of x and y are the same at $t = 0$ and $t = 1.5$, and that the circle is traversed counterclockwise.

11.1.25 Let t be time in seconds, so $0 \leq t \leq 6$ Let $x = 50\cos(\pi/3)t$ and $y = 50\sin(\pi/3)t$. Then since $x^2 + y^2 = 50^2$, the path is a circle of radius 50. Note that the values of x and y are the same at $t = 0$ and $t = 6$, and that the circle is traversed counterclockwise.

11.1.27

Since $t = x - 3$, we have $y = 1 - (x - 3) = 4 - x$, so the line has slope -1. When $t = 0$, we have the point $(3, 1)$.

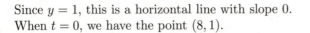

11.1.29

Since $y = 1$, this is a horizontal line with slope 0. When $t = 0$, we have the point $(8, 1)$.

11.1.31 Let $x = x_0 + at$ and $y = y_0 + bt$. Letting $(x_0, y_0) = (0, 0)$, and then finding a and b so that the curve is at the point Q when $t = 1$ yields $x = 2t$, $y = 8t$ for $0 \le t \le 1$.

11.1.33 Let $x = x_0 + at$ and $y = y_0 + bt$. Letting $(x_0, y_0) = (-1, -3)$, and then finding a and b so that the curve is at the point Q when $t = 1$ yields $x = -1 + 7t$, $y = -3 - 13t$ for $0 \le t \le 1$.

11.1.35

Let $x = t$ and $y = 2t^2 - 4$, $-1 \le t \le 5$.

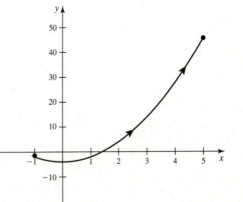

11.1.37

Let $x = -2 + 4t$ and $y = 3 - 6t$, $0 \le t \le 1$, and $x = t + 1$, $y = 8t - 11$ for $1 \le t \le 2$.

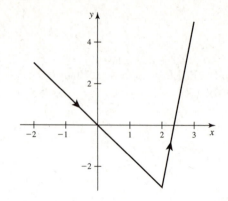

11.1.39

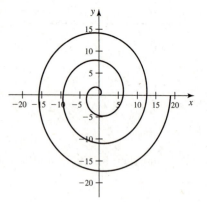

11.1.41

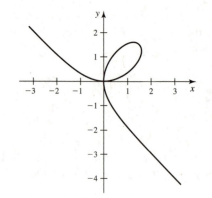

11.1.43

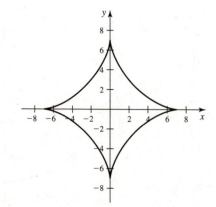

11.1.45

a. $\frac{dy}{dx} = \frac{dy/dt}{dx/dt} = \frac{-8}{4} = -2$ for all t. Since the curve is a line, the tangent line to the curve at the given point is the line itself.

b.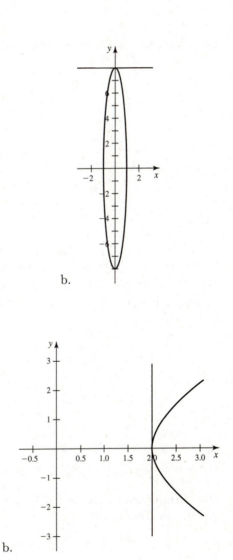

11.1.47

a. $\frac{dy}{dx} = \frac{dy/dt}{dx/dt} = \frac{8\cos t}{-\sin t} = -8\cot t$. At the given value of t, the value of $\frac{dy}{dx}$ is $-8\cot \pi/2 = 0$. The tangent line at the point $(0, 8)$ is thus the horizontal line $y = 8$.

b.

11.1.49

a. $\frac{dy}{dx} = \frac{dy/dt}{dx/dt} = \frac{1+\frac{1}{t^2}}{1-\frac{1}{t^2}} = \frac{t^2+1}{t^2-1}$. At the given value of t, the derivative doesn't exist, and the tangent line is the vertical line $x = 2$, tangent at the point $(2, 0)$.

b.

11.1.51

a. False. This generates a circle in the counterclockwise direction.

b. True. Note that when t is increased by one, the value of $2\pi t$ is increased by 2π, which is the period of both the sine and the cosine functions.

c. False. This generates only the portion of the parabola in the first quadrant, omitting the portion in the second quadrant.

d. True. They describe the portion of the unit circle in the 4th and 1st quadrants.

11.1.53 Let $x = 1 + 2t$ and $y = 1 + 4t$, for $-\infty < t < \infty$. Note that $y = 2(1 + 2t) - 1$, so $y = 2x - 1$.

11.1.55 Let $x = t^2$ and $y = t$, for $0 \le t < \infty$. Note that $x = t^2 = y^2$, and that the starting point is $(0,0)$.

11.1.57

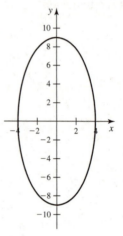

The entire curve is traversed for $0 \le t \le 2\pi$.

11.1.59

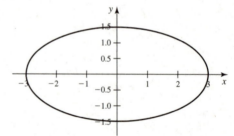

Let $x = 3\cos t$ and $y = \frac{3}{2}\sin t$ for $0 \le t \le 2\pi$. Then the major axis on the x-axis has length $2 \cdot 3 = 6$ and the minor axis on the y-axis has length $2 \cdot \frac{3}{2} = 3$. Note that $\left(\frac{x}{3}\right)^2 + \left(\frac{2y}{3}\right)^2 = \cos^2 t + \sin^2 t = 1$.

11.1.61

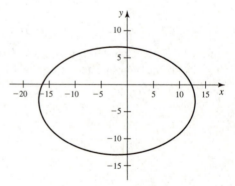

Let $x = 15\cos t - 2$ and $y = 10\sin t - 3$ for $0 \le t \le 2\pi$. Note that $\left(\frac{x+2}{15}\right)^2 + \left(\frac{y+3}{10}\right)^2 = \cos^2 t + \sin^2 t = 1$.

Then the major axis has length 30 and the minor axis has length 20.

11.1.63 The lines **a** and **b**. For line **a**, note that $t = x - 3$, so $y = 4 - 2t = 4 - (2x - 6) = 10 - 2x$. For line **b**, note that $t = \frac{x-3}{4}$, so $y = 4 - 8t = 4 - 2(x - 3) = 10 - 2x$ as well.

For **c**, note that $t^3 = x - 3$, so $y = 4 - t^3 = 4 - (x - 3) = 7 - x$, so this line is not the same as the other two.

11.1.65 Note that $x^2 + y^2 = 4\sin^2 8t + 4\cos^2 8t = 4$, so the curve is $x^2 + y^2 = 4$.

11.1.67 Note that since $t = x$, we have $y = \sqrt{4 - t^2} = \sqrt{4 - x^2}$.

11.1.69 Since $\sec^2 t - 1 = \tan^2 t$, we have $y = x^2$.

11.1.71 $\frac{dy}{dx} = \frac{dy/dt}{dx/dt} = \frac{4\cos t}{-4\sin t} = -\cot t$. We seek t so that $\cot t = -1/2$, so $t = \cot^{-1}(-1/2)$. The corresponding points on the curve are $\left(\frac{-4\sqrt{5}}{5}, \frac{8\sqrt{5}}{5}\right)$ and $\left(\frac{4\sqrt{5}}{5}, \frac{-8\sqrt{5}}{5}\right)$.

11.1.73 $\frac{dy}{dx} = \frac{dy/dt}{dx/dt} = \frac{1+(1/t^2)}{1-(1/t^2)} = \frac{t^2+1}{t^2-1}$. We seek t so that $\frac{t^2+1}{t^2-1} = 1$, which never occurs. Thus, there are no points on this curve with slope 1.

11.1.75 Note that in equation B, the parameter is scaled by a factor of 3. Thus, the curves are the same when the corresponding interval for t is scaled by a factor of $1/3$, so for $a = 0$ and $b = \frac{2\pi}{3}$. In fact, the same curve will be generated for $a = p$, $b = p + 2\pi/3$ where p is any real number.

11.1.77

a. $\frac{dy}{dx} = \frac{dy/dt}{dx/dt} = \frac{2\cos t}{2\cos 2t}$. This is zero when $\cos t = 0$ but $\cos 2t \neq 0$, which occurs for $t = \pi/2$ and $t = 3\pi/2$. The corresponding points on the graph are $(0, 2)$ and $(0, -2)$.

b. Using the derivative obtained above, we seek points where $\cos 2t = 0$ but $\cos t \neq 0$. This occurs for $t = \pi/4$, $3\pi/4$, $5\pi/4$, and $7\pi/4$. The corresponding points on the curve are $(1, \sqrt{2})$, $(-1, \sqrt{2})$, $(-1, -\sqrt{2})$, and $(1, -\sqrt{2})$.

11.1.79

a. Let $\text{sgn}(x) = \begin{cases} 1 & \text{if } x \geq 0 \\ -1 & \text{if } x < 0. \end{cases}$ Let $x = a \cdot \text{sgn}(\cos t) |\cos(t)|^{2/n}$ and $y = b \cdot \text{sgn}(\sin(t)) |\sin(t)|^{2/n}$.

b.

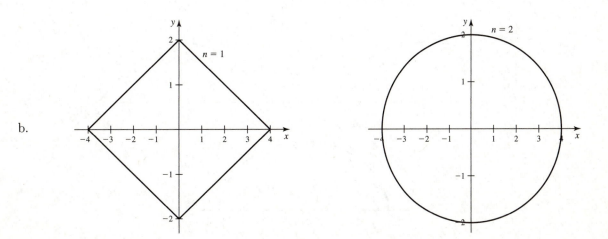

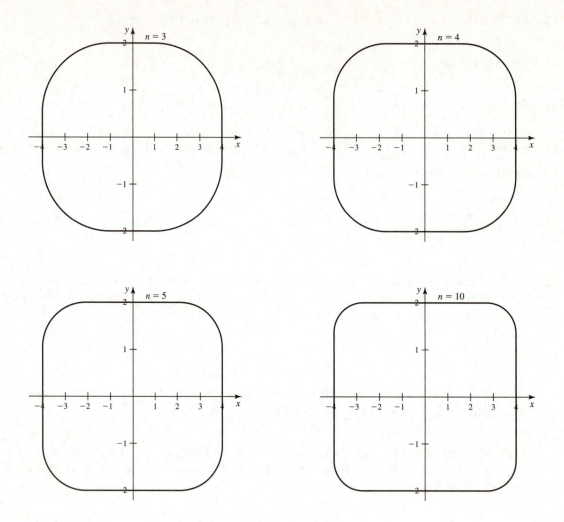

c. As n increases from near 0 to near 1, the curves change from star-shaped to a rectangular shape with corners at $(\pm a, 0)$ and $(0, \pm b)$. As n increases from 1 on, the curves become more rectangular with corners at $(\pm a, \pm b)$.

11.1.81 The first graphic shown is for $a = 1$ and $b = 1$. The second is for $a = 2$, $b = 1$, and the third is for $a = 1$, $b = 2$.

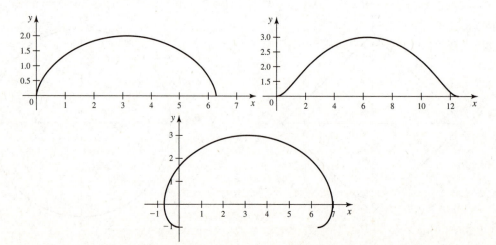

11.1.83

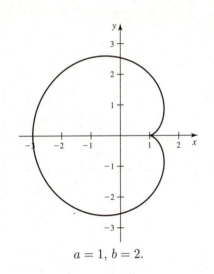

$a = 1$, $b = 2$.

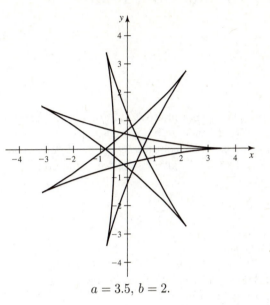

$a = 3.5$, $b = 2$.

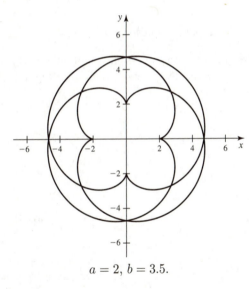

$a = 2$, $b = 3.5$.

Note that for $a < b$, we have cusps pointing inward, while for $a > b$, the cusps point outward.

11.1.85 The package lands when $y = 0$, so when $-4.9t^2 + 4000 = 0$ for $t > 0$. This occurs when $t = \sqrt{\frac{4000}{4.9}} \approx$ 28.57 seconds. At that time, $x \approx 100 \cdot 28.57 = 2857$ meters.

11.1.87 Let $x = 1 + \cos^2 t - \sin^2 t$ and $y = t$, for $-\infty < t < \infty$. Note that since $1 - \sin^2 t = \cos^2 t$, we have $x = 2\cos^2 t$, $y = t$.

11.1.89 Note that $a\cos t + b\sin t = \sqrt{a^2 + b^2}\sin(t + \alpha)$ where $\alpha = \tan^{-1}(a/b)$. This follows because $\sin(t + \tan^{-1}(a/b)) = \sin t \cos(\tan^{-1}(a/b)) + \cos t \sin(\tan^{-1}(a/b)) = \sin t \cdot \frac{b}{\sqrt{a^2+b^2}} + \cos t \cdot \frac{a}{\sqrt{a^2+b^2}}$.

Similarly, $c\cos t + d\sin t = \sqrt{c^2 + d^2} \cdot \cos(t + \beta)$, where $\beta = \tan^{-1}(-d/c)$.

Thus $x^2 + y^2 = (a\cos t + b\sin t)^2 + (c\cos t + d\sin t)^2 = (a^2 + b^2)\sin^2(t + \alpha) + (c^2 + d^2)\cos^2(t + \beta)$. This represents a circle if $a^2 + b^2 = c^2 + d^2$, and if $\alpha = \beta$, which means that $\frac{a}{b} = \frac{-d}{c}$, or $ac + bd = 0$.

11.2 Polar Coordinates

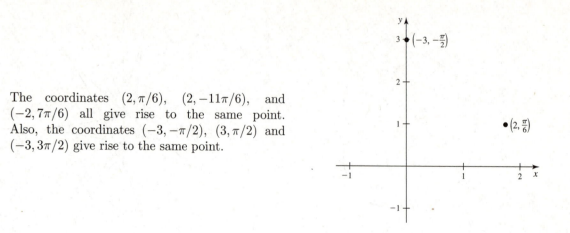

11.2.1 The coordinates $(2, \pi/6)$, $(2, -11\pi/6)$, and $(-2, 7\pi/6)$ all give rise to the same point. Also, the coordinates $(-3, -\pi/2)$, $(3, \pi/2)$ and $(-3, 3\pi/2)$ give rise to the same point.

11.2.3 If a point has cartesian coordinates (x, y) then $r^2 = x^2 + y^2$ and $\tan\theta = y/x$ for $x \neq 0$. If $x = 0$, then $\theta = \pi/2$ and $r = y$.

11.2.5 Since $x = r\cos\theta$, we have that the vertical line $x = 5$ has polar equation $r = \frac{5}{\cos\theta}$.

11.2.7 x-axis symmetry occurs if (r, θ) on the graph implies $(r, -\theta)$ is on the graph. y-axis symmetry occurs if (r, θ) on the graph implies $(r, \pi - \theta) = (-r, -\theta)$ is on the graph. Symmetry about the origin occurs if (r, θ). on the graph implies $(-r, \theta) = (r, \theta + \pi)$ is on the graph.

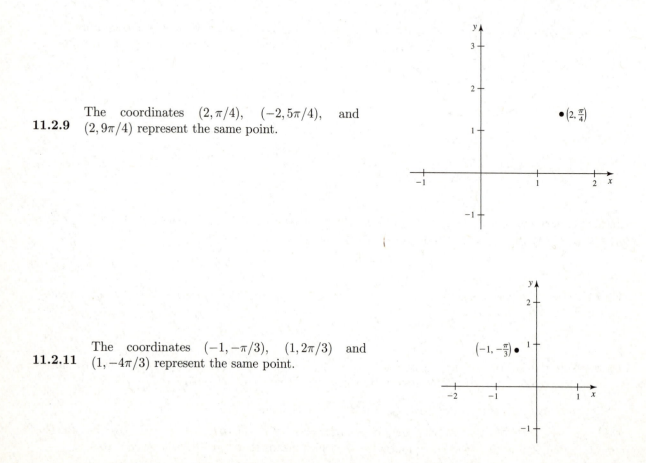

11.2.9 The coordinates $(2, \pi/4)$, $(-2, 5\pi/4)$, and $(2, 9\pi/4)$ represent the same point.

11.2.11 The coordinates $(-1, -\pi/3)$, $(1, 2\pi/3)$ and $(1, -4\pi/3)$ represent the same point.

11.2.13 The coordinates $(-4, 3\pi/2)$, $(4, \pi/2)$ and $(-4, -\pi/2)$ represent the same point.

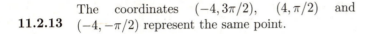

11.2.15 $x = 3\cos(\pi/4) = \frac{3\sqrt{2}}{2}$. $y = 3\sin(\pi/4) = \frac{3\sqrt{2}}{2}$.

11.2.17 $x = \cos(-\pi/3) = \frac{1}{2}$. $y = \sin(-\pi/3) = \frac{-\sqrt{3}}{2}$.

11.2.19 $x = -4\cos(3\pi/4) = 2\sqrt{2}$. $y = -4\sin(3\pi/4) = -2\sqrt{2}$.

11.2.21 $r^2 = x^2 + y^2 = 4 + 4 = 8$, so $r = \sqrt{8}$. $\tan\theta = 1$, so $\theta = \pi/4$, so $(2\sqrt{2}, \pi/4)$ is one representation of this point, and $(-2\sqrt{2}, -3\pi/4)$ is another.

11.2.23 $r^2 = x^2 + y^2 = 1 + 3 = 4$, so $r = \pm 2$. $\tan\theta = \sqrt{3}$, so $\theta = \pi/3$, $4\pi/3$. $(2, \pi/3)$ is one representation of this point, and $(-2, -2\pi/3)$ is another.

11.2.25 $r^2 = 64$, so $r = \pm 8$. $\tan\theta = -\sqrt{3}$, so $\theta = -\pi/3$, $2\pi/3$. One representation of the given point is $(8, 2\pi/3)$, and $(-8, -\pi/3)$ is another.

11.2.27

θ	0	$\pi/6$	$\pi/4$	$\pi/3$	$\pi/2$	$2\pi/3$	$3\pi/4$	$5\pi/6$	π
r	8	$4\sqrt{3}$	$4\sqrt{2}$	4	0	-4	$-4\sqrt{2}$	$-4\sqrt{3}$	-8

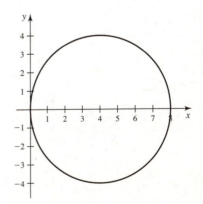

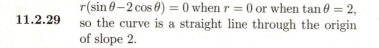

11.2.29 $r(\sin\theta - 2\cos\theta) = 0$ when $r = 0$ or when $\tan\theta = 2$, so the curve is a straight line through the origin of slope 2.

11.2.31 $x = r\cos\theta = -4$, so this is a vertical line $x = -4$ through $(-4, 0)$.

11.2.33 $r\cos\theta = \sin 2\theta = 2\sin\theta\cos\theta$. Note that if $\cos\theta = 0$, then r can be any real number, and the equation is satisfied. For $\cos\theta \neq 0$, we have $x = r\cos\theta = 2\sin\theta\cos\theta$, so $r = 2\sin\theta$, and thus $y = r\sin\theta = 2\sin^2\theta$. Thus $x^2 + y^2 - 2y = 4\sin^2\theta\cos^2\theta + 4\sin^2\theta\sin^2\theta - 4\sin^2\theta = 4\sin^2\theta(\sin^2\theta + \cos^2\theta) - 4\sin^2\theta = 4\sin^2\theta - 4\sin^2\theta = 0$. Note also that $x^2 + y^2 - 2y = 0$ is equivalent to $x^2 + (y-1)^2 = 1$, so we have a circle of radius one centered at $(0, 1)$, as well as the line $x = 0$ which is the y-axis.

11.2.35 $r = 8\sin\theta$, so $r^2 = 8r\sin\theta$, so $x^2 + y^2 = 8y$. This can be written $x^2 + (y-4)^2 = 16$, which represents a circle of radius 4 centered at $(0, 4)$.

11.2.37

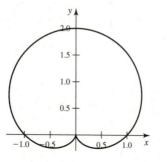

11.2.39

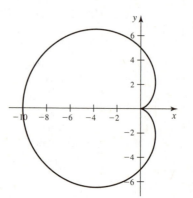

11.2.41

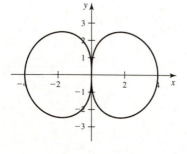

11.2.43

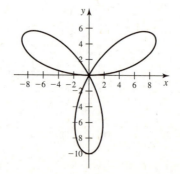

11.2.45 Points B, D, F, H, J, and L have y-coordinate 0, so the graph is at the pole for each of these points. Points E, I, and M have maximal radius, so these correspond to the points at the tips of the outer loops. The points C, G, and K correspond to the tips of the smaller loops. Point A corresponds to the polar point $(1, 0)$.

11.2.47 Points B, D, F, H, J, L, N, and P are at the origin. C, G, K, and O are on the ends of the long loops, while A, E, I, and M are at the ends of the smaller loops.

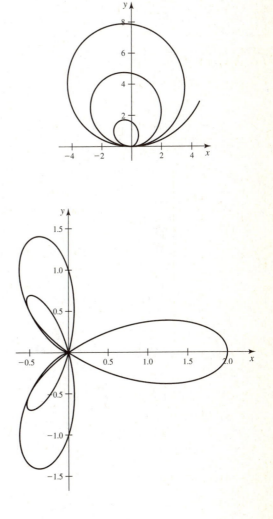

11.2.49 Since $r = \theta \sin \theta > M$ for any real number M (for suitably large θ), no finite interval can generate the entire graph.

11.2.51 The interval $[0, 2\pi]$ generates the entire graph.

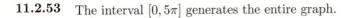

11.2.53 The interval $[0, 5\pi]$ generates the entire graph.

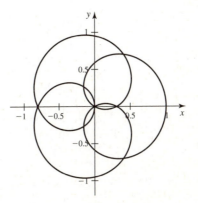

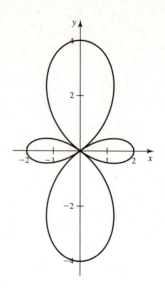

11.2.55 The interval $[0, 2\pi]$ generates the entire graph.

11.2.57

a. True. Note that $r^2 = 8$ and $\tan\theta = -1$.

b. True. Their intersection point (in Cartesian coordinates) is $(4, -2)$.

c. False. They intersect at the polar coordinates $(2, \pi/4)$ and $(2, 5\pi/4)$.

d. False. $3\cos(\pi) \neq 3$.

11.2.59   **11.2.61**

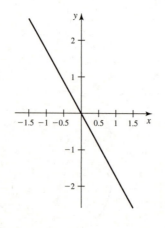   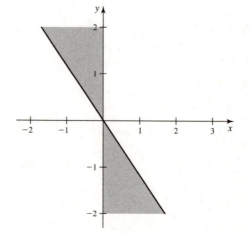

11.2.63

11.2.65

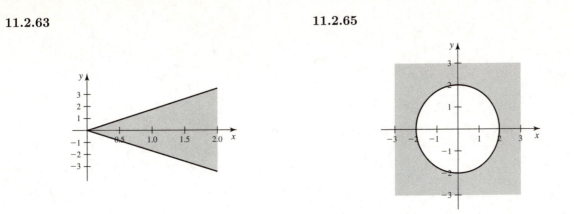

11.2.67 Consider the circle with center $C(r_0, \theta_0)$, and let A be the origin and $B(r, \theta)$ be a point on the circle not collinear with A and C. Note that the length of side BC is R, and that the angle CAB has measure $\theta - \theta_0$. Applying the law of cosines to triangle CAB yields the equation $R^2 = r^2 + r_0^2 - 2rr_0\cos(\theta - \theta_0)$, which is equivalent to the given equation.

11.2.69 In relation to number 67, we have $r_0 = 2$ and $\theta_0 = \pi/3$, and $R^2 - 4 = 12$, so $R^2 = 16$. Thus this is a circle with polar center $(2, \pi/3)$ and radius 4.

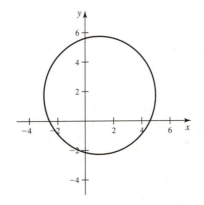

11.2.71 In relation to number 66, we have $a = 2$ and $b = 3$. So $R^2 - a^2 - b^2 = R^2 - 13 = 3$, and thus $R^2 = 16$. Thus we have a circle centered at $(2, 3)$ with radius 4.

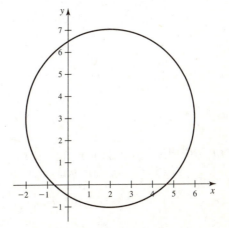

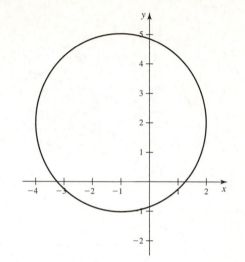

11.2.73 In relation to number 66, we have $a = -1$ and $b = 2$. So $R^2 - a^2 - b^2 = R^2 - 5 = 4$, and thus $R^2 = 9$. Thus we have a circle centered at $(-1, 2)$ with radius 3.

11.2.75

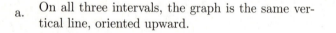

a. On all three intervals, the graph is the same vertical line, oriented upward.

b. For $\theta \neq \frac{2m+1}{2}\pi$ where m is an integer, we have $\cos\theta \neq 0$, so the equation is equivalent to $x = r\cos\theta = 2$. So the graph is a vertical line.

11.2.77 Using problem 76b, this is the line with $r_0 = 3$ and $\theta_0 = \frac{\pi}{3}$. So it is the line through the polar point $(3, \pi/3)$ in the direction of angle $\pi/3 + \pi/2 = 5\pi/6$.

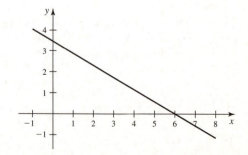

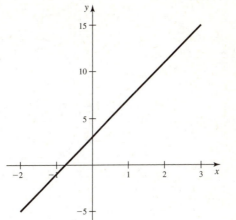

11.2.79 Using problem 76a, this is the line with $b = 3$ and $m = 4$, so $y = 4x + 3$.

11.2.81

a. This matches (A), since we have $|a| = 1 = |b|$, and the graph is a cardiod.

b. This matches (C). This has an inner loop since $|a| = 1 < 2 = |b|$. Note that $r = 1$ when $\theta = 0$, so it can't be (D).

c. This matches (B). This has $|a| = 2 > 1 = |b|$, so it has an oval-like shape.

d. This matches (D). This has an inner loop since $|a| = 1 < 2 = |b|$. Note that $r = -1$ when $\theta = 0$, so this can't be (C).

e. This matches (E). Note that there is an inner loop since $|a| = 1 < 2 = |b|$, and that $r = 3$ when $\theta = \pi/2$.

f. This matches (F).

11.2.83   **11.2.85**

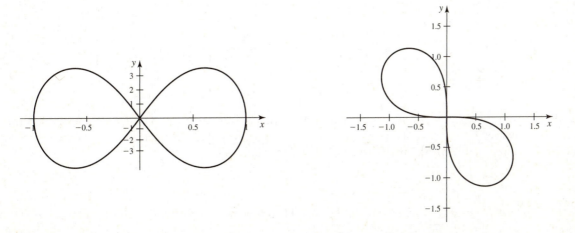

11.2.87   **11.2.89**

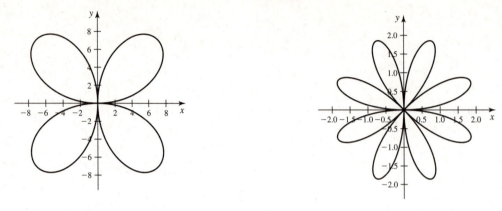

11.2.91 Note that $a \sin m\theta = 0$ for $\theta = \frac{k\pi}{m}$, $k = 1, 2, \ldots, 2m$. Thus the graph is back at the pole $r = 0$ for each of these values, and each of these gives rise to a distinct petal of the rose if m is odd. If m is even, then by symmetry, each petal for $k = 1, 2, \ldots \frac{m}{2}$ is equivalent to one for $k = \frac{m}{2} + 1, \frac{m}{2} + 2, \ldots, m$. (Note that this follows because the sine function is odd.) A similar result holds for the rose $r = a \cos \theta$.

11.2.93 For $a = 1$, the spiral winds outward counterclockwise. For $a = -1$, the spiral winds inward counterclockwise.

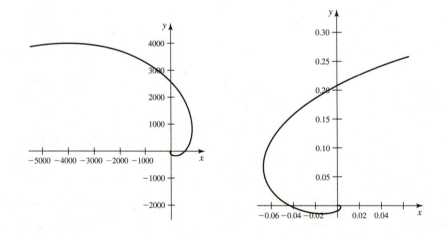

11.2.95 Suppose $2 \cos \theta = 1 + \cos \theta$. Then $\cos \theta = 1$, so this occurs for $\theta = 0$ and $\theta = 2\pi$. At those values, $r = 2$, so the curves intersect at the polar point $(2, 0)$. The curves also intersect when $r = 0$, which occurs for $\theta = \pi/2$ and $\theta = 3\pi/2$ for the first curve and $\theta = \pi$ for the second.

11.2.97 Suppose $1 - \sin \theta = 1 + \cos \theta$, or $\tan \theta = -1$. Then $\theta = 3\pi/4$ or $\theta = 7\pi/4$. So the curves intersect at the polar points $(1 + \sqrt{2}/2, 7\pi/4)$ and $(1 - \sqrt{2}/2, 3\pi/4)$. They also intersect at the pole $(0, 0)$, which occurs for the first curve at $\pi/2$ and for the second curve at π.

11.2.99

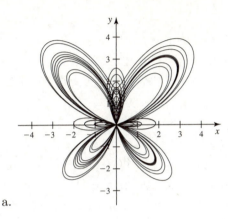

b. It adds multiple layers of the same type of curve as $\sin^5 \theta/12$ oscillates between -1 and 1 for $0 \le \theta \le 24\pi$.

a.

11.2.101

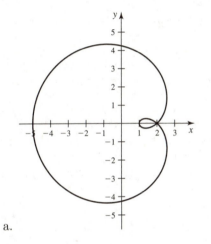

b. $r = 3 - 4\cos \pi t$ is a limaçon, and $x - 2 = r\cos \pi t$ and $y = r\sin \pi t$ is a circle, and the composition of a limaçon and a circle is a limaçon.

a.

11.2.103 With $r = a\cos\theta + b\sin\theta$, we have $r^2 = ar\cos\theta + br\sin\theta$, or $x^2 + y^2 = ax + by$, so $\left(x - \frac{a}{2}\right)^2 + \left(y - \frac{b}{2}\right)^2 = \frac{a^2+b^2}{4}$. Thus, the center is $(a/2, b/2)$ and $r = \frac{\sqrt{a^2+b^2}}{2}$.

11.2.105 Since $\sin(\theta/2) = \sin(\pi - \theta/2) = \sin((2\pi - \theta)/2)$, we have that the graph is symmetric with respect to the x-axis.

11.3 Calculus in Polar Coordinates

11.3.1 Since $x = r\cos\theta$ and $y = r\sin\theta$, we have $x = f(\theta)\cos\theta$ and $y = f(\theta)\sin(\theta)$.

11.3.3 Since slope is given relative to the horizontal and vertical coordinates, it is given by $\frac{dy}{dx}$, not by $\frac{dr}{d\theta}$.

11.3.5 $\frac{dy}{dx} = \frac{-\cos\theta\sin\theta + (1-\sin\theta)\cos\theta}{-\cos^2\theta - (1-\sin\theta)\sin\theta}$. At $(1/2, \pi/6)$, we have $\frac{dy}{dx} = \frac{0}{-1} = 0$. The given curve intersects the origin $r = 0$ for $\theta = \pi/2$. At this point, $\frac{dy}{dx}$ does not exist, and the tangent line is vertical. (It is the line $\theta = \pi/2$.)

11.3.7 $\frac{dy}{dx} = \frac{16\cos\theta\sin\theta}{-8\sin^2\theta + 8\cos^2\theta}$. At $(4, 5\pi/6)$ we have $\frac{dy}{dx} = \frac{-4\sqrt{3}}{4} = -\sqrt{3}$. The given curve intersects the origin $r = 0$ for $\theta = 0$ and $\theta = \pi$. At these points, the derivative is 0, and the tangent line is horizontal, so $\theta = 0$ is the tangent line.

11.3.9 $\frac{dy}{dx} = \frac{-3\sin^2\theta + (6+3\cos\theta)\cos\theta}{-3\cos\theta\sin\theta - (6+3\cos\theta)\sin\theta}$. At both $(3,\pi)$ and $(9,0)$, this doesn't exist. The given curve does not intersect the origin, since $r \geq 3$ for all θ.

11.3.11 $\frac{dy}{dx} = \frac{-8\sin(2\theta)\sin\theta + 4\cos(2\theta)\cos\theta}{-8\sin(2\theta)\cos\theta - 4\cos(2\theta)\sin\theta}$. The tips of the leaves occur at $\theta = 0$, $\pi/2$, π and $3\pi/2$. At 0 and at π, we have that $\frac{dy}{dx}$ doesn't exist. At $\pi/2$ and $3\pi/2$ we have $\frac{dy}{dx} = 0$. The graph intersects the origin for $\theta = \pi/4$, $\theta = 3\pi/4$, $\theta = 5\pi/4$ and $\theta = 7\pi/4$, and thus the two distinct tangent lines are $\theta = \pi/4$ and $\theta = 3\pi/4$.

11.3.13 The curve hits the origin at $\pm\pi/4$, where the tangent lines are given by $\theta = \pi/4$ and $\theta = -\pi/4$. The slopes of those lines are given by $\tan(\pi/4) = 1$ and $\tan(-\pi/4) = -1$.

11.3.15 Note that the curve is at the origin at $\pi/2$, so there is vertical tangent there. Also, $\frac{dy}{dx} = \frac{-4\sin^2\theta + 4\cos^2\theta}{-8\sin\theta\cos\theta} = \frac{1-2\sin^2\theta}{\sin(2\theta)}$. Thus, there are horizontal tangents at $\pi/4$ and $3\pi/4$ (at the polar points $(2\sqrt{2}, \pi/4)$ and $(-2\sqrt{2}, 3\pi/4)$. There is a vertical tangent where $\theta = 0$, at the point $(4,0)$.

11.3.17 Using the double angle identities somewhat liberally:

$$\frac{dy}{dx} = \frac{2\cos(2\theta)\sin\theta + \sin(2\theta)\cos\theta}{2\cos(2\theta)\cos\theta - \sin(2\theta)\sin\theta} = \frac{\sin\theta(\cos 2\theta + \cos^2\theta)}{\cos\theta(\cos(2\theta) - \sin^2\theta)} = \frac{\sin\theta(3\cos^2\theta - 1)}{\cos\theta(1 - 3\sin^2\theta)} = \frac{\sin\theta(3\cos^2\theta - 1)}{\cos\theta(3\cos^2\theta - 2)}.$$

The numerator is 0 for $\theta = 0$ and for $\theta = \pm\cos^{-1}(\pm\sqrt{3}/3)$, so there are horizontal tangents at the corresponding points (all of which have either $r \approx .943$ or $r \approx -.943$). The denominator is 0 for $\theta = \pi/2$ and $3\pi/2$, and for $\theta = \pm\cos^{-1}(\pm\sqrt{6}/3)$, so there are vertical tangents at the corresponding points (all of which have either $r \approx .943$ or $r \approx -.943$.)

11.3.19 The whole curve can be generated by considering values of θ on $[-\pi/4, \pi/4] \cup [3\pi/4, 5\pi/4]$. The curve intersects the origin for the endpoints of the domain, so neither horizontal or vertical tangents exist at the origin. Let $x = r\cos\theta$, so that $x^2 = r^2\cos^2\theta = 4\cos 2\theta\cos^2\theta$ and similarly $y^2 = r^2\sin^2\theta = 4\cos 2\theta\sin^2\theta$. Then $2xx' = -8\sin(2\theta)\cos^2\theta - 8\cos(2\theta)\sin\theta\cos\theta$, and $2yy' = -8\sin(2\theta)\sin^2\theta + 8\cos(2\theta)\sin\theta\cos\theta$. Thus $x' = \frac{4}{r}(-\cos(2\theta)\sin\theta - \sin(2\theta)\cos\theta)$ and $y' = \frac{4}{r}(-\sin(2\theta)\sin\theta + \cos(2\theta)\cos\theta)$. Thus

$$\frac{dy}{dx} = \frac{y'}{x'} = \frac{\sin(2\theta)\sin\theta - \cos(2\theta)\cos\theta}{\cos(2\theta)\sin\theta + \sin(2\theta)\cos\theta} = \frac{\cos\theta}{\sin\theta} \cdot \left(\frac{2\sin^2\theta - (2\cos^2\theta - 1)}{2\cos^2\theta - 1 + 2\cos^2\theta}\right) = \cot\theta \cdot \left(\frac{4\sin^2\theta - 1}{4\cos^2\theta - 1}\right),$$

where we have again made liberal use of the double angle trig identities. We have horizontal tangents on the domain at $\theta = \pm\pi/6$ and $5\pi/6$ and $7\pi/6$. There are vertical tangents on the domain only at $\theta = 0$ and $\theta = \pi$.

11.3.21 $A = \frac{1}{2}\int_0^\pi (8\sin\theta)^2\, d\theta = 32\int_0^\pi \sin^2\theta\, d\theta = 32\int_0^\pi \frac{1-\cos 2\theta}{2}\, d\theta = 32\left(\frac{1}{2}\theta - \frac{\sin\theta\cos\theta}{2}\right)\Big|_0^\pi = 16\pi.$

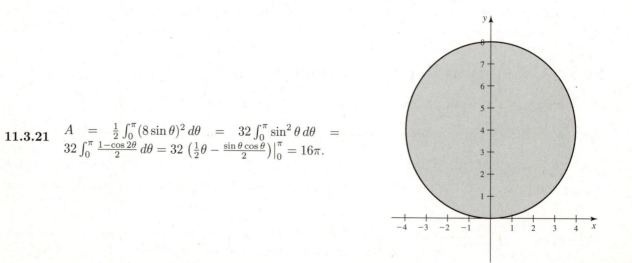

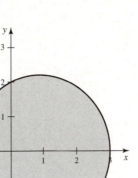

11.3.23 Using symmetry, we have $\frac{1}{2} \cdot 2 \int_0^\pi (2 + \cos\theta)^2 \, d\theta =$
$\int_0^\pi (4 + 4\cos\theta + \cos^2\theta) \, d\theta =$
$\left(4\theta + 4\sin\theta + \frac{1}{2}\theta + \frac{\sin\theta\cos\theta}{2} \right) \Big|_0^\pi = \frac{9\pi}{2}.$

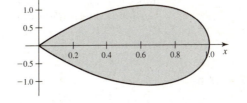

11.3.25 Using symmetry, we compute the area of $1/2$ of one leaf, and then double it. We have
$A = \frac{1}{2} \int_0^{\pi/10} \cos^2(5\theta) \, d\theta = \frac{1}{10} \int_0^{\pi/2} \cos^2 u \, du =$
$\frac{1}{10} \left(\frac{1}{2}u + \frac{\cos u \sin u}{2} \right) \Big|_0^{\pi/2} = \frac{\pi}{40}.$ So the area of one leaf is $2 \cdot \frac{\pi}{40} = \frac{\pi}{20}.$

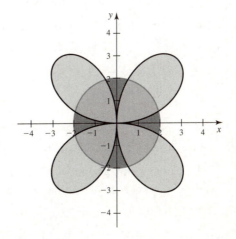

11.3.27 Note that the area inside one leaf of the rose but outside the circle is given by $\frac{1}{2} \int_{\pi/12}^{5\pi/12} (16\sin^2(2\theta) - 4) \, d\theta = (2\theta - \sin(4\theta)) \big|_{\pi/12}^{5\pi/2} = \sqrt{3} + \frac{2\pi}{3}.$
Also, the area inside one leaf of the rose is
$\frac{1}{2} \int_0^{\pi/2} 16\sin^2(2\theta) \, d\theta = \left(4\theta - \sin(4\theta) \big|_0^{\pi/2} \right) = 2\pi.$
Thus the area inside one leaf of the rose and inside the circle must be $2\pi - (\sqrt{3} + \frac{2\pi}{3}) = \frac{4\pi}{3} - \sqrt{3},$
and the total area inside the rose and inside the circle must be $4(\frac{4\pi}{3} - \sqrt{3}) = \frac{16\pi}{3} - 4\sqrt{3}.$

11.3.29 These curves intersect when $\sin\theta = \cos\theta$, which occurs at $\theta = \pi/4$ and $\theta = 5\pi/4$, and when $r = 0$ which occurs for $\theta = 0$ and $\theta = \pi$ for the first curve and $\theta = \pi/2$ and $\theta = 3\pi/2$ for the second curve. Only two of these intersection points are unique: the origin and the point $(3\sqrt{2}/2, \pi/4) = (-3\sqrt{2}/2, 5\pi/4)$.

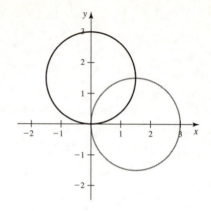

11.3.31 Suppose $4\cos\theta = 1 + 2\cos\theta + \cos^2\theta$. Then $(\cos\theta - 1)^2 = 0$, so $\theta = 0$. At that value, $r = 2$, so the curves intersect at the polar point $(2, 0)$. The curves also intersect when $r = 0$, which occurs for the first curve at $\pi/2$ and $3\pi/2$, and for the second curve at π. Also, the curves intersect when $4\cos\theta = -1 - 2\cos\theta - \cos^2\theta$, which occurs for $\cos^2\theta + 6\cos\theta + 1 = 0$, or (using the quadratic formula) $\theta = \cos^{-1}(-3 + 2\sqrt{2}) \approx 1.743$. This leads to the polar intersection points at approximately $(.8284, \pm 1.743)$.

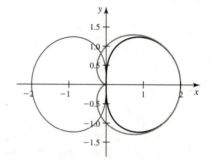

11.3.33

a. False. The area is given by $\frac{1}{2}\int_\alpha^\beta f(\theta)^2\,d\theta$.

b. False. The slope is given by $\frac{dy}{dx}$, which can be computed using the formula $\frac{dy}{dx} = \frac{dy/d\theta}{dx/d\theta} = \frac{f'(\theta)\sin\theta + f(\theta)\cos\theta}{f'(\theta)\cos\theta - f(\theta)\sin\theta}$.

11.3.35 The circles intersect for $\theta = \pi/6$ and $\theta = 5\pi/6$.

The area inside $r = 2\sin\theta$ but outside of $r = 1$ would be given by $\frac{1}{2}\int_{\pi/6}^{5\pi/6}(4\sin^2\theta - 1)\,d\theta = \frac{1}{2}\left(x - \sin(2x)\right)\Big|_{\pi/6}^{5\pi/6} = \frac{\pi}{3} + \frac{\sqrt{3}}{2}$. The total area of $r = 2\sin\theta$ is π. Thus, the area inside both circles is $\pi - \left(\frac{\pi}{3} + \frac{\sqrt{3}}{2}\right) = \frac{2\pi}{3} - \frac{\sqrt{3}}{2}$.

11.3.37 The inner loop is traced out between $\theta = \pi/6$ and $\theta = 5\pi/6$, so its area is given by $\frac{1}{2}\int_{\pi/6}^{5\pi/6}(3 - 6\sin\theta)^2\,d\theta = \frac{1}{2}\int_{\pi/6}^{5\pi/6}(9 - 36\sin\theta + 36\sin^2\theta)\,d\theta = \frac{3}{2}\left(3\theta + 12\cos\theta + 6\theta - 3\sin(2\theta)\right)\Big|_{\pi/6}^{5\pi/6} = 9\pi - \frac{27\sqrt{3}}{2}$.

We can determine the area inside the outer loop by using symmetry and doubling the area of the region traced out between $5\pi/6$ and $3\pi/2$. Thus the area inside the outer region is $2 \cdot \frac{1}{2}\int_{5\pi/6}^{3\pi/2}(3 - 6\sin\theta)^2\,d\theta = 3\left(3\theta + 12\cos\theta + 6\theta - 3\sin(2\theta)\right)\Big|_{5\pi/6}^{3\pi/2} = 18\pi + \frac{27\sqrt{3}}{2}$. So the area outside the inner loop and inside the outer loop is $18\pi + \frac{27\sqrt{3}}{2} - \left(9\pi - \frac{27\sqrt{3}}{2}\right) = 9\pi + 27\sqrt{3}$.

The first horizontal tangent line is at the origin. The next is at approximately $(4.0576, 2.0288)$, and the third at approximately $(9.8262, 4.9131)$.

11.3.39 The first vertical tangent line is at approximately $(1.7206, 0.8603)$, the next is at about $(6.8512, 3.4256)$, and the next at approximately $(12.8746, 6.4373)$.

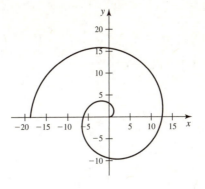

11.3.41

a. $A_n = \frac{1}{2} \int_{(2n-2)\pi}^{(2n-1)\pi} e^{-2\theta}\, d\theta - \frac{1}{2} \int_{2n\pi}^{(2n+1)\pi} e^{-2\theta}\, d\theta = \frac{-1}{4} e^{-(4n-2)\pi} + \frac{1}{4} e^{-(4n-4)\pi} + \frac{1}{4} e^{-(4n+2)\pi} - \frac{1}{4} e^{-4n\pi}.$

b. Each term tends to 0 as $n \to \infty$ so $\lim_{n\to\infty} A_n = 0$.

c. $\dfrac{A_{n+1}}{A_n} = \dfrac{e^{-(4n+2)\pi} + e^{-(4n)\pi} + e^{-(4n+6)\pi} - e^{-(4n+4)\pi}}{e^{-(4n-2)\pi} + e^{-(4n-4)\pi} + e^{-(4n+2)\pi} - e^{-4n\pi}} = e^{-4\pi}$, so $\lim_{n\to\infty} \dfrac{A_{n+1}}{A_n} = e^{-4\pi}$.

11.3.43 One half of the area is given by $\frac{1}{2} \int_0^{\pi/2} 6 \sin 2\theta\, d\theta = \frac{-3}{2} \cos 2\theta \big|_0^{\pi/2} = 3$, so the total area is 6.

11.3.45 The area is given by

$$\frac{1}{2} \int_0^{2\pi} (4 - 2\cos\theta)^2\, d\theta = \int_0^{2\pi} (8 - 8\cos\theta + 2\cos^2\theta)\, d\theta = \left(8\theta - 8\sin\theta + \theta + \frac{1}{2}\sin(2\theta)\right)\Big|_0^{2\pi} = 18\pi.$$

11.3.47 Suppose that the goat is tethered at the origin, and that the center of the corral is $(1, \pi)$. The circle that the goat can graze is $r = a$, and the corral is given by $r = -2\cos\theta$. The intersection occurs for $\theta = \cos^{-1}(-a/2)$.

The area grazed by the goat is twice the area of the sector of the circle $r = a$ between $\cos^{-1}(-a/2)$ and π, plus twice the area of the circle $r = -2\cos\theta$ between $\pi/2$ and $\cos^{-1}(-a/2)$. Thus we need to compute
$A = \int_{\cos^{-1}(-a/2)}^{\pi} a^2\, d\theta + \int_{\pi/2}^{\cos^{-1}(-a/2)} 4\cos^2\theta\, d\theta = a^2\pi - a^2\cos^{-1}(-a/2) + (2\cos\theta\sin\theta + 2\theta)\big|_{\pi/2}^{\cos^{-1}(-a/2)} = a^2(\pi - \cos^{-1}(-a/2)) - \pi - \frac{1}{2}a\sqrt{4 - a^2} + 2\cos^{-1}(-a/2)$. Note that $\pi - \cos^{-1}(-a/2) = \cos^{-1}(a/2)$, so this can be written as $(a^2 - 2)\cos^{-1}(a/2) + \pi - \frac{1}{2}a\sqrt{4 - a^2}$. Note that for $a = 0$ this is 0, and for $a = 2$, this is π, as desired.

11.3.49 Again, suppose that the goat is tethered at the origin, and that the center of the corral is $(1, \pi)$. The equation of the corral fence is given by $r = -2\cos\theta$. Note that to the right of the vertical line $\theta = \pi/2$, the goat can graze a half-circle of area $\pi a^2/2$. Also, there is a region in the 2nd quadrant and one in the 3rd quadrant of equal size that can also be grazed. Let this region have area A, so that the total area grazed will then be $\frac{\pi a^2}{2} + 2A$.

Imagine that the goat is walking "west" from the polar point $(a, \pi/2)$, and is keeping the rope taut until his whole rope is along the fence in the third quadrant. Let ϕ be the central angle angle from the origin to the polar point $(1, \pi)$ to the point on the fence that the goat's rope is touching as he makes this walk. When the goat is at $(a, \pi/2)$, we have $\phi = 0$. When the goat is all the way to the fence, we have $\phi = a$. Then length of the rope not along the fence is $a - \phi$. Thus, the value of A is $\frac{1}{2} \int_0^a (a - \phi)^2\, d\phi = \frac{1}{2}\left(a^2\phi - a\phi^2 + \frac{\phi^3}{3}\right)\Big|_0^a = \frac{a^3}{6}$.

Thus, the goat can graze a region of area $\frac{\pi a^2}{2} + \frac{a^3}{3}$.

11.3.51

a. If $\cot\phi = \frac{f'(\theta)}{f(\theta)}$ is constant for all θ, then $\phi = \cot^{-1}\left(\frac{f'(\theta)}{f(\theta)}\right)$ is constant. Then $\frac{d}{d\theta} \ln(f(\theta)) = \frac{1}{f(\theta)} \cdot f'(\theta) = \cot\phi$ is constant.

b. If $f(\theta) = Ce^{k\theta}$, then $\cot \phi = \dfrac{f'(\theta)}{f(\theta)} = \dfrac{kCe^{k\theta}}{Ce^{k\theta}} = k$.

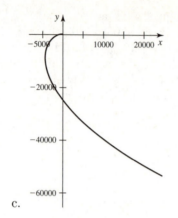

c.

11.4 Conic Sections

11.4.1 A parabola is the set of points in the plane which are equidistant from a given fixed point and a given fixed line.

11.4.3 A hyperbola is the set of points in the plane with the property that the difference of the distances from the point to two given fixed points is a given constant.

11.4.5

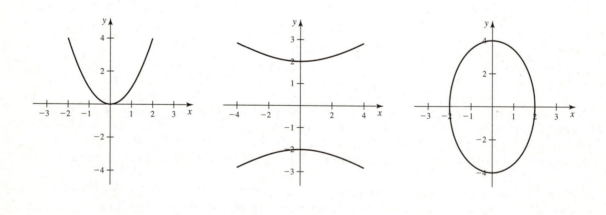

11.4.7 $\left(\dfrac{x}{a}\right)^2 + \dfrac{y^2}{a^2 - c^2} = 1.$

11.4.9 The foci for both are $(\pm ae, 0)$.

11.4.11 The asymptotes are $y = \dfrac{-b}{a} \cdot x$ and $y = \dfrac{b}{a} \cdot x$.

11.4.13 Directrix: $y = -3$. Focus: $(0, 3)$.

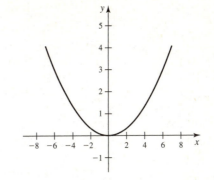

11.4.15 Directrix: $x = 4$. Focus: $(-4, 0)$.

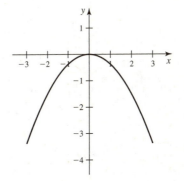

11.4.17 Directrix: $y = \frac{2}{3}$. Focus: $(0, \frac{-2}{3})$.

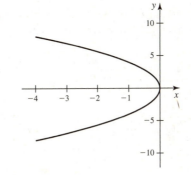

11.4.19

11.4.21

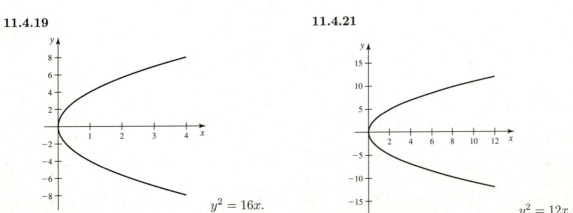

$y^2 = 16x.$

$y^2 = 12x.$

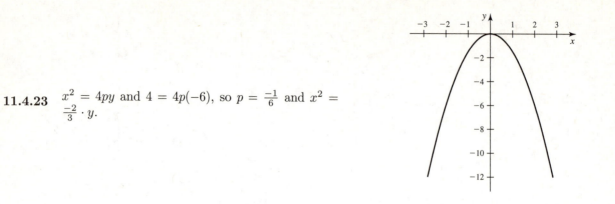

11.4.23 $x^2 = 4py$ and $4 = 4p(-6)$, so $p = \frac{-1}{6}$ and $x^2 = \frac{-2}{3} \cdot y$.

11.4.25 Since the vertex is $(-1, 0)$ and the parabola is symmetric about the x-axis, we have $y^2 = 4p(x+1)$ and since the directrix is one unit left of the vertex, we obtain $p = 1$ and $y^2 = 4(x+1)$.

11.4.27 Vertices are $(\pm 2, 0)$, and the foci are $(\pm\sqrt{3}, 0)$. The major axis has length 4 and the minor axis has length 2.

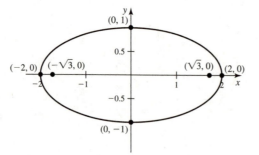

11.4.29 Vertices are $(0, \pm 4)$, and the foci are $(0, \pm 2\sqrt{3})$. The major axis has length 8 and the minor axis has length 4.

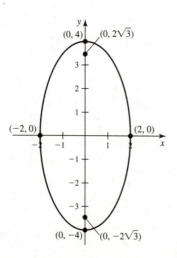

11.4.31 Vertices are $(0, \pm\sqrt{7})$, and the foci are $(0, \pm\sqrt{2})$. The major axis has length $2\sqrt{7}$ and the minor axis has length $2\sqrt{5}$.

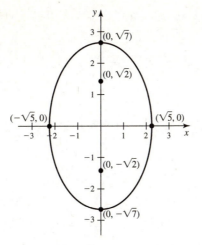

11.4.33 $a = 4$, and $b = 3$, so the equation is $\frac{x^2}{16} + \frac{y^2}{9} = 1$.

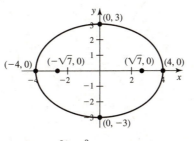

11.4.35 $a = 5$, and the equation is of the form $\frac{x^2}{25} + \frac{y^2}{b^2} = 1$. Since $(4, \frac{3}{5})$ is on the curve, we have $\frac{16}{25} + \frac{9}{25b^2} = 1$, so $b = 1$. The equation is $\frac{x^2}{25} + y^2 = 1$.

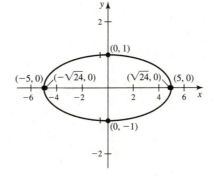

11.4.37 $a = 3$ and $b = 2$, so the equation is $\frac{x^2}{4} + \frac{y^2}{9} = 1$.

11.4.39 The vertices are $(\pm 2, 0)$, and the foci are $(\pm\sqrt{5}, 0)$. The asymptotes are $y = \frac{\pm 1}{2} \cdot x$.

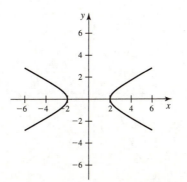

11.4.41 The vertices are $(\pm 2, 0)$, and the foci are $(\pm 2\sqrt{5}, 0)$. The asymptotes are $y = \pm 2x$.

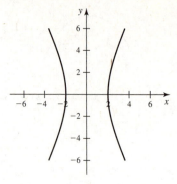

11.4.43 The vertices are $(\pm\sqrt{3}, 0)$, and the foci are $(\pm 2\sqrt{2}, 0)$. The asymptotes are $y = \pm\sqrt{\frac{5}{3}} \cdot x$.

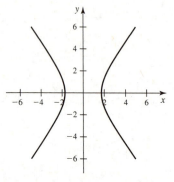

11.4.45 We have $a = 4$ and $c = 6$, so $b^2 = c^2 - a^2 = 20$, so the equation is $\frac{x^2}{16} - \frac{y^2}{20} = 1$.

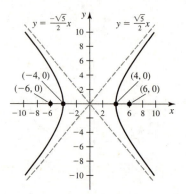

11.4.47 We have $a = 2$, and since the asymptoes are $y = \frac{\pm bx}{a}$, we have that $b = 3$, so the equation is $\frac{x^2}{4} - \frac{y^2}{9} = 1$.

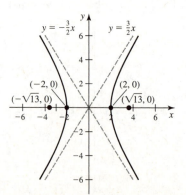

11.4.49 We have $a = 4$ and $c = 5$, so $b^2 = 25 - 16 = 9$, so $b = 3$ and the equation is $\frac{x^2}{16} - \frac{y^2}{9} = 1$.

11.4.51 We have $a = 9$ and $e = \frac{1}{3}$, so $c = ae = 3$, and $b^2 = a^2 - c^2 = 72$, so the equation is $\frac{x^2}{81} + \frac{y^2}{72} = 1$.

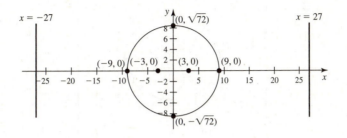

11.4.53 We have $a = 1$ and $e = 3$, so $c = ae = 3$ and $b^2 = c^2 - a^2 = 9 - 1 = 8$. Thus, the equation is $x^2 - \frac{y^2}{8} = 1$.

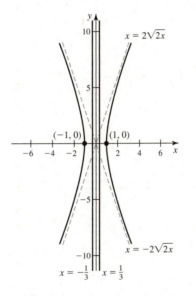

11.4.55 The vertex is $(2, 0)$. The focus is $(0, 0)$, and the directrix is the line $x = 4$.

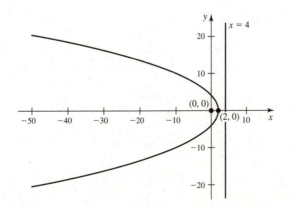

11.4.57 The vertices are $(1, 0)$ and $(-1/3, 0)$. The center is $(1/3, 0)$. The directrices are $x = -1$ and $x = 5/3$.

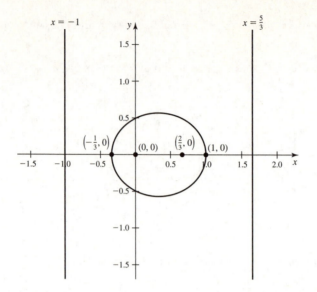

11.4.59 The vertex is $(0, -1/4)$, and the focus is $(0, 0)$. The directrix is the line $y = \frac{-1}{2}$.

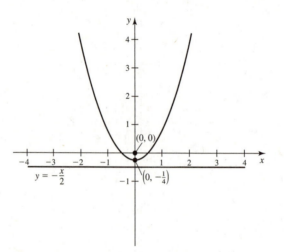

11.4.61 The parabola starts at $(1, 0)$ and goes through quadrants I, II, and III for $\theta \in [0, 3\pi/2]$. It then approaches $(1, 0)$ by traveling through quadrant IV for $\theta \in (3\pi/2, 2\pi)$.

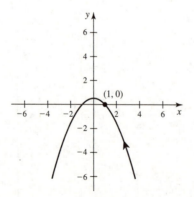

11.4.63 The parabola begins in the first quadrant and passes through the points $(0, 3)$ and then $(-3/2, 0)$ and $(0 - 3)$ as θ ranges from 0 to 2π.

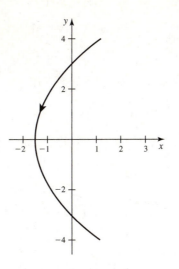

11.4.65 For negative p, the parabola opens to the left and for positive p it opens to the right. As p increases to 0, the parabola opens wider and as p decreases (for $p > 0$), it gets narrower.

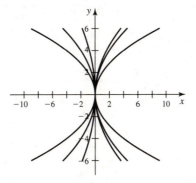

11.4.67

a. True. Note that if $x = 0$, the equation becomes $-y^2 = 9$, which has no solution.

b. True. The slopes of the tangent lines range continuously from $-\infty$ to 0 to ∞ and then back through 0 to $-\infty$ again.

c. True. Given c and d, one can compute a, b, and e. See the summary after Theorem 10.3.

d. True. The vertex is exactly halfway between the focus and the directrix.

11.4.69 Differentiating gives $2x = -6y'$, so at $(-6, -6)$ we obtain $-12 = -6y'$, so $y' = 2$. Thus $y - (-6) = 2(x - (-6))$, or $y = 2x + 6$ is the equation of the tangent line.

11.4.71 Differentiating implicitly, we have $2yy' - \frac{x}{32} = 0$, so at $(6, -5/4)$ we have $\frac{-5}{2}y' - \frac{3}{16} = 0$, so $y' = \frac{-3}{40}$. The equation of the tangent line is $y + \frac{5}{4} = \frac{-3}{40}(x - 6)$, or $y = \frac{-3x}{40} - \frac{4}{5}$.

11.4.73 We have a hyperbola with focal point at the origin and directrix $y = -2$. Furthermore $P = (0, -4/3)$ is a vertex. Thus, $e = \frac{|PF|}{|PL|} = \frac{4/3}{2/3} = 2$, and $r(\theta) = \frac{2(2)}{1 - 2\sin\theta} = \frac{4}{1 - 2\sin\theta}$.

11.4.75 The points on the intersection of the two circles are a distance of $2a + r$ from F_1 and a distance of r from F_2. So for P an intersection point, we have $|PF_1| - |PF_2| = 2a$ for all r, and the set of all such points form a hyperbola with foci F_1 and F_2.

11.4.77 Using implicit differentiation, we have $\frac{2x}{a^2} + \frac{2yy'}{b^2} = 0$, or $y' = \frac{-b^2 x}{a^2 y}$. If (x_0, y_0) is the point of tangency, then $\frac{-b^2 x_0}{a^2 y_0} = \frac{y - y_0}{x - x_0}$, so $\frac{x_0(x - x_0)}{a^2} = \frac{-y_0(y - y_0)}{b^2}$, so $\frac{x_0 x}{a^2} + \frac{y_0 y}{b^2} = \frac{x_0^2}{a^2} + \frac{y_0^2}{b^2} = 1$.

11.4.79 $V_x = \pi \int_{-a}^{a} \left(b^2 - \frac{b^2 x^2}{a^2} \right) dx = \pi b^2 \int_{-a}^{a} \left(1 - \frac{x^2}{a^2} \right) dx = \pi b^2 \left(x - \frac{x^3}{3a^2} \right) \Big|_{-a}^{a} = \frac{4\pi b^2 a}{3}$.

$V_y = \pi \int_{-b}^{b} \left(a^2 - \frac{a^2 y^2}{b^2} \right) dy = \pi a^2 \int_{-b}^{b} \left(1 - \frac{y^2}{b^2} \right) dy = \pi a^2 \left(y - \frac{y^3}{3b^2} \right) \Big|_{-b}^{b} = \frac{4\pi a^2 b}{3}$.

These are different if $a \neq b$. In the case $a = b$, both volumes give $\frac{4\pi a^3}{3}$, the volume of a sphere.

11.4.81

a. $V_x = \pi \int_{a}^{c} \left(\sqrt{\frac{b^2 x^2}{a^2} - b^2} \right)^2 dx = \pi \int_{a}^{c} \left(\frac{b^2 x^2}{a^2} - b^2 \right) dx = \pi b^2 \left(\frac{x^3}{3a^2} - x \right) \Big|_{a}^{c} = \pi b^2 \left(\frac{c^3}{3a^2} - c - \frac{a}{3} + a \right) = \frac{\pi b^2}{3a^2}(c^3 - 3ca^2 + 2a^3) = \frac{\pi b^2}{3a^2}(a-c)^2(2a+c)$.

b. $V_y = 2 \cdot 2\pi \int_{a}^{c} a^2 b \sqrt{\frac{x^2}{a^2} - 1}\, dx = 2\pi \int_{0}^{b^2/a^2} a^2 b \sqrt{u}\, du = 2\pi a^2 b \left(\frac{2}{3} u^{3/2} \Big|_{0}^{b^2/a^2} \right) = 2\pi a^2 b \frac{2b^3}{3a^3} = \frac{4\pi b^4}{3a}$.

11.4.83

a. The slope of a line making an angle θ with the horizontal is $\tan\theta$. The slope of the tangent line at (x_0, y_0) is $y' = \frac{x}{2p}$, so $y' = \frac{x_0}{2p}$, so $\tan\theta = \frac{x_0}{2p}$.

b. The distance from $(0, y_0)$ to $(0, p)$ is $p - y_0$, and $\tan\phi = \frac{\text{opposite}}{\text{adjacent}} = \frac{p - y_0}{x_0}$.

c. Since l is perpendicular to $y = y_0$, we have $\alpha + \theta = \pi/2$, or $\alpha = \frac{\pi}{2} - \theta$, so $\tan\alpha = \cot\theta = \frac{2p}{x_0}$.

d. $\tan\beta = \tan(\theta + \phi) = \frac{\frac{x_0}{wp} + \frac{p - y_0}{x_0}}{1 - \frac{p - y_0}{2p}} = \frac{x_0^2 + 2p^2 - 2py_0}{x_0(p + y_0)}$. Now since $x_0^2 = 4py_0$, we obtain $\tan\beta = \frac{4py_0 + 2p^2 - 2py_0}{x_0(p + y_0)} = \frac{2p(p + y_0)}{x_0(p + y_0)} = \frac{2p}{x_0}$.

e. Since α and β are acute, we have that $\tan\alpha = \tan\beta$, so $\alpha = \beta$.

11.4.85 Assume the two fixed points are at $(c, 0)$ and $(-c, 0)$. Let P be the point $(0, b)$, and note that P is equidistant from the two given points, so we must have $b^2 + c^2 = a^2$ by the Pythagorean theorem. Now let $Q = (u, 0)$ be on the ellipse for $u > c$. Then $u - c + (c + u) = 2a$, so $u = a$. Now let $R = (x, y)$ be an arbitrary point on the ellipse (assume $x > 0$ and $y > 0$ – the other cases are similar.) Using the triangles formed between the foci, R, and the projection of R onto the x-axis, we have $\sqrt{(x + c)^2 + y^2} = 2a - \sqrt{(c - x)^2 + y^2}$. Squaring both sides gives $(x + c)^2 + y^2 = 4a^2 - 4a\sqrt{(c - x)^2 + y^2} + (c - x)^2 + y^2$. Isolating the root gives $\sqrt{(c - x)^2 + y^2} = \frac{1}{4a}\left((c - x)^2 + y^2 - (c + x)^2 - y^2 + 4a^2 \right)$, so $\sqrt{(c - x)^2 + y^2} = a - \frac{c}{a}x$. Squaring again yields $(c - x)^2 + y^2 = a^2 - 2xc + \frac{c^2}{a^2}x^2$, so $c^2 - 2cx + x^2 + y^2 = a^2 - 2cx + \frac{c^2}{a^2}x^2$, or $x^2\left(1 - \frac{c^2}{a^2}\right) + y^2 = a^2 - c^2$. Thus $\frac{x^2}{a^2} + \frac{y^2}{a^2 - c^2} = 1$, which can be written $\frac{x^2}{a^2} + \frac{y^2}{b^2} = 1$, since $b^2 = a^2 - c^2$.

11.4.87 Let the parabola be symmetric about the y-axis with vertex at the origin. Let the circle have radius r and be centered at $(r + a, 0)$, and let the line be $y = -a$. The distance form the point $P(x, y)$ to the line is $u = y + a$. The distance from the point P to the circle is $v = \sqrt{x^2 + (r + a - y)^2} - r$. Setting $u = v$ yields $y + a = \sqrt{x^2 + (r + a - y)^2} - r$, so $y + r + a = \sqrt{x^2 + (r + a - y)^2}$, and squaring gives $y^2 + 2(r + a)y + (r + a)^2 = x^2 + (r + a - y)^2$, so $y^2 + 2(r + a)y + (r + a)^2 = x^2 + (r + a)^2 - 2(r + a)y + y^2$, and thus $4(r + a)y = x^2$, so $y = \frac{1}{4(r + a)}x^2$, the equation of a parabola.

11.4.89 Since the hyperbolas have the same asymptotes, we have that "a" and "b" are interchangeable in the formula. With $c^2 = a^2 + b^2$ and $ae = c$, $bE = c$, we have $e = \frac{\sqrt{a^2 + b^2}}{a}$ and $E = \frac{\sqrt{a^2 + b^2}}{b}$, so $\frac{1}{e^2} + \frac{1}{E^2} = \frac{a^2}{a^2 + b^2} + \frac{b^2}{a^2 + b^2} = 1$.

11.4.91 The latus rectum L intersects the parabola at $x = p$, $y = \pm 2p$. The distance between any point $P(x, y)$ on the parabola to the left of L and L is $p - x$. The distance from F to P is $\sqrt{(x - p)^2 + y^2} = \sqrt{x^2 - 2px + p^2 + 4px} = \sqrt{x^2 + 2px + p^2} = x + p$ (since both x and p are positive.) Thus $D + |FP| = p - x + x + p = 2p$.

11.4.93 Let P be a point on the intersection of the latus rectum and the ellipse. The length of the latus rectum is twice the distance from P to the focus. Let l be the length from P to the focus, and let L be the distance from P to the other focal point. Then $l + L = 2a$, so $L^2 = 4c^2 + l^2$, and thus $(2a - l)^2 = 4c^2 + l^2$, and solving for l yields $l = a - \frac{c^2}{a}$. Since $c^2 = a^2 - b^2$, this can be written as $l = a - \frac{a^2 - b^2}{a} = a - (a - \frac{b^2}{a}) = \frac{b^2}{a}$. The length of the latus rectum is therefore $\frac{2b^2}{a}$. Now since $e = \frac{c}{a}$, we have $\sqrt{1 - e^2} = \sqrt{1 - \frac{a^2 - b^2}{a^2}} = \sqrt{\frac{b^2}{a^2}} = \frac{b}{a}$. The length of the latus rectum can thus also be written as $2b \cdot \frac{b}{a} = 2b\sqrt{1 - e^2}$.

11.4.95 Let the equation of the ellipse be $\frac{x^2}{a^2} + \frac{y^2}{a^2 - c^2} = 1$ and let the equation of the hyperbola be $\frac{x^2}{r^2} - \frac{y^2}{c^2 - r^2} = 1$. Let (x_0, y_0) be a point of intersection. By evaluating both equations at the point of intersection and subtracting, we obtain the result

$$\frac{x_0^2}{a^2} - \frac{x_0^2}{r^2} + \frac{y_0^2}{a^2 - c^2} + \frac{y_0^2}{c^2 - r^2} = 0,$$

which can be written

$$\frac{r_0^2 x_0^2 - a^2 x_0^2}{a^2 r^2} + \frac{(c^2 - r^2)y_0^2 + (a^2 - c^2)y_0^2}{(a^2 - c^2)(c^2 - r^2)} = 0.$$

This equation can be rewritten in the form $\frac{x_0^2}{y_0^2} = \frac{a^2 r^2}{(a^2 - c^2)(c^2 - r^2)}$, which we will use later.

Now implicitly differentiating the equation for the ellipse yields $\frac{2x}{a^2} + \frac{2yy'}{a^2 - c^2} = 0$, and thus the slope of the tangent line to the ellipse at (x_0, y_0) is $y'_e = \frac{-x_0}{y_o} \cdot \frac{a^2 - c^2}{a^2}$. Differentiating the equation of the hyperbola gives $\frac{2x}{r^2} - \frac{2yy'}{c^2 - r^2} = 0$, so the slope of the tangent line to the hyperbola at the point of intersection is $y'_h = \frac{x_0}{y_0} \cdot \frac{c^2 - r^2}{r^2}$.

Now consider the product

$$-1 \cdot y'_e \cdot y'_h = \frac{x_0^2}{y_0^2} \cdot \frac{(a^2 - c^2)(c^2 - r^2)}{a^2 r^2}.$$

By the result of the first paragraph, this is equal to 1, and thus the two curves are perpendicular at the point of intersection.

11.4.97

a. The curve and the line intersect when $x^2 - m^2(x^2 - 4x + 4) - 1 = 0$, which occurs for $\frac{2m^2 \pm \sqrt{1 + 3m^2}}{m^2 - 1}$, assuming $m \neq \pm 1$. So there are two solutions in this case – but if $-1 < m < 1$, one of the solutions is negative (the intersection lies on the other branch of the hyperbola.) If $m^2 = 1$, then the equation becomes $4x - 5 = 0$, and there is only the solution $x = \frac{5}{4}$. So there are two intersection points on the right branch exactly for $|m| > 1$. We have $v(m) = \frac{2m^2 + \sqrt{1 + 3m^2}}{m^2 - 1}$ and $u(m) = \frac{2m^2 - \sqrt{1 + 3m^2}}{m^2 - 1}$.

b. $\lim\limits_{m \to 1^+} u(m) = \lim\limits_{m \to 1^+} u(m) \cdot \frac{2m^2 + \sqrt{1 + 3m^2}}{2m^2 + \sqrt{1 + 3m^2}} = \lim\limits_{m \to 1^+} \frac{4m^4 - 3m^2 - 1}{(m^2 - 1)(2m^2 + \sqrt{1 + 3m^2})} =$
$\lim\limits_{m \to 1^+} \frac{(m^2 - 1)(4m^2 + 1)}{(m^2 - 1)(2m^2 + \sqrt{1 + 3m^2})} = \frac{5}{4}$.

$\lim\limits_{m \to 1^+} v(m) = \lim\limits_{m \to 1^+} v(m) \cdot \frac{2m^2 - \sqrt{1 + 3m^2}}{2m^2 - \sqrt{1 + 3m^2}} = \lim\limits_{m \to 1^+} \frac{4m^4 - 3m^2 - 1}{(m^2 - 1)(2m^2 - \sqrt{1 + 3m^2})} =$
$\lim\limits_{m \to 1^+} \frac{(m^2 - 1)(4m^2 + 1)}{(m^2 - 1)(2m^2 - \sqrt{1 + 3m^2})} = \lim\limits_{m \to 1^+} \frac{(4m^2 + 1)}{(2m^2 - \sqrt{1 + 3m^2})} = \infty$.

c. $\lim\limits_{m \to \infty} u(m) = \lim\limits_{m \to \infty} \frac{2 - \sqrt{\frac{1}{m^4} + \frac{3}{m^2}}}{1 - \frac{1}{m^2}} = 2$.

$\lim\limits_{m \to \infty} v(m) = \lim\limits_{m \to \infty} \frac{2 + \sqrt{\frac{1}{m^4} + \frac{3}{m^2}}}{1 - \frac{1}{m^2}} = 2$.

d. The expression $\lim_{m \to \infty} A(m)$ represents the area of the region bounded by the hyperbola and the line $x = 2$. It is given by $2 \int_1^2 \sqrt{x^2 - 1}\, dx = 2 \left(\dfrac{x}{2} \sqrt{x^2 - 1} - \dfrac{1}{2} \ln(x + \sqrt{x^2 - 1}) \right) \Big|_1^2 = 2\sqrt{3} - \ln(2 + \sqrt{3})$.

11.4.99

a. With $x^2 = a^2 \cos^2 t + 2ab \sin t \cos t + b^2 \sin^2 t$, $y^2 = c^2 \cos^2 t + 2cd \sin t \cos t + d^2 \sin^2 t$, and $xy = ac \cos^2 t + (ad + bc) \sin t \cos t + bd \sin^2 t$, we have $Ax^2 + Bxy + Cy^2 = (Aa^2 + Bac + Cc^2) \cos^2 t + (2Aab + B(ad + bc) + 2Ccd) \sin t \cos t + (Ab^2 + Bbd + Cd^2) \sin^2 t = K$. Thus we have an equation of the desired form as long as there exist A, B, C, and K so that $A(a^2 - b^2) + B(ac - bd) + C(c^2 - d^2) = 0$ and $2Aab + B(ad + bc) + 2Ccd = 0$. This turns out to be the case when $ad - bc \neq 0$. Note that the value of K is $Aa^2 + Bac + Cc^2$.

b. Suppose that $ad - bc \neq 0$, but $ac + bd = 0$. Then $\frac{b}{a} = \frac{-c}{d}$, and $\tan^{-1}(b/a) = \tan^{-1}(-c/d)$.

Note that $x = \sqrt{a^2 + b^2} \cos(t + \tan^{-1}(-b/a))$, $y = \sqrt{c^2 + d^2} \sin(t + \tan^{-1}(c/d))$. This can be seen by applying the trigonometric identities for the sum of two angles. Then $\frac{x^2}{a^2 + b^2} + \frac{y^2}{c^2 + d^2} = \cos^2(t + \tan^1(b/a)) + \sin^2(\tan^{-1}(-c/d)) = 1$.

c. Using the work in part b), we see that the equation is $\frac{x^2}{a^2 + b^2} + \frac{y^2}{c^2 + d^2} = 1$, or $x^2 + y^2 = r^2$, where $r^2 = a^2 + b^2 = c^2 + d^2$.

11.5   Chapter Eleven Review

11.5.1

a. False. For example, $x = r \cos t$, $y = r \sin t$ for $0 \leq t \leq 2\pi$ and $x = r \sin t$, $y = r \cos t$ for $0 \leq t \leq 2\pi$ generate the same circle.

b. False. Since $e^t > 0$ for all t, this only describes the portion of that line where $x > 0$.

c. True. They both describe the point whose cartesian coordinates are $(3\cos(-3\pi/4), 3\sin(-3\pi/4)) = (-3\cos(\pi/4), -3\sin(\pi/4)) = (-3/\sqrt{2}, -3/\sqrt{2})$.

d. False. The given integral counts the inner loop twice.

e. True. This follows because the equation $0 - x^2/4 = 1$ has no real solutions.

f. True. Note that the given equation can be written as $(x - 1)^2 + 4y^2 = 4$, or $\frac{(x-1)^2}{4} + y^2 = 1$.

11.5.3

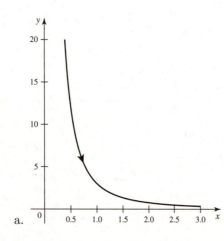

a.

b. $y = 3(e^t)^{-2} = \frac{3}{x^2}$.

c. The curve represents the portion of $\frac{3}{x^2}$ for $x > 0$.

d. $\frac{dy}{dx} = \frac{-6}{x^3}$, so at $(1, 3)$ we have $\frac{dy}{dx} = -6$.

11.5.5

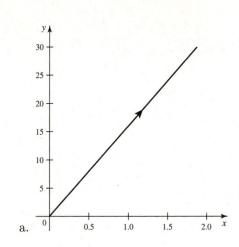

b. Since $\ln t^2 = 2 \ln t$ for $t > 0$, we have $y = 16x$ for $0 \leq x \leq 2$.

c. The curve represents a line segment from $(0,0)$ to $(2, 32)$.

d. $\frac{dy}{dx} = 16$ for all value of x.

a.

11.5.7 $\frac{dy}{dx} = \frac{dy/dt}{dx/dt} = \frac{\sin t}{1 - \cos t}$. At $t = \pi/6$, the slope of the tangent line is $\frac{1}{2 - \sqrt{3}} = 2 + \sqrt{3}$. So the equation of the tangent line is $y - (1 - \sqrt{3}/2) = (2 + \sqrt{3})(x - (\pi/6 - 1/2))$, or $y = (2 + \sqrt{3})x + (2 - \frac{\pi}{3} - \frac{\pi\sqrt{3}}{6})$.

At $t = 2\pi/3$, the slope of the tangent line is $\frac{\sqrt{3}}{3}$, so the equation of the tangent line is $y - \frac{3}{2} = \frac{\sqrt{3}}{3}(x - (\frac{2\pi}{3} - \frac{\sqrt{3}}{2}))$, or $y = \frac{x}{\sqrt{3}} + 2 - \frac{2\pi}{3\sqrt{3}}$.

11.5.9

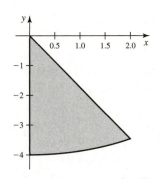

11.5.11 Letting $x = r \cos \theta$, $y = r \sin \theta$, and $r^2 = x^2 + y^2$, we have $x^2 + y^2 + 2y - 6x = 0$, which can be written as $x^2 - 6x + 9 + y^2 + 2y + 1 = 10$, or $(x - 3)^2 + (y + 1)^2 = 10$, so this is a circle of radius $\sqrt{10}$ centered at $(3, -1)$.

11.5.13

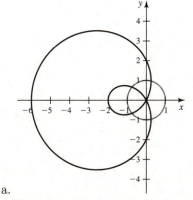

b. Note that $2 - 4\cos\theta = 1$ for $\theta = \cos^{-1}(1/4) \approx 1.32$, and $2 - 4\cos\theta = -1$ for $\theta = \cos^{-1}(3/4) \approx .73$. The points of intersection (in polar form) are approximately $(1, 1.32)$, $(1, 2\pi - 1.32) \approx (1, 4.76)$, $(-1, .73)$, and $(-1, 2\pi - .73) \approx (-1, 5.56)$.

a.

There are 4 intersection points.

11.5.15

a. $\dfrac{dy}{dx} = \dfrac{dy/d\theta}{dx/d\theta} = \dfrac{2\cos\theta\sin\theta + (4 + 2\sin\theta)\cos\theta}{2\cos\theta\cos\theta - (4 + 2\sin\theta)\sin\theta} = \dfrac{4\cos\theta + 4\sin\theta\cos\theta}{2\cos^2\theta - 2\sin^2\theta - 4\sin\theta}.$

This is 0 when $\cos\theta = 0$, and when $4\sin\theta = -4$, so the only solutions are $\theta = \pi/2,\ 3\pi/2$.

The denominator is 0 when $2 - 4\sin^2\theta - 4\sin\theta = 0$ which occurs (using the quadratic formula) for $\sin\theta = \frac{-1}{2} + \frac{\sqrt{3}}{2}$, so there are vertical tangent lines at $\theta = \sin^{-1}(\frac{-1}{2} + \frac{\sqrt{3}}{2})$ and $\theta = \pi - \sin^{-1}(\frac{-1}{2} + \frac{\sqrt{3}}{2})$.

b. The curve is never at the origin.

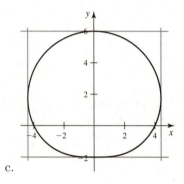

c.

11.5.17

a. Note that the whole curve is generated for $-\pi/4 \le \theta \le \pi/4$, so we restrict ourselves to that domain. Write the equations as $r = \sqrt{2\cos 2\theta}$. Then

$$\frac{dy}{d\theta} = \sqrt{2\cos 2\theta}\cos\theta - \sin\theta\frac{2\sin 2\theta}{\sqrt{2\cos 2\theta}} = \frac{\cos\theta}{\sqrt{2\cos 2\theta}}\left(2\cos 2\theta - 4\sin^2\theta\right) = \frac{\cos\theta}{\sqrt{2\cos 2\theta}}\left(2 - 8\sin^2\theta\right).$$

Also, $\frac{dx}{d\theta} = -\sqrt{2\cos 2\theta}\sin\theta + \cos\theta\frac{2\sin 2\theta}{\sqrt{2\cos 2\theta}} = \frac{\sin\theta}{\sqrt{2\cos 2\theta}}\left(-4\cos^2\theta - 2(\cos 2\theta)\right) = \frac{(2 - 8\cos^2\theta)\sin\theta}{\sqrt{2\cos(2\theta)}}$. Thus $\frac{dy}{dx} = \frac{dy/d\theta}{dx/d\theta} = \cot\theta\left(\frac{1 - 4\sin^2\theta}{1 - 4\cos^2\theta}\right)$.

This expression is 0 on the given domain only for $\sin^2\theta = \frac{1}{4}$, so there are horizontal tangent lines at $\theta = \pm\frac{\pi}{6}$. There are vertical tangent lines on the given domain only for $\theta = 0$ In cartesian coordinates, the lines are $x = \pm\sqrt{2}$.

b. The curve is at the origin for $\theta = \pm\pi/4$, and since $\tan\pi/4 = 1$, the tangent lines have the equations $y = \pm x$.

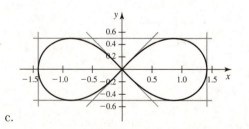

c.

11.5.19 The area is given by $A = \frac{1}{2} \int_0^{2\pi} (3 - \cos\theta)^2 \, d\theta = \frac{1}{2} \int_0^{2\pi} (9 - 6\cos\theta + \cos^2\theta) \, d\theta = \frac{1}{2} \left(9\theta - 6\sin\theta + \frac{1}{2}(\cos\theta\sin\theta + \theta) \right)\Big|_0^{2\pi} = \frac{1}{2}(18\pi + \pi) = \frac{19\pi}{2}$.

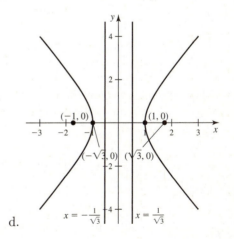

11.5.21 The curves intersect at $\theta = \pm\frac{1}{2}\cos^{-1}(1/16)$. By symmetry the total desired area is $A = 4 \cdot \frac{1}{2} \int_0^{\cos^{-1}(1/16)/2} (4\cos 2\theta - \frac{1}{4}) \, d\theta = 2\left(2\sin 2\theta - \frac{\theta}{4} \right)\Big|_0^{\cos^{-1}(1/16)/2} = \frac{1}{4}\sqrt{255} - \frac{\cos^{-1}(1/16)}{4}$.

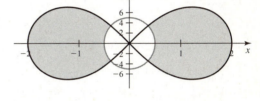

11.5.23

a. This represents a hyperbola with $a = 1$ and $b = \sqrt{2}$.

b. The vertices are $(\pm 1, 0)$, the foci are $(\pm c, 0)$ where $c^2 = a^2 + b^2 = 3$, so they are $(\pm\sqrt{3}, 0)$. The directrices are $x = \frac{\pm a^2}{c} = \frac{\pm 1}{\sqrt{3}}$.

c. The eccentricity is $e = \frac{c}{a} = \sqrt{3}$.

d.

11.5.25

a. This can be written as $\frac{y^2}{16} - \frac{x^2}{4} = 1$. It is a hyperbola with $a = 4$ and $b = 2$.

b. The vertices are $(0, \pm 4)$. The foci are $(0, \pm c)$ where $c^2 = a^2 + b^2 = 16 + 4 = 20$, so they are $(0, \pm\sqrt{20})$. The directrices are $y = \frac{\pm a^2}{c} = \frac{\pm 16}{\sqrt{20}} = \frac{\pm 8}{\sqrt{5}}$.

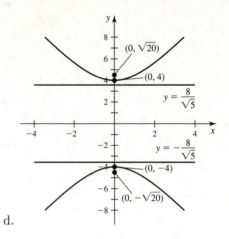

c. The eccentricity is $e = \frac{c}{a} = \frac{\sqrt{20}}{4} = \frac{\sqrt{5}}{2}$.

d.

11.5.27

a. This can be written as $\frac{x^2}{4} + \frac{y^2}{2} = 1$, so it is an ellipse with $a = 2$ and $b = \sqrt{2}$.

b. The vertices are $(\pm 2, 0)$. The foci are $(\pm c, 0)$ where $c^2 = a^2 - b^2 = 4 - 2 = 2$, so they are $(\pm\sqrt{2}, 0)$. The directrices are $x = \frac{\pm a^2}{c} = \frac{\pm 4}{\sqrt{2}} = \pm 2\sqrt{2}$.

c. The eccentricity is $e = \frac{c}{a} = \frac{\sqrt{2}}{2}$.

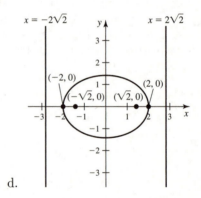

d.

11.5.29 $2y\frac{dy}{dx} = -12$, so at the point in question, $\frac{dy}{dx} = 3/2$. So the equation of the tangent line is $y + 4 = \frac{3}{2}\left(x + \frac{4}{3}\right)$, or $y = \frac{3}{2}x - 2$.

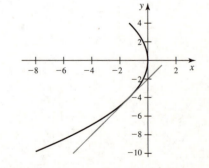

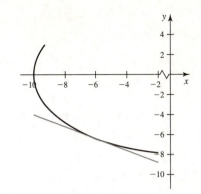

11.5.31 $\frac{x}{50} + \frac{y}{32} \cdot \frac{dy}{dx} = 0$, so at the given point, $\frac{dy}{dx} = \frac{-6}{10} = \frac{-3}{5}$. So the equation of the tangent line is $y + \frac{32}{5} = \frac{-3}{5}(x+6)$, or $y = \frac{-3}{5}x - 10$.

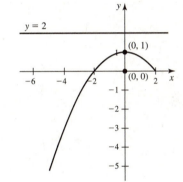

11.5.33 The eccentricity is 1, and the directrix is $y = 2$. The vertex is $(0,1)$ and the focus is $(0,0)$.

11.5.35 The eccentricity is $\frac{1}{2}$, and the directrices are $x = 4$ and $x = \frac{-20}{3}$. The vertices are $\left(\frac{4}{3},0\right)$ and $(-4,0)$ and the foci are $(0,0)$ and $\left(\frac{-8}{3},0\right)$.

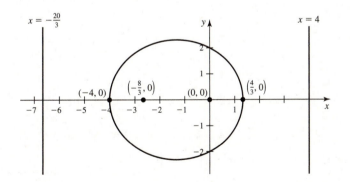

11.5.37

a. Recall that $\cos 2\theta = \cos^2 \theta - \sin^2 \theta$, so $r^2 \cos(2\theta) = 1$ becomes $r^2(\cos^2 \theta - \sin^2 \theta) = x^2 - y^2 = 1$. The curve is a hyperbola.

b. With $a = b = 1$, we have $c^2 = 2$, so the vertices are $(\pm 1, 0)$ and the foci are $(\pm\sqrt{2}, 0)$. The directrices are $x = \pm\frac{a^2}{c} = \pm\frac{1}{\sqrt{2}}$. The eccentricity is $e = \frac{c}{a} = \sqrt{2}$.

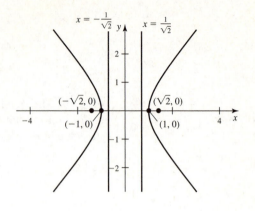

c. It does not have the form as in Theorem 10.4 because it does not have a focus at the origin.

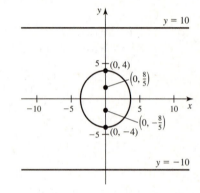

11.5.39 Since the center is halfway between the vertices, it is $(0,0)$. We must have $a = 4$ and since $\frac{a^2}{c} = d = 10$, we have $c = \frac{8}{5}$. So $b^2 = a^2 - c^2 = 16 - \frac{64}{25} = \frac{336}{25}$. The ellipse has equation $\frac{25x^2}{336} + \frac{y^2}{16} = 1$. The eccentricity is $\frac{c}{a} = \frac{8/5}{4} = \frac{2}{5}$.

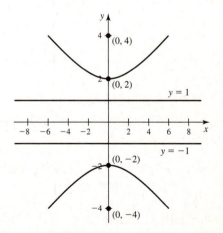

11.5.41 Since the center is halfway between the vertices, it is $(0,0)$. We must have $a = 2$ and since $\frac{a^2}{c} = d = 1$, we have $c = 4$. So $b^2 = 16 - 4 = 12$. The hyperbola has equation $\frac{y^2}{4} - \frac{x^2}{12} = 1$. The eccentricity is $\frac{c}{a} = \frac{4}{2} = 2$.

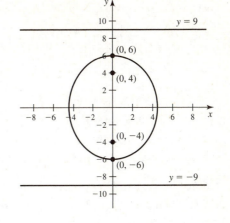

11.5.43 We have $a = 6$, $c = 4$ and $e = \frac{c}{a} = \frac{4}{6} = \frac{2}{3}$. Also, $b^2 = a^2 - c^2 = 36 - 16 = 20$, and the equation is $\frac{y^2}{36} + \frac{x^2}{20} = 1$. The vertices are $(\pm 2\sqrt{5}, 0)$. The directrices are $y = \pm \frac{a^2}{c} = \pm 9$.

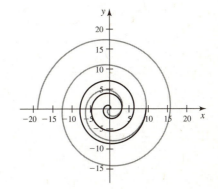

11.5.45 $\sin 2\theta = \theta^2$ when $\theta = 0$. Graphing the functions reveals a root near $\theta = 1$. A CAS reveals the intersection point to be $\theta \approx .9669$. In polar coordinates, the intersection points are $(0,0)$ and $(.9669, .9669)$.

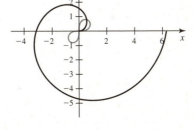

11.5.47 The curves intersect for $\theta = 0$. Note also that when $\theta = k\pi$ for k an odd integer, the curve $r = -\theta$ is at the polar point $(-k\pi, k\pi) = (k\pi, 0)$. And for $\theta = 2k\pi$, the curve $r = \frac{\theta}{2}$ is at the point $(k\pi, 0)$. So the curves intersect at these points.

11.5.49 By symmetry, we can focus on the region in the first quadrant. That area is given by $A = xy$ where $y = \sqrt{b^2 - \frac{b^2}{a^2}x^2}$. So

$$A(x) = x\sqrt{b^2 - \frac{b^2}{a^2}x^2},$$

so

$$A'(x) = \sqrt{b^2 - \frac{b^2}{a^2}x^2} - \frac{b^2 x^2}{a^2 \sqrt{b^2 - \frac{b^2}{a^2}x^2}}.$$

Setting the derivative equal to 0 and clearing denominators yields $\left(b^2 - \frac{b^2}{a^2}x^2\right)a^2 - b^2 x^2 = 0$, and solving for x gives the critical point $x = \frac{\sqrt{2}}{2}a$. Since this is the only critical point and it clearly does not give a minimum (since $A(0) = A(a) = 0$), it must yield a maximum. The whole rectangle has dimensions $\sqrt{2}a \times \sqrt{2}b$, and area $2ab$.

11.5.51 The area of the ellipse in the first quadrant is $\frac{\pi ab}{4}$, so we are seeking θ_0 so that

$$\frac{\pi ab}{8} = \frac{1}{2} \int_0^{\theta_0} \frac{a^2 b^2}{a^2 \sin^2 \theta + b^2 \cos^2 \theta} \, d\theta = \frac{a^2}{2} \int_0^{\theta_0} \frac{\sec^2 \theta}{\frac{a^2}{b^2} \tan^2 \theta + 1} \, d\theta.$$

Let $u = \frac{a}{b} \tan \theta$ so that $du = \frac{a}{b} \sec^2 \theta \, d\theta$. Then we have $\frac{\pi ab}{8} = \frac{ab}{2} \int_0^{\frac{a}{b} \tan \theta_0} \frac{1}{1+u^2} \, du = \frac{ab}{2} \tan^{-1}(\frac{a}{b} \tan(\theta_0))$. Note that this equation is satisfied when $\tan(\theta_0) = \frac{b}{a}$, because then the expression on the right-hand side of that equation is $\frac{ab}{2} \cdot \frac{\pi}{4} = \frac{\pi ab}{8}$. So the desired value of m is $\tan(\theta_0) = \frac{b}{a}$.

11.5.53 Note that $Q = (a\cos\theta, a\sin\theta)$ and $R = (b\cos\theta, b\sin\theta)$, where θ is the angle formed by l and the x-axis. Then $P = (a\sin\theta, b\cos\theta)$ is a point on the ellipse $\frac{x^2}{a^2} + \frac{y^2}{b^2} = 1$, since it satisfies that equation.

11.5.55 The focal point is at the origin, the directrix is $y = -d$, so we have an equation of the form $r = \frac{ed}{1 - e\sin\theta}$. Since c is the distance from the center to the focal point, we have $c = 3/8$, and since a is the distance from the center to a vertex, we have $a = 9/8$. Then we have $e = \frac{c}{a} = \frac{3/8}{9/8} = \frac{1}{3}$, and $d = \frac{a^2}{c} - \frac{3}{8} = 3$. Thus $r = \frac{1}{1 - \frac{1}{3}\sin\theta} = \frac{3}{3 - \sin\theta}$.